T0212622

Communications
in Computer and Information Science 598

Commenced Publication in 2007
Founding and Former Series Editors:
Alfredo Cuzzocrea, Dominik Ślęzak, and Xiaokang Yang

Editorial Board

More information about this series at http://www.springer.com/series/7899

José Braz · Julien Pettré
Paul Richard · Andreas Kerren
Lars Linsen · Sebastiano Battiato
Francisco Imai (Eds.)

Computer Vision, Imaging and Computer Graphics Theory and Applications

10th International Joint Conference, VISIGRAPP 2015
Berlin, Germany, March 11–14, 2015
Revised Selected Papers

 Springer

Editors

José Braz
Escola Superior de Tecnologia do IPS
Setúbal
Portugal

Julien Pettré
Inria-Rennes/MimeTIC Team
Rennes cedex
France

Paul Richard
LISA - ISTIA
University of Angers
Angers
France

Andreas Kerren
Linnaeus University
Växjö
Sweden

Lars Linsen
Jacobs University
Bremen
Germany

Sebastiano Battiato
Università di Catania
Catania, Catania
Italy

Francisco Imai
Research Innovation Center
Canon U.S.A. Inc
San Jose, CA
USA

ISSN 1865-0929   ISSN 1865-0937 (electronic)
Communications in Computer and Information Science
ISBN 978-3-319-29970-9   ISBN 978-3-319-29971-6 (eBook)
DOI 10.1007/978-3-319-29971-6

Library of Congress Control Number: 2016931299

Printed on acid-free paper

This Springer imprint is published by SpringerNature
The registered company is Springer International Publishing AG Switzerland

Preface

This book includes the extended versions of selected papers from VISIGRAPP 2015, the International Joint Conference on Computer Vision, Imaging and Computer Graphics Theory and Applications, which was held in Berlin, Germany, during March 11–14, 2015. The conference was organized by the Institute for Systems and Technologies of Information, Control and Communication (INSTICC), in cooperation with ACM SIGGRAPH and technically co-sponsored by the IEEE Computer Society, IEEE VGMT, and IEEE TCMC.

VISIGRAPP comprises three conferences, namely, the International Conference on Computer Vision Theory and Applications (VISAPP), the International Conference on Computer Graphics Theory and Applications (GRAPP), and the International Conference on Information Visualization Theory and Applications (IVAPP).

VISIGRAPP received 529 paper submissions from more than 50 countries. After a rigorous double-blind evaluation, only 18 % of the papers were accepted and published as full papers. These numbers show that our conference is aiming for the highest scientific standards, and that it can now be considered a well-established venue for researchers in the broad fields of computer vision, image analysis, computer graphics, and information visualization. From the set of full papers, 23 were selected for inclusion into this book. The selection process was based on quantitative and qualitative evaluation results provided by the Program Committee reviewers as well as the feedback on paper presentations provided by the session chairs during the conference. After selection, the accepted papers were further revised and extended by the authors. Our gratitude goes to all contributors and reviewers, without whom this book would not have been possible. Apart from the full papers, 25 % of the papers were accepted for short presentations and 26 % accepted for poster presentations. However, these works were not considered for the present book selection process. We do not expect that each reader is equally interested in all 23 of the selected VISIGRAPP papers. However, the diversity of these papers makes it very likely that all readers can find something of interest in this selection.

As VISAPP 2015 constitutes the largest part of VISIGRAPP with 345 submissions, we decided to select and integrate 15 extended full papers aiming to cover different aspects and areas related to computer vision, such as image formation and pre-processing, image and video analysis and understanding, motion tracking, stereo vision, as well as diverse computer vision applications and services. Here, we would like to mention that when we selected the papers from VISAPP for this book, our intention was to cover and highlight research from different areas and subareas related to computer vision. These papers were mainly competing with other VISAPP papers having similar content, and therefore, we want to explicitly acknowledge that other high-quality papers accepted at the conference could have been integrated in this book if we had more space.

Concerning GRAPP 2015, 93 papers were submitted, and we decided to include four extended full papers in this book. We tried to cover the main areas of computer graphics to make the content of the book similar to the research addressed at the conference.

The four papers selected from 54 submissions to IVAPP 2015 are not only excellent representatives of the field of visualization, but also form a quite balanced representation of the field itself. Above all, they are almost as diverse and exciting as the field of visualization.

VISIGRAPP 2015 also included four invited keynote lectures, presented by internationally renowned researchers, whom we would like to thank for their contribution to reinforcing the overall quality of the conference. They are in alphabetical order: Mauro Barni (Università di Siena, Italy), Andrea Cavallaro (Queen Mary University of London, UK), Gerik Scheuermann (Universität Leipzig, Germany), and Daniel Thalmann (Nanyang Technological University, Singapore).

We wish to thank all those who supported VISIGRAPP and helped to organize the conference. On behalf of the conference Organizing Committee, we would like to especially thank the authors, whose work was the essential part of the conference and contributed to a very successful event. We would also like to thank the members of the Program Committee, whose expertise and diligence were instrumental in ensuring the quality of the final contributions. We also wish to thank all the members of the Organizing Committee, whose work and commitment was invaluable. Last but not least, we would like to thank Springer for their collaboration and help in getting this book to print.

March 2015

José Braz
Sebastiano Battiato
Francisco Imai
Julien Pettré
Paul Richard
Andreas Kerren
Lars Linsen

Organization

Conference Chair

José Braz   Escola Superior de Tecnologia de Setúbal, Portugal

Program Co-chairs

GRAPP

Julien Pettré   Inria Rennes - Bretagne Atlantique, France
Paul Richard   University of Angers, France

IVAPP

Andreas Kerren   Linnaeus University, Sweden
Lars Linsen   Jacobs University, Bremen, Germany

VISAPP

Sebastiano Battiato   University of Catania, Italy
Francisco Imai   Canon USA Inc., Innovation Center, USA

GRAPP Program Committee

Francisco Abad   Universidad Politécnica de Valencia, Spain
Marco Agus   CRS4, Italy
Tremeau Alain   University of Saint Etienne, France
Marco Attene   National Research Council (CNR), Italy
Lilian Aveneau   University of Poitiers, France
Francesco Banterle   Visual Computing Lab, Italy
Dominique Bechmann   University of Strasbourg, France
Gonzalo Besuievsky   Universitat de Girona, Spain
Venceslas Biri   University of Paris Est, France
Jiri Bittner   Czech Technical University in Prague, Czech Republic
Kristopher J. Blom   Virtual Human Technologies, Czech Republic
David Bommes   RWTH Aachen University, Germany
Carles Bosch   Eurecat, Spain
Stephen Brooks   Dalhousie University, Canada
Stefan Bruckner   University of Bergen, Norway
Pere Brunet   Technical University of Catalonia, Spain
Patrick Callet   Laboratoire Mathématiques Appliquées Aux
                                   Systèmes, France
Pedro Cano   University of Granada, Spain

Tolga Capin	Bilkent University, Turkey
Maria Beatriz Carmo	Universidade de Lisboa, Portugal
L.G. Casado	University of Almeria, Spain
Eva Cerezo	University of Zaragoza, Spain
Teresa Chambel	Lasige, University of Lisbon, Portugal
Antoni Chica	Universitat Politecnica de Catalunya, Spain
Hwan-gue Cho	Pusan National University, Korea, Republic of
Ana Paula Cláudio	BioISI– Biosystems and Integrative Sciences Institute, Universidade de Lisboa, Portugal
Sabine Coquillart	Inria, France
António Cardoso Costa	ISEP, Portugal
Carsten Dachsbacher	Karlsruhe Institute of Technology, Germany
Kurt Debattista	University of Warwick, UK
Victor Debelov	Siberian Branch of Russian Academy of Sciences, Russian Federation
John Dingliana	Trinity College Dublin, Ireland
Jean-Michel Dischler	Université de Strasbourg, France
Thierry Duval	Télécom Bretagne, France
Arjan Egges	Utrecht University, The Netherlands
Elmar Eisemann	Delft University of Technology, The Netherlands
Marius Erdt	Fraunhofer IDM@NTU, Singapore
Bianca Falcidieno	Consiglio Nazionale delle Ricerche, Italy
Francisco R. Feito	University of Jaén, Spain
Luiz Henrique de Figueiredo	Impa, Brazil
Pablo Figueroa	Universidad De Los Andes, Colombia
Fabian Di Fiore	Hasselt University, Belgium
Cedric Fleury	Université de Paris-Sud, France
Carla Maria Dal Sasso Freitas	Universidade Federal do Rio Grande do Sul, Brazil
Ioannis Fudos	University of Ioannina, Greece
Alejandro García-Alonso	University of the Basque Country, Spain
Miguel Gea	University of Granada, Spain
Djamchid Ghazanfarpour	University of Limoges, France
Stephane Gobron	HES-SO/HE-Arc/ISIC, Switzerland
Laurent Grisoni	LIFL, France
Jerome Grosjean	LSIIT, France
James Hahn	George Washington University, USA
Peter Hall	University of Bath, UK
Vlastimil Havran	Czech Technical University in Prague, Czech Republic
Nancy Hitschfeld	University of Chile, Chile
Toby Howard	University of Manchester, UK
Ludovic Hoyet	Inria Rennes - Centre Bretagne Atlantique, France
Andres Iglesias	University of Cantabria, Spain
Insung Ihm	Sogang University, Korea, Republic of
Wojciech Jarosz	The Walt Disney Company, Switzerland

Jean-Pierre Jessel	IRIT, Paul Sabatier University, Toulouse, France
Juan J. Jimenez	University of Jaén, Spain
Robert Joan-Arinyo	Universitat Politecnica de Catalunya, Spain
Andrew Johnson	University of Illinois at Chicago, USA
Chris Joslin	Carleton University, Canada
Josef Kohout	University of West Bohemia, Czech Republic
Jaroslav Krivanek	Charles University in Prague, Czech Republic
Torsten Kuhlen	RWTH Aachen University, Germany
David Laidlaw	Brown University - Providence, USA
Miguel Leitão	ISEP, Portugal
Heinz U. Lemke	Foundation for Computer Assisted Radiology and Surgery, Germany
Alejandro León	University of Granada, Spain
Frederick Li	University of Durham, UK
Joaquim Madeira	University of Aveiro, Portugal
Maurizio Mancini	University of Genoa, Italy
Stephen Mann	University of Waterloo, Canada
Michael Manzke	Trinity College Dublin, Ireland
Belen Masia	MPI Informatik, Germany
Oliver Mattausch	University of Zurich, Switzerland
Daniel Meneveaux	University of Poitiers, France
Stéphane Mérillou	University of Limoges, France
Ramon Molla	Universitat Politècnica de València, Spain
Guillaume Moreau	Ecole Centrale Nantes, France
David Mould	Carleton University, Canada
Adolfo Muñoz	Universidad de Zaragoza, Spain
Veronica Costa Orvalho	Face in Motion, FCUP, Spain
Georgios Papaioannou	Athens University of Economics and Business, Greece
Giuseppe Patané	CNR - Italian National Research Council, Italy
Daniel Patel	University of Bergen, Norway
Sumanta Pattanaik	UCF, USA
Nuria Pelechano	Universitat Politecnica de Catalunya, Spain
João Madeiras Pereira	INESC-ID/IST, Portugal
João Pereira	Instituto Superior de Engenharia do Porto, Portugal
Christopher Peters	KTH Royal Institute of Technology, Sweden
Ruggero Pintus	CRS4 - Center for Advanced Studies, Research and Development in Sardinia, Italy
Ronald Poppe	University of Twente, The Netherlands
Nicolas Pronost	Université Lyon 1, France
Anna Puig	University of Barcelona, Spain
Inmaculada Remolar	Universitat Jaume I, Spain
Mickael Ribardière	University of Poitiers, XLIM, France
Tobias Ritschel	MPI Saarbrücken, Germany
María Cecilia Rivara	Universidad de Chile, Chile
Inmaculada Rodríguez	University of Barcelona, Spain
Przemyslaw Rokita	Warsaw University of Technology, Poland

Isaac Rudomin   BSC, Spain
Holly Rushmeien   Yale University, USA
Luis Paulo Santos   Universidade do Minho, Portugal
Rafael J. Segura   Universidad de Jaen, Spain
Etienne de Sevin   Masa Group, France
Ari Shapiro   University of Southern California, USA
Frutuoso Silva   University of Beira Interior, Portugal
A. Augusto Sousa   FEUP/INESC Porto, Portugal
Ching-Liang Su   Da Yeh University, India
Susanne K. Suter   University of Florida, USA
Jie Tang   Nanjing University, China
Marco Tarini   Università degli Studio dell'Insubria, Italy
Matthias Teschner   University of Freiburg, Germany
Daniel Thalmann   Nanyang Technological University, Singapore
Juan Carlos Torres   Universidad de Granada, Spain
Torsten Ullrich   Fraunhofer Austria Research, Austria
Anna Ursyn   University of Northern Colorado, USA
Luiz Velho   IMPA - Instituto de Matematica Pura e Aplicada, Brazil
Daniel Weiskopf   Universität Stuttgart, Germany
Burkhard Wuensche   University of Auckland, New Zealand
Lihua You   Bournemouth University, UK
Jian J. Zhang   Bournemouth University, UK

GRAPP Additional Reviewers

Andrea Brambilla   University of Bergen, Norway
Andrea Cerri   National Council of Research (CNR), Italy
Ismail Khalid Kazmi   Bournemouth University, UK
Gabor Liktor   Karlsruhe Institute of Technology, Germany
Marina Monti   Consiglio Nazionale delle Ricerche, Italy
Cristina Rebollo   Universitat Jaume I, Spain
Tim Reiner   KIT, Germany
Sybren Stüvel   Universiteit Utrecht, The Netherlands
Akemi Galvez Tomida   University of Cantabria, Spain
Giovanna Varni   Università degli studi di Genova, Italy
Zhao Wang   Bournemouth University, UK
Wenshu Zhang   Bournemouth University, UK

IVAPP Program Committee

Wolfgang Aigner   St. Poelten University of Applied Sciences, Austria
Vladan Babovic   National University of Singapore, Singapore
Rita Borgo   Swansea University, UK
David Borland   University of North Carolina at Chapel Hill, USA
Anne Boyer   Loria - Inria Lorraine, France
Massimo Brescia   Istituto Nazionale di AstroFisica, Italy

Ross Brown	Queensland University of Technology, Brisbane, Australia
Maria Beatriz Carmo	Universidade de Lisboa, Portugal
Guoning Chen	University of Houston, USA
R. Jordan Crouser	Smith College, USA
László Czúni	University of Pannonia, Hungary
Christoph Dalitz	Niederrhein University of Applied Sciences, Germany
Robertas Damasevicius	Kaunas University of Technology, Lithuania
Mihaela Dinsoreanu	Technical University of Cluj-Napoca, Romania
Georgios Dounias	University of the Aegean, Greece
Achim Ebert	University of Kaiserslautern, Germany
Chi-Wing Fu	Nanyang Technological University, Singapore
Mohammad Ghoniem	Luxembourg Institute of Science and Technology, Luxembourg
Martin Graham	Edinburgh Napier University, UK
Charles Hansen	University of Utah, USA
Pheng-Ann Heng	Chinese University of Hong Kong, SAR China
Tony Huang	University of Tasmania, Australia
Alfred Inselberg	Tel Aviv University, Israel
Mark W. Jones	Swansea University, UK
Jörn Kohlhammer	Fraunhofer Institute for Computer Graphics Research, Germany
Martin Kraus	Aalborg University, Denmark
Simone Kriglstein	Vienna University of Technology, Austria
Denis Lalanne	University of Fribourg, Switzerland
Chun-Cheng Lin	National Chiao Tung University, Taiwan
Giuseppe Liotta	University of Perugia, Italy
Ross Maciejewski	Arizona State University, USA
Krešimir Matkovic	VRVis Research Center, Austria
Cholwich Nattee	Sirindhorn International Institute of Technology, Thammasat University, Thailand
Steffen Oeltze	University of Magdeburg, Germany
Benoît Otjacques	Luxembourg Institute of Science and Technology (LIST), Luxembourg
Philip J. Rhodes	University of Mississippi, USA
Adrian Rusu	Rowan University, USA
Filip Sadlo	IWR, Heidelberg University, Germany
Giuseppe Santucci	University of Rome, Italy
Angel Sappa	Computer Vision Center, Spain
Tobias Schreck	Graz University of Technology, Austria
Heidrun Schumann	University of Rostock, Germany
Marc Streit	Johannes Kepler Universität Linz, Austria
Yasufumi Takama	Tokyo Metropolitan University, Japan
Levente Tamas	Technical University of Cluj-Napoca, Romania
Sidharth Thakur	Renaissance Computing Institute (RENCI), USA
Huy T. Vo	New York University, USA

Slobodan Vucetic	Temple University, USA
Guenter Wallner	University of Applied Arts Vienna, Austria
Chaoli Wang	University of Notre Dame, USA
Daniel Weiskopf	Universität Stuttgart, Germany
Huub van de Wetering	Technische Universiteit Eindhoven, The Netherlands
Jarke van Wijk	Eindhoven University of Technology, The Netherlands
Kai Xu	Middlesex University, UK
Hsu-Chun Yen	National Taiwan University, Taiwan
Hongfeng Yu	University of Nebraska - Lincoln, USA
Xiaoru Yuan	Peking University, China

IVAPP Additional Reviewers

Bertjan Broeksema	Centre de Recherche Public Gabriel Lippmann, Luxembourg
Andrew Hanson	Indiana University, USA
Jie Liang	Peking University, China
Fintan McGee	CRP - Gabriel Lippmann, Luxembourg
Kawa Nazemi	Fraunhofer IGD, Germany

VISAPP Program Committee

Amr Abdel-Dayem	Laurentian University, Canada
Ilya Afanasyev	Innopolis University, Russian Federation
Tremeau Alain	University of Saint Etienne, France
Vicente Alarcon-Aquino	Universidad de las Americas Puebla, Mexico
Mokhled S. Al-Tarawneh	Mu'tah University, Jordan
Matthew Antone	Massachusetts Institute of Technology, USA
Djamila Aouada	University of Luxembourg, Luxembourg
Pantelis Asvestas	Technological Educational Institute of Athens, Greece
Jamal Atif	Université Paris-Sud 11, France
Lamberto Ballan	Università degli Studi di Firenze, Italy
Hichem Bannour	Marin Software, France
Angelos Barmpoutis	University of Florida, USA
Xavier Baró	Open University of Catalonia, Spain
Arrate Muñoz Barrutia	Universidad Carlos III de Madrid, Spain
Giuseppe Baruffa	University of Perugia, Italy
Mohamed Batouche	University Constantine 2, Algeria
Azeddine Beghdadi	Paris 13 University, France
Saeid Belkasim	Georgia State University, USA
Fabio Bellavia	University of Florence, Italy
Olga Bellon	IMAGO Research Group - Universidade Federal do Paraná, Brazil
Jenny Benois-Pineau	LABRI, University of Bordeaux, France
Neil Bergmann	University of Queensland, Australia
Adrian Bors	University of York, UK

Giosue Lo Bosco	University of Palermo, Italy
Murk Bottema	Flinders University, Australia
Roland Bremond	Institut Français des Sciences et Technologies des Transports, de l'aménagement et des Réseaux (IFSTTAR), France
Marius Brezovan	University of Craiova, Romania
Valentin Brimkov	State University of New York, USA
Alfred Bruckstein	Technion, Israel
Arcangelo R. Bruna	STMicroelectronics, Italy
Xianbin Cao	Beihang University, China
Alice Caplier	GIPSA-lab, France
Barbara Caputo	IDIAP Research Institute, Switzerland
Franco Alberto Cardillo	Consiglio Nazionale delle Ricerche, Italy
M. Emre Celebi	Louisiana State University in Shreveport, USA
Chee Seng Chan	University of Malaya, Malaysia
Satish Chand	NSIT, India
Vinod Chandran	Queensland University of Technology, Australia
Chin-Chen Chang	Feng Chia University, Taiwan
Hang Chang	Lawrence Berkeley National Lab, USA
Jocelyn Chanussot	Grenoble Institute of Technology, France
Chung Hao Chen	Old Dominion University, USA
Samuel Cheng	University of Oklahoma, USA
Michal Choras	University of Technology and Life Sciences Bydgoszcz and ITTI Poznan, Poland
Albert C.S. Chung	The Hong Kong University of Science and Technology, SAR China
Laurent Cohen	Université Paris Dauphine, France
Sara Colantonio	ISTI-CNR, Italy
David Connah	N/A, UK
Donatello Conte	Université François Rabelais Tours, France
Guido de Croon	Delft University of Technology, The Netherlands
Fabio Cuzzolin	Oxford Brookes University, UK
Dima Damen	University of Bristol, UK
Roy Davies	Royal Holloway, University of London, UK
Larry Davis	University of Maryland College Park, USA
Kenneth Dawson-Howe	Trinity College Dublin, Ireland
Emmanuel Dellandréa	Ecole Centrale de Lyon, France
David Demirdjian	Vecna, USA
Joachim Denzler	Friedrich Schiller University of Jena, Germany
Thomas M. Deserno	Aachen University of Technology (RWTH), Germany
Michel Dhome	Institut Pascal, France
Sotirios Diamantas	Athens Information Technology, Greece
Yago Diez	University of Girona, Spain
Jana Dittmann	Otto-von-Guericke-Universität Magdeburg, Germany
Alon Efrat	University of Arizona, USA
Mahmoud El-Sakka	The University of Western Ontario, Canada

Mohan Kankanhalli	National University of Singapore, Singapore
Thomas Paul Karnowski	Oak Ridge National Laboratory, USA
Etienne Kerre	Ghent University, Belgium
Anastasios Kesidis	National Center For Scientific Research, Greece
Sehwan Kim	WorldViz LLC, USA
Nahum Kiryati	Tel Aviv University, Israel
Syoji Kobashi	University of Hyogo, Japan
Sinan Kockara	University of Central Arkansas, USA
Seong Kong	Sejong University, Korea, Republic of
Stephan Kopf	University of Mannheim, Germany
Mario Köppen	Kyushu Institute of Technology, Japan
Andreas Koschan	University of Tennessee, USA
Dimitrios Kosmopoulos	University of Patras, Greece
Constantine Kotropoulos	Aristotle University of Thessaloniki, Greece
Arjan Kuijper	Fraunhofer Institute for Computer Graphics Research & TU Darmstadt, Germany
Paul Kwan	University of New England, Australia
Andreas Lanitis	Cyprus University of Technology, Cyprus
Agata Lapedriza	Universitat Oberta de Catalunya, Spain
Slimane Larabi	U.S.T.H.B. University, Algeria
Mónica G. Larese	CIFASIS-CONICET, National University of Rosario, Argentina
Sébastien Lefèvre	Université de Bretagne Sud, France
Baoxin Li	Arizona State University, USA
Jing Li	Nanchang University, China
Stan Z. Li	Chinese Academy of Sciences, China
Chin-Teng Lin	National Chiao Tung University, Taiwan
Daw-Tung Dalton Lin	National Taipei University, Taiwan
Huei-Yung Lin	National Chung Cheng University, Taiwan
Luis Jiménez Linares	University of Castilla-La Mancha, Spain
Jundong Liu	Ohio University, USA
Xiuwen Liu	Florida State University, USA
Angeles López	Universitat Jaume I, Spain
Jinhu Lu	Chinese Academy of Sciences, China
Rastislav Lukac	Foveon, Inc., USA
Ilias Maglogiannis	University of Piraeus, Greece
Baptiste Magnier	LGI2P de l'Ecole des Mines d'ALES, France
Hanspeter Mallot	University of Tübingen, Germany
Lucio Marcenaro	University of Genoa, Italy
Pere Millan Marco	Universitat Rovira i Virgili, Spain
Emmanuel Marilly	Alcatel Lucent Bell Labs France, France
Jean Martinet	LIFL/CNRS-UMR 8022-University of Lille 1, France
Mitsuharu Matsumoto	The University of Electro-Communications, Japan
Brendan McCane	University of Otago, New Zealand
Javier Melenchón	Universitat Oberta de Catalunya, Spain
Jaime Melendez	Universitat Rovira i Virgili, Spain

Leonid Mestetskiy	Lomonosov Moscow State University, Russian Federation
Jean Meunier	Université de Montréal, Canada
Cyrille Migniot	Université de Bourgogne - le2i, France
Dan Mikami	NTT, Japan
Steven Mills	University of Otago, New Zealand
Pradit Mittrapiyanuruk	Srinakharinwirot University, Thailand
Birgit Moeller	Martin Luther University Halle-Wittenberg, Germany
Thomas B. Moeslund	Aalborg University, Denmark
Bartolomeo Montrucchio	Politecnico di Torino, Italy
Samuel Morillas	Universidad Politécnica de Valencia, Spain
Davide Moroni	Institute of Information Science and Technologies (ISTI)-CNR, Italy
Kostantinos Moustakas	University of Patras, Greece
Lazaros Nalpantidis	Aalborg University, Denmark
Luiz Antônio Neves	UFPR - Universidade Federal do Paraná, Brazil
Mikael Nilsson	Lund University, Sweden
Takahiro Okabe	Kyushu Institute of Technology, Japan
Yoshihiro Okada	Kyushu University, Japan
Anselmo Cardoso de Paiva	Universidade Federal do Maranhao, Brazil
Gonzalo Pajares	Universidad Complutense de Madrid, Spain
Yanwei Pang	Tianjin University, China
Sharathchandra Pankanti	IBM - Exploratory Computer Vision Group, USA
Theodore Papadopoulo	Inria, France
Felipe Pinage	Federal University of Amazonas, Brazil
Edwige Pissaloux	Pierre and Marie Curie University, France
Stephen Pollard	Hewlett Packard Labs, UK
Ramasamy Ponalagusamy	National Institute of Technology, India
Stefan Posch	Martin Luther University Halle-Wittenberg, Germany
Charalambos Poullis	Cyprus University of Technology, Cyprus
Giovanni Puglisi	University of Cagliari, Italy
Xiaojun Qi	Utah State University, USA
Bogdan Raducanu	Computer Vision Center, Spain
Giuliana Ramella	CNR - Istituto per le Applicazioni del Calcolo M. Picone, Italy
Ana Reis	Instituto de Ciências Biomédicas Abel Salazar, Portugal
Huamin Ren	Visual Analysis of People Lab, Aalborg University, Denmark
Alfredo Restrepo	Universidad de Los Andes, Colombia
Phill Kyu Rhee	Inha University, Korea, Republic of
Elisa Ricci	University of Perugia, Italy
Alessandro Rizzi	Università di Milano, Italy
Erik Rodner	Friedrich Schiller University Jena, Germany
Marcos Rodrigues	Sheffield Hallam University, UK
Ramón Ruiz	Universidad Politécnica de Cartagena, Spain

Silvio P. Sabatini	University of Genoa, Italy
Ovidio Salvetti	National Research Council of Italy - CNR, Italy
Andreja Samcovic	University of Belgrade, Serbia
Javier Sánchez	University of Las Palmas De Gran Canaria, Spain
K.C. Santosh	The University of South Dakota, USA
Jun Sato	Nagoya Institute of Technology, Japan
Gerald Schaefer	Loughborough University, UK
Raimondo Schettini	University of Milano - Bicocca, Italy
Mário Forjaz Secca	CEFITEC, FCT/UNL, Portugal
Fiorella Sgallari	University of Bologna, Italy
Shishir Shah	University of Houston, USA
Xiaowei Shao	University of Tokyo, Japan
Lik-Kwan Shark	University of Central Lancashire, UK
Gaurav Sharma	University of Rochester, USA
Maryam Shokri	MaryMas Technologies LLC, USA
Luciano Silva	Universidade Federal do Parana, Brazil
Bogdan Smolka	Silesian University of Technology, Poland
Ferdous Sohel	Murdoch University, Australia
Lauge Sørensen	University of Copenhagen, Denmark
José Martínez Sotoca	Universitat Jaume I, Spain
Ömer Muhammet Soysal	Louisiana State University, USA
Filippo Stanco	Università di Catania, Italy
Liana Stanescu	University of Craiova, Romania
Mu-Chun Su	National Central University, Taiwan
Yajie Sun	Samsung Research America, USA
Shamik Sural	Indian Institute of Technology, Kharagpur, India
David Svoboda	Masaryk University, Czech Republic
Tamás Szirányi	MTA SZTAKI, Hungary
Ryszard Tadeusiewicz	AGH University of Science and Technology, Poland
Norio Tagawa	Tokyo Metropolitan University, Japan
Xue-Cheng Tai	University of Bergen, Norway
Jean-Philippe Tarel	French Institute of Science and Technology for Transport (IFSTTAR), France
Tolga Tasdizen	University of Utah, USA
H.R. Tizhoosh	University of Waterloo, Canada
Yubing Tong	University of Pennsylvania, USA
Yulia Trusova	Dorodnicyn Computing Centre of the Russian Academy of Sciences, Russian Federation
João Vilaça	DIGARC, Polytechnic Institute of Cavado and Ave, Portugal
Muriel Visani	Université de La Rochelle, France
Salvatore Vitabile	University of Palermo, Italy
Frank Wallhoff	Jade University of Applied Science, Germany
Wen-June Wang	National Central University, Taiwan
Yu Wang	Auxogyn, Inc., USA
Zuoguan Wang	3M Company, USA

Toyohide Watanabe	Nagoya Industrial Science Research Institute, Japan
Quan Wen	University of Electronic Science and Technology of China, China
Andrew Willis	University of North Carolina at Charlotte, USA
Christian Wöhler	TU Dortmund University, Germany
Stefan Wörz	University of Heidelberg, Germany
Guoan Yang	Xian Jiaotong University, China
Jucheng Yang	Tianjin University of Science and Technology, China
Vera Yashina	Dorodnicyn Computing Center of the Russian Academy of Sciences, Russian Federation
Hongfeng Yu	University of Nebraska - Lincoln, USA
Yizhou Yu	University of Illinois, USA
Saif Al Zahir	University of Northern British Columbia, Canada
Michalis Zervakis	Technical University of Chania, Greece
Qieshi Zhang	Waseda University, Japan
Yonghui (Iris) Zhao	Xerox Research Center Webster, USA
Huiyu Zhou	Queen's University Belfast, UK
Yun Zhu	UCSD, USA
Zhigang Zhu	City College of New York, USA
Li Zhuo	Beijing University of Technology, China
Peter Zolliker	Empa, Swiss Federal Laboratories for Materials Science and Technology, Switzerland
Ju Jia (Jeffrey) Zou	University of Western Sydney, Australia
Tatjana Zrimec	University of New South Wales, Australia

VISAPP Additional Reviewers

Hassan Afzal	University of Luxembourg, Luxembourg
Michel Antunes	SnT, Luxembourg
Eugene Borovikov	National Library of Medicine, USA
Feng Chen	University of Oklahoma, USA
Rene Grzeszick	TU Dortmund, Germany
Kassem Al Ismaeil	University of Luxembourg, Luxembourg
Michiel Kallenberg	Copenhagen University, The Netherlands
Zhen Lei	Institute of Automation, Chinese Academy of Sciences, China
Guanbin Li	Hong Kong University, SAR China
Shengcai Liao	CASIA, China
Weifeng Liu	University of Copenhagen, Denmark
Pedro Morais	Life and Health Sciences Research Institute (ICVS), Portugal
Antonio Moreira	Minho University, Portugal
Fabricio Batista Narcizo	IT University of Copenhagen, Denmark
Akshay Pai	University of Copenhagen, Denmark
Szilard Vajda	National Institutes of Health, USA
Ruobing Wu	The University of Hong Kong, SAR China

Invited Speakers

Andrea Cavallaro	Queen Mary University of London, UK
Daniel Thalmann	Nanyang Technological University, Singapore
Gerik Scheuermann	Universtät Leipzig, Germany
Mauro Barni	Università di Siena, Italy

Contents

Computer Vision Theory and Applications

Invited Paper

First-Person Palm Pose Tracking and Gesture Recognition in Augmented Reality

Daniel Thalmann[1], Hui Liang[1(✉)], and Junsong Yuan[2]

[1] Institute for Media Innovation, Nanyang Technological University,
50 Nanyang Avenue, Singapore 639798, Singapore
{danielthalmann,lianghui}@ntu.edu.sg
[2] School of Electrical and Electronics Engineering, Nanyang Technological
University, 50 Nanyang Avenue, Singapore 639798, Singapore
jsyuan@ntu.edu.sg

Abstract. We present an Augmented Reality solution to allow users to manipulate and inspect 3D virtual objects freely with their bare hands on wearable devices. To this end, we use a head-mounted depth camera to capture the RGB-D hand images from egocentric view, and propose a unified framework to jointly recover the 6D palm pose and recognize the hand gesture from the depth images. The random forest is utilized to regress for the palm pose and classify the hand gesture simultaneously via a spatial-voting framework. With a real-world annotated training dataset, the proposed method shows to predict the palm pose and gesture accurately. The output of the forest is used to render the 3D virtual objects, which are overlaid onto the hand region in input RGB images with camera calibration parameters to provide seamless virtual and real scene synthesis.

1 Introduction

Augmented Reality (AR) is now widely used in wearable devices such as the Microsoft HoloLens and the Google glasses, which keeps users aware of the real world and provides additional information by synthesizing real visual cues with virtual graphics. Since the traditional input tools like the mouse and keyboard are cumbersome to carry and use with such devices, it would be more favorable if users can use their bare hands to convey commands and inputs to hardware. For instance, a user can move his hands and pose specific gestures for content selection like a traditional mouse [21], or can grasp and manipulate virtual objects for immersive experiences [2].

There have been quite a few AR applications based on vision-based hand tracking and gesture recognition. In [17] the Handvu AR system is proposed to track 2D hand position and recognize key postures with a RGB camera, which are used to interact with the virtual graphical elements, *e.g.*, menu selection, keyboard typing or 3D objects dragging. However, only 2D hand translation is inadequate for fully 3D AR interaction. In [18] the Handy AR system is proposed to track the 3D translation and rotation of the hand by detecting five fingertips

© Springer International Publishing Switzerland 2016
J. Braz et al. (Eds.): VISIGRAPP 2015, CCIS 598, pp. 3–15, 2016.
DOI: 10.1007/978-3-319-29971-6_1

and registering them to a predefined template, which allows users to inspect the 3D virtual objects from different perspectives. The limitation is that the user must keep a fixed hand posture and cannot fully rotate his hand due to occlusion of the fingertips, which leaves a large range of blind angles for 3D inspection. In [2] a head-mounted AR system is proposed, in which the 6D palm motion is tracked with stereoscopic RGB inputs via 3D plane-fitting to the depth cue of the extracted hand region. However, the method is still sensitive to hand posture variations. In [3] the random forest is adopted to predict the normal vector of palm from silhouette images, which is used to manipulate virtual objects. Lacking 3D information, this method can only work in quite limited viewpoints.

Another related field is full degrees of freedom hand pose estimation. This field has gained considerable progresses with the recent advent of low-cost depth cameras [27,33]. However, there still lacks a unified framework for both hand pose tracking and gesture recognition, especially in the AR scenarios. Besides, despite the high flexibility of hand motion, people are usually comfortable with only a set of natural hand postures for interaction [31]. This indicates that 6-DOF palm motion, *i.e.* 3D translation and rotation, with a set of hand postures are sufficient for a lot of applications, which also requires less computation cost to predict.

In this chapter we aim to predict the 6-DOF palm pose and recognize hand gesture simultaneously from egocentric depth images to assist interaction in AR scenarios. Our AR system allows users to manipulate and inspect virtual objects freely from different viewpoints with the recovered 6-DOF palm pose and make color selection with the gesture. Particularly, to get realistic visual feedback, we define a visibility term of the virtual objects based on hand rotation angles to reflect hand-object occlusion, *e.g.* the object becomes more transparent when getting occluded. Technically, the random forest [5] is adopted for both regression of palm pose and classification of hand gestures jointly so that they can be predicted together. Following the previous work on spatial-voting based pose estimation and gesture recognition [15,33], the random forest is learned to map the local features of spatially-distributed voting pixels to the probabilistic votes for either the palm pose or the gesture class. During testing, the per-pixel votes from the spatial-voting pixels are fused for pose and gesture prediction, which proves quite robust against noisy inputs.

2 Literature Review

This section reviews the recent techniques in vision-based hand pose estimation and hand gesture recognition. The generative model-fitting methods and discriminative methods are two main categories of methods for vision-based hand pose estimation. The model-fitting methods are built upon a deformable hand model and seek for the optimal pose by iterative adjustment of pose parameters of the model and compatibility check between model features and input images. In [20] the feasible hand configuration space is discretized and indexed with a KD-tree. The Nelder-Mead simplex algorithm is adopted to search for the hypothesized pose that best matches the input in terms of edge and silhouette similarities. However, no quantitative results are reported. In [30] multiple hand silhouettes are extracted from the images captured with several cameras around the hand, where the background is set using blue boards for easy hand segmentation. A voxel model is generated with the multi-view data and matched to a 3D hand model, and the optimal pose is sought to make the hand model surface stay inside the voxels. In [9] the texture and shading of the skin are captured from input images and synthesized in the hand model, and the illumination sources are controlled in real scenario and simulated during hand modeling. A variational formulation is proposed to estimate the full DOF hand pose. In this way hand pose is recovered quite accurately since matching ambiguity is largely reduced. However, this method is difficult to use in real HCI scenarios. In [22] a Kinect depth camera is adopted to capture the hand image as it can better handle the background clutter and pose ambiguity in monocular color image, the particle swarm optimization algorithm is used to find the optimal pose that best fits the image projection of a 3D hand model to the input depth image and skin silhouette. With the point clouds generated by the depth camera, the iterative closest points algorithm and its extensions to articulated objects are also commonly used for hand pose estimation [23,26], which iteratively build point-to-point correspondences between model and input point cloud and seek for the skeleton transform to minimize the distance between the point pairs.

The discriminative methods infer the hand pose parameters by directly mapping of the image features to pre-indexed templates. Generally, they need to build a large dataset to cover the possible hand postures, and each template in the dataset contains certain features for matching and the associated pose parameters. The dataset are usually indexed for fast search. During testing, the input hand pose is recovered by looking for the templates that share the similar features. In [29] the hand edge image is encoded into a score value vector by matching to a pre-defined set of shape templates, and a multivariate relevance vector machine uses it as the input to retrieve some pose hypotheses. The optimal pose is obtained by a verification stage with the hand model projection. In [11] an isometric self-organizing map is used to learn a nonlinear mapping between image features and pose, which reduces the dataset redundancy by grouping templates with similar features and poses together. The hand edges are captured at only depth discontinuities with a multi-flash camera and encoded into shape context for matching. In [31] a two color camera system is presented to capture 6 DOF palm motion and simple gestures like pinching or pointing for both hands. A pair of hand silhouettes are extracted and coded into binary strings for fast query in the database to retrieve the hand pose. In [33], the random forest is adopted to directly regress for the hand joint angles from depth images. With a pre-trained forest, each pixel casts its votes for the joint positions individually, and the votes from all the pixels are fused to a set of candidates. The optimal one is determined by a verification stage with a hand model. A similar regression forest base method is proposed in [28], with the new characteristic that transfer learning is utilized to handle the discrepancy between synthesized and real-world data. In [16] the authors propose to utilize the regression forest to predict the hand pose. To resolve the ambiguous predictions, their method first finds a set of candidate locations for each joint through mode-seeking, and then applies the bone length constraints to obtain the optimal combination of the different joint locations via Dynamic Programming.

Both types of methods have their pros and cons. The model-fitting methods are sensitive to initialization. The discriminative methods are fast and robust to initialization, but require a large amount of training data and can only produce discrete pose predictions. Therefore, they can be combined to supplement each other so that their advantages can both be exploited. For instance, model-fitting can serve as a verification stage after the discriminative pose retrieval stage [29,33]. On the contrary, pose retrieval can also serve as an initialization stage for model-based fitting. In [32] a human body pose tracking framework based on 3D model fitting is proposed. While the input body size can vary a lot, the random forest classifier provides rough body parsing for fitting the size of the 3D model to the real inputs as well as for initialization and recovering from tracking failure. Both methods can also be used for pose estimation independently and their predictions are finally fused up to certain criteria. In [4] the geodesic extrema are extracted from the depth images, which are used to retrieve the candidate body pose by searching in the database of geodesic extrema templates. Another candidate pose is obtained by fitting a mesh body model to the depth image, and the final prediction is taken to fit to both estimations. In [24], in which the protrusive fingertips are detected by morphological analysis in the depth image. The partial hand pose is recovered from the possible incomplete 3D fingertip positions and used for initialization for the subsequent model-fitting stage, which can help to speed up convergence as well as to avoid local optima.

As hand gestures can be dynamic or static, vision-based gesture recognition also contains two sub-groups: dynamic gesture recognition and static gesture recognition. Dynamic gesture recognition takes both the shapes and motion information of the hands into account during recognition. The motion history image [8] encodes the continuous actions into a single image template and has been adopted to recognize the directional movement of the hand [13]. To better model the dynamics of hand motion, the Hidden Markov Model (HMM) has also been adopted [6], in which a separate HMM is trained for each dynamic gesture respectively. During testing, the gesture is recognized so that the corresponding HMM maximizes the posterior probability conditioned on the video inputs. Static gesture recognition mainly focuses on analyzing the gestural information of the hand shape extracted from the visual images. In [7] the Haar-like features are combined with a cascaded classifier to recognize a set of static gestures. In [10] the statistics of local orientation of each pixel in input images are analyzed by constructing an orientation histogram, which is invariant to both transitions and rotations. A nearest neighbor classifier is adopted to recognize the gestures under various pose variations. In [25] the Kinect is used to get the depth contour of the hand, and a Finger-EarthMover's Distance is proposed to measure the dissimilarity between hand shapes. The method is robust to background clutters and orientation and scale variations. In [34] a histogram of 3D facets is proposed to encode the local 3D shape of the hand surface in the depth image. This descriptor is aggregated via spatial-pooling and the support vector machine is adopted for gesture classification based on the aggregated representation.

3   The Overall Framework

As discussed in Sect. 1, our goal is to recover both the 6D palm motion and the gesture semantics from continuous depth image sequences to assist hand manipulation in Augmented reality. Here the unconstrained 6D palm motion $\boldsymbol{\Phi}$ consists of the translation and rotation in 3D space, which are defined as the Euler angles of pitch, yaw and roll rotations of the hand and the 3D position of the palm center, $i.e.$ $\boldsymbol{\Phi} = (\boldsymbol{\theta}, \boldsymbol{v})$, where $\boldsymbol{\theta} = (\theta_x, \theta_y, \theta_z)$ is the global rotation, and $\boldsymbol{v} = (x_c, y_c, z_c)$ is the palm center position. For gesture recognition, each input hand image is assumed to belong to a predefined alphabet $l \in L$. In our AR system, different hand gestures are used for object color selection, and 6D palm pose is used to change the position and viewpoint of the objects for inspection. Let the image observation be I. Given the sequence of the input depth images, our goal is to obtain the MAP estimation of both the palm pose $\boldsymbol{\Phi}_t^*$ and the gesture label l_t^* at each time t conditioned on all the available frames $I_{1:t}$. The inference problem is formulated as:

$$
\begin{aligned}
\boldsymbol{\Phi}_t^*, l_t^* &= \arg\max_{\boldsymbol{\Phi}_t, l_t} P(\boldsymbol{\Phi}_t, l_t | I_{1:t}) \\
&= \arg\max_{\boldsymbol{\Phi}_t, l_t} P(\boldsymbol{\Phi}_t | I_{1:t}) P(l_t | I_{1:t}) \\
&= \arg\max_{\boldsymbol{\Phi}_t, l_t} P(I_t | \boldsymbol{\Phi}_t) P(I_t | l_t) P(\boldsymbol{\Phi}_t, l_t | I_{1:t-1}),
\end{aligned}
\tag{1}
$$

where we assume the conditional independence of the palm pose and hand gesture on $I_{1:t}$, and they can therefore be estimated separately. The random forest [5] is adopted to predict $\boldsymbol{\Phi}$ and l from single frames. Besides, as the hand gesture only involves a small number of discrete values, we can have a close form solution to predict l_t^* according to formula (1). The optimal Bayesian estimation of l_t^* is obtained by:

$$
\begin{aligned}
l_t^* &= \arg\max_{l_t} P(l_t | I_{1:t}) \\
&= \arg\max_{l_t} P(I_t | l_t) P(l_t | I_{1:t-1}) \\
&= \arg\max_{l_t} P(I_t | l_t) \sum_{l_{t-1} \in L} P(l_t | l_{t-1}) P(l_{t-1} | I_{1:t-1}),
\end{aligned}
\tag{2}
$$

Note that the summation term in formula (2) can be easily calculated by enumerating all the discrete possible gesture classes over $l_{t-1} \in L$. Also, very fast switch between different hand gestures is rare in common HCI applications, such as the manipulative tasks. Therefore, we assign relative big probabilities for transition between the same gestures. Besides, for any transitions between $l_t \neq l_{t-1}$ the probabilities are defined to take equal smaller values. We have:

$$P(l_t|l_{t-1}) = \begin{cases} \beta & if\ l_t = l_{t-1} \\ \dfrac{1-\beta}{|L|-1} & otherwise, \end{cases} \tag{3}$$

where $\beta < 1$ is a constant, $|L|$ is the size of the hand gesture alphabet. In practice we find $\beta = 0.5$ works well enough for the manipulative tasks.

Similarly, the optimal Bayesian estimation of $\boldsymbol{\Phi}_t^*$ with the observations $I_{1:t}$ can be obtained by the following formula:

$$\begin{aligned} \boldsymbol{\Phi}_t^* &= \arg\max_{\boldsymbol{\Phi}_t} P(\boldsymbol{\Phi}_t|I_{1:t}) \\ &= \arg\max_{\boldsymbol{\Phi}_t} P(I_t|\boldsymbol{\Phi}_t)P(\boldsymbol{\Phi}_t|I_{1:t-1}) \\ &= \arg\max_{\boldsymbol{\Phi}_t} P(I_t|\boldsymbol{\Phi}_t) \int_{\boldsymbol{\Phi}_{t-1}} P(\boldsymbol{\Phi}_t|\boldsymbol{\Phi}_{t-1})P(\boldsymbol{\Phi}_{t-1}|I_{1:t-1}). \end{aligned} \tag{4}$$

However, due to the ambiguous per-frame predictions, the pose likelihood function $P(I_t|\boldsymbol{\Phi}_t)$ is generally non-Gaussian and forms multiple peaks. Therefore, there is no close-form solution for the integral term. To this end, we adopt the particle filter [14] to track the continuous palm motion to alleviate such ambiguity based on the single-frame likelihood $P(I|\boldsymbol{\Phi})$. The details are provided in the following sections.

4   Random Forest Prediction

The random forest [5] is an ensemble of T random decision trees, each of which is trained independently with a bootstrap training set. In our algorithm, it is used to map local pixel features to pose and gesture votes during testing, which are then used for final fusion via spatial-voting. The depth context descriptor [19] is adopted as the local pixel feature, which is defined as the depth differences between the current pixel and a set of context points. Figure 1 illustrates the pipeline for palm pose and gesture prediction.

In the random forest, each intermediate node has two children nodes and we store a single vote for palm pose and gesture at each leaf node. Let the vote be $(\bar{\boldsymbol{\theta}}, \bar{\Delta}, \bar{H})$, where $\bar{\boldsymbol{\theta}}$ is the prediction of the rotation angles, $\bar{\Delta}$ is the 3D offset between a pixel and the predicted palm center and $\bar{H}$ is the gesture class distribution. Given a query image I, a set of N_s voting pixels $\{p_i\}$ are first uniformly sampled in the hand region and then cast their pose and gesture votes independently. The pixel p_i branches down each tree in the forest by checking the feature value of its depth context until a leaf node is reached, and thus retrieves in total T votes from the leaf nodes. Let the votes for palm pose and gesture of pixel p_i be $\{\boldsymbol{\Phi}_{ij}, H_{ij}\}_{j=1}^T$, in which $\boldsymbol{\Phi}_{ij}$ is converted from the retrieved relative

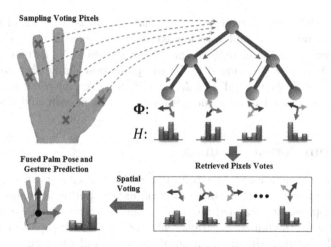

Fig. 1. Palm pose and gesture prediction with the random forest via spatial-voting.

pose vote $(\bar{\boldsymbol{\theta}}_{ij}, \bar{\Delta}_{ij})$ by setting $\boldsymbol{v}_{ij} = \bar{\Delta}_{ij} + \boldsymbol{v}_i$, where $\boldsymbol{v}_i$ is the 3D position of p_i, and H_{ij} is the retrieved probability histogram of the gesture label.

To obtain the final prediction, we need to aggregate the individual votes from all the voting pixels. First we consider the gesture l. Following [15], we define the gesture posterior $P(l|I)$ to be the average of all the per-pixel gesture votes:

$$P(l|I) = \frac{1}{N_s \times T} \sum_{i,j} H_{ij}. \tag{5}$$

Now we consider the palm pose $\boldsymbol{\Phi}$. Since it is unconstrained in 6D space, the different dimensions of $\boldsymbol{\Phi}$ are thus uncorrelated, i.e. $P(\boldsymbol{\Phi}|I) = \prod_{\alpha \in \boldsymbol{\Phi}} P(\alpha|I)$, where $\alpha \in \boldsymbol{\Phi}$ is one dimension of $\boldsymbol{\Phi}$. Similar to [33], we use the Parzen density estimator to evaluate $P(\alpha|I)$:

$$P(\alpha|I) = \sum_i P(\alpha|p_i) = \sum_{i,j} P(\alpha|\alpha_{ij}), \tag{6}$$

where α_{ij} is one dimension of the pixel vote $\boldsymbol{\Phi}_{ij}$. For the palm translation parameters $\alpha \in \boldsymbol{v}$ we adopt the Gaussian kernel for $P(\alpha|\alpha_{ij})$ with an isotropic variance δ_v in all three dimensions. For the palm rotation parameters $\alpha \in \boldsymbol{\theta}$ we use a one-dimensional wrapped Gaussian kernel [12] to model $P(\alpha|\alpha_{ij})$ for these rotation parameters within the range $[0, 2\pi]$, which is basically infinite wrappings of linear Gaussian within $[0, 2\pi]$. However, according to [12], the summation over $z \in [-2, 2]$ approximates the wrapped Gaussian distribution well enough. Thus we have:

$$P(\alpha|\alpha_{ij}) = \begin{cases} \mathcal{N}(\alpha; \alpha_{ij}, \delta_v^2) & \alpha \in \boldsymbol{v} \\ \displaystyle\sum_{-2 \leq z \leq 2} \mathcal{N}(\alpha - 2z\pi; \alpha_{ij}, \delta_\theta^2) & \alpha \in \boldsymbol{\theta}, \end{cases} \tag{7}$$

Based on the above derivation we see that $P(\alpha|I)$ is still sum of Gaussians for both the translation and rotation parameters. Note that for single-frame estimation we can assume uniform priors for Φ_t and l_t. Therefore, the like-lihood functions $P(I_t|\Phi_t)$ and $P(I_t|l_t)$ for palm pose and hand gesture satisfy $P(I_t|\Phi_t) \propto P(\Phi_t|I_t)$ and $P(I_t|l_t) \propto P(l_t|I_t)$. We can combine these two terms into Eqs. 2 and 4 to infer the gesture and palm pose in continuous image sequences.

5   Random Forest Training

To train the random forest, we collect a real-world egocentric depth image dataset with ground truth gestures, rotation angles and palm center positions. The training pixels are sampled uniformly from each training image and associated with the depth context descriptor $\mathcal{D}_j$, the ground truth palm rotation θ_j, the offset Δ_j between the 3D position of the pixel and the ground truth palm center, and the gesture l_j. Each tree of the forest is initialized with an empty root node and built with a bootstrap subset of all the training pixels. Starting from the root node, the training samples are split into two subsets recursively to reduce the prediction errors at the child nodes. To this end, a set of candidate split functions $\{\psi\}$ are randomly generated as the proposals for node splitting at the non-leaf nodes, which takes the form $\mathcal{D}_b \leq \tau$, where $\mathcal{D}_b$ is a randomly selected dimension of $\mathcal{D}$ and τ is a random threshold value to check whether to branch to the left or right children. The optimal split function is selected to maximize a gain measure $G(\psi)$ based on either the palm pose or the gesture distributions of the training samples reaching the node:

$$\psi^* = \arg\max_{\psi} G(\psi)$$

$$= \arg\max_{\psi} \left(H(A) - \sum_{s \in \{l,r\}} \frac{|A_s(\psi)|}{|A|} H(A_s(\psi)) \right), \tag{8}$$

where A denotes the samples reaching the current node and A_l and A_r are the two subsets of A split by ψ. For gesture classification, $H(A)$ is defined as the entropy of gesture label distributions in A for node splitting:

$$H(A) = -\sum_{l \in L} P(l|A) \log P(l|A). \tag{9}$$

For palm pose regression, the function $H(A)$ is defined as the variance of the palm pose among the samples in A to measure the pose uncertainty. As the palm motion is unconstrained, the variance of A can be calculated via $H(A) = \delta_\Phi^2(A) = \delta_\Delta^2(A) + \delta_\theta^2(A)$, where δ_Δ^2 is the variance of the 3D sample offsets and δ_θ^2 is the variance of the 3D rotation angles. Following the assumption that the dimensions of Φ are uncorrelated, we have $\delta_\theta^2 = \sum_{\alpha \in \{\theta_x,\theta_y,\theta_z\}} \delta_\alpha^2$. As the angle follows circular distribution, δ_α^2 is estimated by:

$$\delta_\alpha^2 = 1 - \sqrt{\left[\frac{1}{|A|} \sum_{j \in A} \cos \alpha_j \right]^2 + \left[\frac{1}{|A|} \sum_{j \in A} \sin \alpha_j \right]^2}, \tag{10}$$

At the non-leaf nodes we randomly pick up either the classification mode or the regression mode for node splitting. The training samples are split into left and right branches based on the optimal split function ψ^*. The samples reaching each branch are then used to construct a new tree node by either continuing the splitting procedure or ending up splitting to obtain a leaf node. This is done by checking whether certain stopping criteria are met, *e.g.* the pose and gesture of the samples are pure enough, or the maximum depth is reached. The gesture class distribution $\bar{H}$ stored at the leaf node can be obtained by counting the number of samples belonging to each gesture l in the sample set A reaching it. The pose votes is taken as the mean pose of A:

$$\bar{\Delta} = \frac{1}{|A|} \sum_{j \in A} \Delta_j$$

$$\bar{\alpha} = \mathrm{atan2} \left[\frac{1}{|A|} \sum_{j \in A} \sin \alpha_j, \frac{1}{|A|} \sum_{j \in A} \cos \alpha_j \right].$$

(11)

6   Quantitative Evaluation

We collect a set of 7.2 K depth images with a SoftKinetic DS325 sensor, and annotate each image with hand gesture, palm rotation and center position in a semi-auto way. The gesture set consists of three basic templates, and the fingers are also allowed to move in small ranges for each of them, as illustrated at left of Fig. 2. The resolution of the depth images is 320×240. The forests consist of three trees with maximum depth of 20. During testing, 1000 pixels are uniformly sampled from the hand region for spatial voting. The program is coded in C++/OpenCV, and tested on a PC with Intel i5 750 CPU and 4G RAM. The experiment is based on single-frame evaluation, with 80 % of the images for forest training and the rest 20 % for testing. The error metric for palm center prediction is defined as the Euclidean distance between the prediction and the ground truth. For the palm rotation angles we follow the conventions [1] to define the prediction error between prediction $\tilde{\alpha}$ and ground truth α as their absolute difference:

$$D(\tilde{\alpha}, \alpha) = |(\tilde{\alpha} - \alpha) \mod \pm 180°|. \tag{12}$$

We compare two different modes for forest training, *i.e.* REG: the forest is completely learned for pose regression; HYN: the forest is learned for both pose regression and gesture classification. This is to verify whether simultaneous pose and gesture prediction will degrade the palm pose prediction performance. The results are presented in Fig. 3, which shows the percentage of predictions that are within a threshold of either v_T (cm) or θ_T (Degree) from the ground truth. We can see the pose prediction performances of these two modes are indeed quite close. The right of Fig. 2 shows the gesture recognition confusion matrix with the HYN mode in terms of the absolute number of test images classified to each gesture, and the average accuracy is 97.2 %. The average time cost to process one frame is less than 40 ms, which is sufficient for real-time interaction.

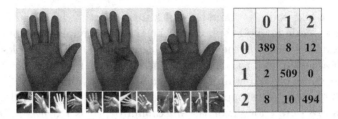

Fig. 2. Left: hand gesture alphabet. The small images show hand viewpoint and shape variations. Right: confusion matrix to recognize gesture 0, 1 and 2.

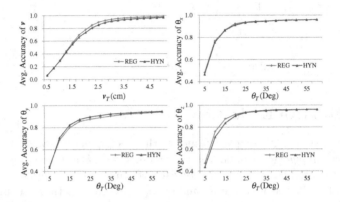

Fig. 3. Palm pose prediction results in REG and HYN modes with respect to different thresholds.

7   Augmented Reality System

With the trained random forest in Sect. 4, we build an AR system to allow users to manipulate the virtual objects with their bare hands, so that the virtual objects are visually put upon the palm for arbitrary poses in the RGB images and the gestures are used to switch among different object colors. To this end, the depth and color cameras of the SoftKinetic sensor are calibrated in advance to get the 3D transformation matrix M_{dc} from the coordinate system centered at the depth camera to that of the color camera. The random forest predicts the palm pose Φ^* and gesture l^* from the depth images, and a Kalman filter is adopted to smooth the pose prediction. Let the transformation matrix of Φ^* be M_Φ. The position and orientation of the virtual object are transformed by $M_\Phi \times M_{dc}$ and projected onto the image plane with OpenGL, which is then overlaid on the RGB image. Besides, we add a visibility term $\zeta \in [0, 1]$ to demonstrate the occlusion between hand and virtual objects, which is defined as:

$$\zeta = \begin{cases} 1 & |\theta_y| \leq 90° \\ 1 - |\theta_y|/\,180° & otherwise, \end{cases} \tag{13}$$

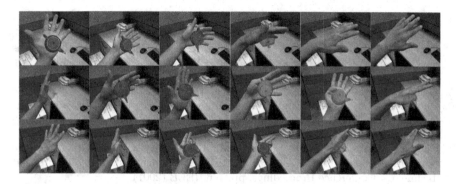

Fig. 4. Virtual object inspection. In several frames the object gets fully transparent due to occlusion.

where θ_y is rounded within $[-\pi, \pi]$. The visibility is implemented via controlling the transparency effect in OpenGL with ζ. That is, the virtual object is fully opaque when the palm is facing the camera, and becomes linearly transparent with θ_y when it rotates backwards. Figure 4 illustrates several exemplar frames of our video demonstration.

8   Conclusions

This chapter presents a unified framework to recover the 6-DOF palm motion and recognize hand gestures in egocentric depth images with the random forest, with an emphasis on its application in first-person viewpoint AR scenarios. The method is tested on a real-world dataset containing large viewpoint variations and different gestures and shows to produce accurate predictions. Based on the output of the random forest and camera calibration parameters, the virtual objects are rendered and overlaid onto the hand in the input RGB images, which provides seamless virtual and real scene synthesis. Especially, we introduce a visibility parameter based on hand rotation angles, so that the user can obtain quite realistic object-hand occlusion feedback.

References

1. Agarwal, A., Triggs, B.: Recovering 3D human pose from monocular images. IEEE Trans. Pattern Anal. Mach. Intell. **28**(1), 44–58 (2006)
2. Akman, O., Poelman, R., Caarls, W., Jonker, P.: Multi-cue hand detection and tracking for a head-mounted augmented reality system. Mach. Vis. Appl. **24**(5), 931–946 (2013)
3. Asad, M., Slabaugh, G.: Hand orientation regression using random forest for augmented reality. In: De Paolis, L.T. (ed.) Augmented and Virtual Reality. LNCS, vol. 8853, pp. 159–174. Springer, Switzerland (2014)

4. Baak, A., Müller, M., Bharaj, G., Seidel, H.-P., Theobalt, C.: A data-driven approach for real-time full body pose reconstruction from a depth camera. In: Fossati, A., Gall, J., Grabner, H., Ren, X., Konolige, K. (eds.) Consumer Depth Cameras for Computer Vision. Advances in Computer Vision and Pattern Recognition, pp. 71–98. Springer, London (2013)
5. Breiman, L.: Random forests. Mach. Learn. **45**(1), 5–32 (2001)
6. Chen, F.-S., Fu, C.-M., Huang, C.-L.: Hand gesture recognition using a real-time tracking method and hidden Markov models. Image vis. comput. **21**(8), 745–758 (2003)
7. Chen, Q., Georganas, N.D., Petriu, E.M.: Real-time vision-based hand gesture recognition using Haar-like features. In: IEEE Instrumentation and Measurement Technology Conference Proceedings, pp. 1–6. IEEE (2007)
8. Davis, J.W., Bobick, A.E.: The representation and recognition of human movement using temporal templates. In: IEEE Conference on Computer Vision and Pattern Recognition, pp. 928–934. IEEE (1997)
9. de La Gorce, M., Fleet, D.J., Paragios, N.: Model-based 3D hand pose estimation from monocular video. IEEE Trans. Pattern Anal. Mach. Intell. **33**(9), 1793–1805 (2011)
10. Freeman, W.T., Roth, M.: Orientation histograms for hand gesture recognition. In: International Workshop on Automatic Face and Gesture Recognition, vol. 12, pp. 296–301 (1995)
11. Guan, H., Feris, R.S., Turk, M.: The isometric self-organizing map for 3D hand pose estimation. In: IEEE International Conference on Automatic Face and Gesture Recognition, pp. 263–268. IEEE (2006)
12. Herdtweck, C., Curio, C.: Monocular car viewpoint estimation with circular regression forests. In: IEEE Intelligent Vehicles Symposium (2013)
13. Hsieh, C.C., Liou, D.H., Lee, D.: A real time hand gesture recognition system using motion history image. In: International Conference on Signal Processing Systems, pp. V2-394–V2-398. IEEE (2010)
14. Isard, M., Blake, A.: Condensation—conditional density propagation for visual tracking. Int. J. Comput. Vis. **29**(1), 5–28 (1998)
15. Keskin, C., Kıraç, F., Kara, Y.E., Akarun, L.: Hand pose estimation and hand shape classification using multi-layered randomized decision forests. In: Fitzgibbon, A., Lazebnik, S., Perona, P., Sato, Y., Schmid, C. (eds.) ECCV 2012, Part VI. LNCS, vol. 7577, pp. 852–863. Springer, Heidelberg (2012)
16. Kirac, F., Kara, Y.E., Akarun, L.: Hierarchically constrained 3D hand pose estimation using regression forests from single frame depth data. Pattern Recogn. Lett. **50**, 91–100 (2014)
17. Kolsch, M.: Vision based hand gesture interfaces for wearable computing and virtual environments. Doctoral thesis, University of California, Santa Barbara (2004)
18. Lee, T., Hollerer, T.: Handy ar: markerless inspection of augmented reality objects using fingertip tracking. In: IEEE International Symposium on Wearable Computers, pp. 83–90 (2007)
19. Liang, H., Yuan, J., Thalmann, D.: Parsing the hand in depth images. IEEE Trans. Multimedia **16**(5), 1241–1253 (2014)
20. Lin, J.Y., Wu, Y., Huang, T.S.: 3D model-based hand tracking using stochastic direct search method. In: IEEE International Conference On Automatic Face and Gesture Recognition, pp. 693–698. IEEE (2004)
21. Lo, R., Chen, A., Rampersad, V., Huang, J., Wu, H., Mann, S.: Augmediated reality system based on 3D camera selfgesture sensing. In: IEEE International Symposium on Technology and Society, pp. 20–31 (2013)

22. Oikonomidis, I., Kyriazis, N., Argyros, A.A.: Efficient model-based 3D tracking of hand articulations using kinect. In: British Machine Vision Conference, vol. 1, p. 3 (2011)
23. Pellegrini, S., Schindler, K., Nardi, D.: A generalisation of the icp algorithm for articulated bodies. In: British Machine Vision Conference, vol. 3, p. 4. Citeseer (2008)
24. Qian, C., Sun, X., Wei, Y., Tang, X., Sun, J.: Realtime and robust hand tracking from depth. In: IEEE Conference on Computer Vision and Pattern Recognition, pp. 1106–1113. IEEE (2014)
25. Ren, Z., Yuan, J., Meng, J., Zhang, Z.: Robust part-based hand gesture recognition using kinect sensor. IEEE Trans. Multimedia 15(5), 1110–1120 (2013)
26. Schroder, M., Maycock, J., Ritter, H., Botsch, M.: Real-time hand tracking using synergistic inverse kinematics. In: IEEE International Conference on Robotics and Automation, pp. 5447–5454. IEEE (2014)
27. Sun, X., Wei, Y., Liang, S., Tang, X., Sun, J.: Cascaded hand pose regression. In: IEEE Conference on Computer Vision and Pattern Recognition (2015)
28. Tang, D., Yu, T.-H., Kim, T.-K.: Real-time articulated hand pose estimation using semi-supervised transductive regression forests. In: IEEE International Conference on Computer Vision, pp. 3224–3231. IEEE (2013)
29. Thayananthan, A., Navaratnam, R., Stenger, B., Torr, P.H., Cipolla, R.: Pose estimation and tracking using multivariate regression. Pattern Recogn. Lett. 29(9), 1302–1310 (2008)
30. Ueda, E., Matsumoto, Y., Imai, M., Ogasawara, T.: Hand pose estimation using multi-viewpoint silhouette images. In: IEEE/RSJ International Conference on Intelligent Robots and Systems, vol. 4, pp. 1989–1996. IEEE (2001)
31. Wang, R., Paris, S., Popović, J.: 6D hands: markerless hand-tracking for computer aided design. In: The Annual ACM Symposium on User Interface Software and Technology, pp. 549–558. ACM (2011)
32. Wei, X., Zhang, P., Chai, J.: Accurate realtime full-body motion capture using a single depth camera. ACM Trans. Graph. 31(6), 188 (2012)
33. Xu, C., Cheng, L.: Efficient hand pose estimation from a single depth image. In: IEEE International Conference on Computer Vision, pp. 3456–3462. IEEE (2013)
34. Zhang, C., Yang, X., Tian, Y.: Histogram of 3D facets: a characteristic descriptor for hand gesture recognition. In: IEEE International Conference and Workshops on Automatic Face and Gesture Recognition, pp. 1–8. IEEE (2013)

Computer Graphics Theory
and Applications

A Sketch-Based Interface for 2D Illustration of Vascular Structures, Diseases, and Treatment Options with Real-Time Blood Flow

Patrick Saalfeld[1]([✉]), Alexandra Baer[1],
Uta Preim[2], Bernhard Preim[1], and Kai Lawonn[1]

[1] Department of Simulation and Graphics, Otto-von-Guericke University,
Magdeburg, Germany
saalfeld@isg.cs.uni-magdeburg.de
[2] Department of Diagnostic Radiology, Municipal Hospital Magdeburg,
Magdeburg, Germany

Abstract. We present a sketching interface, which enables physicians to illustrate various vascular structures, diseases, and treatment options with integrated blood flow. This sketch-based interface provides medical doctors with an effective tool to illustrate different medical scenarios and support patient education. This work integrates methods from sketch-based interfaces and GPU-supported computational fluid dynamics. The usability of the prototype was assessed qualitatively and quantitatively. Additionally, we performed a structured interview with a physician to evaluate the benefits with respect to patient education. The results of the evaluation confirmed the usability of the prototype as well as the usefulness to support physicians during the process of patient education.

Keywords: Sketch-based interface · Vascular diseases · Treatment options · Patient education · Computational fluid dynamics

1 Introduction

The field of vascular diseases causes 31 % of deaths worldwide [1] and has a big impact on economics (€ 196 billion in Europe [2]). In treatment planning, including aspects as prevention, diagnosis, and therapy, physicians discuss treatment options not only with colleagues, but also with patients. In this process of patient education, patients place great value on an understandable presentation of their disease and therapy. Such an appropriate presentation results in various positive aspects [3]:

- the time of treatment may be reduced,
- patients need less medication,
- they are more active in dealing with their diseases and act more responsible, and
- they are more independent from their attending physician.

© Springer International Publishing Switzerland 2016
J. Braz et al. (Eds.): VISIGRAPP 2015, CCIS 598, pp. 19–40, 2016.
DOI: 10.1007/978-3-319-29971-6_2

As also stated by Keulers [3], 42 % of patients feel not adequately informed. Therefore, a method that supports physicians to illustrate and discuss vascular diseases is useful. Such a method are sketches [4,5]. We present a prototype that allows the sketching of different vessel structures, vascular diseases, and various treatment options. This work is an extension of our previous paper [6] and inspired by the work of Zhu et al. [7] about sketching tubular shapes and simulating liquids. Since Zhu et al.'s work refers to wider area of application domains, it has limitations with respect to specifically sketching vascular diseases and treatments. Furthermore, it is not designed for tablet devices, which could be integrated well in the process of patient education. Therefore, we make the following contributions:

- presentation of methods to sketch vessels, vascular diseases, and treatment options,
- plausible integration of real-time blood flow in the sketched vessel structures,
- demonstrate the applicability for physicians with qualitative and quantitative usability measures, and
- showing the usefulness of patient education with a structured interview.

2   Medical Background

This work focuses on the vascular diseases of arteries, i.e., vessels which transporting blood from the heart to the peripheral capillary of the body. A common reason of these diseases is arteriosclerosis, which leads to a hardening of vessels by deposition of blood fat, thrombi and lime [8]. This deposition affects vessels in two problematic ways:

1. A weakening of the vessel wall.
2. A narrowing of the vessel up to a complete occlusion.

The weakening can lead to a dilation of the vessel that may result in an aneurysm, see Fig. 1(a). These are dangerous out of two reasons: first, they can rupture and release blood into the space around the vessel. This is especially critical in the brain, because patients with a ruptured cerebral aneurysm have a mortality rate of 40–60 % [9,10]. Second, the blood can clot inside the aneurysm, which could be carried away and block other arteries. The narrowed vessel, called a *stenosis* (see Fig. 1(b)), can lead to an under supply of involved structures or also cause a clot formation.

For the medical treatment of such vascular diseases, the physician can use several methods. The choice of the treatment depends on parameters such as anatomical access or size and shape of the pathology. In particular, this work focuses on *clipping* as an example of an extravascular treatment method as well as *coiling* and *stenting* as examples of intravascular methods. The *clipping* procedure, e.g., for treating a cerebral aneurysm, starts with a craniotomy to disclose the aneurysm. Afterwards, a titanium clip is placed across the aneurysm neck to stop the blood from entering into the aneurysm [11]. *Coiling* is performed

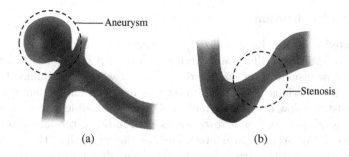

Fig. 1. Illustrations of a saccular aneurysm (a), which is the most common type, and a stenosis (b) caused by a narrowed vessel.

by entering an artery from the *inside*. The coil, a small titanium wire, is used to fill the aneurysm and to induce a thrombus formation [12]. The *stenting* method can be used to treat both, stenosis and aneurysms through an intravascular approach. Similar to the coiling procedure, a catheter is moved to the affected position from the inside. Afterwards, the stent is inflated with a balloon and forces the vessel to expand. For treating aneurysms, stents can help to support the involved vessel during, e.g., a coiling procedure.

Further descriptions of different forms of vascular diseases, their treatments, and possibilities for visualization and exploration are mentioned by Gasteiger [11]. A historical overview of different treatment options can be found in the work of Wong et al. [13].

3 Related Work

This work involves three main topics: computational hemodynamics, flow visualization, and Sketch-based Interfaces (SBIs), a form of user interface (UI) which deals with sketching.

3.1 Computational Hemodynamics

To simulate the behavior of blood, it is necessary to imitate a *non-Newtonian fluid*, i.e., a fluid with varying viscosity. Furthermore, in terms of fluid dynamics, blood is *compressible* and *inhomogeneous*. Such a simulation is complicated and expensive regarding calculation time [14]. Therefore, we consider blood as an incompressible, homogeneous *Newtonian fluid* to achieve a real-time simulation. Examples for methods to calculate non-Newtonian fluids can be found in [15, 16]. To describe the state of a fluid, there are two possibilities: the Lagrangian (particle-based) and the Eulerian (grid-based) description. We use the Eulerian description because the grid-based character is well suited to be calculated with fragment shaders on the GPU. Examples for the Lagrangian description can be found by Müller et al. [17] and Qin et al. [18]. Both deal with particle-based simulation of blood flow in vessels to support surgeons in virtual surgery scenarios.

3.2   Flow Visualization

The simulated flow can be visualized in a variety of ways. Since we use the Eulerian description, we illustrate parameters on each point of the discretized grid. These flow visualizations can be divided in three categories: direct, sparse, and dense visualizations [19,20]. The first two are relevant for our work since they can illustrate flow in a simple and thus, comprehensible way. In *direct flow visualizations*, glyphs such as arrows are used to indicate the flow direction on each grid-point. Additional parameters, such as the strength of the flow, can be visually encoded with the arrow length [21]. Since direct visualization enables the user to get a fast overview of the flow, we use it in our work. *Sparse flow visualization* techniques uses lines to illustrate the flow. The lines are obtained by seeding and tracking particles in the flow. Here, a challenge is to find appropriate seed positions for the particles [20]. We want to enable the physician to control where the particles are seeded. This gives the possibility to show important regions to the patient. Therefore, the seeding points can be placed interactively.

3.3   Sketch-Based Interfaces

The usefulness of SBIs to communicate ideas and concepts is described among others by Jorge and Samavati [22]. They state that the communication of complex issues is possible without the necessity to draw precisely and accurately. However, this is a challenge regarding the automatic recognition of the sketch. As a consequence, several works deal with the interpretation of sketches [22]. The foundations for this were laid by Ivan Sutherland [23] with the program *Sketchpad*. SBIs can be seen as a part of post-WIMP (windows, icons, menus, and pointers) UIs because the sketching is performed with direct input, e.g., a pen or touch [24]. Therefore, the pointer component of the WIMP paradigm is no longer necessary [25]. Xiaogang et al. [26] and Naya et al. [27] showed the advantage of reality-based interfaces (such as SBIs) by comparing WIMP-based interaction with reality-based interaction and presented two findings: first, the users preferred the sketch-based approach and second, they were more efficient with it. SBIs contain three processing steps: *resampling* of the input data, *beautification*, and *recognition*. This paper deals only with the resampling of the input data. Examples for the beautification step can be found, e.g., by Igarashi et al. [28]. There, line segments were analyzed according to geometric constraints such as perpendicularity and parallelism. After that, the program recommends different options of how to interpret the lines. The user chooses an option by clicking on it. The processing step *recognition* describes a procedure where the sketch is compared with an internal representation of symbols. The similarity is expressed with a parameter. If this parameter exceeds a value, the sketch is interpreted as the compared symbol [22]. A simple way to integrate recognition in an application is described with the $1\textrm{\textcent}Recognizer$ [29].

An important use case of SBIs is geometric modeling, i.e., the creation of 3D structures. An example of an SBI for modeling (SBIM) medical structures is described by Pihuit et al. [5]. They describe methods to sketch and model branching vessels. To maintain the sketch-based look of 3D models, it is possible to visualize them with non-photorealistic rendering. A comparison of different line drawing in the medical domain can be found in the work of Lawonn et al. [30]. Contrary to SBIM, the following work addresses the creation of 2D vessels.

4   Methods

Our sketch-based interface consists of different concepts to create vessels with interior blood flow based on intuitive sketch-based interaction. In the following, we describe the methods and the implementation.

4.1   Blood Flow Simulation

We use the Eulerian grid-based method to describe the state of the fluid [14, 31, 32]. For the fluid simulation itself, we use the Navier-Stokes equations for *incompressible* and *homogenous* fluids, which are based on Newton's second law of fluid motion:

$$\frac{\partial u}{\partial t} = -(u \cdot \nabla)u - \frac{1}{\rho}\nabla p + v\nabla^2 u + F, \tag{1}$$

and

$$\nabla \cdot u = 0. \tag{2}$$

Equation 1 (*momentum equation*) describes the behavior of the velocity vector field u under influences such as advection, diffusion, pressure, and external forces, which will be described in more detail in the following. Equation 2 (*continuity equation*) ensures the incompressibility by defining u as a divergence-free vector field.

To achieve a real-time fluid simulation, the grid size is an important factor. While a small grid size accelerates the calculation time, details like whirls may be lost. Another problem are obstacles. To faithfully simulate the behavior of blood flow in the vicinity of obstacles, it is necessary to model the obstacles in the simulation grid. This is achieved by marking grid cells as occupied. Thus, the grid resolution also affects the possible level of detail of the obstacles. To allow a high spatial resolution of grid cells, the simulation is performed on the GPU. For the GPU-based calculation, we used the fragment shader similar to [31].

To solve the equations for the differential operators, the finite difference method is used. Furthermore, Eq. 1 needs to be split up in single terms, which are calculated separately.

An overview of the calculation steps is illustrated in Fig. 2. In the following, we describe the mathematical terms and their effects in the fluid simulation.

The first term $-(u \cdot \nabla)u$ describes the self-advection of the fluid, which is the process of moving the velocity itself through the fluid. Here, the self-advection is realized with semi-Lagrangian advection [32]. Mostly, fluid simulations use the Runge-Kutta method [33] for the integration, which is less error-prone than

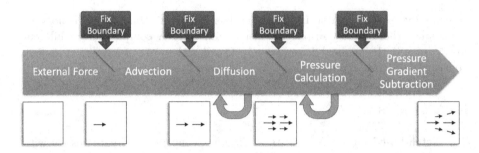

Fig. 2. Pipeline for the fluid simulation including the influence to the underlying vector field. Each step is realized with a separate fragment shader program. The steps provided with self-referencing arrows show the steps which are calculated with an adjustable amount of iterations. Here, a trade-off arises between accuracy (high amount of iterations) and calculation time (low amount of iterations).

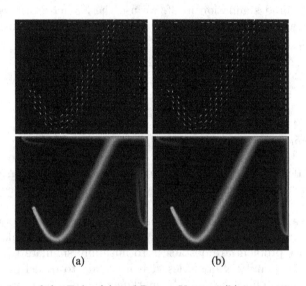

(a)   (b)

Fig. 3. Comparison of the Euler (a) and Runge-Kutta 4 (b) integration methods with a step size of 0.3. Both methods are depicted with a modified arrow plot visualization (top) and a scalar field visualization (bottom), which looks similar to a streakline visualization through the continuous placement of ink during the simulation. Since the differences between both methods are small, the faster Euler method is used.

the simpler and faster Euler method. Interestingly, as illustrated in Fig. 3, the differences between both methods are small. This is due to the small step size used in the application. Therefore, we used the Euler method to decrease the calculation time.

The pressure, a force that gradually spreads from regions with high to regions with low pressure, is described with the second term $\frac{1}{\rho}\nabla p$. The factor ρ is a constant to describe the density. Furthermore, this term also ensures the incompressibility of the velocity vector field, and thus, simultaneously ensures the Eq. 2. To achieve this, the term is calculated at the end. For a description of the derivation, see [14]. To calculate this term, it is necessary to solve a Poisson equation. This is accomplished by the Jacobi approach because it can be mapped directly to GPU facilities. For a discussion of different approaches, see [34].

The third term $v\nabla^2 u$ expresses the physical process of diffusion, i.e., the property of mixing materials without external forces. Here, v is a constant that describes the viscosity. The calculation of the diffusion also requires the solution of a Poisson equation. A disadvantage of diffusion is the resulting smoothing effect on the applied vector field, which causes a loss of details. Therefore, the choice whether diffusion is applied is left to the user.

The last term F describes external forces, which allows the user to influence the simulation dynamically. Normally, such forces are steady, so the influence of the force to the fluid is constant over time. This approach would not represent the pulsating character caused by the contractions of the heart. This pulsation can be imitated by applying a factor to the force, which changes over time. Therefore, we approximate the function measured by an electrocardiogram with the following formula:

$$f(x) = -\frac{3}{4}e^{-\frac{1}{2}(-\frac{1}{2}+x)^2} + 1,12e^{-\frac{x^2}{2}} - \frac{1}{4}e^{-\frac{1}{2}(1+x)^2} + 1. \tag{3}$$

This equation was determined by combining three Gaussian functions with varying heights and widths. In Fig. 4, the difference between a constant and a pulsating force is illustrated.

Finally, boundary conditions are used to simulate the behavior of the fluid at the vessel wall and the boundary of the sketching canvas. These conditions are necessary for the *velocity* vector field and for the scalar field, describing the *pressure*. For the velocity vector field, a Dirichlet boundary condition is used,

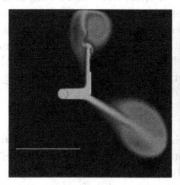
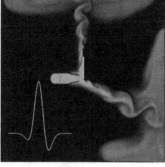

Fig. 4. In (a) a constant force is applied to the vector field. In (b) a heartbeat-like function is applied as a factor, which mimics a more realistic behavior. The corresponding functions are plotted at the left bottom.

which states that the velocity drops to zero at boundaries. For the scalar field, a Neumann condition is used, which states that the derivatives at boundaries are zero. To calculate the derivative, the normal of the boundary is necessary. The normals of the top, right, bottom, and left image boundaries are defined as $(0, -1), (-1, 0), (0, 1)$, and $(1, 0)$, respectively. For determining the normals of arbitrary boundaries, we use the neighborhood of the obstacle. That means that for each grid cell, which is marked as an obstacle, eight neighbor cells are analyzed. Depending on the state of these neighbors (marked as obstacle or no obstacle) the normal is approximated to one of eight possible directions. This approach is described in more detail by Wu et al. [35].

4.2 Blood Flow Visualization

To illustrate the unsteady vector field that represents the blood flow, we implemented two visualization concepts: (1) a direct and (2) a sparse flow visualization, see Fig. 3. The direct visualization is a modified line plot on which fans are drawn on an adjustable grid. Fans are used because they facilitate a fast realization of the flow direction for the user.

Especially for patients, a scientific visualization method for vector fields may be inappropriate. Thus, an additional method is used. It aims to be easily understandable to visualize the behavior of blood in areas such as aneurysms and stenoses in a descriptive way. The used scalar field visualization is inspired by the idea to place *colored ink* in the vector field (also known as *dye injection*). By diffusion and advection, this ink is transported through the vector field. The amount of ink is color-coded with a black-to-hue scale with different colors. This allows using multiple colors, e.g., to show how blood mixes in an aneurysm before and after a treatment, see Fig. 5.

The colors are taken from the *CIELAB* color space, which allows choosing colors that are roughly perceptually linearized regarding hue and brightness. To determine perceptually strongly different colors, the approach of Glaßer et al. [36] is used. If the ink is placed continuously over time, the visualization technique is similar to streaklines. In contrast to streaklines, the placed ink is not connected, but if the amount of placed ink is high enough, there is the impression of connectivity. A difficult task involved in sparse visualization techniques, namely to identify suitable seed point positions, is left to the user. This allows the physician to emphasize specific areas, which supports the patient's understanding.

4.3 Sketching

The obtained data from the input device inherits noise, which is caused by the conversion from the analog to a digital signal (quantization) as well as the imprecise input from the user. Especially the quantization, also depending on the sampling rate, reduces input information during fast input movements from the user. To remove the resulting noise and obtain equidistant input information, the received data of the input device is resampled and smoothed. These steps are commonly applied after the user finishes drawing, which causes abrupt

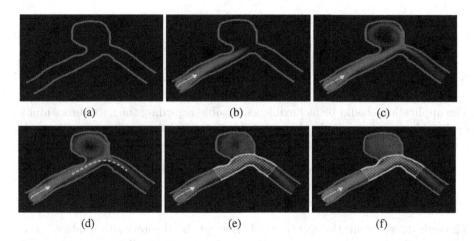

Fig. 5. Illustration of blood flow before and after a stenting treatment. First, the physician sketches a vessel with an aneurysm (a). Afterwards, he manipulates and visualizes the underlying fluid simulation to illustrate the blood flow (b-c). The consequences resulting from the stent treatment (d-e) are illustrated with another color (f).

changes in the sketch. To avoid these sudden changes, we use on-the-fly methods that are applied during sketching. This presents challenges regarding real-time capability. Ideally, the user did not even realize these steps. First, the resampling should reduce the obtained sample points. To achieve this, we use a simple strategy: we ignore all points that are too close to the last accepted point. We use the Euclidean distance to measure the distance of two points, see Fig. 6(a).

A disadvantage of this approach is that it leads to line smoothing and thus, complicates the sketching of zigzag lines. This can be neglected, because vessels, vascular diseases, stents, and coils usually do not have these shapes. To smooth the accepted sample points, a local Gaussian filter is used [37]. More precisely, the 1-neighborhood of the accepted points is used to adjust the points with a

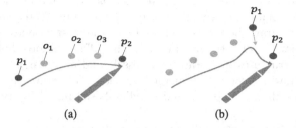

Fig. 6. (a) The new point p_2 has exceeded a specific distance to p_1 and thus, is sampled. The points o_1 - o_3 are too close and thereby omitted. (b) shows the smoothing approach. When p_2 is sampled, the pre-last point p_1 is translated. By transforming the pre-last point, a shrinking of the line is prevented.

Gaussian smoothing. The filter is only applied to the *pre-last* point, which was accepted to prevent the line segments from shrinking, see Fig. 6(b).

5   Application

The application should be as flexible as possible regarding the used direct input device. More precisely, it should be possible to control the application with touch and pen-based input. This implies the following limitations:

- Special functions from pen-based input devices like pressure sensitivity or additional buttons are not supported.
- Multitouch input is not supported.

Through these limitations, the used concept is theoretically usable for a mouse with a left button only. Furthermore, the application is designed according to major usability criteria, such as *suitability for the task* or *self-descriptiveness*. The evaluation of the application revealed possibilities for improvements, which were partially implemented. Improvements based on the quantitative part, the qualitative part, or the interview, are marked and discussed, respectively. The application can be downloaded under www.isg.cs.uni-magdeburg.de/~patrick/application/SketchingVessels.zip.

5.1   Sketch Vessels and Vascular Diseases

Mainly, drawing a vessel requires two lines representing the border of the vessel. Drawing each line separately would lead to strong variations in the resulting vessel structures. Therefore, we use a *create tool* that creates both lines simultaneously by using the sketched path as the *center line* and drawing the vessel wall around it. The general advantage of this process is that uniform vessels can be drawn easier and faster. We fix the width of the vessel to simplify the application. In contrast to Zhu et al. [7], where it is possible to draw vessels under and over already existing ones through a 2,5D sketching canvas, this work limits the sketching area to a *2D canvas*. This decision is motivated by the observation of what could happen if the user sketches over an already created vessel. Besides the possibility to draw the vessel over or under the existing ones, it is possible to *merge* the new vessel with the old ones. This offers the possibility to create more complex structures like branching vessels and aneurysms easily without changing the drawing mode. The merge behavior is illustrated in Fig. 7 and is realized with an already implemented polyline-based functionality in the used framework.

Additionally to the possibility to draw aneurysms, the application offers a possibility to draw stenoses. A *cut tool* is used to allow the user to create irregular non-symmetrical stenoses in a consistent sketch-based way. To prevent problems during cutting, e.g., ambiguity described by Heckel et al. [38], the user sees his sketched contour, which is used for the cutting process. Additionally, the start

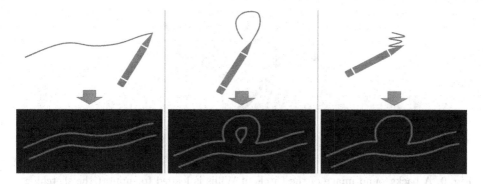

Fig. 7. The merging behavior can be used to draw a vessel containing an aneurysm by simply sketching on top of the previous drawing.

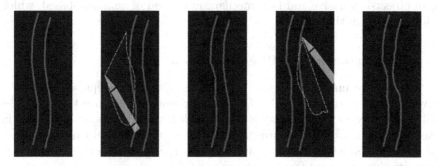

Fig. 8. The *cut tool* is used to remove structures from the already sketched vessel to create a stenosis.

and end point of the sketched contour are connected, and thus, span a cutting area. This area is subtracted from the existing vessels (see Fig. 8).

This tool does not only allow the creation of stenoses: in combination with the *create tool*, the user has a generic sketching tool, which allows the creation of any 2D structure under the usage of only two different modes. To support the medical expert during the sketching process, it is possible to load images in the background. The physician can use this function to load a slice of patient-specific MRI or CT data which contains, e.g., the vessel structures of the patient. In addition, frequently used vessel structures such as the Circle of Willis, can be loaded (see Fig. 9).

During the evaluation, some participants suggested a possibility to load vessel structures. This could help because it allows the physician to not only load standardized structures, but also patient-specific data. The functionality is implemented by loading a monochrome black-and-white image. Every black pixel is interpreted as a vessel and every white pixel as free space, where blood is able to flow. However, this approach has a disadvantage. It is only possible to show blood flow in the loaded vessel structure, but not to use the other tools, e.g.,

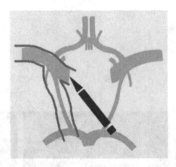

Fig. 9. A background image of the Circle of Willis is loaded to support the sketching process.

cutting or treating the vessel. This disadvantage can be avoided by implementing an object-based save and load mechanism instead of an image-based, which could be added in the future.

5.2   Manipulate and Visualize Blood Flow

The flow can be manipulated with a force term, which is represented with a 2D vector field (recall Sect. 4.1). To allow the user to influence this field in a flexible and easy way, a *direction tool* is provided, which is also implemented in a sketch-based way. After the *direction tool* is selected, the user can sketch lines that are represented by arrows pointing along the draw direction. This arrow represents the influenced force on the vector field.

To transfer the sketched arrow to the vector field, it would be obvious to manipulate only the vectors directly under the sketched arrow. This behavior is not desirable, because it only allows to manipulate the flow in small areas. Instead, the force is applied in a *region* around the sketched arrow. The size of this region is adjustable and is initially set to the width of a vessel. To achieve a natural effect, the force is slightly decreased at the border of the region by applying a Gaussian smoothing, resulting in a strong force at the center and a weak force at the border.

To visualize the flow, the user can use a *dye tool* to place ink (blood) in the fluid region. A circle of ink is placed by just tipping on the canvas or drawing over an area. The ink is interpreted as a source of infinite amount of ink, which is added in every render frame. The width of the ink area is adjustable, but set to the vessel width to allow the user to fill a vessel with ink in a fast way. To allow the physician to show the mixing behavior of fluid in an easy way, a new color is selected (recall Sect. 4.2) after each usage of the *dye tool*.

5.3   Treating Options

To show the patient how to treat aneurysms and stenoses, we implemented the following treatment methods:

- coiling,
- clipping, and
- stenting.

Initially, we implemented the *coiling tool* such that the user should fill out the aneurysm. This coil was illustrated with a line, which occurred during the drawing process. Our user study revealed that generating the coil with this approach is not applicable, since it takes too much effort and is too time-consuming. Therefore, we implemented another approach to improve the coil placement. First, an area is sketched, e.g., on the vessel wall of an aneurysm. After the user raises the pen, every sample point analyzes its neighborhood within a specific adjustable distance. This is performed by sending eight rays in a circular manner, starting at every sample point. Significantly less rays would result in an inaccurate vessel wall detection and more rays would not improve the result. We test every ray for a possible intersection with the vessel wall. If more than one ray collides with the wall, the ray with the shortest distance is chosen. After that, the corresponding sample point is placed on the intersection point of the ray. If no collision occurs, the point keeps the current position. The underlying grid cells are than treated as obstacles and thus, the blood flow changes dynamically corresponding to the drawn coil. Figure 10 illustrates this algorithm.

The *clipping method* was inspired by the line drawing method used in *Sketchpad* [23]. The point where the user starts drawing represents the first point of the clip. As long as the user draws with the pen, the current pen position represents the end point of the clip. These two points are connected with a dashed line to

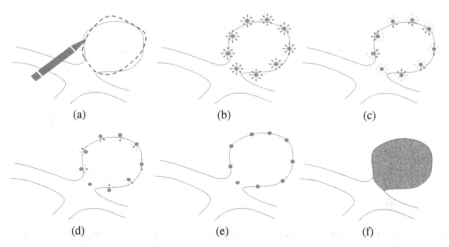

(a)   (b)   (c)

(d)   (e)   (f)

Fig. 10. (a) - (f) shows the algorithm to calculate the coil area. In (a) the user sketches imprecise over the vessel wall of the aneurysm. (b) shows the captured sample points as well as the rays, which are sent in a circular manner from each sample point. In (c) all rays that collide with the vessel wall are highlighted, from which those with the shortest distance are chosen (d). The sample points are moved to the intersection point or remain on their position if no intersection occurred (e). The adjusted sample points build the new area for the coil, which lies precisely on the vessel wall.

give the user a preview of the clipping result. After the user finishes drawing and raises the pen, the clip is placed and the blood flow simulation is affected by it.

The *stent placement* algorithm is inspired by the real treatment. Here, a balloon catheter is inflated to dilate the stent in the vessel. We use this inflating process for providing a stent placement algorithm. The user draws a line in the center of the vessel with the *stenting tool*, which represents the position of the balloon catheter. After the user finishes the sketching process, the application calculates the dilation of the stent. The stent should be dilated in the relevant vessel, but not enter the aneurysm. Since structures are not differentiated semantically (i.e., in vessels and aneurysms), the described behavior must be achieved in a different way. To accomplish this, the surrounding region of the sketched stent is analyzed. The algorithm is described in detail in Fig. 11. This method is more robust according to various input lines. The best results regarding visual aspects are achieved by drawing a line, which is as close as possible to the center of the vessel. Inaccurate lines may lead to an entering of the stent inside the aneurysm. A disadvantage of the described method is that it depends on four control points obtained through the start and the end point. If not all control points could be determined, e.g., if the vessel is too wide, not all normals are calculated and so the stent will not be placed. Similar to the coiling and clipping method, the grid cells under the stent are marked as occupied and thus, influence the fluid simulation.

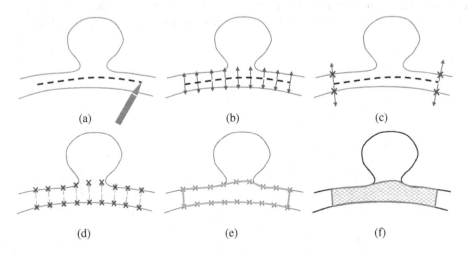

Fig. 11. (a) - (f) illustrate the algorithm to calculate the stent dilation. The user input is illustrated as a sketched line in (a). After the user finishes the stent, the normals of the start and end point are determined (b). The normals have the length of the vessel width. All four normals are tested if they intersect the vessel wall. If not, the stent is discarded. Otherwise, the distance l is determined, which is the longest length of the normal start point to its intersection point. Now, for every point of the sketched line, the normals with length l are determined (c). These normals are tested for intersections with the vessel wall. If they intersect, they are shortened to the intersection point, otherwise they keep the length l (d). The achieved end points of the normals are connected (e) and form the border of the stent, which is illustrated in (f).

5.4   Edit, Delete, and Copy Objects

The possibility to edit and delete objects such as stents and coils, is an important aspect regarding the controllability of the application. This is confirmed by the evaluation (see Sect. 6.1) in which the participants stated that a functionality to delete or edit objects would be useful. To realize this, it is necessary to implement a method to select already created objects. Since a requirement of the application is to control it with a pen without any further buttons, a suitable method is necessary. Besides the use of gestures or a double tap of the pen, the implemented approach is to *press and hold* the pen on the relevant object. Afterwards, the corresponding object is deleted. The possibility to edit objects, which is currently not implemented, can be realized with the same approach.

To provide the physician with an easy and fast possibility to illustrate more treatment options on the same vessel structures, a *copy tool* is added to the application. It allows to select an area, copy its content and paste it at another place. With this possibility, the patient can see different behaviors of the blood-flow, depending on the treatment option. This can help to understand why a specific treatment is chosen or why another is not possible.

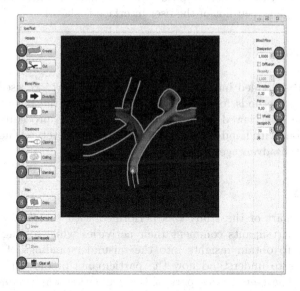

Fig. 12. Screenshot of the prototype. Buttons (1-2) are used to create and edit vessels, (3-4) are used to manipulate and visualize blood flow, (5-7) are used to sketch treatment options, (8) is used to select the *copy tool*, (9a-9b) are used to load a background image to assist the sketching or to load vessel structures out of a monochrome image and (10) is used to reset the whole canvas. The buttons on the right side (11-17) are used to influence the fluid simulation in different ways, e.g., to enable diffusion, change the viscosity, or show the arrow plot visualization.

5.5   User Interface

The structure of the user interface results from two considerations:

- The grid used for the fluid simulation is quadratic and
- a direct input device is used.

The first consideration implies that the canvas, in which the user sketches, is also quadratic. Due to the horizontal format of current displays this means that there is potential free space on the sides of the canvas. This space is used for the menus. In detail, the space is divided in a left and a right region and used for a semantic differentiation of the functionality. On the left side, menus and buttons are placed, which are used to create and manipulate vessels, to sketch treatment options, and to visualize and control blood flow. On the right side, there are menus to control simulation parameters such as the number of iterations for solving or to activate the diffusion process. The second consideration (using a direct input device) leads to the following requirements with respect to interaction. To achieve an easy interaction, all buttons have a bigger size and there are no sub-menus. Furthermore, classical WIMP input elements like spin boxes were omitted since exact inputs would be hard to enter with a pen or touch. In Fig. 12, a screenshot of the user interface is shown.

6   Evaluation

The evaluation is divided into two parts: in the first part, we used *qualitative* and *quantitative* methods for assessing the usability of the prototype. In the second part, we interviewed a physician to compare the procedure of typical patient education (with hand-drawn sketches) with the prototype and investigate advantages and disadvantages.

6.1   Usability

The *qualitative* part of the study was performed with the *think aloud protocol*, where the participants comment their activities while solving a problem. This is helpful to obtain insights into the misunderstandings of the participants as well as to understand how the participants predict the behavior of the prototype. The *quantitative* part of the evaluation was conducted with a questionnaire. The questions are modeled after the questionnaire for ergonomic principles from ISO 9241-110 (suitability for learning, suitability for the task, self-descriptiveness, conformity with user expectations, controllability, and error tolerance) [39]. The single questions were categorized in the different principles and rated with a 7-point Likert scale.

The evaluation started with a short introduction of the prototype on a SMARTBoard, a 70″ screen which allows pen interaction. All features were demonstrated and we asked to think-aloud and noted the spoken comments of the participants. Initially, the participants were asked to perform several tasks that were handed out in written form. For example, they should draw a vessel with a trifurcation. Then, they should use the *cut tool* to change the vessel to a bifurcation. Furthermore, they were asked to draw an aneurysm and treat it with the *clipping*, *coiling*, and *stenting tool*. Finally, they should create and visualize the blood flow in a specific direction. Afterwards, the participants were asked to fill out the questionnaire.

The evaluation was conducted with 14 researchers with medical visualization knowledge. We had three female and eleven male participants, aged from 25 to 44 with an average of 31 years. The participants are experienced computer users with an experience of 20 years on average (minimum: 14, maximum: 30). Ten participants were experienced with pen interaction. It took about 15 to 20 min to fill out the questionnaire. The statements of the participants are denoted with [P#].

Think-Aloud Method. Mostly, the participants were satisfied with the prototype, e.g., one stated that "it is possible to create vessels and aneurysms according to my own ideas" [P13]. Regarding the different tools, the majority of the participants had no problems using them. For example, it was stated that the *cut tool* "is more precise than conventional eraser tools" [P6]. Unfortunately, the *cut tool* leads to misunderstandings during the first use. This was caused by different expectations of its functionality, e.g., some participants thought that it can be used as a conventional eraser. After some practice, the functionality was understood and all participants rated this tool as positive. Another example of a positively rated tool is the *stenting tool*. Especially the automatic expanding behavior was noted as useful. Some participants highlight the alignment of the tools, as it supports the typical workflow (generate vessel, treatment, and visualize blood flow). In general, they stated that this prototype allows a fast and easy generation of vessels with simulated blood flow.

Questionnaire. The results of the questionnaire were determined by calculating the average of every answer. For this, we assigned the symbols of the 7-Point Likert scale $(- - -$ to $+ + +)$ to the values from -3 to 3. Figure 13 depicts the average for every category. The category *error tolerance* has the lowest rates. Here, the users were asked if the effort to correct an inadvertently drawn error is significantly high. Mostly, the participants stated that it should be possible to delete drawn objects and the effort is high to manually correct them by redrawing. Thus, we added the possibility to delete drawn objects like stents and coils individually, see Sect. 5.4. However, with an average of 1.68, this category was rated well. In summary, the ratings of all categories were positive and the participants were satisfied with the functional range. For refinement, we used the results of the evaluation and improved the prototype according to the suggestions of the participants.

6.2   Structured Interview

We performed a structured interview of about 20 min with a physician, which has 12 years of experience in the field of vascular diseases and patient education.

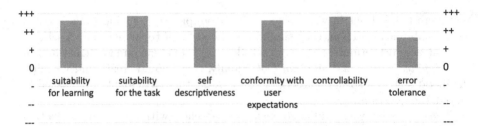

Fig. 13. The average results of the questionnaire for each category of the usability principles is shown.

The audio of the interview was recorded and analyzed after the interview. First, we asked the physician to explain typical patient education with an example and outline the typical education procedure. During typical patient education, the patient receives textual and image-based templates with respect to his disease. Additionally, the physician explains the intervention verbally and supports this with hand-drawn sketches (see Fig. 14). These can help to illustrate specific medical cases and answer individual questions from the patient. The description of the physician revealed the following disadvantages of the *hand-drawn* sketches:

- To correct errors, the physician draws over the existing sketch, which results in a cluttered image.
- The blood flow and the implications of a treatment are usually not drawn, because it is hard to illustrate these and could lead to a confusing image.
- The physician only uses one color to draw all structures, which makes it difficult to distinguish between different elements such as vessels and a clip.
- Some patients have problems to understand the sketch, because they find it difficult to imagine diseases and treatment options.

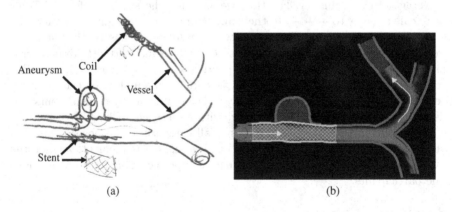

Fig. 14. On the left side is the hand sketch of a physician with additional labels. On the right side are the same structures made with the application, which the physician assessed clearer and more understandable.

After that, we explained the prototype and asked to perform the patient education again using our application. Thereby, we let the physician compare his hand-drawn images with sketches from the prototype which showed the following:

- By using different colors, symmetrical, consistent, and uniform structures (e.g., for the vessels or stents), the result is more clear, descriptive and plastic than the hand sketching, and thereby, the perception is supported.
- Due to the illustration of blood flow, the vessel structures, the effects of different intervention methods and possible complications are more understandable and imaginable.

Furthermore, the physician stated additional advantages of the prototype, e.g., due to the clearer resulting images, other persons, who were not involved in the sketching process, are able to understand the sketch. This is difficult with hand sketches, because of their cluttering nature. The physician rated the tool as easy to learn and use, which matches the results of our usability evaluation.

7   Conclusions and Future Work

This work provides methods to sketch vessels, vascular diseases, and treatment options. It is possible to interactively create and manipulate blood flow, which adapts in real-time depending on the sketch.

The usability of the prototype was assessed with *qualitative* and *quantitative* methods. The positive feedback of the evaluation indicates that the proposed concept and prototype are suitable for sketching vascular structures and treatment options. The *structured interview* with the physician revealed further benefits compared to hand-drawn sketches and confirmed the idea to improve patient education and intelligibility by integrating animated blood flow.

A limitation of this work is the representation of the vessels and blood flow in 2D. The representation of vessels and fluids in 3D is a challenge with respect to visualization and interpretation [40]. However, advantages of the 2D representation are its easier intelligibility and lower computational effort, which allows the calculation of a more detailed fluid simulation in real-time.

An improvement of our system would be to automatically differentiate the sketched vascular structures semantically, i.e., distinguish between healthy vessels, stenoses, and aneurysms. With this differentiation, placing a stent in narrowed vessels could be implemented more easily. The sketched stent could inflate a stenosis only, without affecting the surrounding, healthy vessels. By distinguishing between vessels and aneuryms, an entering of the stent into the aneurysm could be prevented as well. Beneath this, we want to investigate how the described concept and prototype can be used for collaboration between physicians. This collaboration can happen at different places where each medical doctor interacts with an instance of the application. The different instances could be connected and mirror the input from one place to another. Additionally, the communication could be supported with voice chat and webcams. Another way of collaboration is possible with only one application, where several physicians are sketching at the same time in front of a big screen and discuss a medical case.

Further analysis could investigate the possibility to realize the described concept in 3D. This leads to challenges and questions regarding a real-time 3D fluid simulation and visualization as well as creating the vessels. For 3D interaction, there are input and output devices which are more suited than pen or touch interaction. By lifting the application in the third dimension, the behavior of blood in vessels could be simulated more accurately, and thus, make the application more relevant for scenarios like operation planning and training.

Acknowledgements. This work was partially funded by the German Federal Ministry of Education and Research (BMBF) within the research campus STIMULATE under grant number '13GW0095A'.

References

1. Mendis, S., Puska, P., Norrving, B.: Global Atlas on Cardiovascular Disease Prevention and Control. Nonserial Publications Series. World Health Organization in collaboration with the World Heart Federation and the World Stroke Organization (2011)
2. European Heart Network: European cardiovascular disease statistics (2012). http://www.ehnheart.org/cvd-statistics.html (last visited: 19 August 2015)
3. Keulers, B.: Computer-based Patient Education: Its Potential in General and Plastic Surgery. Ph.D. thesis, University Nijmegen (2008)
4. Lidal, E.M.: Sketch-based Storytelling for Cognitive Problem Solving. Ph.D. thesis, Department of Informatics, University of Bergen, Norway (2013)
5. Pihuit, A., Cani, M.P., Palombi, O.: Sketch-based modeling of vascular systems: a first step towards interactive teaching of anatomy. In: Proceedings of the Sketch-Based Interfaces and Modeling Symposium, pp. 151–158 (2010)
6. Saalfeld, P., Baer, A., Preim, U., Preim, B., Lawonn, K.: Sketching 2D vessels and vascular diseases with integrated blood flow. In: Proceedings of the International Conference on Computer Graphics Theory and Applications (GRAPP), pp. 379–390 (2015)
7. Zhu, B., Iwata, M., Haraguchi, R., Ashihara, T., Umetani, N., Igarashi, T., Nakazawa, K.: Sketch-based dynamic illustration of fluid systems. In: Proceedings of the SIGGRAPH Asia Conference, pp. 134:1–134:8. ACM (2011)
8. Kutikhin, A., Brusina, E., Yuzhalin, A.E.: A hypothesis of virus-driven atherosclerosis. Viruses and Atherosclerosis. SpringerBriefs in Immunology, vol. 4, pp. 1–3. Springer, New York (2013)
9. Bederson, J.B., Connolly, E.S., Batjer, H.H., Dacey, R.G., Dion, J.E., Diringer, M.N., Duldner, J.E., Harbaugh, R.E., Patel, A.B., Rosenwasser, R.H.: Guidelines for the management of aneurysmal subarachnoid hemorrhage. Stroke **40**, 994–1025 (1994)
10. Neugebauer, M., Diehl, V., Skalej, M., Preim, B.: Geometric reconstruction of the ostium of cerebral aneurysms. In: Proceedings of VMV 2010-Vision, Modeling, Visualization. pp. 307–314 (2010)
11. Gasteiger, R.: Visual Exploration of Cardiovascular Hemodynamics. Ph.D. thesis, Otto-von-Guericke University Magdeburg (2014)

12. Teitelbaum, G.P., Higashida, R.T., Halbach, V.V., Larsen, D.W., McDougall, C.G., Dowd, C.F., Hieshima, G.B.: Flow-directed use of electrolytically detachable platinum embolization coils. J. Vasc. Interv. Radiol. **5**, 453–456 (1994)
13. Wong, G.K., Tan, H.B., Kwan, M.C., Ng, R.Y., Yu, S.C., Zhu, X.L., Poon, W.S.: Evolution of intracranial aneurysm treatment: from Hunterian ligation to the flow diverter. Surg. Pract. **15**, 16–20 (2011)
14. Bridson, R.: Fluid Simulation for Computer Graphics. A K Peters/CRC Press, Boca Raton (2008)
15. Ciarlet, P., Glowinski, R., Lions, J.: Numerical Methods for Non-Newtonian Fluids: Special Volume. Handbook of Numerical Analysis. Elsevier, New York (2011)
16. Ferziger, J.H., Perić, M.: Compressible flow. Computational Methods for Fluid Dynamics, pp. 309–328. Springer, Heidelberg (2002)
17. Müller, M., Schirm, S., Teschner, M.: Interactive blood simulation for virtual surgery based on smoothed particle hydrodynamics. Technol. Health Care **12**, 25–31 (2004)
18. Qin, J., Pang, W.M., Nguyen, B.P., Ni, D., Chui, C.K.: Particle-based simulation of blood flow and vessel wall interactions in virtual surgery. In: Proceedings of the Symposium on Information and Communication Technology, pp. 128–133 (2010)
19. Hansen, C., Johnson, C.: The Visualization Handbook. Butterworth-Heinemann, Waltham (2005). Referex Engineering
20. Weiskopf, D.: Vector field visualization. GPU-Based Interactive Visualization Techniques, pp. 81–159. Springer, Heidelberg (2007)
21. Boring, E., Pang, A.: Directional flow visualization of vector fields. In: Proceedings of the Conference on Visualization, pp. 389–392 (1996)
22. Jorge, J., Samavati, F.: Sketch-based Interfaces and Modeling. Springer, London (2011)
23. Sutherland, I.E.: Sketchpad, A Man-Machine Graphical Communication System. Outstanding Dissertations in the Computer Sciences. Garland Publishing, New York (1963)
24. Preim, B., Dachselt, R.: Interaktive Systeme: User Interface Engineering, 3D-Interaktion, Natural User Interfaces, vol. 2. Springer, Heidelberg (2015)
25. van Dam, A.: Post-WIMP user interfaces. Commun. ACM **40**, 63–67 (1997)
26. Xu, X., Liu, W., Jin, X., A, Z.S.: Sketch-based user interface for creative tasks. In: Proceedings of Asia Pacific Conference on Computer Human Interaction, pp. 560–570 (2002)
27. Naya, F., Contero, M., Aleixos, N., Company, P.: ParSketch: a sketch-based interface for a 2D parametric geometry editor. In: Jacko, J.A. (ed.) HCI 2007. LNCS, vol. 4551, pp. 115–124. Springer, Heidelberg (2007)
28. Igarashi, T., Matsuoka, S., Kawachiya, S., Tanaka, H.: Interactive beautification: a technique for rapid geometric design. In: Proceedings of ACM Symposium on User Interface Software and Technology, pp. 105–114 (1997)
29. Herold, J., Stahovich, T.F.: The $1^¢$; recognizer: a fast, accurate, and easy-to-implement handwritten gesture recognition technique. In: Proceedings of the International Symposium on Sketch-Based Interfaces and Modeling. SBIM 2012, Eurographics Association, pp. 39–46 (2012)
30. Lawonn, K., Saalfeld, P., Preim, B.: Illustrative visualization of endoscopic views. In: Bildverarbeitung für die Medizin (BVM) pp. 276–281 (2014)
31. Harris, M.J.: GPU GEMS Chapter 38, Fast Fluid Dynamics Simulation on the GPU. Pearson Higher Education, New York (2004)
32. Stam, J.: Stable fluids. In: Proceedings of the Conference on Computer Graphics and Interactive Techniques (ACM SIGGRAPH), pp. 121–128 (1999)

33. Butcher, J.C.: A history of Runge-Kutta methods. Appl. Numer. Math. **20**, 247–260 (1996)
34. Krüger, J., Westermann, R.: Linear algebra operators for GPU implementation of numerical algorithms. ACM Trans. Graph. **22**, 908–916 (2003)
35. Wu, E., Liu, Y., Liu, X.: An improved study of real-time fluid simulation on GPU: research articles. Comput. Animat. Virtual Worlds **15**, 139–146 (2004)
36. Glaßer, S., Lawonn, K., Preim, B.: Visualization of 3D cluster results for medical tomographic image data. In: Proceedings of Conference on Computer Graphics Theory and Applications (VISIGRAPP/GRAPP), pp. 169–176 (2014)
37. Taubin, G.: Curve and surface smoothing without shrinkage. In: Proceedings of the International Conference on Computer Vision, pp. 852–857. IEEE Computer Society (1995)
38. Heckel, F., Moltz, J.H., Tietjen, C., Hahn, H.K.: Sketch-based editing tools for tumour segmentation in 3D medical images. Comput. Graph. Forum **32**, 144–157 (2013)
39. Prmper, J.: Der Benutzungsfragebogen ISONORM 9241/10: Ergebnisse zur Reliabilität und Validität. In: Software-Ergonomie: Usability Engineering: Integration von Mensch-Computer-Interaktion und Software-Entwicklung, pp. 254–262 (1997)
40. Lawonn, K., Gasteiger, R., Preim, B.: Adaptive surface visualization of vessels with animated blood flow. Comput. Graph. Forum **33**(8), 16–27 (2014)

GPU Accelerated Computation of Geometric Descriptors in Parametric Space

Anthousis Andreadis$^{(\boxtimes)}$, Georgios Papaioannou, and Pavlos Mavridis$^{(\boxtimes)}$

Department of Informatics, Athens University of Economics and Business,
76 Patission Street, 10434 Athens, Greece
{anthousis,gepap,pmavridis}@aueb.gr
http://graphics.cs.aueb.gr

Abstract. We present a novel generic method for the fast and accurate computation of geometric descriptors. While most existing approaches perform the computations directly on the geometric representation of the model, our method operates in parametric space, decoupling the computational complexity from the underlying mesh geometry. In contrast to other parametric space approaches, our method is not restricted to specific descriptors or parameterisations of the surface. By using the parametric space representation of the mesh geometry, we can trivially exploit massive parallel GPU architectures and achieve interactive computation times, while maintaining high accuracy. This renders the method suitable for computations involving large areas of support and animated shapes.

Keywords: Geometric descriptors · Parametric space · Mesh geometry · GPU acceleration

1 Introduction

The computation of *geometric descriptors*, like curvature, is central in a wide range of applications, including object retrieval, registration, texture synthesis, stylized rendering and many more. The computation of these fundamental metrics is usually performed by algorithms that operate directly on discrete polygonal representations of the continuous surface. In the case of static meshes, these geometric descriptors can be computed once without worrying about the performance. In contrast, in the case of moderately dense meshes with large areas of support and especially in the case of animated or dynamic meshes, performance becomes critical and this computation process becomes a challenging task.

In this work we focus on the general class of metrics with finite local support, whose computation depends on the local neighbourhood of an arbitrary point $\mathbf{p}$ on the object's surface. Robustness in the presence of noise is achieved by performing computations at multiple scales [30]. The computation of these types of descriptors often relies on data structures that encode the adjacency information and allow efficient discovery of the neighbouring points on the surface. This is especially true for algorithms that operate on meshes. The computational complexity of such *object-space* approaches is directly proportional to the geometric density and quadratic with respect to the extent (i.e. radius) of the local area of support.

© Springer International Publishing Switzerland 2016
J. Braz et al. (Eds.): VISIGRAPP 2015, CCIS 598, pp. 41–61, 2016.
DOI: 10.1007/978-3-319-29971-6_3

Fig. 1. (a) "Lucy" model (200 K) colourized with mean curvature, computed in 49 ms, (b) geometry (position) buffer (normalized for visualization), (c) surface normal buffer, (d) polygon chart identifiers (colourized for clarity) along with the adjacent chart identifiers on border pixels, (e) mean curvature in parametric space (colourized for visualization).

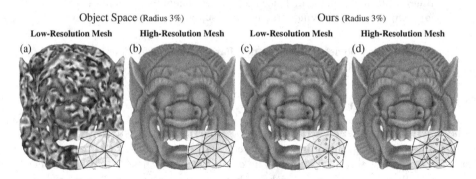

Fig. 2. (a) Using an *object-space* approach on a sub-sampled surface (low-resolution mesh) with a relative small support radius (3 % of the object's diagonal), results in inaccuracies. (b) The same computations applied on the densely sampled surface (high-resolution mesh). (c) Using our parametric-space approach on the low-mesh we obtain accurate results without any extra effort, due to the linear interpolated samples of the surface. (d) Results of our approach used on the high-resolution mesh.

Despite the fact that computing the metric for independent surface points is an inherently parallel task, the use of complex data structures for storing the adjacency information, prevents a trivial and efficient mapping of these computations to massively parallel stream processors, like commodity GPUs, at arbitrary scales. For these reasons, real-time computation is often limited to meshes with relatively low geometric complexity and 1-ring vertex neighbourhoods [6].

In order to alleviate these limitations, we shift all computations from *object-space* to *parametric space*, by transferring all the geometric data of the object to a two-dimensional layout, along with extra adjacency information that allows us to reconstruct the object-space local neighbourhood of a given point on the fly.

While this choice is similar to *Geometry Images* [7], we do not restrict our method by requiring a specific parameterisation of the surface, but rather develop a scheme that handles any underlying parameterisation, including multi-chart layouts. The benefits of parametric-space computations are twofold: First, sampling the geometry at arbitrarily large areas of support is much more efficient in parametric space, since the samples can be directly indexed in contrast to a geometry-based estimation, where the traversal of a surface patch is performed via the connectivity information of the vertices. Second, the parametric space computations are directly mapped to the GPU/many-core computing paradigm in a very efficient manner, rendering the approach suitable for real time calculations over deformable or animated objects. Another gain that stems from the utilization of GPUs, is that we have access to linear interpolated data between all the sample points of the surface with minimal impact on the performance, as linear interpolation is natively supported by the hardware. This is very useful, especially in cases of sub-sampled surfaces, where *object-space* methods give inaccurate results for small support radius (see Fig. 2).

2   Related Work

Most of the existing methods in the bibliography concentrate on the computation of a specific geometric descriptor, and do not try to generalize their framework. For our overview, we do not focus on the specific descriptors used in the existing works, but rather focus and classify methods, based on how they sample the geometric information of the object. Existing methods can be classified to those that sample the geometry in *object-space*, in *screen space*, from a *volumetric representation* and those that operate in *parametric space*. In the remainder of this section we will review the main representatives for each category.

Object-space methods operate directly on the discrete mesh representations of a surface. [18,29] generalize the differential-geometry-based definition of curvatures to discrete meshes but their computations are limited to 1-ring neighbourhoods, which renders them sensitive to noise. Similarly [23] estimate the curvature over meshes using essentially a 2-ring neighbourhood. For efficient arbitrary neighbourhoods, *object-space* methods require a data structure that encodes the adjacency information between the triangles of the mesh, such as the *half-edge* [2] or a kd-tree data structure. However, as discussed in the introduction, a mapping of this data structure to the GPU is neither trivial nor optimal. Most of the existing methods belong to this category and thus operate on the CPU. GPU-based methods, have been proposed for the computation of specific descriptors, like curvature [6], but these methods do not generalize to the sampling of arbitrary neighbourhoods.

Screen space methods sample the geometric information of a mesh from a 2D pixel buffer, where each pixel encodes the projected surface position of the mesh from a specific point of view. In this form of representation, adjacency information is implied by the pixel grid, and therefore sampling is trivial and can be efficiently mapped to GPUs. This efficiency in sampling is also the main motivation behind our method. The main disadvantage of screen-space methods is that computations are limited to the surface points visible from a particular view, resulting in inaccuracies near occluded points and at the screen-space silhouettes of the object. Such screen-space methods have been proposed for curvature estimation in real-time stylized rendering [13,17]. Our method retains most of the sampling efficiency of the screen-space methods, but avoids the view-dependence of the results by moving all the computation to the parametric space.

Volumetric data and algorithms can be also employed for the computation of descriptors. In this case, the input mesh is initially converted to a volumetric representation, such as a *level set*, and geometric descriptors are computed by sampling this representation, instead of the original mesh. Finally, the results of these calculations can be mapped back to the original mesh. The advantage of this approach is that the computational complexity does not depend on the underlying geometry but rather on the new volumetric representation, where sampling a local neighbourhood around a surface point is often more efficient than sampling the same neighbourhood on the original geometry. Features, like curvature, can be quickly approximated using the gradient field of the object, as described in the OpenVDB [19] or by using convolutions, which can be accelerated using FFT as shown in [22]. The disadvantage of this approach is that an efficient voxelisation method is required, additional memory is consumed for the storage of the volumetric format and most importantly, the computations are based on a volumetric discretisation, which is a more rough representation of the original surface than the triangular mesh. Furthermore, certain descriptors when computed on volumetric data, are incompatible with the results of the respective surface-based measurements, especially for non-manifold surfaces.

Finally, **parametric space** methods have also been proposed. Methods of this category rely on a the unwrapping of the model's surface on a 2D plane. Using this representation, computational complexity is decoupled from the underlying geometry and additionally, several image analysis techniques can be applied intuitively to 3D data. To our knowledge, so far there has been no practical and generic approach that would allow both geometric and image space descriptors to be computed efficiently, as existing methods focus on applying image space techniques only. [20] propose a method for corner and edge detection that requires a user-driven single chart parameterisation. Furthermore, to handle points lying near the perimeter of charts, the authors construct complementary parameterisations, for which boundary regions are then mapped to internal chart locations. [10] describe another method that locates extrema using a scale space representation. This approach method relies on a specialised conformal mapping and expects pre-computed per-vertex values of mean-curvature and geodesic distance. In contrast, our method does not rely on a specific parameterisation approach, nor does it require any pre-computed descriptors.

3 Methodology

Our method operates on fully parameterised geometry but does not rely on a specific method for this task. Initially, we perform a pre-processing step in order to locate the surface edges of the polygonal representation, which are mapped to discontinuous regions in parametric space. This is usually part of the model loading process. In real-time, we create the parametric-space representation of the geometry, augmented by the adjacency information and perform the computation of discrete locations in parametric space, i.e. on a texture buffer. During this step, we utilize the information stored in our geometry and adjacency buffers in order to index arbitrary surface samples in the neighbourhood of a point $\mathbf{p}$, regardless of its parametric mapping. The measured metrics can be then queried per vertex, using standard texture look-up operations, or used directly in image space, e.g. to extract salient features and local image-space descriptors. In the rest of this section we each one of the above steps in detail.

3.1 Surface Parameterisation

Surface parameterisation as explained in [5], can be viewed as a one-to-one mapping from a suitable domain to a surface. The parameter domain is also a surface and thus the procedure maps one surface onto another. Our method expects fully parameterised geometry in a normalized 2D domain. This procedure is also known as (bijective) *uv-mapping* and the resulting surface patches are referred to as *charts* or *uv-islands* (see Figs. 1*(d)*, and 3). The area of surface parameterisation has been extensively researched in the past years, [5,26] and the minimization of stretch distortion has been the goal of several works, such as that of [25,31,32]. Therefore, we do not address this part in our work, but rather rely on existing methods and solutions.

3.2 Pre-processing Operations

The estimation of a local descriptor, requires the calculation of an operator $F(\mathbf{p}, S(\mathbf{p}))$ at a point $\mathbf{p}$, given a neighbourhood $\mathbf{x} \in S(\mathbf{p})$, where $\mathbf{x}$ satisfies a set of criteria, such as a maximum Euclidean or geodesic distance from $\mathbf{p}$. Finally another option, is to use the n-ring adjacency of $\mathbf{x}$ to $\mathbf{p}$ (max. n vertex graph distance), but due to the imposed limitation of uniform triangulation of the surface, this is often impractical. These relations in geometric space are easily represented using data structures with topology. For a review of the existing geometric data representations, see the work of [4].

On the other hand, when operating in 2D parametric space, the connectivity information is implied by the adjacency of neighboring pixels. However, this is not true on the borders of charts, where adjacent geometry is mapped to discontinuous locations in parametric space (see example in Fig. 4). In this case, additional information should be stored at the border pixels to keep track of the hops to geometrically-adjacent pixels in different charts.

In order to appropriately annotate the chart pixels, mesh vertices located at the borders of charts must be first identified and the link to the geometrically adjacent vertices on different charts has to be stored on the affected vertices. Details regarding the information stored can be found in Sect. 3.3. The complexity of this step is equivalent to the pre-processing stage of all object-space approaches for the adjacency information generation (e.g. half-edge data structure) and even for large meshes, it only takes a few seconds to complete. This stage needs to be performed only once, as the adjacency information for topologically unchanging geometry can be stored in the 3D model file itself.

Fig. 3. The "bunny" model with two parameterisations, resulting in different set of *charts.*

3.3 Data Buffer Generation

The computation of geometric descriptors requires a set of attributes per sampled surface location, such as the coordinates of **p** in the object's local space and the respective normal vector **n**. These data must be transferred to the parametric space and stored in appropriate buffers, i.e. a set of textures that correspond to the normalized parametric space of the unwrapped geometry. The buffers also store the identifier of the polygon chart that **p** belongs to. The object-space position of surface points is stored in a *geometry buffer* $P(u,v)$, the normal vectors are placed in a *normal buffer* $N(u,v)$, whereas the chart identifiers are registered in an ID channel in the geometry buffer ($ID(u,v)$). Another set of textures, comprising the *adjacency buffer*, equal in size to the geometry buffer, store the identifier of the adjacent chart, the *local metric distortion* of the parameterisation (see below) and the corresponding (u,v) coordinates on the adjacent chart. An example of the data channels for the position, normal and current and adjacent chart identifiers is shown in Fig. 1.

The buffer generation process is performed in two steps. First, the geometric information is efficiently generated in the GPU by rasterising the object triangles using orthographic projection, where the normalized texture coordinates $(u,v,1)$ are used as the vertex coordinates of the mesh. The chart ID is passed as a vertex attribute and copied for all points inside the triangles of a chart. Similarly to [24] we also rasterise each chart's boundary edges in order to avoid the generation of disconnected regions.

In the second step, we compute the *local metric distortion* factors that will be used for the anisotropic adjustment of scale and sampling direction in various

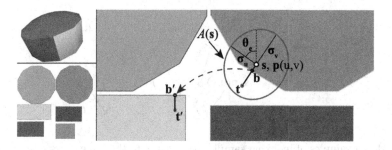

Fig. 4. Indexing a sample inside the neighbourhood of a point. Sample **t** does not lie inside the chart of **s**. Locate the boundary point **b**, read adjacency buffers and relocate sample to adjacent chart.

calculations. In order to do so we use the *Jacobian* $J_P = (P_u, P_v)$, where P_u and P_v the partial derivatives of the surface. The left-singular vectors of J_P are used in order to get the θ_e angular distortion of the anisotropic ellipse while the singular values of J_P σ_u, σ_v are the stretch factors in the u and v direction. Due to the fact that the singular value decomposition is a tedious task, we use the equivalent eigen decomposition of the 2×2 first fundamental form matrix:

$$J_P^T J_P = \begin{bmatrix} E & F \\ F & G \end{bmatrix}, \tag{1}$$

where $E = (\partial P(u,v)/\partial u)^2$, $G = (\partial P(u,v)/\partial v)^2$ and
$\quad\;\; F = (\partial P(u,v)/\partial u) \cdot (\partial P(u,v)/\partial v)$.
For more information see [9].

Additionally, in this second pass we also store the rest of the adjacency data. These attributes are calculated when setting up the triangle connectivity and are simply copies to the adjacency buffers for the chart border pixels. While for static objects the buffer generation step could be performed only once, we focus on a method suitable for deformable/animated objects, and treat it as a per-frame event. Therefore, the reported timings in the remaining text include the data buffer generation overhead.

3.4 Sampling a Point's Neighbourhood

In order to be able to perform the calculation of a feature $F(\mathbf{p}, S(\mathbf{p}))$ in parametric space, we need to establish a procedure for drawing individual samples from the neighbourhood $S(\mathbf{p})$ of $\mathbf{p}$. Our approach estimates $F(\mathbf{p}, S(\mathbf{p}))$ in image space and therefore, for every pixel (i, j) with a corresponding parameter pair $\mathbf{s} = (u, v)$, $\mathbf{p}(u, v)$ is first retrieved from the geometry buffer: $\mathbf{p} = P(u, v)$. Then, assuming a maximum radius of support r_{max} for the local feature estimator in *object-space* units, a sample $\mathbf{t} = (u', v')$ is generated in a region $A(\mathbf{s})$ in parametric space so that $\mathbf{x} = P(\mathbf{t})$ satisfies the neighbourhood criterion. $A(\mathbf{s})$ is calculated as an ellipse of radii $(1/\sigma_u(\mathbf{s}), 1/\sigma_v(\mathbf{s})) \cdot r_{max}$ in the parametric domain

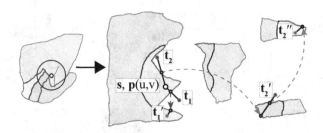

Fig. 5. More examples of indexing samples. t_1 returns to the same chart after a jump. t_2 parametric location is located using two jumps. In the right part, chart adjacencies are colored across the borders.

(upper distance bound) rotated by θ_e, in order to account for local angular distortion and scale, and $\mathbf{x}$ is acquired with rejection sampling according to the selected neighbourhood criterion (Fig. 4). The exact pattern or random distribution with which the samples are generated is specific to the feature estimator and the generic sampling approach presented here is agnostic to it. Also, since we perform a random sampling of the neighbourhood of $\mathbf{s}$, no assumption is made about the chart's convexity.

Since the ellipse $A(\mathbf{s})$ may extend beyond the boundary of the chart containing $\mathbf{p}$ (Fig. 4), a more elaborate mechanism is required to handle the samples that fall outside the chart. These samples obviously contribute to the result and should not be discarded. Identifying whether the sample $\mathbf{x}$ at $\mathbf{t}$ lands on the same chart as $\mathbf{p}$ is trivially resolved by checking their respective chart identifiers $ID(u, v)$ and $ID(u', v')$.

In the case where $\mathbf{t}$ lands outside the chart of $\mathbf{p}$, we utilize the parametric adjacency data stored in our buffers to find its true location. Initially, we march along the direction $\overrightarrow{\mathbf{st}}$ in pixel-sized increments to locate the first pixel with the chart ID as $\mathbf{p}$ (boundary point $\mathbf{b}$). The adjacency buffer for a border pixel $\mathbf{b}$ of a chart contains the ID of the adjacent chart and the parametric location $\mathbf{b}'$ of the corresponding point on it. For samples across seams of the same chart, the ID of the adjacent buffer is identical to that of $\mathbf{p}$, but the parametric location $\mathbf{b}'$ points safely to the corresponding location on the same chart (see Fig. 5-t_1). The adjacency buffer contains also the relative chart edge rotation $\theta(\mathbf{b} \rightarrow \mathbf{b}')$ between $\mathbf{b}$ and $\mathbf{b}'$. Finally, a non-uniform scale factor $s(\mathbf{b} \rightarrow \mathbf{b}')$ can be calculated, corresponding to the relative scale of the two charts in parametric space at their border locations $\mathbf{b}$ and $\mathbf{b}'$ (this scale factor may vary across a chart):

$$s(\mathbf{b} \rightarrow \mathbf{b}') = \left(\frac{\sigma_u(\mathbf{b}')}{\sigma_u(\mathbf{b})}, \frac{\sigma_v(\mathbf{b}')}{\sigma_v(\mathbf{b})} \right). \tag{2}$$

Therefore, we can adjust the location of $\mathbf{t}$ according to the following parametric space transformation to obtain the relocated sampling position $\mathbf{t}'$ on the adjacent chart:

$$\mathbf{t}' = \mathbf{b}' + \mathbf{R}_{\theta(\mathbf{b}\rightarrow\mathbf{b}')}\mathbf{S}_{s(\mathbf{b}\rightarrow\mathbf{b}')}(\mathbf{t} - \mathbf{b}), \tag{3}$$

where $\mathbf{R}_{\theta(\mathbf{b} \to \mathbf{b}')}$ is the rotation matrix of angle $\theta(\mathbf{b} \to \mathbf{b}')$ and $\mathbf{S}_{s(\mathbf{b} \to \mathbf{b}')}$ is the non-uniform scale matrix of factor $s(\mathbf{b} \to \mathbf{b}')$. In case $\mathbf{t}'$ lands outside the expected chart, the same search is performed similarly in the $\overrightarrow{\mathbf{sb}'}$ direction (see Fig. 5-$\mathbf{t_2}$). The sample relocation procedure is shown in Fig. 4. Note that the full non-rigid transformation of $\mathbf{t}$ corresponds to the adaptation of the initial sampling ellipse to the new charts. Therefore, if no severe stretching is present, $S(\mathbf{p})$ is properly covered.

A useful side-effect of the parametric-space computation is that feature estimation can take into account displacement and normal mapping. In the special case of displacement mapping, our method could be easily adopted in order to handle the changes in the geometry that could break the neighbourhood estimation heuristic. Points lying within the initial Euclidean neighbourhood that stretch out of it due to the displacement are automatically handled by measuring the Euclidean distance from $\mathbf{p}$. The problem arises when point with initial location outside the Euclidean neighbourhood of $\mathbf{p}$ fall within r_{max} after displacement. By scaling σ_u and σ_v with the maximum expected displacement distortion, which is usually a user defined parameter, the method successfully handles these points as well.

Finally, we need to clarify that if our method focused only on single chart parameterisations such as *Geometry images* [7] we could avoid highly irregular transitions and in this way reduce the complexity of our operations. On the other hand, multi-chart parameterisations offer an added flexibility that can be used to reduce distortion, particularly for shapes with long extremities, high genus, or disconnected components [24] (see Fig. 13).

4   Estimating Integral Geometric Descriptors

Central to many geometric descriptor computations is the estimation of surface and volume integrals in the neighbourhood of $\mathbf{p}$. Integral invariant features for instance, are often used in the formulation of local descriptors [11], or provide the means to estimate differential invariants such as the mean curvature H (see [3, 15]). Another typical example of an integral operator in $S(\mathbf{p})$) is the estimation of Gaussian curvature K, which can be efficiently computed via the local geodesic area at $\mathbf{p}$.

We estimate integrals in a neighbourhood $S(\mathbf{p})$) using Monte Carlo integration in parametric space and in Sect. 5 we use this approach to compute a variety of integral and differential features interactively for arbitrary feature scales.

4.1   Monte Carlo Integration

In parametric space, the generated data buffers hold not only the vertex information but also all internal polygon samples, generated by the GPU through linear interpolation during rasterisation. An approach that would use all the per-pixel information inside $S(\mathbf{p})$ would be unnecessarily exhaustive and computationally expensive. Utilizing Monte Carlo integration with a uniform distribution in

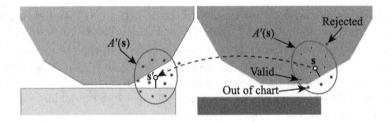

Fig. 6. Monte Carlo Sampling in current and adjacent charts described in Sect. 4.1.

the parametric domain, any integral $I(\mathbf{p})$ of a function $g(\mathbf{p})$ over $S(\mathbf{p})$ can be approximated by:

$$\langle I \rangle (\mathbf{p}) = \frac{A'(\mathbf{s})}{N} \sum_{i=1}^{N} g(P(\mathbf{t}_i)), \tag{4}$$

where $A'(\mathbf{s})$ is the portion of the elliptical sampling area $A(\mathbf{s})$ centered at parameter pair $\mathbf{s}$ corresponding to the central point $\mathbf{p} = P(\mathbf{s})$ after rejection sampling with the criterion of neighbourhood $S(\mathbf{p})$ (e.g. Euclidean distance of $P(\mathbf{t})$ to $\mathbf{p}$) and N is the number of valid samples. While performing a similar sampling on the geometry itself would require area-weighted probabilities, the parametric-space values can be sampled uniformly, assuming of course a low-distortion parameterisation.

Random samples are generated uniformly using a stratification scheme. Uniform samples in the cells of a planar grid are transformed to disk samples using the concentric mapping of [27]. The disk samples are anisotropically scaled along the u and v axes to form the elliptical region $A(\mathbf{s})$, according to the distortion factors discussed in Sect. 3.3. The same samples are used at each pixel, randomly rotating them to avoid statistical noise.

The elliptical region $A(\mathbf{s})$ is an approximation that favors fast computations. A more refined but rather more computationally expensive approach would be to pre-compute the maximal distortion for discretised polar coordinates at each pixel and subsequently anisotropically scale each random sample according to the closest distortion term from its conversion to polar coordinates. Nevertheless, as demonstrated in the experiments, the elliptical approximation proved to be both robust and efficient, even for large neighbourhoods.

Given a point $\mathbf{p}$ and its location in parametric space $\mathbf{s}$, initially we perform computations only for the samples that lie on the same chart as $\mathbf{p}$ (Fig. 6 - right). At the same time, for all parametric-space samples that fall outside the chart, we mark the ID of the chart they land on. Subsequently, we compute for each marked chart the transformed parametric position $\mathbf{s}'$ of the central parametric pair $\mathbf{s}$ and repeat the sampling procedure on the new location, using the entire sampling pattern (Fig. 6 - left). Only samples falling within the new chart are accounted for and contribute to the final integral. The marking of charts and the central point transformation is done according to the procedure described in Sect. 3.4.

The sampling scheme described above is generic and can be implemented for an arbitrary number of jumps, excluding each time the already visited charts. In our experiments we noticed that, no more than one jump per sample point was typically required, even for large-scale local feature neighbourhoods. Of course, this also depends on the size of the charts produced by the parameterisation. For example, in Fig. 13, where the bunny model is shown in two different parameterisations, for the left one we reported the first missing sample using a support area of 10 % the object's diagonal. Conversely, for the one on the right we did not report any missing samples even for neighbourhoods larger than 16 % the object's diagonal.

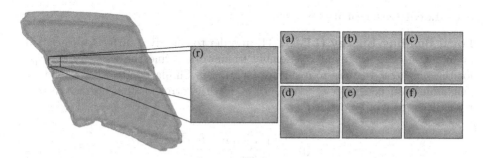

Fig. 7. Comparison of mean curvature for Full and Adaptive Sampling. (r) Reference (a), (b), (c) Full Sampling using 64, 100 and 256 samples in respect. (d), (e), (f) Adaptive Sampling using 32/64, 64/128 and 128/256 samples.

4.2   Adaptive Sampling

Since $g(\mathbf{x})$ is a function of the surface geometry, smooth areas of the objects, i.e. areas with smaller variance of the evaluated function $g(\mathbf{x})$, give satisfactory results even when lowering the sampling rate significantly. Given the fact that in our approach the computation time is proportional to the total number of samples drawn, we speed-up our method by exploiting adaptive sampling.

Typically, adaptive sampling methods continue to draw random samples, until the variance of the computed quantity falls below a certain threshold. In our method however, we perform a simplified, two-step adaptive sampling, instead of waiting for the variance to converge: Initially, we compute the integral with $N/2$ samples and measure the variance. For points $\mathbf{p}$ with variance of $g(\mathbf{x})$ greater than a predetermined threshold, we use an additional set of $N/2$ random samples. Using a fixed, two-stage adaptive sampling creates exactly two different GPU execution loads, generally coherent across the output buffer, thus maximizing shader core utilization and performance.

Our experiments show that as the number of samples increases, the difference of % Absolute Error (% AE) between the full and adaptive sampling declines, while at the same time the performance savings increase (see Table 1 and Fig. 7).

5 Performance and Quality Evaluation

In this section we present a number of local geometric descriptor operators using our method and provide a qualitative comparison against respective reference *object-space* CPU algorithms that operate directly on the polygonal geometry using the *Halfedge* data-structure (HE) [2]. Furthermore, we showcase the use of our method in order to exploit 2D image-space interest point detectors over 3D geometry without the need for a specific data structure or implementation. Initially, we briefly present each of the local descriptors used and subsequently evaluate the results against various factors.

5.1 Local Descriptors

Local Bending Energy (LBE). [11] in order to classify a surface as fractured or intact in their fragment reassembly framework define the LBE term $e_k(\mathbf{p})$ for the k nearest vertices to a surface location $\mathbf{p}$. Similarly, given an Euclidean neighbourhood $\mathbf{q}_i \in S(\mathbf{p}, r) : \|\mathbf{q}_i - \mathbf{p}\| \leq r$ with corresponding normal vectors $\mathbf{n}_i$, LBE $e_r(\mathbf{p})$ can be defined as:

$$e_r(\mathbf{p}) = \frac{1}{N} \sum_{i=1}^{N} \frac{\|\mathbf{n} - \mathbf{n}_i\|^2}{\|\mathbf{p} - \mathbf{q}_i\|^2}, \tag{5}$$

where $\mathbf{n}$ is the normal at the central point $\mathbf{p}$ and N is the number o samples taken in the $S(\mathbf{p}, r)$ neighbourhood.

Table 1. Computation Time and % Absolute Error for Full and Adaptive Sampling over the same metric. Error in comparison to reference *object-space* implementation.

Samples	Full		Adaptive	
	Time	% AE	Time	% AE
64	17.57 ms	1.172	15.94 ms	1.331
100	22.17 ms	1.035	19.54 ms	1.110
256	50.54 ms	1.005	41.44 ms	1.007
400	74.21 ms	0.789	61.75 ms	0.824

Sphere Volume. [16] presented a stochastic solid angle computation for the approximation ambient occlusion in the hemisphere above a point $\mathbf{p}$. Inspired by this idea, we extend it to a full sphere and compute a fast approximation of the unoccupied volume of a sphere of radius r centered at $\mathbf{p}$. Assuming a smoothly varying tangential elevation around $\mathbf{p}$, the vector $\mathbf{q}_i - \mathbf{p}$ from the central point to any sample $\mathbf{q}_i$ within the Euclidean neighbourhood $S(\mathbf{p}, r)$ approximates the horizon in this direction with respect to the normal vector $\mathbf{n}$ at $\mathbf{p}$ at a distance scale equal to $\|\mathbf{q}_i - \mathbf{p}\|$. Taking a uniform rotational and radial distribution of

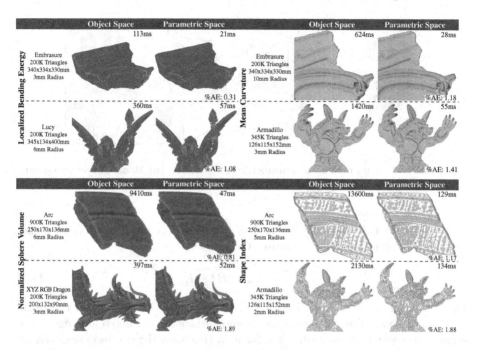

Fig. 8. Comparative visualization, timings and % Absolute Error for the implemented geometric features (Sect. 5.1).

samples (direction and scale) $\mathbf{q}_i$ in $S(\mathbf{p}, r)$, we can approximate the open volume $V_o(\mathbf{p})$ above $\mathbf{p}$ by:

$$V_o(\mathbf{p}) = \frac{4\pi r^3}{3N} \sum_{i=1}^{N} \frac{(\mathbf{q}_i - \mathbf{p})\mathbf{n}}{\|\mathbf{q}_i - \mathbf{p}\|}. \tag{6}$$

The *sphere volume integral invariant*, i.e. the part of the sphere volume of radius r "inside" the surface at $\mathbf{p}$ [21] is the complement of the above integral quantity.

$$V_r(\mathbf{p}) = \frac{4\pi r^3}{3} - V_o(\mathbf{p}). \tag{7}$$

Mean Curvature (MC). [12] derive the relation of MC to the sphere volume integral invariant as:

$$V_r(\mathbf{p}) = \frac{2\pi}{3}r^3 - \frac{\pi H}{4}r^4 + O(r^5), \tag{8}$$

from which we can directly compute MC H at $\mathbf{p}$ for a given radius r.

Shape Index (SI). Introduced by [14], SI is a local descriptor that combines the principal curvatures (PC) in order to classify the locale shape of the surface. SI is a normalized descriptor and for a given surface point $\mathbf{p}$ is defined as:

$$S(\mathbf{p}) = \frac{2}{\pi} \arctan \frac{K_2(\mathbf{p}) + K_1(\mathbf{p})}{K_2(\mathbf{p}) - K_1(\mathbf{p})}, \tag{9}$$

where $K_1(\mathbf{p})$, $K_2(\mathbf{p})$ are the principal curvatures at $\mathbf{p}$.

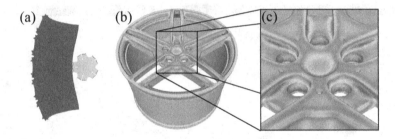

Fig. 9. Genus 20 Rim model. (a) Parametric space charts. (b) Mean Curvature colourized (generated in 58 ms). (c) Zoom to detail.

In order to calculate the K_1 and K_2, we rely on their relation to mean curvature H and Gaussian curvature (GC) K:

$$K_{1,2} = H \pm \sqrt{H^2 - K}. \tag{10}$$

The computation of H was discussed earlier. For the GC we rely on the work of [1] that relates K with the perimeter and surface area of a geodesic disk on a surface. In particular, we utilize the formula that uses the geodesic area GA of distance r:

$$K = 12 \frac{\pi r^2 - GA^r}{\pi r^4}. \tag{11}$$

The only unknown parameter now is the geodesic area GA^r at a given distance r. In the case of the geometric evaluation, we sum the Voronoi area of each vertex within a neighbourhood of geodesic distance r. For the parametric-space computation of GA^r, we first draw a number of samples N_{tot} in the Euclidean neighbourhood of $\mathbf{p}$ (see Sect. 4.1). Then, for each sample $\mathbf{q}_i$ at parametric location $\mathbf{t}_i$, the geodesic distance to $\mathbf{p}$ is approximated by a sum of chords at $P(\mathbf{s}_j)$, i.e. at the intermediate parametric space coordinates $\mathbf{s}_j = \mathbf{t}_i + j(\mathbf{s} - \mathbf{t}_i)/N_{steps}$, where $\mathbf{s}$ are the uv coordinates of $\mathbf{p}$ and N_{steps} is the number of chords. Depending on the local distortion of the parameterisation, $P(\mathbf{s}_j)$ may not reside exactly on the same plane. According to the computed geodesic distance between $\mathbf{q}_i$ and $\mathbf{p}$, a final set of N_g samples is retained, $N_g \leq N_{tot}$, and GA^r is estimated by:

$$GA^r = \frac{N_g}{N_{tot}} EA^r, \tag{12}$$

where EA^r is the Euclidean area.

Euclidean area can be approximated in the following way. Let P_{tot} be the total number of pixels in the elliptical region. Given the ratio of the samples that satisfy the Euclidean criterion to the total samples $N_{A(s)}/N_{tot}$, and $A_{\mathbf{q}_m}$ the mean area represented by each sample in $A(s)$, we approximate EA^r as:

$$EA^r = P_{tot} \frac{N_{A(s)}}{N_{tot}} A_{\mathbf{q}_m}, \tag{13}$$

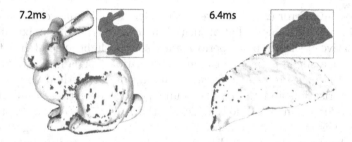

Fig. 10. Harris corner points computed over the Normalized Sphere Volume descriptor and painted (red) on the mesh vertices (Color figure online).

$A_{\mathbf{q}_m}$ is given by the formula:

$$A_{\mathbf{q}_m} = \frac{1}{N_{A(s)}} \sum_{i=1}^{N_{A(s)}} A_{\mathbf{q}_i}, \tag{14}$$

where $A_{\mathbf{q}_i}$ is the product of the distortion factors $l_u(u, v)$, $l_v(u, v)$.

5.2   Image Descriptors

Harris Corner Detection. Harris and Stephens [8] describe a mathematical operator for the computation of corner points of interest (features) on images, based on the change of intensity. These feature-points are invariant under rotational and intensity changes and can be used for matching. The same computation has been used on 3D meshes to generate feature points for object registration and retrieval [28]. The mathematical formulation of Harris corner response is:

$$R = \det(M) - k \cdot \operatorname{trace}(M)^2, \tag{15}$$

where k is constant ($k \in [0.04, 0.06]$) and M is given by:

$$M = \sum_{x,y} w(x, y) \begin{bmatrix} I_x^2 & I_x I_y \\ I_x I_y & I_y^2 \end{bmatrix}. \tag{16}$$

I_x, I_y are the image derivatives, and $w(x, y)$ a Gaussian window function. Using the parametric-space indexing scheme (see Sect. 3.4) we compute the image-space derivatives and the Gaussian window function. Of course, using our indexing scheme, any other image space descriptor can also be applied over the 3D data.

5.3   Results and Discussion

We have tested our method using a large variety of objects, ranging from simple geometric shapes to complex and detailed 3D scanned models. Indicative results can be seen in Figs. 8 and 9, where we report an average of 49× acceleration and

1.245 % Absolute Error (AE) relative to the reference CPU *object-space* method described below. Please note that in such comparisons, reporting maximum error is not indicative of the method's performance, since a slight mismatch in the representation at a single point due to parameterisation can cause an isolated but inconsequential measurement difference. Timings of our method do not include the parameterisation and the charts boundary edge detection. Similarly, timings of the reference method do not include the *Half-Edge (HE)* data structure generation. It is important to mention here that while geometric algorithms for computing features operate on discretised values at a vertex or triangle level, the parametric space calculations can exploit interpolated values at arbitrary surface locations. Therefore, the measurement deviations that are reported here as errors, mostly stem from the different approximation and sampling of the underlying surface (see Fig. 2). Finally in Fig. 10 we show *Harris corner points* detection over 3D data. Timings for the GPU parametric implementation are shown for an NVIDIA GTX 670 GPU. We use 1024×1024 floating point texture buffers, while metrics are computed over a 512×512 buffer with 256 samples per pixel unless stated otherwise. The reference geometric algorithm results are shown for a Corei7-3820 system (4 cores @ 3.60 GHz, 8 threads). Our implementation uses the *OpenMP API* and takes advantage of the current generation multi-core CPU's.

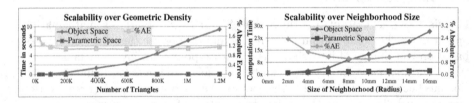

Fig. 11. (Left Graph) Computations using the same metric and neighbourhood size (Left axis). (Right Graph) Computations using the same metric and geometric complexity (Left axis). Green line shows the % AE of the parametric method (Right axis) (Color figure online).

The efficiency of our method is attributed to the shift of the computations from a topology-detail-dependent representation to two dimensions with application-controlled (sampling) quality settings, which enables very good scaling for multi-core and many-core architectures. The proposed implementation is tailored for (but not limited to) commodity GPUs.

Geometric Detail. In Fig. 11(Left) we present comparative results computed over a fixed neighbourhood size (4 % of object's diagonal) for a single model (Embrasure) decimated at different geometric detail levels. For small resolutions (25 K, 50 K triangles) we observe similar computation times between geometric and parametric space approaches, while the % AE is high in comparison to higher detail versions of the mesh. This is expected as the parametric method uses the position samples as interpolated by the GPU resulting in smoother and therefore slightly different results than the CPU method (see Fig. 2). For larger resolutions, we report an acceleration of 3× for the 100 K model to 137× for the 1000 K model, with a steady AE. Finally, for the original scanned object resolution of 1200 K, we report a 181× faster computation with a slight increase in the % AE. This is also expected and attributed to the relative small buffer size for the dense geometric detail. However, this can be trivially addressed by increasing the geometry buffer resolution.

Neighborhood Size. In the measurements of Fig. 11(Right) we shift the focus from the geometric detail to neighbourhood size. Results are for the same model (Embrasure) and metric (mean curvature) at 600 K resolution. We notice that for small neighbourhoods the % AE is higher. This deviation between the para-metric and geometric domain results are due to the inadequate discrete representation of the area of support in the geometric solution. While in the parametric domain due to the interpolation of values we mentioned earlier, an increasing neighbourhood size is directly reflected in a wider selection of samples, the geometric neighbourhood expands in discrete steps, which is actually a deficiency (see Fig. 2). For very large neighbourhoods we notice also an increase to the % AE, this time, due to the one jump per sample approach of our implementation (see end of Sect. 4.1), which starts missing samples. Performance-wise, the parametric space method scales very well and is not significantly affected by the 8× growth of neighbourhood size. More specifically, the computation time for the parametric domain feature estimator grows by 2.25 times in contrast to the 26.45× factor reported by the geometric approach.

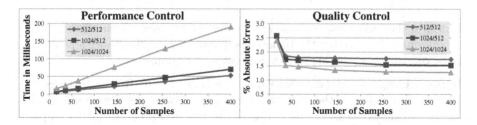

Fig. 12. (Left) Average performance over several models using different number of samples, buffers size and size of texture over which computations are performed. (Right) Average quality over several models using different number of samples, buffers size and size of texture over which computations are performed. Legends show Buffers Texture/Computations Texture Size in square format.

Performance and Quality Control. The number of samples per pixel, buffer size and size of the texture over which computations are performed, are para-meters that control the quality/performance of our method. As we can see in Fig. 12(Left), increasing the number of samples reduces the % AE and has linear impact on the computation time, regardless of the buffer resolution. The same

Table 2. Average Computation time and % Absolute Error over a set of models for the same metric over different resolutions and buffer precision. Error in comparison to the reference CPU implementation.

Buffer size				
	512x512		1024x1024	
Precision	Time	% AE	Time	% AE
Full-Float	26.5 ms	1.662	74.9 ms	1.307
Half-Float	24.4 ms	1.725	66.8 ms	1.384

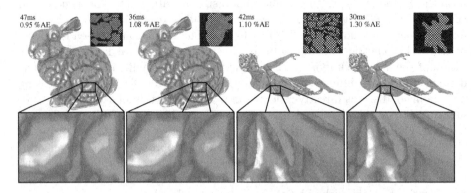

47ms 0.95 %AE   36ms 1.08 %AE   42ms 1.10 %AE   30ms 1.30 %AE

Fig. 13. Mean Curvature (colourized) computed using different parameterisations. Multiple charts result in increased computation times, but smaller error, due to the smaller distortion of the generated charts.

effect have the buffer size and the size of the texture over which computations are performed (Fig. 12(Right)). Using these parameters, performance and quality can be controlled depending on the application requirements.

Memory Usage - Texture Size and Precision. Four RGBA textures are used (see Sect. 3.3). All the presented results so far were performed using half-float precision textures. In order to evaluate the performance/quality impact of full-float-precision textures (FF), which double the memory requirements, we performed experiments using both resolutions (Table 2). FF buffers present an 8 % and 11 % performance degradation on 512×512 and 1024×1024 buffers respectively, while the corresponding improvement in AE is 4 % and 6 %. We can conclude that the minor quality improvement does not justify the performance drop and the doubled memory requirements.

UV Parameterization. In order to evaluate how our method is affected by the underlying parameterisation in terms of speed and quality, we performed several tests. When operating on maps coming from global surface parameterisation (single chart) techniques, we notice faster times, and increased error rates (see Fig. 13) compared to multi-chart parameterisations opting for minimal stretching. Single charts, minimize branching operations but at the same time result in greater distortion and less uniform sampling leading thus in loss of representation and measurement accuracy.

6 Limitations

Due to the fact that parameterisation of the objects surface is required, the method is limited to mesh geometries and it cannot be directly applied on point-clouds. Still, most of the local descriptors rely on the notion of a connected neighbourhood. While simple distance queries (without connectivity) can be used in the case of point-clouds, the resulting computations will have inaccuracies, especially when using large areas of support.

7 Conclusion

We presented a novel generic parametric-space approach for the computation of geometric descriptors in multiple scales, that can also be used to trivially apply computer vision algorithms on 3D data. Our method, decouples the computational complexity from the underlying geometry and by taking advantage of modern multi-core architectures (GPUs), achieves real-time computations even for large areas of support, rendering the method suitable for deformable and animated objects. Finally, despite the focus of our method on efficiency, computations are accurate and equivalent to those of the traditional object-space approaches as shown by our experiments.

Acknowledgements. This work was supported by EC FP7 STREP Project PRE-SIOUS, grant no. 600533. Armadillo, Lucy, Bunny and XYZ RGB Dragon models are from Stanford 3D Scanning Repository. Angel model is from the Large Geometric Models Archive of Georgia Institute of Technology. Rim model is from TurboSquid. All other models used are from the PRESIOUS project data collection.

References

1. Bertrand, J., Diquet, C., Puiseux, V.: Démonstration d'un théorème de Gauss. J. Math. **13**, 80–90 (1848)
2. Campagna, S., Kobbelt, L., Seidel, H.P.: Directed edges — a scalable representation for triangle meshes. J. Graph. Tools **3**(4), 1–11 (1998)
3. Connolly, M.L.: Measurement of protein surface shape by solid angles. J. Mol. Graph. **4**(1), 3–6 (1986)
4. De Floriani, L., Hui, A.: Data structures for simplicial complexes: an analysis and a comparison. In: Proceedings of the Third Eurographics Symposium on Geometry Processing. SGP 2005. Eurographics Association (2005)
5. Floater, M., Hormann, K.: Surface parameterization: a tutorial and survey. In: Dodgson, N.A., Floater, M.S., Sabin, M.A. (eds.) Advances in Multiresolution for Geometric Modelling. Mathematics and Visualization, pp. 157–186. Springer, Heidelberg (2005)

6. Griffin, W., Wang, Y., Berrios, D., Olano, M.: GPU curvature estimation on deformable meshes. In: Symposium on Interactive 3D Graphics and Games, I3D '11, pp. 159–166. ACM (2011)

7. Gu, X., Gortler, S.J., Hoppe, H.: Geometry images. In: Proceedings of the 29th Annual Conference on Computer Graphics and Interactive Techniques. SIG-GRAPH '02, pp. 355–361. ACM (2002)

8. Harris, C., Stephens, M.: A combined corner and edge detector. In: Proceedings of the 4th Alvey Vision Conference, pp. 147–151 (1988)

9. Hormann, K., Polthier, K., Sheffer, A.: Mesh parameterization: theory and practice. In: ACM SIGGRAPH ASIA 2008 Courses. SIGGRAPH Asia '08, pp. 12: 1–12: 87. ACM (2008)

10. Hua, J., Lai, Z., Dong, M., Gu, X., Qin, H.: Geodesic distance-weighted shape vector image diffusion. IEEE Trans. Vis. Comput. Graph. **14**(6), 1643–1650 (2008)

11. Huang, Q.X., Flöry, S., Gelfand, N., Hofer, M., Pottmann, H.: Reassembling fractured objects by geometric matching. ACM Trans. Graph. **25**(3), 569–578 (2006)

12. Hulin, D., Troyanov, M.: Mean curvature and asymptotic volume of small balls. Am. Math. Monthly **110**(10), 947–950 (2003)

13. Kim, Y., Yu, J., Yu, X., Lee, S.: Line-art illustration of dynamic and specular surfaces. In: ACM SIGGRAPH Asia 2008 Papers. SIGGRAPH Asia '08, pp. 156: 1–156: 10. ACM (2008)

14. Koenderink, J.J., van Doorn, A.J.: Surface shape and curvature scales. Image Vis. Comput. **10**(8), 557–565 (1992)

15. Manay, S., Hong, B.-W., Yezzi, A.J., Soatto, S.: Integral invariant signatures. In: Pajdla, T., Matas, J.G. (eds.) ECCV 2004. LNCS, vol. 3024, pp. 87–99. Springer, Heidelberg (2004)

16. McGuire, M., Osman, B., Bukowski, M., Hennessy, P.: The alchemy screen-space ambient obscurance algorithm. In: Proceedings of the ACM SIGGRAPH Symposium on High Performance Graphics. HPG '11, pp. 25–32. ACM (2011)

17. Mellado, N., Barla, P., Guennebaud, G., Reuter, P., Duquesne, G.: Screen-space curvature for production-quality rendering and compositing. In: ACM SIGGRAPH 2013 Talks. SIGGRAPH '13, pp. 42: 1–42: 1. ACM (2013)

18. Meyer, M., Desbrun, M., Schrder, P., Barr, A.: Discrete differential-geometry operators for triangulated 2-manifolds. In: Hege, H.-C., Polthier, K. (eds.) Visualization and Mathematics III. Mathematics and Visualization, pp. 35–57. Springer, Heidelberg (2003)

19. Museth, K.: Vdb: high-resolution sparse volumes with dynamic topology. ACM Trans. Graph. **32**(3), 27:1–27:22 (2013)

20. Novatnack, J., Nishino, K.: Scale-dependent 3D geometric features. In: IEEE 11th International Conference on Computer Vision, 2007. ICCV 2007, pp. 1–8. IEEE, October 2007

21. Pottmann, H., Wallner, J., Huang, Q.X., Yang, Y.L.: Integral invariants for robust geometry processing. Comput. Aided Geom. Des. **26**(1), 37–60 (2009)

22. Pottmann, H., Wallner, J., Yang, Y.L., Lai, Y.K., Hu, S.M.: Principal curvatures from the integral invariant viewpoint. Comput. Aided Geom. Des. **24**(8), 428–442 (2007)

23. Rusinkiewicz, S.: Estimating curvatures and their derivatives on triangle meshes. In: Proceedings of the 2nd International Symposium on 3D Data Processing, Visualization, and Transmission. 3DPVT '04, pp. 486–493. IEEE Computer Society (2004)

24. Sander, P.V., Wood, Z.J., Gortler, S.J., Snyder, J., Hoppe, H.: Multi-chart geometry images. In: Proceedings of the 2003 Eurographics/ACM SIGGRAPH Symposium on Geometry Processing. SGP '03, pp. 146–155. Eurographics Association (2003)

25. Sander, P.V., Snyder, J., Gortler, S.J., Hoppe, H.: Texture mapping progressive meshes. In: Proceedings of the 28th Annual Conference on Computer Graphics and Interactive Techniques. SIGGRAPH '01, pp. 409–416. ACM (2001)

26. Sheffer, A., Praun, E., Rose, K.: Mesh parameterization methods and their applications. Found. Trends. Comput. Graph. Vis. 2(2), 105–171 (2006)

27. Shirley, P., Chiu, K.: A low distortion map between disk and square. J. Graph. Tools 2(3), 45–52 (1997)

28. Sipiran, I., Bustos, B.: Harris 3D: a robust extension of the harris operator for interest point detection on 3D meshes. Vis. Comput. 27(11), 963–976 (2011)

29. Taubin, G.: Estimating the tensor of curvature of a surface from a polyhedral approximation. In: Proceedings of the Fifth International Conference on Computer Vision. ICCV '95, pp. 902–907. IEEE Computer Society (1995)

30. Yang, Y.L., Lai, Y.K., Hu, S.M., Pottmann, H.: Robust principal curvatures on multiple scales. In: Symposium on Geometry Processing, pp. 223–226 (2006)

31. Yoshizawa, S., Belyaev, A., Seidel, H.P.: A fast and simple stretch-minimizing mesh parameterization. In: Proceedings of the Shape Modeling International. SMI '04, pp. 200–208. IEEE Computer Society (2004)

32. Zhou, K., Synder, J., Guo, B., Shum, H.Y.: Iso-charts: stretch-driven mesh parameterization using spectral analysis. In: Proceedings of the 2004 Eurographics/ACM SIGGRAPH Symposium on Geometry Processing. SGP '04, pp. 45–54. ACM (2004)

Interference Shader for Multilayer Films

Fukun Wu[1,2](✉) and Changwen Zheng[1]

[1] Science and Technology on Integrated Information System Laboratory,
Institute of Software, Chinese Academy of Sciences, Beijing, China
`iscassucess@gmail.com, cwzheng@ieee.org`
[2] University of Chinese Academy of Sciences, Beijing, China

Abstract. To visualize the interference effects of objects with multi-layer film structures such as soap bubbles, optical lenses and Morpho butterflies is challenging and valuable in the physics-based framework, a novel multilayer film interference shader is constructed. The multi-beam interference equation is applied to effectively simulate the multiple reflection and transmission inside films, and calculate the composite reflectance and transmittance to model the amplitude and phase variations related to interference. The absorption of photons due to the film materials is accounted for by the Fresnel coefficients used for metallic and dielectric films. In addition, the irregularity of multilayer film microstructures is incorporated into the iridescent illumination model to explain the isotropic and anisotropic optical properties. The new wave bidirectional scattering distribution function is proposed and integrated into the existing ray tracer in the form of the material plugin to further enhance the photorealistically rendering capabilities. The experiments show that our interference shader gives accurate results in both visual and numerical quality.

Keywords: Interference effects · Interference shader · Multi-beam interference equation · Fresnel coefficients · Isotropic and anisotropic

1 Introduction

Photorealistic rendering is the main rendering technology of the existing modeling softwares such as Maya, 3Dmax and Blender. Different from the non-photorealistic rendering technology, it is involved in the physical simulation of interaction between light and objects where the illumination information can be accurately modeled. The ray tracing is generally used for modeling the propagation process of light in space. Specially, it calculates the reflectivity and transmissivity from surfaces by acquiring the material properties of objects, and eventually obtains the radiant energy of each ray arriving at the imaging plane after recursively tracing rays to generate the realistic images.

The thin film interference is an important part of photorealistically rendering. However, the existing graphical development platform or softwares lack the ability of describing the phase of light. To construct precise interference

© Springer International Publishing Switzerland 2016
J. Braz et al. (Eds.): VISIGRAPP 2015, CCIS 598, pp. 62–74, 2016.
DOI: 10.1007/978-3-319-29971-6_4

model to model the interaction between light and multilayer film structures in order to visualize iridescent colors of objects coated with multilayer films such as soap bubble, beetles and butterflies is a significant but challenging research task. In computer graphics, many wave models have been developed to render the wave phenomena generated by these multilayer film structures. Gondek et al. [1], for example, used a wavelength-dependent bidirectional reflectance distribution function and a virtual goniospectrophotometer to analyze and generate the reflection spectrum of thin films and pearl materials. Hirayama et al. [2,3] constructed a series of multilayer dielectric and metallic film models to visualize the richer interference effects through the iterative calculation of multi-beam reflection and transmission. Sun [4] applied the analytical calculation and the numerical simulation methods to implement an iridescent shading process to render the biological iridescences. These method can approximately describe the wave properties of films, but rarely consider the microstructure or material characteristics of surfaces, which are not applicable for the accurate simulation of the anisotropic iridescent colors.

For the sake of accurately rendering the optical phenomena of diverse film materials, this paper constructs a general multilayer film interference shader in the ray-based framework. It adopts the multi-beam interference equation and Fresnel formulas to account for the multiple reflection, interference and absorption of light. Fresnel coefficients for dielectric and metallic films are introduced to visualize interference due to the complex refractive indexes and photon absorption. In addition, the irregularity of multilayer film microstructures is incorporated into the iridescent illumination model to accurately describe the isotropic and anisotropic optical properties. The new wave bidirectional scattering distribution function is proposed and integrated into the PBRT [5] in the form of the material plugin, which has become a practical technology by applying the existing modeling software to render complex interference optical effects.

2   Related Work

In computer graphics, to solve the problem of wave rendering in the physics-based rendering framework, multiple classical technologies have been developed where the wave bidirectional scattering distribution function is applied to simulate the behavior of light on surfaces [1,6–8]. For instance, Moravec [9] used the wave theory of light to solve the global illumination problem and applied wave model to computer graphics based on the phase tracking technology. Kajiya [10] developed a bidirectional reflectance distribution function to model the anisotropic spectral reflectance by numerically calculating Kirchhoff integral. Later, Stam [11] implemented a general diffraction shaders, followed by the works of Agu [12], Sun [13] and Wu [14,15], to render iridescent colors from periodic structures such as compact discs in a ray-based renderer. The solutions above, however, are constructed to model diffraction effects that are a part of wave rendering, not applicable for rendering film interference effects due to the lack of ability of encapsulating phase variations into transmitted radiant energy.

To construct the accurate interference model to model the interactive behavior of light and surfaces in order to visualize the iridescent appearance of objects coated with multilayer films such as soap bubble, beetles and butterflies has attracted a lot of attentions. There exist many models used for rendering these interference effects generated by multilayer films. Icart et al. [16], for example, constructed a physics-based bidirectional reflectance model for multilayer systems consisting of homogeneous and isotropic thin films with rough boundaries, which can account for interference, diffraction and polarization effects. Hirayama et al. [2,3,17] constructed a comprehensive multilayer film interference model to model scattering characteristics of rough multilayer surfaces. Sun [4] implemented an iridescent shading process for rendering the biological iridescences of butterflies and beetles due to multilayer interference based on analytical calculation and numerical simulation. Few of the previous models, however, takes into account specific geometrical properties of multilayer films or other sub-wavelength microstructures. They also lack the ability of modeling the back-scattering and anisotropic properties for photorealistic renderings of Morpho butterfly. Okada et al. [18] applied the nonstandard finite-difference time-domain method to numerically solve Maxwell's equations for brilliant iridescences. This approach can gain accurate results, but depends on a fine defined numerical grid.

3   Iridescent Illumination Model

The key to rendering iridescent colors of multilayer films in the ray-based framework is to account for the interaction between the films with the periodic structure and light with amplitude and phase. Therefore, this paper builds on the multilayer film interference theory and incorporates the geometry of rough surfaces to construct an accurate film interference shader in order to model isotropic and anisotropic iridescent effects, which is further integrated in Maya modeling software to improve its practicality. The iridescent colors from objects coated with similar multilayer films can be efficiently visualized where the refractive indexes, thicknesses and amount of alternative arrangement of films and the incident direction of light source play an important role.

3.1   Multi-beam Interference

When light interacts with multilayer films, it undergoes multiple reflection, transmission and absorption inside films. It is desired to develop a more general model that take complex interactions into consideration. In this work, we analytically compute multilayer film interference based on the recursive composition method [2,4] to visualize optical properties of multilayer structures.

As an example, consider a pair of film and air layers as shown in Fig. 1. Given a thickness H and a refractive index $n_j, j = 0, 1, 2$. The r_1, r_2, t_1 and t_2 denote the reflection and transmission coefficients of light propagating from air to film, and the $r_1^{'}$ and $t_1^{'}$ denote the reflection and transmission coefficients of

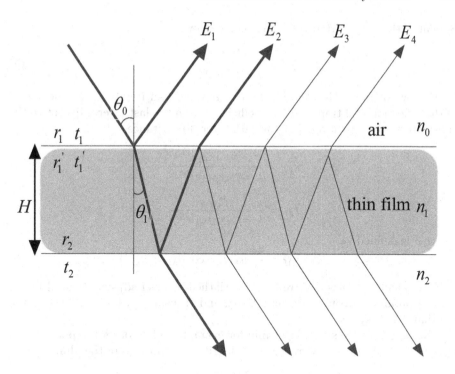

Fig. 1. Interference modeling from a single layer film structure of ridge.

light propagating from film to air, which are derived using the Fresnel equations. The refractive angle complies with Snell's law. The indices of refraction of the air and the film are denoted as n_0 and n_1 respectively where $n_0 = n_2$.

Hence the reflectivity of light from a pair of film and air layers corresponding to Fig. 1 are formulated as

$$
\begin{aligned}
E_1^{(r)} &= r_1 E_0^{(r)} \\
E_2^{(r)} &= t_1 r_2 t_1' E_0^{(r)} e^{i\delta} \\
E_3^{(r)} &= t_1 r_2 (r_1' r_2) t_1' E_0^{(r)} e^{i2\delta} \\
E_4^{(r)} &= t_1 r_2 (r_1' r_2)^2 t_1' E_0^{(r)} e^{i3\delta}
\end{aligned}
\tag{1}
$$

$$\vdots$$

where $\delta = \frac{4\pi}{\lambda} n_1 H \cos\theta_1$ denotes the phase difference of two adjacent reflected or transmitted light propagating through the film.

Referring to the interference theory of multilayer films [2,19–21], the composite reflectivity $\bar{r}$ and transmissivity $\bar{t}$ of this single layer film can be further formulated as

$$
\bar{r} = \frac{E_1^{(r)} + E_2^{(r)} + \cdots}{E_0^{(r)}} \approx \frac{r_1 + r_2 e^{i\delta}}{1 + r_1 r_2 e^{i\delta}}
\tag{2}
$$

Similarly, the transmitted coefficient is given by

$$\bar{t} \approx \frac{t_1 + t_2}{1 + r_1 r_2 e^{i\delta}} \tag{3}$$

For two or more M-layer thin film system, we can iterative the calculation of the reflection and transmission coefficients from the last layer adjacent to the substrate to the first layer. For the Mth layer, for instance,

$$\bar{r}_M = \frac{r_M + r_{M+1} e^{i\delta_M}}{1 + r_M r_{M+1} e^{i\delta_M}} \tag{4}$$

$$\bar{t}_M = \frac{t_M t_{M+1}}{1 + r_M r_{M+1} e^{i\delta_M}} \tag{5}$$

where δ is defined as

$$\delta_M = \frac{4\pi}{\lambda} n_M H_M \cos\theta_M \tag{6}$$

The calculation process is repeated until the first layer adjacent to air. Finally, we can obtain the composite reflectivity and transmissivity coefficients of the multilayer film system.

Taking the above single layer film for example and ignoring the polarization of the light, we can get the reflectance distribution function of the film, namely,

$$R_{fresnel} = \frac{r_1^2 + r_2^2 + 2r_1 r_2 cos\delta}{1 + r_1^2 r_2^2 + 2r_1 r_2 cos\delta} \tag{7}$$

The transmittance distribution derivation of the film is written as

$$T_{fresnel} = \frac{n_2 cos\theta_1}{n_0 cos\theta_0} \frac{t_1^2 t_2^2}{1 + r_1^2 r_2^2 + 2r_1 r_2 cos\delta} \tag{8}$$

The thin film interference is one of most simple structural colors and widely exists in nature. Its most remarkable characteristics is that the reflected wave is selective. Namely, a specific wavelength plays a determinant role in a specific direction. Based on the Eq. 7, the construction interference condition of reflected wave is given by

$$2n_1 H cos\theta_1 = m\lambda \tag{9}$$

Referring to the above constructive equation, it is clear that the wavelength leading to higher reflectivity changes to a shorter wavelength with the increase of the incident angle. The result is that the color changes with viewing angle. For example, the color of Morpho butterflies changes from blue to purple as the viewing angle is increased that will be verified in the following experiment.

3.2   Fresnel Coefficient

According to the above section, the composite reflection and transmission coefficients on film surfaces play a key role in producing iridescent colors. They affect the spatial distribution of radiant energy by changing the amplitude and phase of light, which are determined by Fresnel Eq. 10. In experiments, light is assumed to be unpolarized and randomly oriented. Hence, the reflectance of multilayer film structure is approximated as the average of squares of the parallel and perpendicular polarization Fresnel terms.

$$
\begin{aligned}
r_j^{\|} &= \frac{n_j cos\theta_{j-1} - n_{j-1} cos\theta_j}{n_j cos\theta_{j-1} + n_{j-1} cos\theta_j} \\
t_j^{\|} &= \frac{2n_{j-1} cos\theta_{j-1}}{n_{j-1} cos\theta_j + n_j cos\theta_{j-1}} \\
r_j^{\perp} &= \frac{n_{j-1} cos\theta_{j-1} - n_j cos\theta_j}{n_{j-1} cos\theta_{j-1} + n_j cos\theta_j} \\
t_j^{\perp} &= \frac{2n_{j-1} cos\theta_{j-1}}{n_j cos\theta_j + n_{j-1} cos\theta_{j-1}}
\end{aligned}
\tag{10}
$$

where $r^{\|}$ and $r^{\perp}$ denote the Fresnel coefficients for parallel polarized light, $t^{\|}$ and $r^{\perp}$ denote the coefficients for perpendicular polarized light, and n_{j-1} and n_j denote the refractive indexes of incident and transmitted medium respectively. The transmitted angle complies with Snell's law [2,5].

The applied Fresnel term, namely Eq. 10, depends on the assumption that the potential polarization states of the light are not considered. This is an approximation, as the reflected parallel term can lead to a phase shift, or, in the case of total reflection, become purely imaginary, therefore leading again to a phase delay. On the other hand, the perpendicular Fresnel term does not affect the phase. These situations, which may be important for interference, are neglected by averaging them.

3.3 Regularity and Irregularity of Structures

For the multilayer film structures with a certain amount of irregularity such as Morpho butterflies, the occlusion, shadowing and interreflection of light among ridges may lead to the uneven spatial distribution of energy as illustrated in Fig. 2. For example, Kinoshita and Yoshioka [23,24] have demonstrated that the behavior of light is the result of joint action of the regularity and irregularity of multilayer film structures.

It is necessary to incorporate the irregularity of the film structure to model the diffusive nature where isotropic Phong exponent [25] is commonly used [4]. However, some film structures often show backscattering and anisotropic spectral characteristics. Many geometrical models have been developed [26,27]. They work by statistically modeling the scattering of light, where the wave-like properties are ignored. As an extension, we incorporate the Ashikhmin microfacet scattering shader [28], namely Eq. 11, into the wave BSDF illumination model to describe the local anisotropic effects.

$$
D_{facet} = \frac{\sqrt{(e_x + 2)(e_y + 2)}}{2\pi} (\omega_h \cdot n)^{e_x \cos^2 \phi + e_y \sin^2 \phi}
\tag{11}
$$

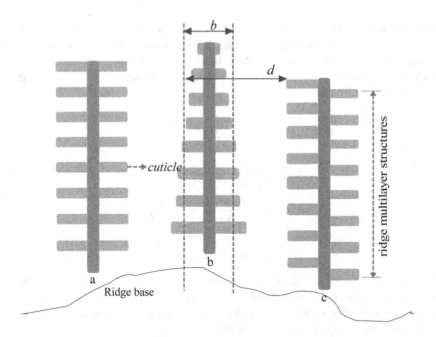

Fig. 2. The approximated geometry of ridge films where the film width b is set to 300 nm and the film separation d is set to 675 nm based on the measurements of Platter [22].

where ω_h denotes the half angle vector for incident direction ω_i and outgoing direction ω_o, n denotes the surface normal, ϕ denotes the orientation angle, and e_x and e_y denote the exponents of anisotropic distribution along the x and y axes respectively.

This paper develops a new wave bidirectional scattering distribution function that provides an efficient way in accurately rendering the interference appearance of films with periodic structures, written as

$$\text{BSDF}_\lambda = c_a I_{diffuse} + c_b \frac{D_{facet} F_{fresnel} G(\omega_o, \omega_i)}{4 \cos \theta_o \cos \theta_i} \tag{12}$$

where $G(\omega_o, \omega_i)$ denotes a geometric attenuation term [5,26], $I_{diffuse}$ denotes the diffuse effects due to the surface irregularities, $F_{fresnel}$ is determined by Eqs. 7 and 8, and c_a and c_b is the weighted coefficient.

3.4   Rendering Equation

In this work, we construct a multilayer film interference model to describe the interaction between light and film structures, where the wave bidirectional scattering distribution function is constructed to accurately model the amplitude and phase information of light in the physics-based framework [5]. With the wavelength λ, the wave rendering equation is defined as:

$$L_o(p,\omega_o,\lambda) = L_e(p,\omega_o,\lambda) + \int_{\delta^2} \text{BSDF}_\lambda(p,\omega_i,\omega_o)L_i(p,\omega_i,\lambda)|\cos\theta|d\omega_i \quad (13)$$

where $L_o(p,\omega_o,\lambda)$ is the reflected radiance of wavelength λ in the direction ω_o at the point P, $L_e(p,\omega_o,\lambda)$ is the self-emitted irradiance, $L_i(p,\omega_i,\lambda)$ is the incident irradiance with wavelength λ in the direction ω_i at the point P, and θ is the angle between the incident direction ω_i on the sphere δ^2 and the surface normal. $\text{BSDF}_\lambda(p,\omega_i,\omega_o)$ consisting of the reflectance BRDF and the transmittance BTDF denotes the our proposed wave model that model the anisotropic optical property of multilayer film structures.

4   Simulations

We implemented our wave model for rendering iridescent colors of objects coated with multilayer thin films by creating a new material plugin for the PBRT system [5]. All of the images in this work were produced using Maya software on a Dell T7600 workstation with a 2.40 GHz Intel Xeon CPU E5-2609 and a NVIDIA Quadro 6000. We focused on the visible spectrum (350~750 nm) and showed physics-based renderings as interference examples for the multilayer structures.

Figure 3 visualizes the optical phenomena of a single layer dielectric film coating on a glass sphere due to the multi-beam interference where refractive indices of the glass and dielectric materials are set to 1.5 and 2.0 respectively. Note how refraction through the transmissive object distorts the scene behind it and how the left mirror reflects the interference effect.

In Fig. 4, our model is further applied to render the iridescent patterns of objects coated with a 450 nm dielectric film, whose interference effects clearly appear on the near mirror surface. The indices of refraction of substrate and dielectric are set to 1.5 and 2.0. Compared with Fig. 3, the color shows a shift to red with the increase of the film thickness which is in agreement with the experiments.

This paper also implements the film interference patterns of opaque objects as illustrated in Fig. 5 where Blinn [27] isotropic exponent is used. The indices of refraction of substrate and dielectric are also set to 1.5 and 2.0. In addition, the approach proposed in this article is applicable for other cases of iridescence rendering. For example, Figs. 6 and 7 are two examples of anisotropic renderings of opaque objects based on the proposed approach in the physics-based PBRT [5], where the parameters denoting the film thickness and the microfacet roughness used for each object can be easily adjusted as needed. We addressed the effects of the surface roughness and anisotropy on the visual optical appearance where Ashikhmin [28] anisotropic functions is used as the basis.

The iridescent objects can be biological or nonbiological. With the help of the optics and electric microscopes, researchers have extensively reported the tree-like periodic structures of ridges on Morpho butterfly wings as shown in Fig. 2 where ridges consist of alternate cuticle film and air. Our proposed model can be used for generate iridescent colors of butterflies as illustrated in Fig. 8. The four butterflies with different structural parameter values are rendered,

Fig. 3. Renderings from a mirror sphere with perfect specular reflection (left) and a glass sphere coated with the 420 nm dielectric film (right) (Colour figure online).

Fig. 4. Renderings from a mirror sphere coated with perfect specular reflection (left) and a glass sphere with the 450 nm dielectric film (right) (Colour figure online).

where the thickness of film layer is set to 80 nm, 100 nm, 120 nm, and 135 nm respectively. From left to right, the rendered colors of wings are approximate violet, blue, yellow and red. Comparing these cases, a color shift from the violet to the red happens. A detailed comparison with the work of Sun [4] is further illustrated in Fig. 9. This renderings also agree with the observed iridescences and experimental measurements of Morpho butterflies [29–31]. This experimental measurements provide us with a basis to apply the multilayer interference model to visualize the iridescent colors reflected by the biological structures.

(a) 400nm   (b) 600nm

Fig. 5. Interference renderings of opaque objects coated with 400 nm and 600 nm dielectric films respectively.

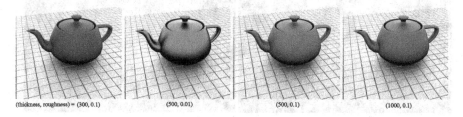

(thickness, roughness) = (300, 0.1)   (500, 0.01)   (500, 0.1)   (1000, 0.1)

Fig. 6. Interference renderings of teapots with the different surface roughnesses and thin film thicknesses.

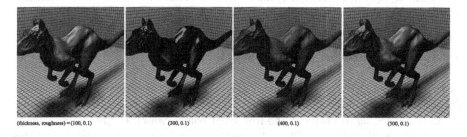

(thickness, roughness) = (100, 0.1)   (300, 0.1)   (400, 0.1)   (500, 0.1)

Fig. 7. Interference renderings of kangaroos with the different surface roughnesses and thin film thicknesses.

(a) 80 nm   (b) 100 nm   (c) 120 nm   (d) 135 nm

Fig. 8. Rendered biological iridescences of Morpho butterflies consisting of tree-like ridge structures with cuticle films of different thicknesses (left to right: 80 nm, 100 nm, 120 nm and 135 nm) (Colour figure online).

[Sun]   [Our work]

Fig. 9. Comparison of rendered biological iridescences of Morpho butterfly wings with 90 nm cuticle layer thickness using Sun butterfly shader [4] (left) and our proposed multilayer interference model (right) respectively.

5   Conclusion and Future Work

In the photorealistic rendering field, a lot of attentions are paid to the wave properties of multilayer film structures. This paper constructs an interference illumination model to visualize the iridescent colors caused by the interaction of light and layered structures where the indices of refraction, thicknesses and the irregular geometry of films play an important role. In ray tracers, This model creates a wavelength-dependent bidirectional scattering distribution function to describe the spatial spectrum distribution of light. The multi-beam interference equations have been introduced to represent the multiple reflection and transmission inside films, whose values are applied to the weight-based Monte Carlo sampling to realistically render local illumination. The Fresnel formulas for dielectric and metallic films are also described which are applied to trace the amplitude and phase variations. In addition, the microfacet scattering coefficient is incorporated to consider the optical characteristics from rough surfaces for the sake of accurately exhibiting backscattering and anisotropic phenomena. Compared with experimental measurements, we have shown that this model suffices to describe the optical effects, and have facilitated its practical application in Maya software.

However, there still exist many work for future. For example, how to handle the polarized effects of light. Due to the complexity of film structures, it is desirable to gain the measured appearance data to improve accuracy of wave rendering. In addition, our proposed model can be applied to render other objects exhibiting structural colors such as optical lenses, beetles and birds.

Acknowledgements. We sincerely acknowledge all anonymous reviewers for their valuable comments. This work was funded by National High Technology Research and Development Program of China (2012AA011206 and 2009AA01Z303).

References

1. Gondek, J.S., Meyer, G.W., Newman, J.G.: Wavelength dependent reflectance functions. In: Proceedings of SIGGRAPH 1994, pp. 213–219 (1994)
2. Hirayama, H., Kaneda, K., Yamashita, H., Monden, Y., Yamaji, Y.: Visualization of optical phenomena caused by multilayer films with complex refractive indices. In: 7th Pacific Conference on Computer Graphics and Application, pp. 128–137 (1999)
3. Hirayama, H., Yamaji, Y., Kaneda, K., Yamashita, H., Monden, Y.: Rendering iridescent colors appearing on natural objects. In: Proceedings of the 8th Pacific Conference on Computer Graphics and Application, pp. 15–22 (2000)
4. Sun, Y.: Rendering biological iridescences with RGB-based renderers. ACM Trans. Graph. **25**(1), 100–129 (2006)
5. Pharr, M., Humphreys, G.: Physically Based Rendering: From Theory to Implementation. Morgan Kaufmann, Burlington (2010)
6. Smits, B.E., Meyer, G.W.: Newton's colors: simulating interference phenomena in realistic image synthesis. In: Proceedings of Eurographics Workshop on Photosimulation, Realism and Physics in Computer Graphics, pp. 185–194 (1990)
7. Dias, M.: Ray tracing interference color. IEEE Comput. Graph. Appl. **11**(2), 54–60 (1991)
8. Jakob, W., D'Eon, E., Jakob, O., Marschner, S.: A comprehensive framework for rendering layered materials. In: Proceedings of SIGGRAPH 2014 (2014)
9. Moravec, H.P.: 3D graphics and the wave theory. In: Proceedings of SIGGRAPH 1981, pp. 289–296 (1981)
10. Kajiya, J.T.: The rendering equations. In: Proceedings of SIGGRAPH 1986, vol. 20, pp. 143–150 (1986)
11. Stam, J.: Diffraction shaders. In: Proceedings of SIGGRAPH 1999, pp. 101–110 (1999)
12. Agu, E.: Diffraction shading models for iridescent surfaces. In: Proceedings of IASTED VIIP (2002)
13. Sun, Y., Fracchia, F.D., Drew, M.S., Calvert, T.W.: Rendering iridescent colors of optical disks. In: Eurographics Workshop on Rendering, pp. 341–352 (2000)

14. Wu, F.-K., Zheng, C.-W.: A comprehensive geometrical optics application for wave rendering. Graph. Models **75**(6), 318–327 (2013)

15. Wu, F.-K., Zheng, C.-W.: Microfacet-based interference simulation for multilayer films. Graph. Models **78** (2015)

16. Icart, I., Arquès, D.: A physically-based BRDF model for multilayer systems with uncorrelated rough boundaries. In: Rendering Techniques, pp. 353–364 (2000)

17. Hirayama, H., Kaneda, K., Yamashita, H., Monden, Y.: An accurate illumination model for objects coated with multilayer films. Comput. Graph. **25**(3), 391–400 (2001)

18. Okada, N., Zhu, D., Cai, D.S., Cole, J.B., Kambe, M., Kinoshita, S.: Rendering Morpho butterflies based on high accuracy nano-optical simulation. J. Optics **42**(1), 25–36 (2013)

19. Born, M., Wolf, E.: Principles of Optics: Electromagnetic Theory of Propagation, Interference and Diffraction of Light. Cambridge University Press, London (2005)

20. Goodman, J.W.: Introduction to Fourier Optics. Roberts & Co., Englewood (2005)

21. Liang, Q.T.: Physical Optics (in Chinese). Publishing House of Electronics Industry, Beijing (2008)

22. Plattner, L.: Optical properties of the scales of Morpho rhetenor butterflies: theoretical and experimental investigation of the back-scattering of light in the visible spectrum. Comput. Graph. Forum **1**, 49–59 (2004)

23. Kinoshita, S., Yoshioka, S., Fujii, Y., Okamoto, N.: Photophysics of structural color in the morpho butterflies. Forma **17**, 103–121 (2002)

24. Kinoshita, S., Yoshioka, S.: Structural colors in nature: the role of regularity and irregularity in the structure. Chemphyschem. **6**, 1442–1459 (2005)

25. Phong, B.T.: Illumination for computer generated pictures. Commun. ACM **18**(6), 311–317 (1975)

26. Torrance, K.E., Sparrow, E.M.: Theory of off-specular reflection from roughened surfaces. J. Opt. Soc. Am. **57**(9), 1105–1112 (1967)

27. Blinn, J.F.: Models of light reflection for computer synthesized pictures. In: Proceedings of SIGGRAPH 1977, pp. 192–198 (1977)

28. Ashikhmin, M., Shirley, P.: An anisotropic Phong BRDF model. J. Graph. Tools **5**(2), 25–32 (2000)

29. Fox, D.L.: Animal Biochromes and Structural Colours. University of California Press, Berkeley (1976)

30. Simon, H.: The Splendor of Iridescence of Structural Colors in the Animal World. Dodd Mead & Company, New York (1971)

31. Vukusic, P., Sambles, J.R., Lawrence, C.R., Wootton, R.J.: Quantified interference and diffraction in single Morpho butterfly scales. Proc. R. Soc. B Biol. Sci. **266**, 1403–1411 (1999)

Enhancement of Direct Augmented Reality Object Selection by Gravity-Adapted Target Resizing

Daniela Markov-Vetter[1,3(✉)], Vanja Zander[2], Joachim Latsch[2],
and Oliver Staadt[3]

[1] German Aerospace Center (DLR), Institute of Aerospace Medicine,
Linder Höhe, 51147 Cologne, Germany
daniela.markov-vetter@dlr.de
[2] Institute of Cardiology and Sports Medicine, German Sport University
Cologne, Am Sportpark Müngersdorf 6, 50933 Cologne, Germany
{v.zander,j.latsch}@dshs-koeln.de
[3] Institute of Computer Science, University of Rostock, Albert-Einstein-Str. 22,
18051 Rostock, Germany
oliver.staadt@uni-rostock.de
http://dlr.de/me

Abstract. Direct object selection in an Augmented Reality environment that is coded outside of human body frame of reference is deteriorated under short-term altered gravity. As countermeasures we developed a gravity-adapted resizing technique based on the Hooke's law that resulted in two techniques of target and interface deformation (compression, elongation). To prove the concept of this resizing approach we initially conducted two experiments under simulated hypergravity conditions. While during the first study hypergravity was induced by a long-arm human centrifuge, in the second study hypergravity was simulated by additional arm weightings that were balanced and attached to the participants' pointing arm. We investigated the difference of the task performance with respect to the pointing frequency, response time, pointing speed and accuracy, when participants performed a visuomotor task under the resizing conditions compared to the unchanged condition. During the second study we additionally evaluated the speed-accuracy tradeoff of the resizing techniques according to Fitts' law and the physiological workload by cardiac responses analyzing the heart rate variability. Both experiments showed that the online adaption of the present gravity load to targets' size and distance influences the performance of direct AR direct pointing. The results revealed that the pointing performance benefits from elongation target deformation by increased target sizes and distances, while pointing towards compressed targets mostly decreases the physiological workload under increased gravity conditions.

Keywords: Augmented reality · Interaction · Direct pointing · Usability

© Springer International Publishing Switzerland 2016
J. Braz et al. (Eds.): VISIGRAPP 2015, CCIS 598, pp. 75–96, 2016.
DOI: 10.1007/978-3-319-29971-6_5

1 Introduction

The application of Augmented Reality [1] to intravehicular space operations could support astronauts in their task performance at complex technical facilities aboard the International Space Station [2, 3]. Beside of innovative technologies intended to make Augmented Reality (AR) ready-to-use in smart technical environments, the human performance decides finally on the usability. While AR as interface technology claims being in the physical reality, resulted AR systems should be optimized for the environmental conditions to which the AR interface will be applied [12]. Current research on human factors of handling AR interfaces presumes the application under normogravity (1 *g*) conditions on Earth. Changes in gravity otherwise, causes sensorimotor disruptions in human motor coordination and eye movements that interfere with astronauts' work and thus can also deteriorate their task performance during space operations.

In previous work [14, 15] we investigated the impact of altered gravity on direct AR object selection for symbolic-input and control tasks. Therefore we performed two experiments under parabolic flight conditions to find out which interface alignment is the most efficient one and preserves human's hand-eye coordination under altered gravity. We compared three alignment conditions (see Fig. 1) of the virtual pointing interface, whereby each condition had its own interrelation between the human body frame of reference and the support of haptic cues. The results showed that aimed pointing movements for direct AR selection under altered gravity benefits from targets with haptic cues and targets that are coded inside the human body frame of reference (e.g., attached to hand), while targets placed outside the body frame deteriorate the pointing performance.

However, given that the future main application of an AR supported guidance system, is predominantly coded outside of the user's body frame, the maintenance of user performance in object selection requires the introduction of appropriate countermeasures. Supposing that an outside coded AR interface that the astronaut has to operate in space, is affected in the same way as the astronaut by gravity changes, it is conceivable that the hand-eye coordination will be improved. Whether detecting labels and annotations by gaze control, or pointing movements during symbolic tasks, an adequate transformation of the AR interface could overcome the human sensorimotoric disruptions under gravity changes. As stated by Fitts' Law, timing effects of the selection performance can be affected by targets' transformation in size and distance [6]. Thereby a supportive approach could be the dynamical transformation of the AR interface with respect to the active gravity. Before conducting expensive experiments under simulated weightlessness conditions (e.g., parabolic flight), we initially performed a proof-of-concept (POC) study under simulated hypergravity (+Gz) conditions. Therefore we developed a gravity-adapted strategy for interface resizing that we studied by AR selection using a direct touch interface under increased gravity conditions. The resizing approach not only affects the size of a pointing target (or label), but also affects the position, i.e. in cases of more than one target, also the distance between them is affected. This paper extends our previous work [16].

Outside coded target	**Inside coded target**	**Inside coded target**
- Physical surface alignment	- Body alignment (hand-held)	- HMD alignment (HUD)
- Head and body movements	- Bimanual handling	- No haptic feedback

Fig. 1. Placement conditions of Augmented Reality Interfaces for control and symbolic input tasks previously studied under short-term hyper and microgravity during parabolic flights.

2 The Force-Based Resizing Approach

For improving the performance of aimed pointing movements towards virtual targets under altered gravity conditions, we developed a force-based approach for dynamic target transformation. Force-based approaches are typically used for automated positioning of labels and annotations, e.g. in 3D information visualization [8, 18]. Depending on the active gravity load we calculated a corresponding force affecting target's size and position. The approach for target resizing and positioning was derived from the elastic behaviour of soft bodies, which are proportional deformed to the applied gravity load G_{sim}, similar to Hooke's law (Eq. 1). Therefore, we calculated the axial (Eq. 2) and transversal (Eq. 3) strain of the target using empirical values for the modulus of elasticity E and Poisson's ratio v. Thereby, we distinguished between two techniques of target sizing – *sizing by compression* (SC, Eq. 4) and *sizing by elongation* (SE, Eq. 5). For evaluation purposes their output was compared with the unmodified sizing technique (SU) as baseline condition that did not affect the targets. For initially experimentation we limited the evaluated parameters by automated target resizing without the transversal strain Δw, but applied the axial strain Δh proportionally to target's height and width. Figure 2 shows the resulted sizing techniques that we have investigated. We also applied the gravity-based changes to the complete interface, i.e. to targets' position, that resulted in a larger target distance with the SE technique and in

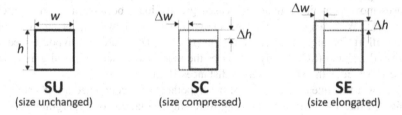

Fig. 2. Resulted resizing methods applied to the pointing targets and the overall interface.

smaller distances with the compressed SC technique. While the SC sizing technique provides smaller targets and benefits from smaller target distances, the SE technique offers larger target size at larger distance.

$$\sigma = E\varepsilon \text{ with } \varepsilon = \frac{\Delta h}{h} \text{ and } \sigma = \frac{F}{A} \tag{1}$$

$$\Delta h = \frac{F * h}{A * E} \text{ with } A = h^2 \text{ and } F = G_{sim} \tag{2}$$

$$\Delta w = -w * v \frac{\Delta h}{h} \tag{3}$$

$$h_{SC} = h - \Delta h \ , \ w_{SC} = w - \Delta w \tag{4}$$

$$h_{SE} = h + \Delta h \ , \ w_{SE} = w + \Delta w \tag{5}$$

To our knowledge, gravity-adapted target sizing was not reported until now. For designing the normal sized targets (SU) we followed the recommended ergonomic size range for push buttons [5] and used a squared target of 15 mm width and height.

3 Research Objective

Focused on sensorimotor hand-eye coordination we investigated the effect of gravity-adapted interface resizing on performance during an AR visuomotor task presented by a head-mounted display under increased gravity conditions. In accordance with the active gravity load, we expected that the variations of the pointing performance (e.g., response time, speed, accuracy; etc.) are correlating with the resulted AR interface that was affected by the resizing approach and led to the following question: Does gravity-adapted target resizing affect the performance and workload of direct AR pointing under altered hypergravity conditions? [Q1].

The resulted sizing conditions using the gravity-adapted sizing approach interrelate to the characteristics of Fitts' law that predicts longer movement times at greater distances, as well as at smaller targets [6]. While the elongated method (SE) provides greater targets at larger distances, the compressed method (SC) provides smaller targets at shorter distances. Even though elongated sizing will cause the largest distance, we expect that fast pointing movements under increased gravity conditions benefit from greater targets. Therefore we hypothesized that elongated resizing (SE) mostly decreases movement times under increased gravity load, because it provides greater targets [H1]. With respect to the physiological workload, we expect an increase in physical effort for increased target distances (SE). Therefore we hypothesized that compressed resizing (SC) mostly decreases the workload under increased gravity load, because it provides the shortest target distances [H2].

To answer this question and to test the hypotheses we conducted a POC study that was divided into two experiments using different simulations of hypergravity. Firstly, we performed a case study where +Gz load was induced by a long-arm human centrifuge (LAHC). Secondly, we conducted an experiment under normogravity and simulated +Gz loads by additional arm weighting [7]. Using a visuomotor task, we aimed to investigate the impact of increased gravity loads on size and distance of a given target interface evaluated on direct object selection by the performance and physiological workload. For evaluating the performance we used common measures, such as the frequency of correct and incorrect pointing, the accuracy, the response time and the pointing speed. Since the physiological factor is essential in sensorimotor coordination, we recorded and evaluated the physiological strain by assessing the heart rate variability (HRV) [17, 21]. This is an immanent expression of sympathetic and parasympathetic influences of the function of the heart [20]. HRV recording has only been applied during the Weight-Study. The studies did not only differ in the way of hypergravity simulation and workload assessment, they were also distinct in their experimental task. For the LAHC-Study we used the experiment task that has already been used during the parabolic flight studies (pointing towards an AR soft keyboard). Because we additionally evaluated the task performance related to Fitts' law during the Weight-Study, we considered the international standard for pointing devices [9] and this time we chose to use the multi-directional tapping task [13], with eight targets arranged in a ring. There have been only few studies applying Fitts' law on evaluation of AR interaction [19]), or on head-mounted Mixed Reality pointing [11].

4 First Experiment: Case Study by Long-Arm Human Centrifuge

To proof the concept of the gravity-adapted approach initially, we were allowed to perform a case study under +Gz load induced by a long-arm human centrifuge (LAHC, see Fig. 3). Human centrifuges enable research in medicine and human physiology during altered +Gz load and are also used to train pilots and astronauts. The case study was performed with one participant. The male participant (51 years old, space engineer) is very experienced under altered +Gz load (human centrifuge, parabolic flight) and familiarized with the used AR pointing system and task.

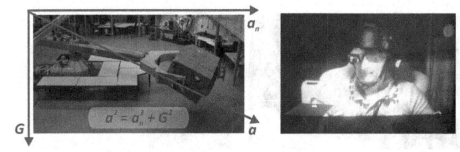

Fig. 3. Used LAHC (5 m radius) with centrifugal acceleration a_n. The cabin is swinging out during the rotation with resulted acceleration a in line with subject's long body axis.

4.1 Apparatus

We used a right-sided monocular optical see-through head mounted display (OST HMD, Shimadzu dataGlass2/a), which has a semi-transparent LCD display with a resolution of 800 × 600 pixels and a diagonal field of view (FOV) of 30 degrees (see Fig. 4, left). The HMD was connected to the data processing unit (Lenovo ThinkPad T420 s, 2.8 GHz CPU, NVIDIA Quadro NVS 4200 M), which was installed under the participant's seat in the centrifuge cabin. For optical inside-out marker tracking we equipped the HMD with an optical sensor (Microsoft HD 5000 webcam with 66 degree diagonal FOV). To compute the position of participant's eye relative to the optical sensor, the participant had to perform a self-calibration [10]. To realize pointing with haptic feedback we used a panel that was installed in front of the participant and was equipped with a multi-marker configuration. For the pointing purpose a single marker was attached to the participant's fingertip at the dominant hand. The pose data were captured with a mean frame rate of 38.74 fps (SD = 10.05) by the optical sensor at constant artificial light conditions.

4.2 Experiment Task

In response to visual stimuli the participant should point towards virtual targets under altered +Gz loads while wearing an optical see-through head-mounted display (OST HMD). Pointing in response to visual stimuli was done based on the visuomotor task used for parabolic flight studies. By using a soft AR keyboard with squared keys of 15 mm width and height [5], the participant was requested to enter prescribed random pseudo-letters on a virtual keyboard (see Fig. 4, right). Entering letters onto the keyboard was determined by collision tests of a virtual ray ranging from the origin of the fingertip marker to the top of the index finger. The requested letter was signalled in green, hitting a correct key was highlighted in red and then the next key was signalled. Because the data processing unit was installed in the cabin of the centrifuge, the participant needed to start the experiment with a virtual start button displayed above the keyboard and hidden afterwards.

Fig. 4. Participant sitting in the LAHC cabin, wearing the OST-HMD and pointing toward the panel (left). The soft-AR keyboard (right).

4.3   Experiment Design and Procedure

We conducted the study in three experimental sessions, on three successive days. For the visuomotor task performance we specified four target pools that were counterbalanced presented per sizing technique. we defined a target pool as a pre-randomized series of keys. The completion time for pointing towards the keys of one target pool was predefined by 25 s. Within one centrifugation the participant performed the task for two sizing techniques that resulted in a total pointing time of 200 s per +Gz load. To adjust the duration of a key pool and the Gz loads, we used the first day for pilot testing. Because it was quite exhausting for the participant to perform arm movements in series for 200 s, we decide to reduce the operation time of a target pool to 20 s (in total 160 s per centrifugation). Pilot testing was done under 1.5 g and 2 g. For better differentiation between the Gz loads we decided to perform the experiment under 1.5 g, 1.8 g and 2.3 g. To avoid transition effects between the target pools and the method changes, we did not recorded the first and the last signaled key. For physiological regeneration and to limit learning effects there was a 10 min break between the changes of the +Gz loads. For the experiment sessions of the second day, within one centrifugation the participant performed the task under one +Gz load using the unchanged method (SU) and one of the methods with force-based target sizing (SC, SE). Thereby the sizing technique was changed after one target pool. We performed the experiment for SU and SC on the second day, and for SU and SE on the third day under 1.5 g, 1.8 g and 2.5 g. Thereby the sizing conditions and the Gz loads were systematically counterbalanced.

4.4   Results

For comparison of the sizing conditions we analyzed the frequencies of correct and false target hits, the percentage error rate, the response time and the pointing speed. Thereby a false target hit constitutes that the participant has pointed toward a wrong key. Because the number of resulted target hits was variable by a predefined completion time, we consider the percentage error rate calculated by dividing the total number of triggered targets by the total number of false target hits. The response time mirrors the time elapsed between the visual stimulus onset and motoric response onset, while the speed was calculated by the response time and the Euclidean distance between the centre of the last target and the centre of the present target. For statistical analysis we compared the sizing methods across all Gz loads and on same stage of Gz load. For comparing count data we assumed a Poisson distribution and applied a general linear mixed model (SAS® 9.4 PROC GLIMMIX) with the logarithm as link function. For repeated measures analysis we assume a normal distribution and used PROC MIXED (SAS® 9.4) as linear mixed model with simulated adjustment to keep the experiment-wise error rate $\alpha = 0.05$. To compare the percentage error rate between the sizing conditions we transformed the data values by the natural logarithm. Aware that the experiment was conducted by only one participant, this case study should prove whether the resizing approach generally effects the performance under altered hypergravity conditions.

The participant completed 46 target pools (with 20 s per pool) across all Gz loads (1.5 *g*, 1.8 *g* and 2.5 *g*) and accomplished 566 correct target hits in total with a mean of 12.30 (SD = 1.39) per target pool with 42 false target hits with a mean of 0.91 (SD = 1.28) per target pool. For analyzing the time that the participant required to respond to a visual stimulus we considered only data sets where the number of false target hits was zero and the response time was less than 3000 ms. Therefore we analyzed 460 valid trials for the response time and the resulted pointing speed. In Table 1 we present measures of the central tendency and the variability of the dependent variables for the studied sizing techniques per Gz load. As also shown in Fig. 5 (left), pointing towards elongated targets (SE) resulted in average mostly correct target hits under the gravity levels 1.5 *g* and 1.8 *g*, while the compressed sizing technique (SC) led to the lowest number of correct target hits under all Gz loads. But

Table 1. Measures of frequency of correct target hits, percentage error rate, response time and pointing speed.

Resizing	Correct Hits [ms]	Error [%]	Response time [ms]	Speed [mm/ms]
	Mean ± SD	Mean ± SD	Mean ± SD	Mean ± SD
Gz = 1.5 *g*				
SU	12.63 ± 1.77	06.44 ± 5.87	1367.25 ± 217.33	0.063 ± 0.013
SC	10.75 ± 0.96	13.14 ± 9.05	1483.18 ± 260.69	0.040 ± 0.008
SE	**13.25** ± 0.96	**00.00** ± 0.00	**1341.46** ± 081.74	**0.075** ± 0.005
Gz = 1.8 *g*				
SU	13.13 ± 1.46	02.03 ± 03.79	**1300.01** ± 166.84	0.064 ± 0.007
SC	11.50 ± 0.58	07.28 ± 10.11	1338.24 ± 181.74	0.044 ± 0.006
SE	**13.25** ± 0.96	01.67 ± 03.33	1312.80 ± 133.50	**0.077** ± 0.008
Gz = 2.5 *g*				
SU	**12.14** ± 1.07	04.49 ± 06.11	1322.43 ± 256.35	0.062 ± 0.007
SC	11.00 ± 0.00	19.41 ± 03.49	**1239.52** ± 168.98	0.048 ± 0.007
SE	11.75 ± 0.96	14.02 ± 16.23	1389.13 ± 166.74	**0.074** ± 0.012

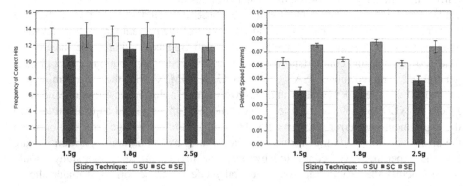

Fig. 5. Mean frequency of correct target hits and pointing speed (with CI = 95 %) of the sizing techniques grouped by the Gz loads.

comparing the sizing conditions (see Table 2) only revealed significant differences across all Gz loads, with a higher mean frequency of correct hits with SU ($M = 12.65$, $SD = 1.47$) and with SE ($M = 12.75$, $SD = 1.14$) compared to SC ($M = 11.09$, $SD = 0.70$). The same applied to the percentage error rate that revealed the lowest error rate with SE under 1.5 g and 1.8 g and always the highest with SC, but did not show significant differences.

Because the elongated method provides larger and the compressed method shorter distances we expected longer response times with the elongated method (SE) and shorter response times with the compressed method (SC). But the performance did not show meaningful variations of the mean response times. Therefore it is more reasonable to analyze the pointing speed, because the size and distance of the target varies with the presented sizing method. The mean speed values (see Fig. 5, right) showed most prominent variations for the elongated method (SE) with the highest speed and for the compressed method (SC) with the lowest speed under all Gz loads. The comparison of the pointing speed (see Table 2) resulted in significant differences between the sizing methods on the same level of gravity load and between the sizing methods across all Gz loads with significant highest speed under the SE conditions ($M = 0.076$ mm/ms, $SD = 0.008$) followed by the SU condition ($M = 0.063$ mm/ms, $SD = 0.010$), and with the slowest speed under the SC condition ($M = 0.044$ mm/ms, $SD = 0.007$).

Table 2. Significant differences of comparison of correct hits, response time and pointing speed between the sizing techniques on the same level of Gz load using SAS® 9.4 PROC GLIMMIX (dist = poisson; link = log) and PROC MIXED (adjust = simulate).

Dependent Variable	+Gz	Resizing		*Estimate*	*Stderr*	*DF*	*Adj. p-Val*
Correct Target Hits	–	SU	SC	1.5613	0.4555	43	0.0039
	–	SC	SE	−1.6591	0.5186	43	0.0075
Reponse Time [ms]	1.5	SU	SC	−123.58	39.8470	166	0.0058
	1.5	SC	SE	131.95	43.6318	166	0.0075
	2.5	SU	SC	119.92	40.1622	110	0.0093
	2.5	SC	SE	−149.61	44.9627	110	0.0034
Pointing Speed [mm/ms]	–	SU	SC	0.0194	0.0011	457	<.0001
	–	SU	SE	−0.0126	0.0009	457	<.0001
	–	SC	SE	−0.0320	0.0012	457	<.0001
	1.5	SU	SC	0.0223	0.0021	166	<.0001
	1.5	SU	SE	−0.0125	0.0018	166	<.0001
	1.5	SC	SE	−0.0348	0.0023	166	<.0001
	1.8	SU	SC	0.0205	0.0015	175	<.0001
	1.8	SU	SE	−0.0129	0.0013	175	<.0001
	1.8	SC	SE	−0.0334	0.0017	175	<.0001
	2.5	SU	SC	0.0138	0.0021	110	<.0001
	2.5	SU	SE	−0.0123	0.0019	110	<.0001
	2.5	SC	SE	−0.0261	0.0024	110	<.0001

In conclusion, the case study showed that gravity-adapted target resizing and positioning significantly impacts aimed pointing performance under increased Gz loads and showed a tendency to improve the pointing performance using elongated targets (SE), particularly taking into account the significant increase in the pointing speed. That means, it seems that pointing under increased gravity benefits from greater targets at larger distances. Because we could only test one participant on the long-arm centrifuges, we looked for alternative approaches to simulate hypergravity.

5   Second Experiment: User Study by Arm Weightings

To verify the observed effect of the case study using the LAHC we performed a subsequent experiment under normogravity condition. For simulation the +Gz loads we used corresponding weightings [7] that were balanced attached to the participant's dominant forearm (see Fig. 6). The extended arm weights (see Table 3) were calculated (Eq. 6) for each participant as follows:

$$m_{add} = (G_{sim} - G) * \frac{m_{body}}{100} * 5.38\% \tag{6}$$

with G_{sim} for the simulated gravity force, m_{body} for the body weight of the participant and 5.38 % as averaged percentage arm weights introduced by Clauser et al. [4].

The LAHC-Study has shown that the performance under 2.5 g was strongly influenced by physical demand. Therefore we decided to change the used Gz loads, so that the user study by arm weightings was performed under 1.5 g, 2 g and 2.3 g. In this study we additionally investigated the effect of gravity-adapted resizing on the physiological workload that was assessed by cardiac responses using the heart rate variability (HRV).

Fig. 6. Participants wearing the arm weightings to simulate different hypergravity loads.

Table 3. Weights of body, arm and the resulted weights that was added to participants arm.

Participant	m_{body} [kg]	m_{arm} [kg]	m_{add} [kg]		
			1.5 g	2 g	2.3 g
S1	80.0	4.3	2.2	4.3	5.6
S2	78.0	4.2	2.1	4.2	5.5
S3	75.0	4.0	2.0	4.0	5.3
S4	80.0	4.3	2.2	4.3	5.6
S5	65.0	3.5	1.8	3.5	4.6
S6	69.0	3.7	1.9	3.7	4.8
S7	60.0	3.3	1.7	3.3	4.3
S8	78.0	4.2	2.1	4.2	5.5

5.1 Apparatus

We used the same HMD setup as for the LAHC-Study (see Sect. 4.1). All participants also performed an eye-sensor calibration [10] immediately before the experiment. To perform the task of pointing towards outside coded targets, the participant stood in front of a wall with 50 cm distance. Depending on participant's body height the multi-pattern was individually aligned in the horizontal position, thus the target area was at the participant's eye level. The optical sensor captured the pose data with a mean frame rate of 38.52 fps (SD = 12.54). For the assessment of the physiological workload by the HRV, the participants were equipped with a wireless eMotion HRV sensor from Mega Electronic.

5.2 Experiment Task

To evaluate the speed-accuracy tradeoff related to Fitts' law we decided to use an appropriate task and designed a multi-directional pointing task as proposed by the ISO/DIS 9241-9 standard [9]. Therefore we used eight squared targets with a default size of $a = 15.0\,mm$ (see Fig. 7). The targets were arranged in a circle with a default diameter of $d = 82.5\,mm$. Like the LAHC task, the participants should point towards the targets in response to visual stimuli. For evaluation purposes by Fitts' law we defined "true" target connections of $0°$, $45°$, $90°$ that implied the same target distance and involve horizontal and vertical arm movements. The remaining target connections were used for pointing transition only.

5.3 Participants

Participants were 6 male and 2 female aged between 24 and 51 years (20–31 years: 4 participants, 40–51 years: 4 participants, $M = 37.25$, $SD = 10.55$). Seven participants have had experiences with AR interfaces in terms of participation in previous studies, while one participant was a novice. They came from backgrounds in biology, physiology, aerospace and medicine. All participants had a right-dominant arm that was used for the pointing task.

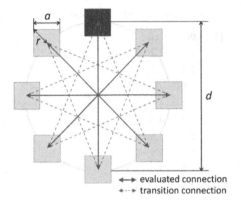

◄─► evaluated connection
◄∙∙► transition connection

Fig. 7. The multi-directional pointing task used during the Weight-Study.

5.4 Experiment Design and Procedure

The study followed a repeated measure design with two independent variables containing three levels for gravity-based resizing (SU, SC, SE) and three levels for Gz load (1.5 g, 2 g, 2.3 g). Thereby the SU level constituted the baseline condition. In a within-subject design, each participant performed the test series for all resizing methods under all gravity loads, resulting in a factorial design of 3 × 3. The repetition rate for each method amounted to five target pools per Gz load. Thereby a target pool was specified as a predefined series of randomized target connections for the multi-directional pointing tasks. Pointing towards the targets of one target pool should be completed by the participants in 25 s and constituted one test series. Overall each participant performed 45 test series. Because we did not compare the gravity loads neither on the same level of sizing methods, nor across the sizing methods, was the multi-directional task performed in a fixed order of Gz loads (2.3 g, 1.5 g, 2 g). But we used systematic variations of the presentation order of the sizing methods per Gz load. To avoid transition effects between pool changes the first and the last signaled targets performance were not recorded. Between changes of the Gz load and the sizing techniques, the participant had a five minute break for physiological regeneration. To be familiar with the pointing task and to check the integrity of the tracking operation, the participants undertook a short training session before starting the first condition. Before conducting the experiment sessions each participant performed the experiment without added arm weightings under the SU condition as baseline for the workload assessment by HRV. This condition we did not use for evaluating the performance.

5.5 Results

Table 4 shows the resulted target sizes a with its surrounding radius (Eq. 7) and targets' distances d calculated by the force-based resizing approach using the active Gz load. The distance reflects the pointing range between two "true" target connections. While pointing towards normal sized targets (SU) always resulted in same target sizes and

Table 4. Resulted target size a, radius r_s and distance d.

Resizing	+Gz	a [mm]	r_S [mm]	d [mm]
SU	–	15.00	10.61	082.50
SC	1.5	11.67	08.25	064.17
	2.0	10.56	07.45	058.06
	2.3	09.89	06.99	054.39
SE	1.5	18.33	12.96	100.83
	2.0	19.44	13.75	106.94
	2.3	20.11	14.22	110.61

distances, the elongated sizing (SE) resulted in increased sizes and distances on increased Gz loads and contrary for the compressed sizing (SC).

$$r_s = \frac{a\sqrt{2}}{2} \tag{7}$$

As for the LAHC-Study the pointing performance was assessed by the frequency of correct target hits, the percentage error rate, the response time and the pointing speed depending on the resulted target distances. To compare the percentage error rate between the sizing conditions we transformed the data values by the natural logarithm. In addition we evaluated the Euclidean distance between the target's center and the final intersection point and the percentage accuracy depending on the resulted target sizes. We compared the sizing conditions for each dependent variable on same level of Gz load and across all Gz loads using SAS® 9.4 PROC MIXED with simulated adjustment to keep the experiment-wise error rate $\alpha = 0.05$. For comparing count data we assumed a Poisson distribution and applied a general linear mixed model (SAS® 9.4 PROC GLIMMIX) with the logarithm as link function. Additionally we evaluated the speed-accuracy tradeoff of the sizing methods according to Fitts' law and present the throughput (TP) and the resulted movement models predicting the movement times of the studied sizing conditions. For workload assessment we evaluated the physiological strain by the R-R distance that was obtained from the heart rate variability (HRV). The R-R distance is the interval between two heartbeats in milliseconds.

Task Performance

Pointing Frequency. The data revealed that the participants pointed towards 6708 targets in total in a correct way with a mean frequency of correct target hits of 19.39 (SD = 3.37) per target pool and pointed towards 102 targets in a wrong way with a mean frequency of false target hits of 0.27 (SD = 0.66) per target pool. Figure 8 shows the mean frequency of correct target hits and the log transformed mean percentage error rate of the sizing conditions across all Gz loads and per Gz load. With respect to the frequency of correct target hits the comparison of the sizing conditions on same level of Gz load revealed no significant differences, but comparing the sizing conditions over all Gz loads showed the highest frequency of correct target hits for pointing towards elongated targets SE (M = 20.00, SD = 2.53) that significantly differed (see Table 5) from SU (M = 19.05, SD = 3.52) and SC (M = 18.82, SD = 4.09). With respect to the

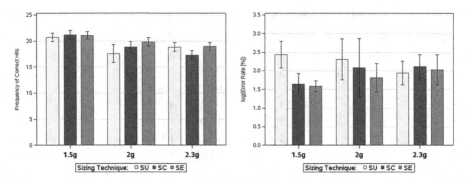

Fig. 8. Mean frequency of correct target hits and log(error rate[%]) (with CI = 95 %) of the sizing methods grouped by Gz load.

Table 5. Significant differences of comparison of frequency of correct target hits and percentage error rate between the sizing techniques across all Gz loads and on same level of Gz load using SAS® 9.4 PROC GLIMMIX (dist = poisson; link = log) and PROC MIXED (adjust = simulate).

Dependent Variable	+Gz	Resizing		Estimate	Stderr	DF	Adj. p-Val
Correct Target Hits	–	SU	SE	1.0523	0.4262	338	0.0373
	–	SC	SE	−1.1584	0.4131	338	0.0141
log(Error Rate [%])	1.5	SU	SC	0.7974	0.1449	10	0.0005
	1.5	SC	SE	0.8524	0.1415	10	0.0003

log transformed error rate the data only yielded significant differences (see Table 5) under 1.5 g with a significant lower error rate with SU compared to SC, and with a significant lower error rate under SE compared to SC.

Response Time and Pointing Speed. For analyzing time effects we only considered target hits with "true" target connections resulting in the same pointing distance per Gz load and sizing technique, as well as data sets where the number of false target hits was zero and the response time was less than 3000 ms. This resulted in 2055 valid trials. Overall the participants pointed with a mean response time of 1038.60 ms (SD = 179.26). The mean response times of the sizing conditions across all Gz loads and per Gz load are presented in Fig. 9 (left) and show that in principle the response times increased as the gravity load increased. But the mean response times did not show prominent variations between the sizing conditions across all Gz loads and per Gz load. This is contrary to our expectation of significant slower response times at larger target distances (SE) and significant faster response times at shorter distances (SC). Because targets' size and distance vary with the used sizing technique and the Gz load, analyzing the pointing speed was more meaningful than the response time. The pointing speed was calculated by the distance between the targets (see Table 4) divided by the response time. Overall the participants pointed with a mean speed of 0.083 mm/ms (SD = 0.024). The mean pointing speed by the sizing techniques across all gravity loads and per Gz load is presented in Fig. 9 (right).

Prominent mean variations yielded from the elongated sizing technique (SE) with highest speed under all Gz loads, while using compressed targets (SC) always revealed lowest speeds. The comparison of the sizing technique (see Table 6) yielded significant differences ($p < .0001$) between all conditions across all Gz loads and grouped per Gz load. The result related to the pointing speed mean that hypothesis H1 can be accepted.

Pointing Accuracy. The pointing accuracy reflects the precision of target pointing and was measured by the Euclidean distance d_{ED} and the surrounding radius r_s of the targets (see Table 4). For evaluating the pointing accuracy we initially analyzed the Euclidean distance relative to the resulted target sizes using 6708 correct target hits.

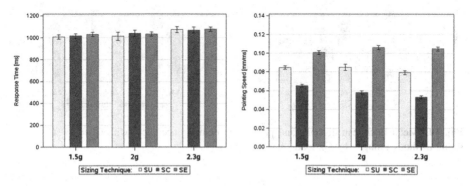

Fig. 9. Mean response time and pointing speed (with CI = 95 %) of the sizing methods grouped by Gz load.

Table 6. Significant differences of comparison of frequency of pointing speed between the sizing techniques across all Gz loads and on same level of Gz load using SAS® 9.4 PROC MIXED (adjust = simulate).

Dependent Variable	+Gz	Resizing		*Estimate*	*Stderr*	*DF*	*Adj. p-Val*
Pointing Speed [mm/ms]	–	SU	SC	−0.0242	0.00091	2046	<.0001
	–	SU	SE	0.0209	0.00088	2046	<.0001
	–	SC	SE	−0.0451	0.00081	2046	<.0001
	1.5	SU	SC	−0.0195	0.00131	2046	<.0001
	1.5	SU	SE	0.0161	0.00126	2046	<.0001
	1.5	SC	SE	−0.0355	0.00128	2046	<.0001
	2.0	SU	SC	−0.0269	0.00183	2046	<.0001
	2.0	SU	SE	0.0211	0.00175	2046	<.0001
	2.0	SC	SE	−0.0480	0.00150	2046	<.0001
	2.3	SU	SC	−0.0262	0.00155	2046	<.0001
	2.3	SU	SE	0.0256	0.00152	2046	<.0001
	2.3	SC	SE	−0.0518	0.00143	2046	<.0001

Thereby the Euclidean distance d_{ED} was the distance between the centre of the target and the intersection point within the target. As presented in Fig. 10 (left), the variations of the mean distances show a proportional ratio between the distance and target's size, i.e. the pointing distance was greater with the increment of target's size and vice versa. Statistical analyzing of the Euclidean distance confirmed this observation by significant differences ($p < .0001$) between the sizing techniques across all Gz loads and per Gz load (see Table 7). The test revealed that pointing towards SC targets resulted in the

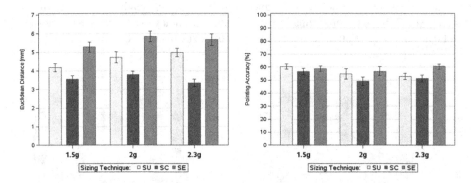

Fig. 10. Mean Euclidean distance and percentage pointing accuracy (with CI = 95 %) of the sizing methods grouped by Gz load.

Table 7. Significant differences of comparison of the Euclidean distance and log(accuracy) between the sizing techniques across all Gz loads and on the same level of Gz load using SAS® 9.4 PROC MIXED (adjust = simulate).

Dependent Variable	+Gz	Resizing		*Estimate*	*Stderr*	*DF*	*Adj. p-Val*
Euclidean Distance [mm]	–	SU	SC	1.2840	0.07675	6705	<.0001
	–	SU	SE	−1.1159	0.07405	6705	<.0001
	–	SC	SE	−2.3999	0.07161	6705	<.0001
	1.5	SU	SC	1.0259	0.1162	2685	<.0001
	1.5	SU	SE	−0.9865	0.1136	2685	<.0001
	1.5	SC	SE	−2.0123	0.1128	2685	<.0001
	2.0	SU	SC	1.3637	0.1569	1862	<.0001
	2.0	SU	SE	−1.1426	0.1469	1862	<.0001
	2.0	SC	SE	−2.5063	0.1365	1862	<.0001
	2.3	SU	SC	1.5742	0.1335	2152	<.0001
	2.3	SU	SE	−1.1634	0.1310	2152	<.0001
	2.3	SC	SE	−2.7375	0.1258	2152	<.0001
log(Accuracy [%])	–	SC	SE	0.1095	0.02084	6705	<.0001
	2.0	SC	SE	−0.2210	0.03607	1862	0.0178
	2.3	SU	SE	−0.1522	0.03608	2152	0.0072
	2.3	SC	SE	−0.3099	0.03462	2152	0.0436

significantly shortest distances and towards SE targets in significant largest pointing distances. Contrary to the Euclidean distance, the percentage accuracy (Eq. 8) mirrors the percentage ratio of the distance d_{ED} to the target size expressed by the radius r_s.

$$accuracy = 100 - \left(\frac{d_{ED}}{r_s} * 100 \right) \tag{8}$$

The participants pointed with an overall mean percentage accuracy of 56.4 %. The variations of the mean accuracy are presented in Fig. 10 (right). Per sizing technique over all Gz loads data revealed that relative to target's size, participants pointed most precisely using the elongated method (SE) with 58.7 % accuracy (SU: 56.7 %, SC: 53.2 %). Statistical analyzing (see Table 7) revealed that pointing towards elongated targets (SE) enabled a significant improvement over all Gz loads compared to pointing towards compressed targets (SC). On the same Gz level, SE yielded significant increased accuracy under 2 g than SE, and under 2.3 g than SC and SU.

Speed-Accuracy-Trade-off (Fitts' Law). In designing Human-Computer-Interfaces the assessment of ergonomics is mainly determined by Fitts' model of movement time (Eq. 7) [6] that a human needs to point at a target of a given size and distance. Fitts' law predicts longer movement times at larger distances, as well as at smaller targets. The sizing approach interrelates these characteristics to each other, whereby the elongated method (SE) provides larger targets at larger distances, while the compressed method (SC) results in smaller targets at smaller distances. We used Fitts' law to evaluate the speed-accuracy trade-off of the studied sizing techniques related to direct pointing affected by added arm weightings. The metric for comparing the performance is the Throughput TP (Eq. 10), in bits per second (bps) calculated by the Index of Difficulty ID and mean movement time MT (Eq. 9) as time to hit a target (in milliseconds) with a for the intercept and b for the slope of measured mean response time by the target width W. The ID measures the tasks difficulty in bits using target size and distance. Because we used squared targets, we calculated the ID only by the targets' width. For computing the ID (Eq. 11) we used the Welford formulation [22]. To reflect the observed pointing performance of the participants, we used the effective target width W_e (Eq. 12) [13, 22] as the central 96 % of the spatial distribution with SD_x as standard deviation of the mean pointing accuracy. Table 8 shows the resulting Fitts' parameter for the three sizing methods per +Gz load. Because the target size and distance increase with an increase in gravity, the SE method resulted in the most difficult targets with the highest ID_e under 2 g and 2.3 g, but also in the highest throughput (TP).

$$MT = a + bID_e \tag{9}$$

$$TP = \frac{ID_e}{MT} \tag{10}$$

$$ID_e = \log_2 \left(\frac{A}{W_e} + 0.5 \right) \tag{11}$$

Table 8. Fitts' resulted parameters: targets' distance (A), target width (W), effective target width (W_e), mean measured movement time (MT), effective Index of Difficulty (ID_e), and Throughput (TP).

Resizing	+Gz	A [mm]	W [mm]	W_e [mm]	MT [ms]	ID_e [bits]	TP [bps]
SU	1.5	082.50	15.00	14.47	1010.61	**2.63**	**2.61**
	2.0	082.50	15.00	11.04	1052.83	**2.99**	**2.85**
	2.3	082.50	15.00	11.78	1071.96	**2.91**	**2.71**
SC	1.5	064.17	11.67	08.47	1021.53	**3.02**	**2.96**
	2.0	058.06	10.56	11.12	1032.42	**2.52**	**2.44**
	2.3	54.39	09.89	09.09	1085.23	**2.69**	**2.49**
SE	1.5	100.83	18.33	13.68	1029.36	**2.98**	**2.89**
	2.0	106.94	19.44	13.39	1035.30	**3.09**	**2.98**
	2.3	110.61	20.11	13.18	1089.69	**3.15**	**2.89**

$$W_e = 4.133 * SD_x \tag{12}$$

The compressed sizing method (SC) yielded the highest index of difficulty under 1.5 g, while under 2 g and 2.3 g yielded most the simple targets, but with the lowest *TP*. Pointing towards normal sized targets (SU) yielded an increased *ID_e*, as well as a growing throughput with the increment of gravity. The resulting Pearson's correlation coefficient r and the regression equations of Fitts' movements' model for the sizing conditions are presented in Eqs. 13, 14 and 15. While the movement time and the index of difficulty were very strongly positively correlated ($r > 8.0$) for the SU and SE conditions, a moderate negative correlation was revealed by the SC condition. The Fitts' model of movement time of the SU (Eq. 13) and SE (Eq. 15) sizing conditions provides good descriptions of the observed pointing behaviour and predicted an increase in response time with an increased target difficulty with the fastest increase under the SE condition. In contrast, the model of the compressed sizing technique SC (Eq. 14) resulted in a model with the highest intercept and lowest negative slope, i.e. that the Movement Time will slow decrease with an increase in the Index of Difficulty.

$$SU: \ r = 0.856, \ MT = 638 + 142ID_e \tag{13}$$

$$SC: \ r = -0.321, \ MT = 1163 - 43ID_e \tag{14}$$

$$SE: \ r = 0.842, \ MT = 81 + 315ID_e \tag{15}$$

Physiological Workload. Analyzing the R-R interval obtained by HRV shows the impact to the cardiovascular system on a certain workload. Larger workload causes a larger impact in the cardiovascular system and therefore causes a higher heart frequency and subsequently a shorter R-R interval between the heartbeats. The cardiovascular parameters were assessed during all phases of the experiment. The 1 g SU output was used as reference measurement and showed the lowest impact on the

Table 9. Assessed HRV parameter: R-R distance in [ms] median and SD across all participants.

Gz	SU	SC	SE
	Mean ± SD	Mean ± SD	Mean ± SD
1.0	723.86 ± 156.88	-	–
1.5	674.24 ± 114.76	**680.45** ± 120.29	665.09 ± 102.56
2.0	642.11 ± 119.43	**649.57** ± 117.03	625.28 ± 088.07
2.3	641.97 ± 105.77	**648.31** ± 119.25	645.24 ± 126.40

cardiovascular system. Since the physiological workload respectively grows with the increment of gravity, the R-R distance decreased during the experiment under 2 *g* and even more under 2.3 *g* (see Table 9). The data revealed the lowest R-R for all sizing conditions under high Gz load since the weight attached to the participants' arm constituted the major part of the workload. While the SE conditions yielded the lowest values for the R-R distances under all Gz loads, the SC condition always resulted in the highest values for the R-R distance. This suggests that pointing towards elongated targets (SE, largest target distances) mostly increased the workload, while pointing towards compressed targets (SC, shortest target distances) decreased the workload the most. Statistical analysis showed a significant main effect of the Gz loads on the R-R distance ($F_{2,4} = 27.69$, $p < .05$), but did not show effects by applying the sizing methods. This result indicates that hypothesis H2 can be conditionally accepted, i.e. that compressed resizing (SC) mostly decreases the workload under increased gravity load, although the difference to the other sizing conditions (SE, SU) was not statistically significant.

6   Discussion

The results of the LAHC-Study has already shown that pointing towards outside coded targets during AR selection is affected by gravity-adapted resizing under increased Gz loads and revealed an upcoming trend for the elongated sizing method (SE) by a significant increase in the pointing speed. The Weight-Study confirmed the findings of the LAHC-Study and also showed that the workload during direct AR selection towards outside coded targets is influenced by gravity-adapted resizing with significant differences between affected targets (SE, SC) and unaffected targets (SU). In this way we can affirm that gravity-adapted target resizing affects the performance and workload of direct AR pointing under altered hypergravity conditions (Q1). The results of the Weight-Study also confirmed the observed trend of an improved performance of pointing towards targets that are influenced by elongated resizing (SE) implicating greater target sizes and larger distances between the targets. An elongated target interface not only significantly increases the pointing frequency across all Gz loads, it also shows a decreasing tendency in the error rate. In contrast to elongated targets, the compressed sizing technique (SC) yields the small targets at short distances that always caused significant closer hits to the targets' centre. But relative to the target size, pointing towards elongated targets (SE) significantly provided the most precise pointing across all Gz loads than pointing towards compressed targets (SE), and even under 2.3 g it revealed significant more accurate pointing than SE and the unchanged baseline condition (SU). With respect to the response time we was astonished that the resulted target distances did not affect the time that was needed to respond to visual stimuli. We excepted the shortest times for pointing towards compressed targets with shortest distances, and the longest times for pointing towards elongated targets with the longest distances. But considering the distance covered yielded the significantly fastest pointing speed by the longest distances using the elongated sizing method (SE). Also the analysis of the speed-accuracy tradeoff related to Fitts' law showed that pointing towards elongated targets (SE) yielded in an increase in throughput with increased gravity load. The HRV based workload assessment showed an effect caused by changed gravity with attached arm weights and the alternation of the workload. The mostly increased workload was caused by pointing towards elongated targets (SE) that could be due to the distance that gets larger with the increase in gravity load. In contrast, the mostly decreased workload was provided by pointing towards the compressed targets with the shortest distances. In conclusion, pointing towards outside coded target interfaces will benefit from elongated resizing that provides an increase in pointing frequency, most accurate pointing and fastest pointing speed, but causes the mostly increased physiological workload. Conversely, using compressed targets will deteriorate the pointing performance, but provided an decrease in physiological workload.

7 Summary

The performance of Augmented Reality direct object selection coded outside of the human body frame of reference is impaired under short-term altered gravity. Therefore we looked for adequate countermeasures and introduced a gravity-adapted resizing approach that dynamically modify the size and position of the pointing targets related to the active gravity load. Before conducting experimentations under microgravity conditions, we started studying how gravity-adapted resizing affects visuomotor coordination under altered hypergravity conditions. Applying the resizing approach can affect the pointing interface in two different ways that either results in elongated targets or in compressed targets. This means that with an increase in gravity elongated interfaces will increase in targets' size and distance, while compressed interfaces will decrease in sizes and distances. We conducted a proof-of-concept study under simulated hypergravity conditions to investigate the influence of gravity-adapted resizing on the performance and workload of visuomotor coordination during AR selection towards outside coded interfaces. The study was divided into two experiments, where simulated hypergravity was induced, firstly by long-arm human centrifugation (LAHC) and then by added arm weightings under normogravity. The workload was assessed by cardiac response using the heart rate variability (HRV). Summarizing the results of both studies showed that AR selection towards outside coded targets is effected by gravity-adapted resizing and proofed evidence that pointing towards outside coded target interfaces under increased gravity benefits from elongated resizing, while compressed resizing deteriorates the pointing performance. But contrary, pointing towards elongated targets increases the physiological workload, while pointing towards compressed targets provides a decrease in workload.

In future work we will investigate the effect of gravity-adapted resizing on performance and workload under microgravity conditions. Therefore the resizing approach needs to be correspondingly adjusted to the changed condition. Further research could also investigate the effect of gravity-adapted resizing on the view management of an AR supported assistant system for space operation. It is conceivable that analogous to aimed pointing movements the gravity-adapted resizing approach affects the gaze control during the detection of labels and annotations. Related to the workload assessment by HRV, future work should consider the separation of physical and cognitive workload. This can be done by additional evaluation of the muscular activity measured by electromyogram (EMG). By doing so, it could even more precisely assess the workload during pointing and targeting.

References

1. Azuma, R.T.: A survey of augmented reality. Presence Teleoperators Virt. Environ. **6**(4), 355–385 (1997)
2. Agan, M., Voisinet, L.A., Devereaux, A.: NASA's Wireless Augmented Reality Prototype (WARP). In: Proceedings of AIP 1998, pp. 236–242 (1998)
3. Scheid, F., Nitsch, A., König, H., Arguello, L., De Weerdt, D., Arndt, D., Rakers, S.: Operation of European SDTO at Col-CC. SpaceOps 2010 Conference (2010)
4. Clauser, C.E., McConville, J.T., Young, J.W.: Weight, volume, and center of mass of segments of the human body. AMRL Technical Report 69-70 (1969)
5. Department of Defense.: Design criteria standard, human engineering. Technical Report MIL-STD-1472F
6. Fitts, P.M.: The information capacity of the human motor system in controlling the amplitude of movement. J. Exper. Psychol. **47**, 381–391 (1954)
7. Guardiera, S., Schneider, S., Noppe, A., Strüder, H.K.: Motor performance and motor learning in sustained +3Gz acceleration. Aviat. Space Environ. Med. **79**(9), 852–859 (2008)
8. Hartmann, K., Ali, K., Strothotte, T.: Floating labels: applying dynamic potential fields for label layout. In: Butz, A., Krüger, A., Olivier, P. (eds.) SG 2004. LNCS, vol. 3031, pp. 101–113. Springer, Heidelberg (2004)
9. ISO/DIS 9241-9.: Ergonomic requirements for office work with visual display terminals (VDTs) - Part 9: Requirements for non-keyboard input devices. International Standard, International Organization for Standardization (2000)
10. Kato, H., Billinghurst, M.: Marker Tracking and HMD calibration for a video-based augmented reality conferencing system. In: Proceedings of IWAR 1999, pp. 85–94 (1999)
11. Kohli, L., Whitton, M.C., Brooks, F.P.: Redirected touching: the effect of warping space on task performance. In: Proceedings of 3DUI 2012, pp. 105–112 (2012)
12. Li, N., Duh, H.B.L.: Cognitive issues in mobile augmented reality: an embodied perspective? In: Huang, T., Alem, L., Livingston, M. (eds.) Human Factors in Augmented Reality Environment, pp. 109–135. Springer Science + Business, New York (2013)
13. MacKenzie, I.S.: Fitts' law as a research and design tool in human-computer interaction. J. Hum. Comput. Interact. **7**, 91–139 (1992)
14. Markov-Vetter, D., Moll E., Staadt O.: Evaluation of 3D selection tasks in parabolic flight conditions: pointing task in augmented reality user interfaces. In: Proceedings of VRCAI 2012, pp. 287–293 (2012)

15. Markov-Vetter, D., Zander, V., Latsch, J., Staadt O.: The impact of altered gravitation on performance and workload of augmented reality hand-eye-coordination: inside vs. outside of human body frame of reference. In: Proceedings of JVRC 2013, pp. 65–72 (2013)
16. Markov-Vetter, D., Zander, V., Latsch, J., Staadt, O.: The influence of gravity-adapted target resizing on direct augmented reality pointing under simulated hypergravity. In: Proceedings of GRAPP 2015, pp. 401–411 (2015)
17. Oehme, O., Schmidt, L., Luczak, H.: Comparison between the strain indicator HRV of a head based virtual retinal display and LC-mounted displays for augmented reality. In: Proceedings of WWDU 2002, pp. 387–389 (2002)
18. Pick, S., Hentschel, B., Tedjo-Palczynski, I., Wolter, M., Kuhlen, T.: Automated positioning of annotations in immersive virtual environments. In: Proceedings of JVRC 2010, pp. 1–8 (2010)
19. Rohs, M., Oulasvirta, A., Suomalainen, T.: Interaction with magic lenses: real-world validation of a Fitts' Law model. In: Proceedings of CHI 2011, pp. 2725–2728 (2011)
20. Scheid, F., Nitsch, A., König, H., Arguello, L., De Weerdt, D., Arndt, D., Rakers, S.: Operation of European (2010)
21. SDTO at Col-CC. SpaceOps 2010 Conference
22. Task Force of the European Society of Cardiology and the North American Society of Pacing and Electrophysiology. Heart Rate Variability: standards of measurement, physiological interpretation and clinical use. Circulation 1996 (93): 1043–1065 / Eur. Heart J. 17(3), 354–381 (1996)

Information Visualization Theory and Applications

A Linear Time Algorithm for Embedding Arbitrary Knotted Graphs into a 3-Page Book

Vitaliy Kurlin[1,2]([⊠]) and Christopher Smithers[2]

[1] Microsoft Research Cambridge, 21 Station Road, Cambridge CB1 2FB, UK
vitaliy.kurlin@gmail.com
[2] Durham University, Durham DH1 3LE, UK
christopher.smithers@durham.ac.uk
http://kurlin.org

Abstract. We introduce simple codes and fast visualization tools for knotted structures in complicated molecules and brain networks. Knots, links and more general knotted graphs are studied up to an ambient isotopy in Euclidean 3-space. A knotted graph can be represented by a plane diagram or a Gauss code. First we recognize in linear time if an abstract Gauss code represents a graph embedded in 3-space. Second we design a fast algorithm for drawing any knotted graph in the 3-page book, which is a union of 3 half-planes along their common boundary. The complexity of the algorithm is linear in the length of a Gauss code. Three-page embeddings are encoded in such a way that the isotopy classification for graphs in 3-space reduces to a word problem in finitely presented semigroups.

1 Introduction: Motivation and Problems on Knotted Structures

This is an extended version of the conference paper [13] with extra Appendices B, C, D that describe key stages of the full algorithm for drawing 3-page embeddings.

Knotted structures are common in nature. For example, microscopic lines in liquid-crystals [18] or Reeb graphs of complex shapes [2] can be knotted. Figure 1 shows large brain neurons with many branching points. These structures are usually huge and more complicated than simple closed curves studied in classical knot theory.

Pictures of knots can be attractive for humans, but robots would prefer a smaller form or codes representing the same knotted object. Such codes are needed for automatic analysis, however a final output is also important to visualise. We summarise our requirements for processing knotted structures in the following 3 problems.

- **Modeling:** find a mathematical model for all possible knotted structures in $\mathbb{R}^3$.

© Springer International Publishing Switzerland 2016
J. Braz et al. (Eds.): VISIGRAPP 2015, CCIS 598, pp. 99–122, 2016.
DOI: 10.1007/978-3-319-29971-6_6

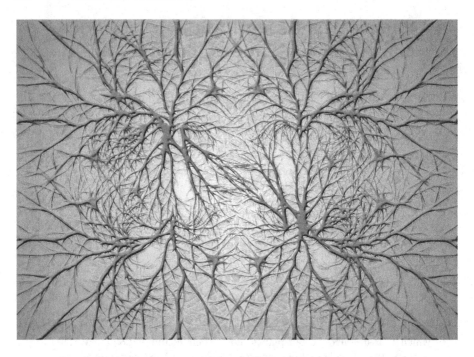

Fig. 1. Neurons in the brain form a large knotted graph with many branching points in 3-space.

Fig. 2. Plane diagrams (projections) of the trefoil, the Hopf link and a simple knotted graph.

- **Encoding:** represent any knotted structure by a simple code in a computer memory.
- **Visualization:** design a fast algorithm to visualize knotted structures given by codes.

Our suggested model for knotted structures is a possibly disconnected graph with branching vertices and multiple edges that might be knotted in 3-space, see Definition 1. For instance, any knot in 3-space is a non-self-intersecting closed curve or a loop.

Knots live in 3-space, but it is easier to draw their planar projections with double crossings. Such plane diagrams are usually represented by Gauss codes that specify the order of overcrossings and undercrossings along a knot. We will

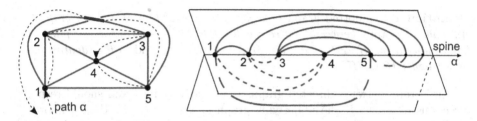

Fig. 3. Straighten a path α to build a 3-page embedding of a graph $K_5 \subset \mathbb{R}^3$ from its diagram.

extend classical Gauss codes of knots and links to arbitrary knotted graphs in Definition 4.

A random code (of a required form) may not represent a real knotted graph, because a planar drawing may need extra crossings. We solve this planarity problem for Gauss codes of knotted graphs in Theorem 9. Our algorithm checks if a Gauss code is realized by a graph in 3-space with a linear time complexity in the length of the given code.

Starting from any realizable Gauss code, we draw a corresponding graph in the *3-page book*, see Theorem 12. This book consists of 3 half-planes attached along their common boundary α called the *spine*. It is well-known that any graph can be topologically embedded in the 3-page book [1, Theorem 5.4]. However, an embedded graph may cross many times the spine of the book. It is only known that $O(|E| \log |V|)$ spine crossings suffice for embedding a graph with $|V|$ vertices and $|E|$ edges [6].

We largely strengthen the former result by designing a linear time algorithm to continuously move any graph embedded in 3-space to a graph within 3 pages. We review other related work throughout the paper. Figure 3 is a high-level illustration of the fast algorithm for a 3-page embedding of the graph $K_5 \subset \mathbb{R}^3$. Appendix A contains more details on the following advantages of 3-page embeddings over plane diagrams.

- Theorem 13 encodes 3-page embeddings of all knotted graphs in 3-space by easy linear codes that form a finitely presented semigroup.
- Theorem 14 decomposes any topological equivalence between 3-page embeddings of knotted graphs into finitely many local relations between 3-page codes.

2   Key Concepts on Knotted Graphs and Isotopy in 3-Space

A *homeomorphism* between spaces is a bijection that is continuous in both directions. An *embedding* of one space into another is a continuous function $f : X \to Y$ that induces a homeomorphism between X and its image $f(X) \subset Y$.

We study embeddings of undirected finite graphs, possibly disconnected and with loops or multiple edges. The concept of a knotted graph extends the classical theory of knots to arbitrary graphs considered up to isotopy in 3-space $\mathbb{R}^3$.

Definition 1. *A knotted graph $G \subset \mathbb{R}^3$ is an embedding of a finite graph G. An ambient isotopy between knotted graphs $G, H \subset \mathbb{R}^3$ is a continuous family of homeomorphisms $f_t : \mathbb{R}^3 \to \mathbb{R}^3$, $t \in [0, 1]$ such that $f_0 = \mathrm{id}$ is the identity map on $\mathbb{R}^3$ and $f_1(G) = H$.*

An isotopy between directed graphs is similarly defined and should respect directions of edges. If the underlying graph G is a circle S^1, then a knotted graph is a *knot*. If G is a disjoint union of several circles, $G \subset \mathbb{R}^3$ is a *link*. A link isotopic to a union of disjoint circles in $\mathbb{R}^2$ is *trivial*. The simplest non-trivial knot is the *trefoil* in the 1st picture of Fig. 2. The simplest non-trivial link is the *Hopf* link in the middle of Fig. 2.

If an ambient isotopy keeps a small neighborhood of each vertex of a knotted graph in one moving plane, the graph is called *rigid*. Rigid knotted graphs with vertices of only degree 4 are sometimes called *singular knots*, because they consist of one or several circles intersecting each other at singular points.

Definition 2. *A plane diagram D of a knotted graph $G \subset \mathbb{R}^3$ is the image of G under a projection $\mathbb{R}^3 \to \mathbb{R}^2$ from 3-space $\mathbb{R}^3$ to a horizontal plane $\mathbb{R}^2$. In a general position we assume that all intersections of a plane diagram D are double crossings so that the crossings and the projections of all vertices of G are distinct. For each crossing of D, we specify one of two intersecting arcs that crosses over another arc.*

The key problem in knot theory is to efficiently classify knots and graphs up to ambient isotopy. The first natural step is to reduce the dimension from 3 to 2. Any isotopy of knotted graphs can be realized by finitely many moves on plane diagrams. The following result extends Reidemeister's theorem from knots to any knotted graphs.

Theorem 3. *[8] Two plane diagrams represent isotopic knotted graphs in 3-space $\mathbb{R}^3$ if and only if the diagrams can be obtained from each other by an isotopy in $\mathbb{R}^2$ and finitely many Reidemeister moves in Fig. 4. (The move R5 is only for rigid graphs, the move R5' is only for non-rigid graphs.)*

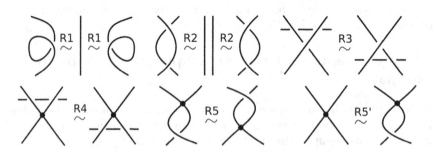

Fig. 4. These Reidemeister moves on diagrams generate any isotopy of graphs in $\mathbb{R}^3$.

The move R4 is shown in Fig. 4 only for a degree 4 vertex, moves for other degrees are similar. The move R5 turns a small neighborhood of a vertex in the plane upside down. So a cyclic order of edges at vertices is preserved in rigid graphs. The move R5′ can arbitrarily reorder all edges at a vertex. Theorem 3 formally includes all symmetric images of moves in Fig. 4.

The Reidemeister moves or their analogs on Gauss codes are not local as they involve distant parts of a graph or a Gauss code. This non-locality is a key obstacle for simplifying codes of knots. That is why we later consider 3-page embeddings that allow only finitely many local moves, see Theorems 12, 13, 14.

3   Gauss Codes of Knotted Graphs and Abstract Gauss Codes

A standard way to encode a plane diagram of a knot is to write down labels of crossings in a Gauss code. The Gauss code of a link has several words corresponding to all connected components of the link. We extend this classical concept to any knotted graphs $G \subset \mathbb{R}^3$. If a component of a knotted graph $G \subset \mathbb{R}^3$ is a circle without vertices, we add a *base point* (a degree 2 vertex) to this circle.

Definition 4. *Let $D \subset \mathbb{R}^2$ be a plane diagram of a knotted graph G with vertices $A, B, C, \ldots$ We fix directions of all edges of G and arbitrarily label all crossings of D by $1, 2, \ldots, n$. Then each crossing of D has the sign locally defined in Fig. 5.*

The Gauss code W of the diagram D consists of all words W_{AB}, where each word W_{AB} is associated to a directed edge from a vertex A to a vertex B as follows:

- *W_{AB} starts with A, finishes with B and has the labels of all crossings in AB;*
- *if AB goes under another edge at a crossing i with a sign $\varepsilon \in \{\pm\}$ as in Fig. 5, we add the superscript ε to i and get the symbol i^ε with the sign ε in W_{AB}.*

In the plane diagram the edges at each vertex A of the graph G are clockwisely ordered in $\mathbb{R}^2$, so the Gauss code also specifies a cyclic order of all edges at the vertex A.

If G is a knot, Definition 4 requires at least one degree 2 vertex (a base point) on the circle G. Then we may ignore degree 2 vertices and consider W as a *cyclic* word.

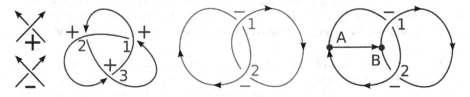

Fig. 5. Local rules for assigning signs of crossings in plane diagrams of knotted graphs.

In Fig. 5 the diagram of the blue trefoil has the cyclic Gauss code $12^+31^+23^+$. The diagram of the red knotted graph has the Gauss code $W = \{AB; A1^-2A; B12^-B\}$ with the cyclic orders of edges at vertices $(AB, 2A, A1^-)$ and $(AB, B1, 2^-B)$. In this example each edge is denoted by the pair of its endpoints. In general, if there are multiple edges with the same endpoints, we use distinct labels for all different edges.

A Gauss code of any undirected graph depends on a choice of extra degree 2 vertices, directions of edges, an order of crossings. If a plane diagram of a knotted graph corresponds to a Gauss code, then this diagram is unique up to isotopy in the plane. We explicitly construct a plane diagram from a Gauss code in the proof of Theorem 9.

Here is a naive approach to drawing a plane diagram represented by a Gauss-like code W. We can plot vertices $A, B, C, \ldots$ and crossings $1, 2, \ldots, n$ anywhere in $\mathbb{R}^2$. Since W specifies the cyclic order of edges at each vertex A in Definition 4, we may draw short arcs around A in a correct cyclic order. Now we should connect all vertices and crossings that have adjacent positions in the code W by continuous non-intersecting arcs in the plane.

The last step fails for the word 12^+1^+2 that does not encode any plane diagram. Indeed, if we try to draw a closed curve with 2 self-intersections as required by 12^+1^+2, we have to add a 3rd intersection (a virtual crossing) to make the curve closed. This obstacle can be resolved if we draw a diagram on a torus as in Fig. 7, because we can hide a virtual crossing by adding a handle. Another approach is to embrace virtual crossings, which has led to virtual knots.

If we study properly embedded graphs, we need to recognize planarity of Gauss codes, namely we will determine if a Gauss code W represents the plane diagram of some knotted graph $G \subset \mathbb{R}^3$. So we first introduce abstract Gauss codes in Definition 5 and then recognize their planarity in the general case of knotted graphs in Theorem 9.

Definition 5. *Let the alphabet consist of m letters $A, B, C, \ldots$ and $3n$ symbols i, i^+, i^- for $i = 1, \ldots, n$. An* abstract Gauss code W *is a set of words such that*

- *the first and last symbols of each word in the code W are letters (of vertices),*
- *the set of symbols in all words (apart from the initial and final letters) contains, for each $i = 1, \ldots, n$, the symbol i and exactly one symbol from the pair i^+, i^-.*

Each of the m letters defines a cyclic order of all symbols adjacent to this letter. The length $|W|$ *is the total length of all words minus the number of words.*

The Gauss code of any plane diagram of a knotted graph G from Definition 4 satisfies the conditions above. Indeed, the letters $A, B, C, \ldots$ denote (projections of) vertices of G. Then every edge contains crossings labeled by i, i^+ or i^- for $i = 1, \ldots, n$.

The clockwise order of edges around any vertex A in the plane diagram of G in $\mathbb{R}^2$ defines the cyclic order of vertices and crossings adjacent to A. If a component of G is a circle, we may remove its vertices of degree 2 and write the remaining symbols as in the cyclic code $12^+31^+23^+$ of the trefoil in Fig. 5. The total number of these symbols equals the double number of crossings.

4 Planarity Criterion for Gauss Codes of Knotted Graphs

The planarity problem is to determine whether it is possible to draw a plane diagram represented by an abstract Gauss code W. To avoid potential self-intersections, we shall draw a diagram not in the plane, but in the Gauss surface $S(W)$ defined below.

First we introduce the abstract graph $G(W)$ describing the adjacency relations between symbols in a Gauss code W. Then we attach disks to $G(W)$ to get the surface $S(W)$ containing a required diagram without self-intersections. The criterion of planarity will check if the surface $S(W)$ is a topological sphere S^2.

Definition 6. *Any abstract Gauss code W with m letters $A, B, \ldots$ and $2n$ symbols from $\{i, i^+, i^- \mid i = 1, \ldots, n\}$ gives rise to the Gauss graph $G(W)$ with $m+n$ vertices labeled by $A, B, \ldots$ and $1, 2, \ldots, n$.*

We connect vertices p, q by a single edge in $G(W)$ if p, q (possibly with signs) are adjacent symbols in W. Below when we travel along an edge from p to q, we record our path by $(p, q)_+$ if q follows p in the code W (in the cyclic order), otherwise by $(p, q)_-$.

We define unoriented cycles in the graph $G(W)$ by going along edges and turning at vertices according to the following rules illustrated in Fig. 6:

- *if we came to one of the vertices $A, B, C, \ldots$ from its neighbor, then we turn to the next neighbor in the clockwise order specified in the Gauss code W;*
- *at each vertex labeled by $i \in \{1, \ldots, n\}$ we turn to the next edge by one of the rules below for a unique possible choice of $\delta \in \{+, -\}$ and both $\varepsilon \in \{+, -\}$*

$$(p, i)_+ \to (i^\delta, q)_\delta, \quad (p, i)_- \to (i^\delta, q)_{-\delta}, \quad (p, i^+)_\varepsilon \to (i, q)_{-\varepsilon}, \quad (p, i^-)_\varepsilon \to (i, q)_\varepsilon.$$

We stop traversing cycles when every edge was passed once in each direction. The Gauss surface $S(W)$ is obtained from $G(W)$ by gluing a disk to each cycle.

The number of edges in the graph $G(W)$ equals the length $|W|$ of the code W. The rules for traversing cycles in Definition 6 geometrically mean that at each vertex or crossing we turn left to a unique edge and can pass every edge exactly once in each direction. Hence the Gauss surface of any abstract Gauss code is a compact orientable surface without boundary. From now on we assume that all diagrams, Gauss graphs and surfaces are connected. Otherwise each connected component is considered separately.

Lemma 7. *For the Gauss code W of any connected plane diagram of a knotted graph $G \subset \mathbb{R}^3$, the Gauss surface $S(W)$ is homeomorphic to a topological sphere S^2.*

Proof. We assume that the given diagram D is contained in a sphere S^2 instead of a plane $\mathbb{R}^2$. Then the Gauss graph $G(W)$ can be identified with the diagram D, though $G(W)$ was introduced as an abstract graph not embedded into any space. When we traverse the cycles in $D = G(W)$ from Definition 6, we pass over the boundaries of all connected components of $S^2 - D$. Indeed, each time

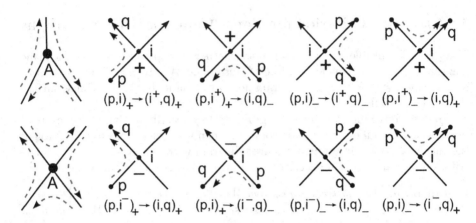

Fig. 6. Interpretation of the 'turning-left' rules for traversing cycles in the Gauss graph $G(W)$.

we turn left in the diagram $D \subset S^2$ according to the geometric rules in Fig. 6. Hence the Gauss surface $S(W)$ can be identified with the sphere S^2 containing the diagram $D = G(W)$. $\qquad\square$

Example 8. *We construct the Gauss surface of the abstract Gauss code* $W = 12^+1^+2$, *whose diagram with one virtual crossing is in Fig. 7. For simplicity, we removed the degree 2 vertex from the circle and consider the word* 12^+1^+2 *in the cyclic order.*

Then 4 pairs 12^+, 2^+1^+, 1^+2, 21 *of adjacent symbols in the code* W *lead to the Gauss graph* $G(W)$ *whose 2 vertices with labels* $1, 2$ *are connected by 4 edges with labels* $(1, 2^+)$, $(2^+, 1^+)$, $(1^+, 2)$, $(2, 1)$, *see Fig. 7. Recall that the edges labeled by* $(2, 1)$ *and* $(2^+, 1^+)$ *meet at a non-avoidable virtual crossing in the plane, but the abstract Gauss graph* $G(W)$ *has only 2 vertices.*

If we start traveling from the edge $(1, 2^+)_+$ *in the same direction as in* W, *the next edge should be* $(2, 1^+)_-$ *by the rule* $(p, i^+)_\varepsilon \to (i, q)_{-\varepsilon}$, *where* $p = 1$, $i = 2$, $\varepsilon = +$ *uniquely determine the next symbol* $q = 1^+$ *from the code* W *(going from 2 in the opposite direction). After passing the second edge* $(2, 1^+)_-$, *we return to the first edge* $(1, 2^+)_+$ *by the same rule* $(p, i^+)_\varepsilon \to (i, q)_{-\varepsilon}$ *for* $p = 2$, $i = 1$, $\varepsilon = -$, $q = 2^+$.

So the 1st cycle consists of 2 edges $(12^+)_+$ *and* $(2, 1^+)_-$. *The 2nd cycle consists of 6 edges* $(1^+, 2)_+ \to (2^+, 1^+)_+ \to (1, 2)_- \to (2^+, 1)_- \to (1^+, 2^+)_- \to (2, 1)_+$. *Both cycles of* $G(W)$ *are shown by red dashed closed curves in Fig. 7. The resulting Gauss surface* $S(W)$ *with 2 vertices, 4 edges, 2 faces has the Euler characteristic* $\chi = 2 - 4 + 2 = 0$ *and should be a torus as expected from the 2nd picture in Fig. 7.*

The *Euler characteristic* of a surface subdivided by a graph with $|V|$ vertices and $|E|$ edges into $|F|$ faces (topological disks) is defined as $\chi = |V| - |E| + |F|$ and is invariant up to a *homeomorphism* (a bijection continuous in both directions).

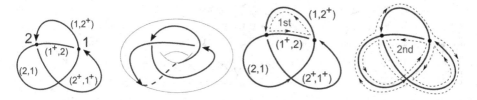

Fig. 7. The code $W = 12^+1^+2$ is realizable on a torus, $G(W)$ has two red dashed cycles (Color figure online).

Any orientable connected compact surface of a genus g (the number of handles) and b boundary components (circles) has $\chi = 2 - 2g - b \leq 2$. Hence a sphere S^2 with $\chi = 2$ is detectable by the Euler characteristic among connected compact surfaces.

Theorem 9 extends [12, Algorithm 1.4] from links to arbitrary knotted graphs.

Theorem 9. *Given an abstract Gauss code W of a length $|W|$, an algorithm of time complexity $O(|W|)$ can determine if the given Gauss code W represents a plane diagram of a knotted graph $G \subset \mathbb{R}^3$.*

Proof. The Gauss surface $S(W)$ of any abstract Gauss code W contains the diagram D encoded by W due to the geometric interpretation of the rules in Fig. 6. We assume that $S(W)$ is connected, otherwise we separately consider each connected component below. This surface has the maximum Euler characteristic χ among all orientable connected compact surfaces S that contain the diagram D and have no boundary.

Indeed, after cutting the underlying graph of the diagram $D \subset S$, the surface S splits into several components. The Euler characteristic of S is maximal when all these components are disks as in the Gauss surface. The disk has $\chi = 1$, which is maximal among all compact surfaces whose boundary is a circle.

To decide the planarity of the Gauss code W, it remains to determine if the Gauss surface $S(W)$ is a sphere S^2, which is detectable by the Euler characteristic $\chi = 2$ in the class of all orientable connected compact surfaces S without boundary. For computing the Euler characteristic χ, we use the Gauss graph $G(W)$, which splits the Gauss surface $S(W)$ into topological disks by Definition 6.

Namely, the surface $S(W)$ has $m + n$ vertices, $|W|$ edges and the number of faces equal to the number of cycles. We count all cycles in the graph $G(W)$ in time $O(|W|)$ by a double traversal of W according to the rules in Fig. 6. Hence in time $O(|W|)$ we compute $\chi = m + n - |W| + \#(\text{cycles})$ and determine if the Gauss surface $S(W)$ is homeomorphic to a topological sphere S^2.   □

5   Embedding Any Knotted Graph into a 3-Page Book

Our algorithm will draw a 3-page embedding of a knotted graph G, which is usually represented by a plane diagram or by a Gauss code. Even for knots,

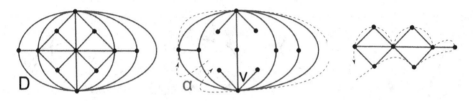

Fig. 8. A path α through all vertices of any planar graph meets every edge at most once.

an abstract Gauss code may not represent a closed curve in 3-space. That is why we first solve the planarity problem for Gauss codes of knotted graphs in Theorem 9.

If we know that a given Gauss code represents a plane diagram D of a knotted graph G, the next step in Theorem 11 is to draw the diagram D in a 2-page book as defined below. After that we upgrade this topological 2-page embedding of D to a 3-page embedding of G in linear time, see Theorem 12.

Definition 10. *The k-page book consists of k half-planes with a common boundary line α called the* spine *of the book. An embedding of an undirected graph G into the k-page book is* topological *if the intersection of G with the spine α is finite and includes all vertices of G. A* spine point *of the embedded graph G is any non-vertex point in the spine α. If G has no spine points, so every edge of G is contained in a single page, then the k-page book embedding of G is called* combinatorial.

A graph D is *planar* if D can be embedded in $\mathbb{R}^2$. Any undirected planar graph has a combinatorial 4-page embedding [20]. Figure 8 shows a non-hamiltonian maximal planar graph that can not be combinatorially embedded into 2 pages [1, Sect. 5]. Any topological 2-page embedding of this graph will have spine points. The linear time algorithm below guarantees at most two spine points per edge.

Theorem 11. *[5, Theorem 1] Given a planar undirected graph $D \subset \mathbb{R}^2$ with $|V|$ vertices, an algorithm of linear time complexity $O(|V|)$ can draw a topological embedding of the graph D in the 2-page book with at most two spine points per edge.*

Two more pictures in Fig. 8 illustrate the key idea how we can construct a non-self-intersecting path α that passes through each vertex once and intersects each edge at most once. By an isotopic deformation of $\mathbb{R}^2$, the path α can be converted into a straight spine, which splits the plane into 2 pages. Since all vertices and crossings of D are in the spine α, we get a required topological 2-page embedding of D.

We are not going to minimize the number of bends of edges in a 2-page embedding of a plane diagram D, because we shall construct 3-page embeddings of original knotted graphs with a linear number $O(|W|)$ of total bends in the length of a Gauss code W.

Table 1. Necessary local upgrades of crossings from 2 to 3 pages, see Lemma 20 in Appendix D.

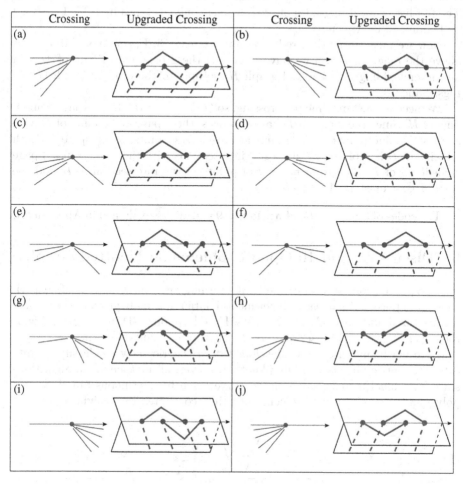

Theorem 12. *Given an abstract Gauss code W, an algorithm of time complexity $O(|W|)$ determines if W represents a plane diagram of a knotted graph $G \subset \mathbb{R}^3$ and then draws a topological 3-page embedding of a graph H isotopic to G. Moreover, the graph H has at most $12|W|$ intersections with the spine of the book.*

Proof. We first apply the linear time algorithm from Theorem 9 to determine if the code W represents a plane diagram D of a knotted graph G. If yes, we draw a 2-page embedding of the diagram $D \subset \mathbb{R}^2$ in linear time using the algorithm of Theorem 11.

At every crossing in the diagram D, we mark a short red arc that crosses over another arc in D. The centers of all these marked arcs are all crossings of D, which are already in the straight spine α of the 2-page book. We may slightly deform the embedding of D by pushing the marked red arcs into the spine α. The full list of local upgrades of crossings has only 10 types in Table 1 justified by Lemma 20 in Appendix D.

Now we push all marked red arcs into the extra 3rd page attached along α above the diagram D. So we have upgraded the 2-page embedding of D to a 3-page embedding of a knotted graph H isotopic to the original graph G, see Figs. 9 and 10.

We need a constant time per crossing, so $O(|W|)$ in total, for a 3-page embedding of H. Since the diagram D has $|W|$ edges, the 2-page embedding of D with at most 2 spine points per edge has at most $3|W|$ points in the spine α. Each crossing of D is replaced by at most 4 intersections with the spine α in a 3-page embedding of H. The total number of points in the intersection of H and the spine α is at most $12|W|$. $\qquad\square$

The codes of 3-page embeddings in Fig. 9 and 10 are explained in Appendix A.

6   Discussion and 10 Open Problems on Knotted Graphs

We now discuss our results in the light of a huge gap between real-life experiments and pure mathematics. Experimental data are usually given in the form of unstructured and noisy clouds of points. If we have only 2D images as in Fig. 1, then we also need to extract a knotted structure in a suitable form.

Pure mathematicians have developed deep theories how to classify complicated geometric objects including knots. However, all mathematical algorithms start from ideal models, say a closed curve given by continuous functions or a polygonal curve given by a sequence of points connected by straight edges.

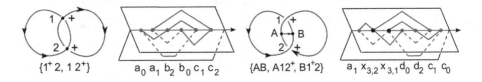

Fig. 9. Hopf link and Hopf graph with Gauss codes and 3-page embeddings

The key challenge is to convert any unstructured experimental data into an ideal theoretical model that can be rigorously analyzed by existing mathematical methods. The first advance in this direction is computing the fundamental group of a knot complement from a point cloud in [4]. We state open problems relating practice and theory for knotted graphs. We are open to collaboration on these and any related projects.

1. State and prove a criterion of planarity of Gauss codes of knotted graphs using combinatorial invariants like sums of signs similarly to [12, Theorem 3.6].

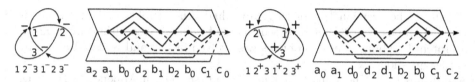

$1\,2^-3\,1^-2\,3^-$ $a_2\,a_1\,b_0\,d_2\,b_1\,b_2\,b_0\,c_1\,c_0$ $1\,2^+3\,1^+2\,3^+$ $a_0\,a_1\,d_0\,d_2\,d_1\,b_2\,b_0\,c_1\,c_2$

Fig. 10. Left-handed and right-handed trefoils with Gauss codes and 3-page embeddings.

2. Let a link of n components be given as an unordered union of $m \geq 2n$ open arcs (or sequences of points). How can we 'correctly' join corresponding endpoints of the arcs to form n closed curves in $\mathbb{R}^3$?

3. When drawing pictures on a tablet, a few intersecting curves can be represented by several sequences of 2D points sampled along the curves. Under what conditions on the curves and sample, can we quickly reconstruct the curves using only the sample?

4. Design a fast algorithm to convert an unstructured 3D point cloud sampled around an unknown knotted structure into a Gauss code W of a knotted graph.

5. Design an algorithm to convert a 2D image of a knotted graph into a Gauss code W.

Our current work on visualizing Gauss codes is an important step in the hard problems above. First, we may try to recognize small patches of vertices and crossings in a 2D image of a knotted graph, but after that we should combine them in a Gauss code whose planarity can be quickly checked by Theorem 9.

Second, if we need to visualize any noisy cloud sampled from an unknown knot $K \subset \mathbb{R}^3$, we may draw a knot isotopic to K using its Gauss code and Theorem 12. Even more importantly we often wish to get a simplified (minimal) version of a knot.

The state-of-the-art simplification algorithm for recognizing trivial knots available at http://www.javaview.de/services/knots is based on 3-page embeddings. We remind how to extend this approach to graphs in Appendix A and state more problems below.

6. Design an algorithm to untangle diagrams of graphs isotopic to planar graphs.

7. Extend the algorithm for drawing knotted graphs in 3 pages to drawing 2-dimensional surfaces in a universal 3-dimensional polyhedron (the *hexabasic book*) from [9].

8. Decide if the problem to find a 3-page embedding of a knotted graph $G \subset \mathbb{R}^3$ having the minimum number of intersections with the spine α is NP-hard.

9. Use the computed invariants to build a database of isotopy classes of knotted graphs similarly to the Knot Atlas at http://katlas.math.toronto.edu.

10. Define a kernel [16] on point clouds representing knotted graphs so that one can use tools of machine learning for automatic recognition of real-life knotted structures.

The earlier version [13] of this paper had other Problems 1 and 8 about knotted graphs $G \subset \mathbb{R}^3$ given as sequences of points, say positions of atoms in a protein backbone. These problems were solved in [14] and replaced above by harder questions.

Algorithms from Theorems 9, 11 and 12 are described in Appendices B, C, D, respectively. A C++ code will be on the webpage http://kurlin.org of the first author, who thanks EPSRC for funding his secondment at Microsoft Research Cambridge. More examples are included in the forthcoming MSc thesis [17] of the second author.

A   Semigroups for Classifying Graphs up to Isotopy

We remind how to encode 3-page embeddings of all knotted graphs by words in a simple alphabet. Since edges with vertices of degree 1 can be easily unknotted by isotopy in 3-space, for simplicity we consider below only graphs without degree 1 vertices.

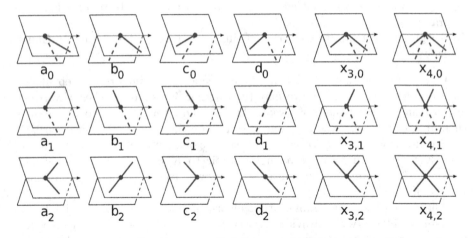

Fig. 11. Local 3-page embeddings for the generators of the semigroups from Theorem 14.

To explain the 3-page encoding of knotted graphs, let us deform any 3-page embedding so that all arcs are monotonically projected to the spine α. Then the 3-page embedding can be uniquely reconstructed from its thin neighborhood around α.

Namely, if we know only directions of arcs going from all spine intersections, we can uniquely join these arcs in each of 3 pages. Hence we can encode any 3-page embedding by the ordered list of local embeddings at all intersections in the spine α.

Theorem 13 [11, Theorem 1.6a]. *Any 3-page embedding of a knotted graph G with vertices up to degree n can be encoded by a word in the alphabet consisting* *of the letters a_i, b_i, c_i, d_i and $x_{k,i}$ for each degree $k = 3, \ldots, n$, where $i = 0, 1, 2$, see Fig. 11.*

Figure 11 shows 12 local embeddings a_i, b_i, c_i, d_i, $i \in \mathbb{Z}_3 = \{0, 1, 2\}$, which are sufficient for encoding 3-page embeddings of knots and links. The notation a_i emphasizes that all a_i can be obtained from each other by a rotation around the spine α.

For encoding graphs in Theorem 13, we can make sure that at each vertex the spine separates one or two arcs from others. Then only 3 local embeddings are enough for each degree, see 3 neighborhoods $x_{3,0}$, $x_{3,1}$, $x_{3,2}$ of a degree 3 vertex in Fig. 11.

The 3-page embedding $K_5 \subset \mathbb{R}^3$ in Fig. 3 can be represented by the 3-page code $w = a_1 d_1 (a_1 b_1 x_{4,1})^2 (a_1 d_1 x_{4,1}) d_1 (x_{4,1} d_1 c_1)(x_{4,1} b_1 c_1) d_2 c_1 c_2$.

The following result completely reduces the topological classification of knotted graphs up to isotopy in $\mathbb{R}^3$ to a word problem in finitely presented semigroups. The cases of 3-regular graphs and 4-regular graphs (singular knots) appeared in [10, 15].

Theorem 14 [11, Theorems 1.6 and 1.7]. *There is a finitely presented semigroup whose all central elements are in a 1-1 correspondence with all isotopy classes of knotted graphs with vertices of degree up to n. An algorithm of a linear complexity $O(|w|)$ decides if an element w of the semigroup is central, i.e. commutes with all other elements.*

So two knotted graphs $G, H \subset \mathbb{R}^3$ are isotopic in 3-space if and only if their corresponding central elements w_G, w_H are equal in the semigroup. A stronger result in [9] says that all isotopies between 3-page embeddings of arbitrary knotted graphs are realizable in the *hexabasic book* $U \times [0, 1]$, where U is the union of the 3-page book $\mathbb{P}_0 \cup \mathbb{P}_1 \cup \mathbb{P}_2$ (with the common boundary line α) and a plane $\mathbb{P}_3$ orthogonal to α. Theorem 14 has been extended to the isotopy classification of surfaces in $\mathbb{R}^4$ [9].

There are two semigroups: RSG_n for rigid knotted graphs with vertices up to degree n and NSG_n for non-rigid graphs. Both semigroups have 12 generators a_i, b_i, c_i, d_i, $i \in \{0, 1, 2\}$, and $3(n - 2)$ generators for vertices up to degree n, so 3 generators for each degree from 3 to n, see Fig. 11. The operation in the semigroups is the concatenation of words. The unit is the empty word $\emptyset$. The generators $a_i, c_i, x_{k,i}$ are not invertible, while b_i, d_i are inverses of each other. In the case of links for $n = 2$, the semigroup has 48 relations (1)–(4), where the index $i \in \mathbb{Z}_3 = \{0, 1, 2\}$ is considered modulo 3.

(1) $d_0 d_1 d_2 = 1$ and $b_i d_i = 1 = d_i b_i$;
(2) $a_i = a_{i+1} d_{i-1}$, $b_i = a_{i-1} c_{i+1}$, $c_i = b_{i-1} c_{i+1}$, $d_i = a_{i+1} c_{i-1}$;
(3) $w(d_i c_i) = (d_i c_i) w$ for $w \in \{\, c_{i+1}, \; b_i d_{i+1} d_i \,\}$;
(4) $uv = vu$, where $u \in \{\, a_i b_i, \; b_{i-1} d_i d_{i-1} b_i \,\}$, $v \in \{\, a_{i+1}, \; b_{i+1}, \; c_{i+1}, \; b_i d_{i+1} d_i \,\}$.

One of the 7 relations in (1) is superfluous as it follows from the remaining 6. The generators a_i, b_i, c_i, d_2 can be expressed only in terms of d_0, d_1, but the resulting relations between d_0, d_1 will be longer. All defining relations of the semigroups represent elementary isotopies between 3-page embeddings, see [13, Appendix].

For knotted graphs with vertices of only degree 3, any non-rigid isotopy can be made rigid, because we can keep 3 short arcs at any vertex in a moving plane. Hence both semigroups for rigid and non-rigid isotopies from Theorem 14 are the same for $n = 3$. In this case the extra relations in addition to (1)–(4) are (5)–(9), see [10, 11]:

(5) $x_{3,i-1} = d_{i-1}x_{3,i}d_{i+1}$;
(6) $x_{3,i}b_i(d_i^2 d_{i+1}^2 d_{i-1}^2) = (d_i d_{i+1} d_{i-1})x_{3,i}b_i$;
(7) $x_{3,i}d_i = a_i(x_{3,i}d_i)c_i$, $\quad b_i x_{3,i}b_i = a_i(b_i x_{3,i}b_i)c_i$;
(8) $ux_{3,i+1} = x_{3,i+1}u$ for any word u from $\{\, a_i b_i,\ d_i c_i,\ x_{3,i}b_i,\ b_{i-1}d_i d_{i-1}b_i \,\}$;
(9) $(x_{3,i}b_i)v = v(x_{3,i}b_i)v$ for any word v from $\{\, a_{i+1},\ b_{i+1},\ c_{i+1},\ b_i d_{i+1}d_i \,\}$.

Knotted graphs that have only vertices of degree 4 and are considered up to rigid isotopy are often called *singular knots*. Each singular point remains a transversal intersection of two arcs during a rigid isotopy, so the cyclic order of all arcs at any degree 4 vertex is invariant. The semigroup of Theorem 14 for singular knots has 15 generators $a_i, b_i, c_i, d_i, x_{4,i}$, relations (1)–(4) above and relations (10)–(14) below, see [11, 15]:

(10) $x_{4,i-1} = b_{i+1}x_{4,i}d_{i+1}$;
(11) $(d_i x_{4,i}b_i)(d_i^2 d_{i+1}^2 d_{i-1}^2) = (d_i^2 d_{i+1}^2 d_{i-1}^2)(d_i x_{4,i}b_i)$;
(12) $d_i x_{4,i}d_i = a_i(d_i x_{4,i}d_i)c_i$, $\quad b_i x_{4,i}b_i = a_i(b_i x_{4,i}b_i)c_i$;
(13) $wx_{4,i+1} = x_{4,i+1}w$ for any word w from $\{\, a_i b_i,\ d_i c_i,\ d_i x_{4,i}b_i,\ b_{i-1}d_i$
$\quad d_{i-1}b_i \,\}$;
(14) $(d_i x_{4,i}b_i)v = v(d_i x_{4,i}b_i)v$ for any word v from $\{\, a_{i+1},\ b_{i+1},\ c_{i+1},\ b_i d_{i+1}d_i \,\}$.

The hard part of Theorem 14 says that any isotopy between graphs decomposes into finitely many elementary isotopies involving *a small part* of a 3-page code. This is the main advantage of the 3-page encoding over plane diagrams and Gauss codes. Indeed, Reidemeister moves in Fig. 4 and their analogues on Gauss codes are not local.

The linear time algorithm for detecting a central element w checks if the arcs corresponding to all letters of w properly meet each other in every page to form an embedding of a graph without hanging edges. For example, the letter a_2 doesn't encode any knotted graph, but $a_2 c_2$ does, because the arcs of a_2, c_2 meet and form a closed curve.

The 3-page code of a knotted graph commutes with any other element w in the semigroups from Theorem 14. For instance, a trivial knot has the code $a_2 c_2$ and can be isotopically moved in $\mathbb{R}^3$ to another side of the 3-page embedding represented by w.

B   Algorithm 1 for Checking Planarity of any Gauss Code

The input is an abstract Gauss code W from Definition 5. The output is a plane diagram (if it exists) having the same Gauss code W. The plane diagram will be obtained as the Gauss graph $G(W)$ with topological disks attached to certain cycles of $G(W)$.

Stage 1.1: Simplifying a Gauss code W by Reidemeister moves I, see Fig. 4.

We go along a Gauss code W and check all pairs of two successive symbols. If the pair is one of $(k, k^+), (k, k^-), (k^+, k), (k^-, k)$ corresponding to the same k-th crossing, we remove this pair from W and continue from the symbol before the removed pair. If W is cyclic (for a circle), at the end we compare the 1st and last symbols of W.

Stage 1.2: Building the abstract Gauss graph $G(W)$ of a Gauss code W.

The Gauss graph $G(W)$ was introduced in Definition 6. The nodes of $G(W)$ are all different vertices and crossings extracted from the Gauss code W after forgetting all signs. We store the graph $G(W)$ in memory by using three arrays NodeTypes, EdgeList, WedgeList. NodeTypes[r] is 0 if r is a vertex, 1 if r is a positive crossing, -1 if r is a negative crossing. The EdgeList consists of ordered pairs (i, j), where i, j are indices of nodes in NodeTypes. The WedgeList for each node n contains indices of edges attached to n and ordered according to the cyclic order from W (starting from any edge).

Stage 1.3: Subdividing the Gauss graph $G(W)$ to remove multiple edges.

It will be convenient to avoid multiple edges of $G(W)$ for Algorithm 2 building a 2-page embedding. For each node n, we check all attached edges. If we find two edges with the same endpoint $k \neq n$, we add a midpoint to one of these edges.

Stage 1.4: Splitting the Gauss graph into different connected components.

We split the subdivided Gauss graph into connected components by using the Boost algorithm [3] based on a Breadth First Search. From now on we assume that $G(W)$ is a connected graph without loops and double edges.

Stage 1.5: Finding cycles in the Gauss graph $G(W)$ and checking planarity.

We initialise the boolean *PassList* whose 2 halves can be viewed as two (forward and backward) lists indexed by edges of $G(W)$. Every bit in each of the halves indicates whether we have passed the corresponding edge in the forward or backward direction whilst building cycles. Each entry in *PassList* is initially false. We shall keep track of the index *least* in this list so that all edges with indices less than *least* are passed.

Step 1. Starting at the index *least*, check each entry of the edge list until a false entry is found. If we reach the end of the list, we have found all cycles.

Step 2. Change *least* to the found index i of the next edge that wasn't passed yet. Set $PassList[i] = true$ and start a new cycle from this i-th oriented edge e. Repeat the following substeps until the next passed edge is once again e.

2a. For each edge, use *EdgeList*, *WedgeList* to find the node being moved towards.

2b. To make the "left-turn" for traversing a cycle from Definition 6 and Fig. 6, we take the next edge from the list of cyclically ordered edges at the node from the substep 2a.

2c. Change the boolean entry in *PassList* for the new passed edge to true.

Step 3. Store the cyclic order of the edges in the found cycle, and move back to Step 1.

Step 4. Compute the Euler characteristic of the Gauss surface $S(W)$ by the formula $\chi = V - E + F$. Here V, E are the numbers of nodes and edges, respectively, in $G(W)$. The number of cycles (or disks attached to the graph) is denoted by F.

Step 5. If $\chi \neq 2$, the given Gauss code W doesn't correspond to any knotted graph as shown in the proof of Theorem 9. Otherwise, we output an embedding $G(W) \subset S^2$ as the graph $G(W)$ with its *NodeTypes*, *EdgeList*, *PassList* and all found cycles.

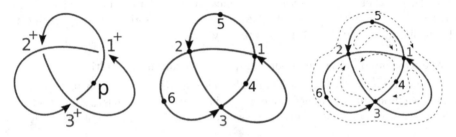

Fig. 12. Left: Trefoil with signs of crossings and a base point p for $W_{Tref} = 12^+31^+23^+$. **Middle:** subdivided Gauss graph without double edges from Stage 1.3 of Algorithm 1. **Right:** 5 cycles found in the subdivided graph at Stage 1.5 of Algorithm 1, so $\chi = 6 - 9 + 5 = 2$.

C   Algorithm 2 for Drawing a 2-page Embedding

The input is an embedding $G \subset S^2$ of an abstract graph G with the boundary cycles of all faces (connected components of $S^2 - G$). The output is a topological 2-page embedding $G \subset \mathbb{P}_0 \cup \mathbb{P}_1$ (of a graph isotopic to the given one in S^2). Here the half-planes $\mathbb{P}_0$ and $\mathbb{P}_1$ have the common boundary (spine) α containing all nodes of G.

Stage 2.1: Extending a given graph G to a maximal planar graph $\bar{G}$.

We triangulate each face whose boundary has more than 3 nodes as follows.

Step 1. Pick a node v and use *WedgeList, EdgeList* to find its neighbors u, w.

Step 2. If u, w are connected by an edge (outside the current face), v cannot be connected to any other node $x \neq u, v, w$ of the face, so we add all such edges

(v, x). Otherwise, we add the edge (u, w) cutting the triangle (u, v, w) from the current face.

Step 3. Start again with a new face if it still has more than 3 nodes.

For the embedded Gauss graph of the trefoil code W_{Tref}, the only non-triangular face in Fig. 12 is $\{5, 1, 4, 2, 6, 3\}$. To triangulate this face, we begin with node 1, and check if its neighbors in the face are adjacent in the graph. Node 1 has neighbors 4 and 5 in the face, which a quick check revels is not an edge in our list and so we add edge (4,5), and create new cycles $\{5, 1, 4\}$ and $\{4, 2, 6, 3, 5\}$. Similarly edges (4,6) and (6,5) are added, in order to get a triangulation from the graph $\bar{G}(W_{Tref})$.

Stage 2.2: Building a canonical ordering on the nodes of the maximal planar graph $\bar{G}$.

A graph G is *k-connected* if G has at least $k + 1$ nodes and the removal of any $k - 1$ or fewer nodes with all their incident edges keeps the graph connected.

Definition 15. *Given a maximal planar graph $G \in \mathbb{R}^2$ on $n \geq 3$ nodes, an ordering of the nodes $v_1, v_2, \ldots, v_n$ of G is called* canonical *if for each $3 \leq k < n$ the subgraph G_k induced by the nodes $v_1, \ldots, v_k$ has the following properties (Fig. 13).*

(a) The subgraph G_k is 2-connected and its external face F_k has the edge $v_1 v_2$.
(b) The neighbors of v_{k+1} in G_k form a path on the boundary of the face F_k.

A canonical ordering exists by [7, Proposition 3] and is implemented in [3].

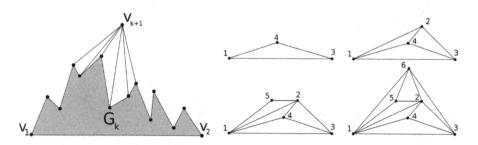

Fig. 13. Left: illustration of Definition 15. **Right:** drawing $\bar{G}(W_{Tref})$ using a canonical ordering.

Stage 2.3: Drawing a 2-page Embedding of the Maximal Planar Graph $\bar{G}$.

We implemented the algorithm from [5] drawing a topological 2-page embedding $G \subset \mathbb{P}_0 \cup \mathbb{P}_1$ of any maximal planar graph whose nodes $v_1, \ldots, v_n$ are added to the spine according to their canonical ordering. We put the edge between the first 2 nodes v_1, v_2 into the lower page $\mathbb{P}_1$. For each next node v_k, we follow the steps below.

Step 1. Find the embedded neighbors $w_1, \ldots, w_l$ of v_k, where w_1 is the leftmost in α.

Step 2. Embed v_k into α just to the right of w_1, and embed the edge (v_k, w_1) in $\mathbb{P}_1$.

Step 3. We connect v_k with its neighbor w_2 according to the substeps below.

3a. If v_k and w_2 are consecutive in the spine, we embed the edge (v_k, w_2) in $\mathbb{P}_1$.

3b. If v_k, w_2 are not consecutive, we embed (v_k, w_2) with 2 extra spine intersections p, q, namely (v_k, w_2) splits into 3 subedges $(v_k, p) \subset \mathbb{P}_1$, $(p, q) \subset \mathbb{P}_0$, $(q, w_2) \subset \mathbb{P}_1$.

Step 4. For each neighbor w_i, $3 \le i \le l$, we embed the edge (v_k, w_i) as in substep *3b*.

Step 5. Update *EdgeList* when we subdivide an edge into 3 subedges in *Steps 3b, 4*.

We explain the steps above by drawing the trefoil graph $\bar{G}(W_{Tref})$ from the last picture in Fig. 14. This graph has a canonical ordering of nodes $\{1, 3, 4, 2, 5, 6\}$. We embed the node $v_3 = 4$ between the first 2 nodes $v_1 = 1$ and $v_2 = 3$ by Steps 2 and 3a in the 2nd picture of Fig. 14. We embed the node $v_4 = 2$ with 3 already embedded neighbors $1, 4, 3$ by Steps 2, 3, 4 above in the 3rd picture of Fig. 14.

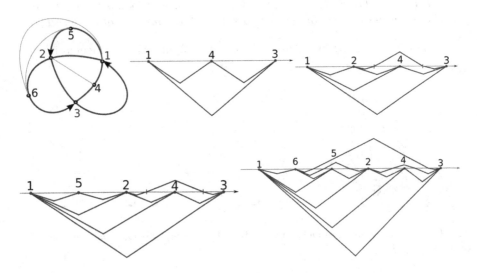

Fig. 14. A maximal planar graph $\bar{G}(W_{Tref})$ and drawing its 2-page embedding in Stage 2.3.

We introduce the dual graph below in order to show in Proposition 19 that the above algorithm produces an embedding isotopic to an original maximal planar graph.

Definition 16. *Given a planar graph $G \subset \mathbb{R}^2$, the* dual graph $G^* \subset \mathbb{R}^2$ *has a node for each face of G and a continuous arc connecting nodes from any adjacent faces of G.*

Theorem 17 (Whitney [19, Theorem 11]). *A 3-connected planar graph G with a fixed external face has a unique dual graph $G^* \subset \mathbb{R}^2$ up to isotopy in the plane.*

Corollary 18. *Any 3-connected planar graph without loops and multiple edges has a unique planar embedding (up to isotopy) with a choice of external face of G.*

Proof. Let $G_1, G_2 \subset \mathbb{R}^2$ be planar embeddings of the same abstract graph with the same external face. By Theorem 17, the duals G_1^* and G_2^* are isotopic. By Definition 16 the double dual G_1^{**} is isotopic to G_1 and G_2^{**} is isotopic to G_2. Since $G_1^* \sim G_2^*$, we conclude that $G_1^{**} \sim G_2^{**}$, hence $G_1 \sim G_1^{**} \sim G_2^{**} \sim G_2$.

Proposition 19. *For any maximal planar graph $\bar{G} \subset \mathbb{R}^2$ with a fixed external face, Stage 2.3 outputs a 2-page embedding isotopic to the original embedding $\bar{G} \subset \mathbb{R}^2$.*

Proof. We may assume that a given maximal planar graph G has at least 4 nodes, then G is 3-connected. Since G has a fixed external face, no loops or double edges, G has a unique planar embedding by Corollary 18. Since the 2-page embedding from Stage 2.3 has the same face, the output is isotopic to the original embedding of G. $\qquad\qquad\square$

Stage 2.4: Restricting a 2-page embedding of a maximal planar graph $\bar{G}$ to G.

At Stage 2.1 we extended a planar graph G to a maximal planar graph $\bar{G}$ in order to use a canonical ordering for drawing a 2-page embedding. Erase the edges added at Stage 2.1 to get a 2-page embedding of G, see the 1st picture in Fig. 15.

D   Algorithm 3 for Drawing a 3-page Embedding

Algorithm 1 in Appendix B starts from a Gauss code W and outputs an embedded (possibly, subdivided) Gauss graph $G(W) \subset S^2$ (if it exists). Then we fix an external face to get an embedding $G(W) \subset \mathbb{R}^2$. Algorithm 2 in Appendix C outputs an isotopic 2-page embedding $G(W) \subset \mathbb{P}_0 \cup \mathbb{P}_1$. In Algorithm 3 the input is the 2-page embedding with signs of crossings coming from the Gauss code W. The output will be a 3-page embedding $K(W) \subset \mathbb{P}_0 \cup \mathbb{P}_1 \cup \mathbb{P}_2$ of a knotted graph isotopic to a graph given by W.

Stage 3.1: Upgrading crossings in a 2-page embedding to get a 3-page embedding.

Lemma 20. *There are exactly 10 types of crossings in $\mathbb{P}_0 \cup \mathbb{P}_1$ obtained by Algorithm 2.*

Proof. According to the steps of Stage 2.3, all 4 (sub)edges incident to each crossing are in the lower page $\mathbb{P}_1$. Each of these 4 edges goes either to the left or to the right of the crossing. There are 5 ways to split 4 edges into 2 groups of left and right edges. For each of 5 ways, there are 2 types of crossings, see all 10 types of in Table 1.   □

We upgrade each crossing in a 2-page embedding $G(W) \subset \mathbb{P}_0 \cup \mathbb{P}_1$ to a local 3-page embedding according to Table 1. The resulting global embedding defines a knotted graph $K(W) \subset \mathbb{P}_0 \cup \mathbb{P}_1 \cup \mathbb{P}_2$, because in every page all arcs join each other.

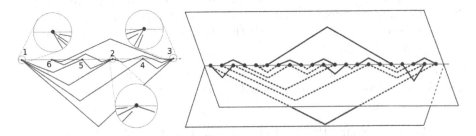

Fig. 15. Left: 2-page embedding of the Gauss graph $G(W_{Tref})$ from Stage 2.4 with resolved crossings for Stage 3.1. **Right**: 3-page embedding of the trefoil $G(W_{Tref})$ obtained at Stage 3.1.

Stage 3.2: Computing the element of a knotted graph $K(W)$ in a 3-page semigroup.

The 3-page embedding obtained at Stage 3.1 may contain spine intersections with both arcs in the same page see how to normalize such spine intersections in Fig. 16. All the letters from the 3-page alphabet in Fig. 11 have only local embeddings where arcs around every point in the spine occupy exactly 2 pages. Examples in Fig. 16 show how to get a 3-page embedding encoded by an element in a semigroup from Appendix A. After these upgrades the trefoil $K(W_{Tref})$ from Fig. 15 has the 3-page code $a_0 a_1 d_0 b_1 a_1 b_1 d_1 d_1 d_1 b_1 b_1 d_1 b_0 c_1 d_0 d_1 d_1 b_1 b_1 b_1 b_0 d_1 c_0 c_1$.

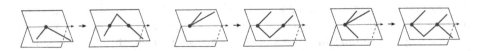

Fig. 16. Normalizations of 3-page embeddings around points with arcs in the same page.

Stage 3.3: Local simplifications for shortening elements in a 3-page semigroup.

Relations (1)–(2) in the semigroups from Appendix A are illustrated in Fig. 17 and allow us to locally simplify the element obtained at Stage 3.2. The trefoil

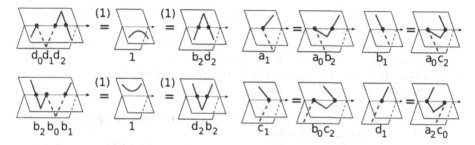

Fig. 17. A geometric interpretations of relations (1) and (2) in the semigroups from Appendix A.

from Fig. 15 will have the shorter code $a_0a_1d_0b_1a_1d_1b_0c_1d_0d_2d_1c_0c_1$ in the 1st picture of Fig. 18. The 2nd picture shows a minimal 3-page embedding with a shortest code $a_2a_1d_0d_2d_1d_0d_2c_1c_0$ by using more relations in the semigroups from Appendix A.

Stage 3.4: Computing 3D coordinates for a straight-line 3-page embedding of $K(W)$.

The spine α is identified with the x-axis. The upper page $\mathbb{P}_0$ is in the (x, z)-plane. The pages $\mathbb{P}_1, \mathbb{P}_2$ are obtained from $\mathbb{P}_0$ by the rotations through the angles $\pm\frac{2\pi}{3}$. If a 3-page code of a knotted graph $K(W) \subset \mathbb{P}_0 \cup \mathbb{P}_1 \cup \mathbb{P}_2$ has n letters, we embed corresponding points in the spine at $(j, 0, 0)$, $1 \leq j \leq n$. After finding which points are connected in each page $\mathbb{P}_i$, we embed all edges as broken lines with 2 straight segments.

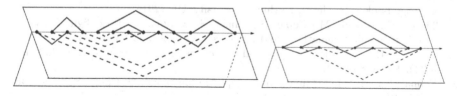

Fig. 18. Left: The 3-page embedding of a trefoil after local simplifications at Stage 3.3. **Right**: The minimal 3-page embedding of a trefoil after global simplifications by relations from Appendix A.

The MSc thesis [17] explains in more details why all stages in Appendices B, C, D require a linear time and memory in the length of a Gauss code.

References

1. Bernhart, F., Kainen, P.: The book thickness of a graph. J. Comb. Theory B **27**, 320–331 (1979)
2. Biasotti, S., Giorgi, D., Spagnuolo, M., Falcidieno, B.: Reeb graphs for shape analysis and applications. Theoret. Comput. Sci. **392**, 5–22 (2008)
3. Boost C++ Libraries (version 1.59.0). http://www.boost.org

4. Brendel, P., Dlotko, P., Ellis, G., Juda, M., Mrozek, M.: Computing fundamental groups from point clouds. Appl. Algebra Eng. Commun. Comp. **26**, 27–48 (2015)
5. Di Giacomo, E., Didimo, W., Liotta, G., Wismath, S.: Curve-constrained drawings of planar graphs. Comput. Geom. **30**, 1–23 (2005)
6. Enomoto, H., Miyauchi, M.: Lower bounds for the number of edge-crossings over the spine in a topological book embedding of a graph. SIAM J. Discrete Math. **12**, 337–341 (1999)
7. De Fraysseix, H., Pach, J., Pollack, R.: How to draw a planar graph on a grid. Combinatorica **10**(1), 41–51 (1990)
8. Kauffman, L.: Invariants of graphs in three-space. Trans. AMS **311**, 697–710 (1989)
9. Kearton, C., Kurlin, V.: All 2-dimensional links live inside a universal 3-dimensional polyhedron. Algebraic Geom. Topology **8**(3), 1223–1247 (2008)
10. Kurlin, V.: Dynnikov three-page diagrams of spatial 3-valent graphs. Funct. Anal. Appl. **35**(3), 230–233 (2001)
11. Kurlin, V.: Three-page encoding and complexity theory for spatial graphs. J. Knot Theory Ramifications **16**(1), 59–102 (2007)
12. Kurlin, V.: Gauss paragraphs of classical links and a characterization of virtual link groups. Math. Proc. Cambridge Philos. Soc. **145**(1), 129–140 (2008)
13. Kurlin, V.: A linear time algorithm for visualizing knotted structures in 3 pages. In: Proceedings of Information Visualization Theory and Applications, IVAPP 2015, pp. 5–16 (2015)
14. Kurlin, V.: Computing invariants of knotted graphs given by sequences of points in 3-dimensional space. In: Proceedings of Topology-Based Methods in Visualization, TopoInVis 2015 (2015)
15. Kurlin, V., Vershinin, V.: Three-page embeddings of singular knots. Funct. Anal. Appl. **38**(1), 14–27 (2004)
16. Schölkopf, B., Smola, A.: Learning with Kernels. MIT Press, Cambridge (2002)
17. Smithers, C.: A linear time algorithm for embedding arbitrary knotted graphs into a 3-page book. MSc thesis, Durham University, UK (2015)
18. Tkalec, U., Ravnik, M., Copar, S., Zumer, S., Musevic, I.: Reconfigurable knots and links in chiral nematic colloids. Science **333**(6038), 62–65 (2011)
19. Whitney, H.: Congruent graphs and the connectivity of graphs. Am. J. Math. **54**(1), 150–168 (1932)
20. Yannakakis, M.: Embedding planar graphs in four pages. J. Comp. Syst. Sci. **38**, 36–67 (1989)

Leaf Glyphs: Story Telling and Data Analysis Using Environmental Data Glyph Metaphors

Johannes Fuchs$^{(\boxtimes)}$, Dominik Jäckle, Niklas Weiler, and Tobias Schreck

Data Analysis and Visualization Group, University of Konstanz, Universitätstr. 10,
78457 Konstanz, Germany
`fuchs@dbvis.inf.uni-konstanz.de`

Abstract. In exploratory data analysis, important analysis tasks include the assessment of similarity of data points, labeling of outliers, identifying and relating groups in data, and more generally, the detection of patterns. Specifically, for large data sets, such tasks may be effectively addressed by glyph-based visualizations. Appropriately defined glyph designs and layouts may represent collections of data to address these aforementioned tasks. Important problems in glyph visualization include the design of compact glyph representations, and a similarity- or structure-preserving 2D layout. Projection-based techniques are commonly used to generate layouts, but often suffer from over-plotting in 2D display space, which may hinder comparing and relating tasks.

Inspired by contour and venation shapes of natural leafs, and their aggregation by stems, we introduce a novel glyph design for visualizing multi-dimensional data. Motivated by the human ability to visually discriminate natural shapes like trees in a forest, single flowers in a flowerbed, or leaves at shrubs, we design a flexible leaf-shaped data glyph, where data controls main leaf properties including leaf morphology, leaf venation, and leaf boundary shape. Our basic leaf glyph can map to more than a dozen of numeric and categorical variables. We also define custom visual aggregation schemes to scale the glyph for large numbers of data records, including prototype-based, set-based, and hierarchic aggregation. We show by example that our design is effectively interpretable to solve multivariate data analysis tasks, and provides effective data mapping. The design provides an aesthetically pleasing appearance, and lends itself easily to storytelling in environmental data analysis problems, among others. The glyph and its aggregation schemes are proposed as a scalable multivariate data visualization design, with applications in data visualization for mass media and data journalism, among others.

Keywords: Glyph visualization and layout · Nature-inspired visualization · Leaf shape · Multi-dimensional data analysis · Data aggregation

1 Introduction

Glyph-based data visualization has a long tradition in Information Visualization research and application. The basic idea in glyph visualization is to map data

© Springer International Publishing Switzerland 2016
J. Braz et al. (Eds.): VISIGRAPP 2015, CCIS 598, pp. 123–143, 2016.
DOI: 10.1007/978-3-319-29971-6_7

properties to visual properties of some appropriately designed visual structure. By the interplay of the different visual properties, each glyph then represents a data record. Many data records can be compared by appropriately laid out glyph displays. Glyph visualization, like other areas in Information Visualization, can be considered both a science and an art. Specifically, the design of glyphs may be inspired intuitively by common, well-known shapes or icons. For example, Chernoff faces were inspired by face properties, and sticky figures by abstraction of human body shapes.

A subset of the designs studied in Information Visualization to date has been inspired by nature. For example, tree structures have inspired hierarchical node-link diagrams. As another example, the notion of information landscapes or terrains is also borrowed from nature. There is reason to believe that the human visual sense, due to long evolutionary processes, is highly trained in recognizing, distinguishing and comparing natural forms. These visual recognition processes typically work well even in low illumination conditions, or in presence of partial occlusion of natural objects. By background knowledge and experience, humans are able to efficiently recognize natural shapes, also often in cases where only parts of the shape or their boundary are visible.

Based on this motivation, we investigate the design space for leaf shapes as natural metaphors for data glyphs. From observing leaves in nature, it is clear that there is a large variability in the different types and forms of leaves that exist. Overall leaf shape, shape boundary, and shape interior all comprise several visual parameters that can in principle, be used to map data to generate glyphs. To the best of our knowledge, this is the first work to systematically study the design space of leaf-based glyph visualization, and identify an encompassing set of leaf variables to map data to. In conjunction with appropriate glyph layouts (based e.g., on projection), and visual aggregation techniques, effective and intuitive data displays can be realized. Our rationale for using leaf-based data visualization is two-fold. First, the design space is large, giving ample opportunities for the visualization expert to map data variables to visual variables. As will be discussed, our variable space amounts to more than 20 different visual variables that can be controlled. While we have not formally evaluated the effectiveness of these variables or their combinations, we presume this is a large design space from which appropriate effective selections can be found. Second, we propose that nature-inspired designs, by their potential aesthetic appearances and familiarity, can be suited to spark interest in visual data analysis for wider audiences, e.g., for use in mass media. Also, it resonates well with visualization of environmental data, as has been previously demonstrated, e.g., by a respective infographic used by OECD (see Sect. 2.2).

The remainder of this paper is structured as follows. In Sect. 2, we discuss glyph-based and nature-inspired data visualization approaches. Section 3 defines the design space for leaf glyphs, based on identification of main visual leaf properties which are candidates for data mapping. Then, in Sect. 4, we define several visual aggregation schemes to scale 2D glyph layouts for large numbers of data points. Section 5 then applies our design to several data sets. By exemplary data analysis cases, we demonstrate the principal applicability of our approach. Finally, Sect. 6 summarizes our work and outlines future research in the area.

2 Related Work

Our work extends the design space of two existing branches of research by introducing a compact data representation making use of environmental cues. The related work is, therefore, split into two parts. The first part covers the area of space efficient visualization techniques, namely, data glyphs. The second part addresses research using environmental cues to convey data. We do not address research in the area of computer graphics, since this work mainly focuses on photo-realistic representation of the environment. We refer the interested reader to a summary work about this topic by Deussen and Lintermann [1].

2.1 Glyphs

In the literature, there exists a large variety of glyph designs. Elaborate summaries can be found in [2,3]. To come up with a comprehensive categorization we make use of Ward's classification of data glyphs [3]. In his research he distinguishes between three different ways a data point can be mapped to a glyph representation.

First, Many-to-One Mapping: All data dimensions and their respective value are mapped to a common visual variable. Therefore, these designs can be systematically created by choosing the most effective visual variable for a certain task. Additional guidance is given by Cleveland et al. with a ranking of visual variables [4]. Well-known examples making use of a position/length encoding are star glyphs [5], whisker and fan plots [6,7], or profile glyphs [8]. The designs just differ in their layout of the dimensions (i.e., circular or linear) and some minor variations like the presence or absence of a surrounding contour line. Other glyph designs make use of color encodings to represent the data value. Clock glyphs [9] map the dimensions in a radial fashion, whereas pixel-based glyph designs [10] layout the dimensions linearly. Of course, color cannot convey the data as accurate as a position/length encoding [11], however, for certain tasks like spotting outliers the color encoding is a reasonable choice. There is even a design mapping the data values to the angle of its rays. Sticky figures [6] use the visual variable orientation, which is not so accurate in communicating exact data values. However, when used as an overview visualization the designs convey individual shapes, which are perceived as a whole nicely approximating the underlying data point.

Second, One-to-One Mapping: Each dimension is mapped to a different visual variable. Probably, the most well-known representations here are Chernoff faces [12]. The single data values are mapped to face characteristics, like the size of the nose or the angle of the eyebrows. Other more exotic designs are bugs [13] (changing the shape, length or color of wings, tails and spikes), or hedgehogs [14] (manipulating the spikes by changing the orientation, thickness and taper). The major drawback of these kinds of glyph representations is that they are often sensitive to the order by which the data dimensions are mapped to visual variables. Variation of the order could significantly change the final glyph representation and its visual perception by users. Additionally, measuring differences between single dimension values within a data point is typically a difficult task, as the analyst has to compare different kinds of visual variables with each other (e.g., compare length with saturation or angle, etc.).

Third, One-to-Many Mapping: The dimensions are represented by two or more visual variables. This redundant mapping can be useful to strengthen the perception of individual dimensions. For example, in star or profile glyphs the dimensions can be additionally encoded by coloring the single data rays. Clock glyphs can make use of an additional length encoding for the single colored slices to encode the underlying data values more accurately.

Metaphoric Glyph Designs: Another category of glyph representations are metaphors for communicating domain specific data. A well-known example are Chernoff faces [12], which were already introduced in the one-to-one mapping category. In two quantitative experiments conducted by Jacob and Flury et al. these faces were compared against other visual representations like polygons or simple digits. In both evaluations data from human beings like anthropometric variables [15] or medical patient information [16] had to be encoded. The results indicate that metaphors outperform the more abstract designs. In addition, also other metaphoric glyph designs like clock glyphs [11] or car glyphs [17] have been subject to quantitative experiments yielding similar results.

As can be seen from these experiments metaphoric designs seem to be superior for specific domains compared to more abstract representations. This insight is an interesting starting point to think about designs for visualizing environmental data.

2.2   Environmental Cues

Visualizations making use of environmental cues need not necessarily be glyph representations. Stefaner uses an abstract tree layout to show the editing history of Wikipedia entries represented as single branches [18]. The branches grow to the right whenever people decided to delete an article or to the left in the other case. The resulting tree nicely summarizes 100 articles with the longest discussion whether to keep them or not. Another tree-based approach in combination with leaves visualizes poems in a more artistic way [19]. The branches of the tree are invisible just dealing as an anchor point to arrange the glyphs. Each word in the poem is represented with a leaf glyph and attached along the tree structure.

The work is not eligible of representing the text data accurately but tries to illustrate a creative unique picture or fingerprint of the underlying poem.

A more data-driven glyph design is the botanical tree [20], which again uses a 3D tree layout to represent hierarchical information. The single nodes are represented as fruits. The authors argue that people can more easily identify single nodes in this visualization compared to a more abstract representation because they are used to detect fruits or leaves on shrubs or trees. A 2D visualization using a botanical tree metaphor are so-called ContactTrees [21] which show relationships in data, e.g., contacts between persons. The branches consist of single lines representing an attribute in the data, e.g., a longer line refers to an older tie between people. Finally, fruits or leaves are added to the tree according to some data property, e.g., the kind of relation between people (friends, co-workers etc.). However, the fruits and leaves are highly abstract representations (mainly colored dots) and their shape does not change according to some data characteristics. The OECD's Better Life Index visualization [22], on the other hand, systematically changes the appearance of the single flower glyphs used to represent data. Stefaner uses such environmental cues to visualize multi-dimensional data about country characteristics. Each country is represented by one flower. The petals encode the different economic branches with varying sizes and lengths for the corresponding values. The flowers are arranged according to their weighted rank across all dimensions. People can change the layout by changing the weights of the dimensions or simply focusing on just one dimension.

We contribute to this body of existing work with the definition of a highly detailed leaf glyph, which closely follows the main morphological and functional variations among leaves. It is able to effectively map data variables. We also provide a custom aggregation scheme to scale leaf layouts for large number of records.

3   Environmental Glyph

According to Biological literature, leaves may be categorized by their function or usage in the environment [23]. For our purposes, we divide leaves according to their shape (or morphology). The overall appearance of a leaf consists of the combination of (1) the overall shape type, (2) the boundary details, and (3) the leaf venation. We consider these three aspects as the main dimensions for controlling the leaf glyph by mapping data. As a result we come up with a design space structured along the overall leaf shape, which we discuss next.

3.1   Leaf Shape Design Space

Following Palmer who pointed out: "Shape allows a perceiver to predict more facts about an object than any other property" [24], this visual variable should be used for the most important data dimension. In the environment, there exists a nearly endless amount of different leaf shapes since each leaf is unique. However, it is possible to distinguish leaves according to their overall shape [1]. A first categorization can be done between conifer and deciduous leaves.

Conifer leaves can be found for example at fir or pine trees and have a thin long needle-like shape. Therefore, they do not offer much space for a venation pattern, which we want to use later for mapping additional attributes (e.g., Acicular leaves). Since the differences in shape are quite small for the different kinds of this group and the provided area is limited due to the distorted aspect ratio, we do not consider them in our design space.

Deciduous leaves cover a large group of different shapes and can again be further divided into four sub-categories [1].

Pinnate and *palmate* compound leaves are shapes, which consist of several smaller leaflets attached to a shared branch (e.g., Alternate, or Odd and Even Pinnate leaves etc.). In order to avoid any misinterpretation between single leaflets at a branch and individual leaves, we discard this group from our final design space. However, these kinds of leaves seem an appropriate representation to visually summarize multiple data points where one leaflet corresponds to a single leaf.

Lance-like leaves have a parallel venation and are thin and long, similar to conifer leaves. Therefore, it is difficult to distinguish different kinds of these leaves since the differences in the overall shape are limited. Like the conifer leaves, we do not keep them in our design space because of the limited area to map a venation pattern, and because of possible confusion of different lance-like shapes.

Leaves with *net veins* or *reticulate* venation patterns encompass the largest group of deciduous leaves with a big diversity in shape. We restrict ourselves to the most common leaf shapes for this category to avoid misinterpretation of intermediate structures, which could not clearly be distinguished. Additionally, we focus on leaves with a big surface to show venation patterns and small stems to save space. Leaves similar to Flabellate, Unifoliate, etc. will, therefore, not be considered.

The most important requirement for shapes in visualizations is that they should be easily distinguishable. Therefore, our final design space covers elliptic (e.g., Ovate, Obtuse, Obcurdate etc.), circular (e.g., Orbicular), triangular (e.g., Deltoid), arrow-like (e.g., Hastate, Spear-shaped etc.), heart-like (e.g., Cordate, Deltoid etc.), two variations of tear-drop like (e.g., Acuminate, Cuneate etc.), wave-like (e.g., Pinnatisect), and star-like (e.g., Palmate, Pedate, etc.) shapes. Figure 1 illustrates the nine different leaf shape categories covered by our design space. In Sect. 5 we will introduce a heuristic to map data points to leaf shapes, based on the idea of representing outlying points by the more jagged leaf shapes; conversely, non-outlying points will be represented by the more regular or smooth leaf shapes.

We take these categories as a starting point and further extend them by mapping additional attribute dimensions to the width and the height of the glyph, scaling the overall shape. Therefore, similar shapes according to a certain data characteristic can look different because of the varying aspect ratio. However,

Fig. 1. Leaf shapes: Selected from our overall design space, these are the shapes used in our final glyph design. From left to right: Wave-like, circular, triangular, heart-like, arrow-like, tear drop up, tear drop down, elliptic, and star-like shapes.

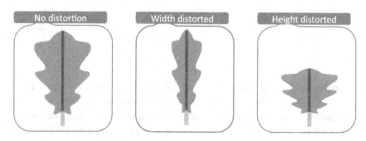

Fig. 2. Leaf scaling: The Lobate leaf shape is scaled using either the width (middle), or the height (right) of the glyph. Even after scaling, the glyph can still be recognized as a wave-like leaf, although the precise environmental reference to the Lobate leaf is reduced.

the individual shape categories can still be distinguished (Fig. 2). Because of this decision, we will deviate from the precise environmental reference, where leaves typically show a homogeneous aspect ratio. However, we thereby are able to encode additional data dimensions. Note that we do not want to represent leaves as accurate as possible (or even photo realistic), but use their expressiveness to visualize data.

3.2   Leaf Boundary Design Space

Basically, the boundary (or margin) of a leaf can be described as either serrated or unserrated. *Unserrated* boundaries have a smooth contour adapting to the overall leaf shape. *Serrated* boundaries are toothed with slight variations depending on the size of teeth, their arrangement along the boundary, and their frequency. Of course, there are more detailed differences and variations in nature. However, especially in overview visualizations (the major domain of data glyphs), distinguishing between small variations of the contour line of a leaf shape is nearly impossible. We therefore focus on just the two main boundary categories of teethed or smooth (serrated or unserrated). For mapping data values to the leaf boundary, we distinguish between a smooth and a toothed contour line and vary the width, height, and frequency of the teeth according to the underlying data value (Fig. 3).

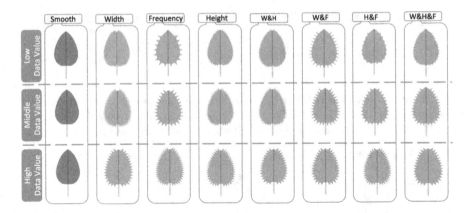

Fig. 3. Leaf boundary: Modifying the boundary in our design is realized by changing the frequency, the height, or the width of the boundary serration (teeths). Combinations of these three variables are possible and increase the expressiveness of the glyph. The figure illustrates all possible combinations for low, middle, and high data values for an elliptically shaped leaf glyph.

3.3 Leaf Venation Design Space

We also control the leaf venation pattern as to map additional data variables to the glyph. Several main leaf venation patterns exist, which differ in their overall structure within the leaf. A rough distinction can be made between single, not intersecting (e.g., Parallel), paired (e.g., Pinnate), or net-like (e.g., Reticulate) veins. The venation is perceived as an additional texture for the glyph and further increases the glyph expressiveness. Since it is hard to find a natural order within this texture, we propose to use the venation type for visualizing qualitative (or categorical) data, similar than the overal leaf shapes discussed in Sect. 3.1. Within a given venation type, we may also encode numeric data. This works as follows. Generally, the leaf is split in the middle by a main vein, with small veins growing from there in a given direction (angle). For mapping numerical data, we may either control this *angle of the veins* branching out from the main vein. An alternative is to control the *number of veins* shown on the surface Fig. 4. As a result, we come up with a venation texture able of encoding categorical and numerical data.

3.4 Summary

Besides modifying the leaf shape given by morphology, boundary and venation, further dimensions can be assigned to the color hue or saturation of the glyph. Of course, the designer has to pay attention to the contrast between the venation texture and the background color. Additionally, orientation of the glyph in the display can be used to encode further numeric information. We draw a short stem to each leaf shape, showing its orientation. Finally, it is also possible to modify the stem's width or height as well.

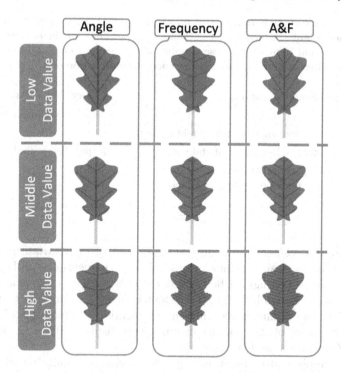

Fig. 4. Leaf venation: The texture for the venation system can either be created by mapping data values to the angle or frequency of the veins separately, or by combining the two. The figure illustrates all possible combinations for low, middle, and high data values for a wave-like leaf shape.

This represents a comprehensive design space for mapping data to leaf glyphs, controlled by 12 categorical and 14 numeric parameters, summing up to 26 variables altogether (see Table 1 for an overview of all variables.) We propose this design space as a toolbox from which the designer may select visual variables as appropriate. The number of 26 parameters is considered more a theoretical upper limit of data variables that we can show. We expect not all visual parameters in this design space to be of the same expressiveness; but some variables may be more effective than others, and may not all be orthogonal to each other. Careful choice should be done in selected and prioritizing the variables. An option is of course always, to redundantly code data variables to different glyph variables, to emphasize perception of important data variables. In Sect. 5, we will illustrate by practical examples, how glyph variables can be combined to form data displays.

4   Leaf Glyph Aggregation

When visualizing large data sets, leaf glyphs, like many other glyphs, are prone to overlap in the display, reducing the effectiveness of perceiving data from

Table 1. Summary of the parameters of our glyph design. It comprises 14 numeric and 12 categorical variables, which form the theoretic upper limit for the expressiveness of our glyph. Note that in practice, these variables are expected to not all be orthogonal, and comprise different perceptual performance, depending also on the data.

Leaf Design	Numeric Variables	Categorical Variables
Shape	2 (x/y scale)	9 (selected morphologies)
Boundary	3 (frequency, width, height of teeth)	–
Venation	2 (number, angle of child veins)	3 (parallel, paired, net)
Other	7 (hue, saturation, orientation, x/y position, stem width/height)	–
Sum	14	12

individual glyphs. Generally, an increasing amount of multivariate points in a visualization produces significant clutter resulting in perceptual problems – the user is not able to distinguish between data points properly anymore. This is mainly due to our design intention to use larger shapes for adding e. g., venation patterns. Next, we discuss three different aggregation techniques, to help cope with large numbers of data points in our glyph display: Alpha Compositing, Prototype Generation, and Abstraction.

First, we apply transparency in Fig. 5 to provide a visually pleasing representation that also reveals differences between data points. In some cases, the application of transparency is not enough. For example, if multiple data points share the same position, the opacity might sum up until no difference is perceivable. Therefore, we propose two different aggregation techniques that build on top of transparency and the application of a grid-based aggregation. Specifically, we place a user-defined grid on top of the visualization. All data points sharing the same cell are aggregated (see Fig. 6).

These effects can at the same time be perceived in nature: leaves can overlap or coincide with others. We adapt the proposed aggregation techniques and extend them in order to find a representative aggregate glyph which summarizes multiple leaf glyphs.

In Figs. 5 and 6 we point out the application of the aggregation techniques – Alpha Compositing, Prototype Generation, and Abstraction – with respect to nature. We next explain them in terms of their counterpart in nature, and apply them to our visualization of leaf glyphs.

4.1   Alpha Compositing

We use Alpha Compositing [25] to reveal details on overlapping glyphs by applying transparency. This technique describes the process of combining multiple, separately rendered images in order to provide a transparent appearance. The result of the application of transparency to the glyphs is shown in Fig. 5.

As mentioned in Sect. 3, different leaf shapes and characteristics need to be taken into account. In nature, leaves own the characteristic that even when

Fig. 5. Aggregation by Alpha Compositing. When multiple leaves overlap or coincide, we are not able to distinguish properly between their shapes and related characteristics. To overcome this issue, we propose to apply alpha compositing. It reveals details by applying transparency to the leaves.

multiple leaves overlap, we perceive differences due to their diverse shape and color. To support this, we apply transparency to the leaves. Figure 5 presents the first results. The application of transparency works well, in our experience, for a limited amount of leaf glyphs. When too many leaves overlap, perceptional problems can arise: Since the transparency also aggregates, from a certain extent on, the glyphs can become occluded and not be distinguishable anymore. For this reason, we propose two additional aggregation techniques we observed in nature: Prototype Generation and Abstraction.

4.2 Prototype Generation

As mentioned above, transparency may not be enough when aggregating multiple glyphs. Therefore, we propose to additionally generate a prototype glyph that aggregates the characteristics of all considered glyphs. We apply a grid and aggregate all leaves the calculated center point of which fall into the same grid cell; the cell dimensions are user defined. The glyph representing each cell can be given either by (1) a single glyph, determined by statistical aggregates of the member element dimensions, e.g., the mean or median values, or (2) a visual aggregate combining small multiples of the member elements, by a connecting structure (so-called bouquet glyph, inspired by combinations of different flower types). Figure 6 shows the result of both techniques, visualization of the median as well as the visualization in form of a bouquet. For both techniques, the transparency is preserved to be able to distinguish between different attribute values that determine the shape of a leaf glyph.

Our first proposed prototype is the representation of the median. We therefore create a new leaf glyph that has a simple appearance by means of its shape.

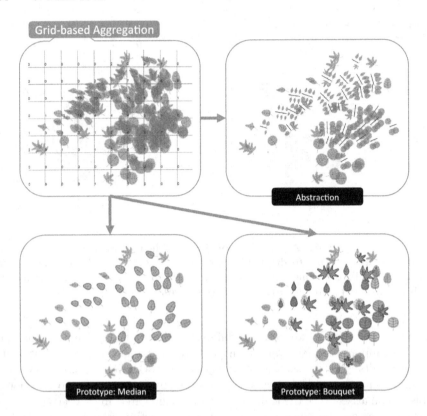

Fig. 6. Grid-based aggregation. We apply a grid to the visualization and calculate the center point of each leaf glyph, and aggregate all glyphs whose center points coincide within the same cell. Two different aggregations can be used: *Prototype Generation* and *Abstraction*. The first determines a representative glyph for the corresponding cell in the form of a median glyph or a bouquet glyph. The second creates (similar to what we observe in nature), a branch with multiple leaves based on the attributes of the considered leaves.

We use the median venation, margin, and shape in order to describe a set of leaves that coincide in one cell.

Similar to a bouquet, we derive our second proposed prototype by combining and aligning all contained leaf glyphs. First, all leaf glyphs sharing the same shape are stacked using transparency as described in Sect. 4.1. Second, stacked leaf glyphs are aligned in a radial manner according to their shape. This means, while in the first step glyphs are stacked according to their shape, in the second step they are radially moved and aligned according to the shape classes as pointed out in Sect. 3. As a result, we get a representation similar to a bouquet.

4.3   Abstraction by Visual Aggregation

Based on the grid aggregation, we need to address issues that emerge when too many glyphs fall into one cell. Prototype generation may fail, if too many glyphs along too many different shapes are aggregated, and the visualized prototype may then suffer from clutter. Therefore, we propose abstraction by visual aggregation. We describe the new visual representation for an aggregated set of glyphs. Similar to growth characteristics of leaves we observe in nature, this aggregation technique represents an aggregated set of leaf glyphs as a new branch with multiple leaves on it. All leaf glyphs are aligned side-by-side along a branch according to Fig. 6.

4.4   Hierarchical Aggregation

The previously introduced aggregation techniques are not only suitable to visualize dense areas in 2D projections. Another design alternative is to use hierarchical arrangements, which can convey aggregate information and therefore, help with scalability. The relevant concept is that of a dendrogram (see Fig. 7). Each parent node in a dendrogram may be represented by an aggregate prototype showing properties of the represented data partitions. Basic hierarchical visualizations can, therefore, be enriched with additional information like the composition of data points for individual clusters.

In Fig. 7 we clustered the Iris dataset from the UCI Machine learning repository and represented the hierarchical structure in a radial dendrogram. The class attribute is used to assign different leaf shapes to the data. Other visual features like color, venation, and margin represent different attribute dimensions of the dataset. In each level, the nodes have been replaced with aggregated leaf glyphs using alpha composition together with a position bundling. The leaf glyph positioned in the middle of the visualization (*#1*) aggregates the dimension values of all nodes in the diagram. It, therefore, contains many different sub-clusters as can be seen in Fig. 7. When traversing the single branches to the lower levels (from inside out) the prototype representations of lower aggregate levels are getting more homogeneous. For example, after the first hierarchical split two main clusters are separated (*a* and *b*). The node labeled with *b* shows only green ovate leaves thus representing a homogeneous group of data points. The other aggregated prototype labeled with *a* seems to be more heterogeneous showing two different kinds of leaf shapes (hastate leaves and maple leaves). However, after descending to the next hierarchy level these two sub-clusters are separated. The inner node labeled with *#2* represents only maple leaves, whereas the other node labeled with *#3* contains hastate leaves. By traversing along the different branches the inner node is getting more and more homogeneous (e.g., similar colored leaves). Step by step different sub-clusters are divided till the lowest level of the hierarchy is reached.

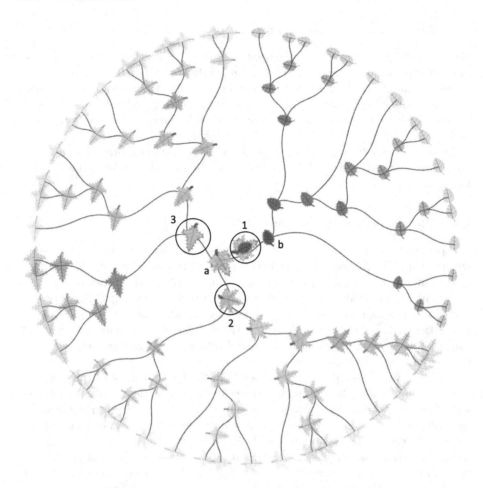

Fig. 7. Enhanced dendrogram: A selection of data points from the iris dataset have been hierarchically clustered and their structure represented in a radial dendrogram. Leaf glyphs are used to visualize the groups and individual data points along the hierarchy. As can be seen, the visual structure of the leaf glyph is getting more and more precise when approaching the leaf nodes illustrating the homogeneity of the lower levels in the dendrogram (Color figure online).

5   Story Telling and Data Analysis

We defined an encompassing scheme to generate leaf glyph-based data visualizations for large data sets. We implemented the above described designs in an interactive system. We here exemplify results we obtained for analyzing the forest fire data set, showcasing the applicability of our approach. Note that a formal comparison against alternative glyph designs and user testing remain future work.

To facilitate memorizing the visual mappings we explain our design choices step by step (see Figs. 9, 10, 11, and 12). Such a story telling approach guides the audience through a use case scenario, which analyzes complex data structures combining multi-dimensional characteristics with time-series data. Whenever possible metaphoric features are used to represent data dimensions. As studies suggest such an approach will help to better understand the underlying data.

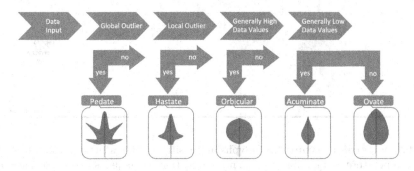

Fig. 8. Shape categories: Based on the results of the clustering we assign different leaf shape templates according to the data characteristics.

Forest Fire: The *forest fire* data set is available in the UCI machine learning repository [26]. It contains data about burned areas of forests in Portugal on a daily basis for one year.

Additionally, weather information is included, e.g., temperature, humidity, rain and wind conditions at respective points in time. This data set does not contain any categorical data which could be directly mapped to the leaf shape. Therefore, we initially clustered the data points with the DBSCAN algorithm [27] and assign local or global outliers to different glyph shapes (Fig. 8). Our idea is to map outliers to the more jagged leaf shapes, while non-outlier points get mapped to more regular or smooth shapes, thereby providing a first visual assessment of the degree of outlyingness for the data. Our analysis task is to find similarities between burned areas to be able to predict fires due to certain weather conditions.

First, we wanted to get an idea about the data distribution. We used one data glyph for each data point and positioned the leaf glyphs in a common scatterplot layout. The x-axis is reflecting the temperature and the y-axis the humidity. By intention, we swapped the y-axis showing low data values at the top and high data values at the bottom. This reflects our background knowledge that possible indicators for forest fires are a high temperature and a low humidity. Potentially vulnerable areas are, therefore, positioned at the top right corner of the scatterplot. Figure 9 allows a first view of the data. There seems to be a positive correlation between temperature and humidity. However, because of the high number of data points, substantial information is lost due to overplotting.

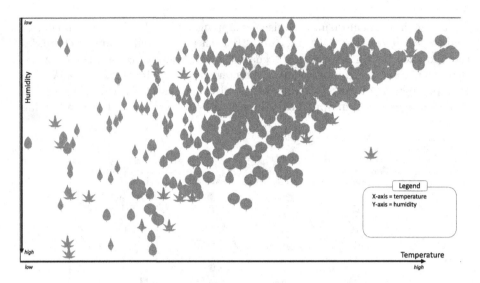

Fig. 9. Scatterplot layout: Leaf glyphs are positioned in a scatterplot according to their temperature and humidity. Since no aggregation technique is applied on the data a lot of overplotting occurs.

As a next step, we applied transparency to the data points and also use color to show temporal information and orientation to encode the wind speed. The alpha composition technique helps to detect some more leaf shapes, however, especially in the dense area on the diagonal still a lot of overplotting exists. For the color encoding, we decided to use a metaphoric approach to help understand the encoding without a color legend. We try to associate the seasons (i.e., winter, spring, summer, autumn) with the leaves. During winter and autumn, the leaves in nature have a brownish or reddish color, whereas the color hue changes during spring and summer getting more green. Therefore, we colored our leaf glyphs accordingly. As can be seen in Fig. 10 the data points are divided into 2 main clusters. Brown and red leaf glyphs are located above the diagonal and the more greener leaves are positioned on the diagonal. It seems as if humidity and temperature are both lower during autumn and winter times compared to spring or summer.

Another metaphoric approach was used to represent the magnitude of wind. The orientation of the leaf glyphs is changing according to the wind speed. Data points with low speed are oriented to the left. With an increasing wind speed the angle changes pointing right. The idea was to simulate a blast blowing from left to right catching all leaves and changing their direction accordingly. However, no additional visual pattern can be perceived. The leaf glyphs are pointing in various directions showing no correlation between wind magnitude and temperature, humidity, or time.

To find similarities between burned forest areas, we map the size of the burned regions to the size of the glyphs. While this encoding is not strictly a

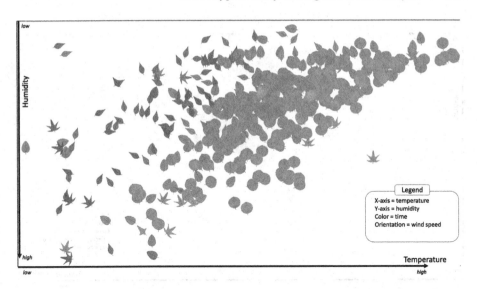

Fig. 10. Alpha Composition: Transparency is used to better perceive the data in cluttered areas. Since too many data points are located in the dense regions this aggregation technique does not provide the best view on the data (Color figure online).

metaphoric representation, it does help to associate the information with the respective visual dimension. When inspecting Fig. 11, it appears all leaf glyphs are reduced in size, and differences according to size cannot be perceived. This is surprising, since we would expect the size of burned forest areas to be different. One possible explanation is that some data points with different size are located in the cluttered area on the diagonal.

To get a different perspective on the data, and to further reduce overplotting, we switch to an alternative aggregation technique to better understand the highly cluttered area (Fig. 12). Due to the design of the bouquet prototype generation, the visual attribute of orientation is lost, and therefore, we cannot map the wind magnitude to this variable anymore. In the highly cluttered area in the middle of the plot, several different maple leaf shapes become apparent. These refer to outliers detected by our previous clustering algorithm. However, more interesting are the two big maple leaf shapes located at the top right corner. They represent huge areas of burned forests during the summer time with high temperature and low humidity. When switching to Fig. 11, and keeping in mind the concrete location of these data points, we can further extract the wind magnitude, which seems to be medium. With this understanding of the data, it is plausible why the burned forest areas are large. High temperature, medium winds, and low humidity all support the spread of forest fires. However, since there are more smaller data points with similar data characteristics, these features are not necessarily an indication for large forest fires. Perhaps other factors, e.g., the area or the coverage of fire stations, which are not covered in the data visualization discussed, may constitute additional factors.

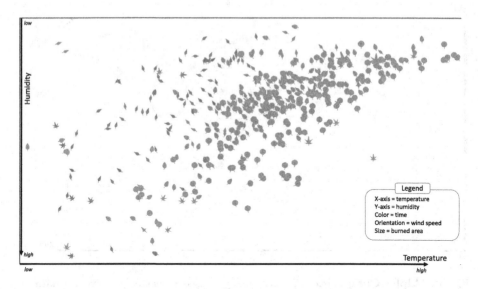

Fig. 11. Forest fire data set: We applied alpha compositing as an aggregation technique to get a first overview of the data set. We used the following mapping to represent the multi-dimensional data: Shape ≙ local/global outlier, x-position ≙ temperature, and y-position ≙ humidity, color hue/saturation ≙ time (i.e., month), size ≙ area of burned forests, orientation ≙ magnitude of wind.

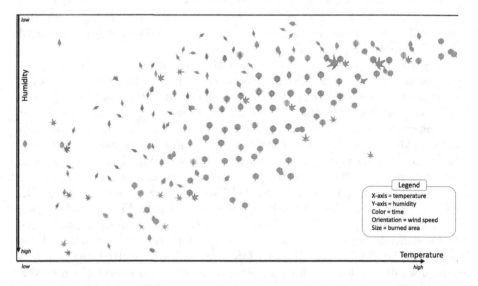

Fig. 12. Forest fire data set: We applied a prototype aggregation technique to reveal insights to the highly cluttered areas in the plot. Interesting to note are the relatively big outlier leaf shapes, which were not visible beforehand.

Of course, these findings would need to be substantiated by additional data considerations. Further information, e.g., the amount of firemen fighting the fire, the exact kind and amount of trees, or the time until the fire was recognized are important side factors not covered within the used data. However, with our new glyph approach, we were able to easily identify timely patterns, outliers, and similar behavior of data points. Other glyph designs (i.e., star glyphs etc.) might also be suitable to represent the data, however, our leaf glyph technique helps to easily associate the appearance of the data point with its attribute dimensions.

6   Conclusion and Future Work

We introduced Leaf Glyph, a novel glyph design inspired by an environmental metaphor. Due to its natural and familiar appearance, we expect users are likely to be able to discriminate data by its visual properties. The glyph is based on a naturally prominent shape, which should connect well to human perception, supposedly also under conditions of partial overlap. We systematically structured the leaf glyph design space. Specifically, we mapped data to the main properties of the leaf glyph: Leaf morphology, leaf venation, and leaf boundary. Furthermore, we defined visual aggregations including set-oriented and hierarchical aggregation, to scale the glyph display for large numbers of data records, based on inspirations from nature. Finally, we exemplified the applicability and effectiveness of our approach in a multivariate data analysis task, showing its strengths in illustrative storytelling using a consistent metaphor.

This work is a first step in studying the effectiveness of nature-oriented data visualization. While we believe leaf glyphs can form intuitive and effective data glyphs, more thorough evaluation is needed. Specifically, we want to compare the leaf glyph against alternative glyphs from the literature, such as Chernoff faces, and pixel-oriented glyphs. This should also include user-studying of effectiveness and efficiency of the technique. We also believe our approach is aesthetically pleasing and may spark interest by a wider audience, for use, e.g., in mass media communication. The leaf glyph by design may fit well e.g., to visualization of environment survey data. Also, this should be evaluated by qualitative consideration.

As a next step, we will combine our multi-dimensional leaf glyph representation with related botanical tree metaphors to extend the design space with a hierarchical layout. A natural combination would be to pair it with the botanical tree layouts proposed in [20]. We assume the combination of the two will support people with no computer science background more easily in understanding complex data structures due to the environmental reference. We further want to test this in a controlled environment against more abstract hierarchical representations such as TreeMaps.

Acknowledgements. This work has been supported by the Consensus project and has been partly funded by the European Commission's 7th Framework Programme through theme ICT-2013.5.4 ICT for Governance and Policy Modelling under contract no. 611688.

References

1. Deussen, O., Lintermann, B.: Digital Design of Nature. Springer, Heidelberg (2005)
2. Borgo, R., Kehrer, J., Chung, D.H., Maguire, E., Laramee, R.S., Hauser, H., Ward, M., Chen, M.: Glyph-based visualization: foundations, design guidelines, techniques and applications. In: Proceedings of Eurographics, Eurographics, pp. 39–63 (2012)
3. Ward, M.: Multivariate data glyphs: principles and practice. Handbook of Data Visualization. Springer Handbooks Computer Statistics, pp. 179–198. Springer, Heidelberg (2008)
4. Cleveland, W., McGill, R.: Graphical perception: theory, experimentation, and application to the development of graphical methods. J. Am. Stat. Assoc. **79**, 531–554 (1984)
5. Siegel, J., Farrell, E., Goldwyn, R., Friedman, H.: The Surgical Implications of Physiologic Patterns in Myocardial Infarction Shock. Surgery **72**, 126 (1972)
6. Pickett, R.M., Grinstein, G.G.: Iconographic displays for visualizing multidimensional data. In: Proceedings of the Conference on Systems, Man, and Cybernetics, pp. 514–519. IEEE (1988)
7. Ware, C.: Information Visualization: Perception for Design. Morgan Kaufmann, Waltham (2012)
8. Du Toit, S.H., Steyn, A.G.W., Stumpf, R.H.: Graphical Exploratory Data Analysis. Springer-Verlag, New York (1986)
9. Kintzel, C., Fuchs, J., Mansmann, F.: Monitoring large IP spaces with clockview. In: Proceedings Symposium on Visualization for Cyber Security, p. 2. ACM (2011)
10. Levkowitz, H., Herman, G.: Color scales for image data. IEEE Comput. Graph. Appl. **12**, 72–80 (1992)
11. Fuchs, J., Fischer, F., Mansmann, F., Bertini, E., Isenberg, P.: Evaluation of alternative glyph designs for time series data in a small multiple setting. In: Proceedings Human Factors in Computing Systems (CHI), pp. 3237–3246. ACM (2013)
12. Chernoff, H.: The use of faces to represent points in k-dimensional space graphically. J. Am. Stat. Assoc. **68**, 361–368 (1973)
13. Chuah, M.C., Eick, S.G.: Information rich glyphs for software management data. IEEE Comput. Graph. Appl. **18**, 24–29 (1998)
14. Klassen, R.V., Harrington, S.J.: Shadowed hedgehogs: a technique for visualizing 2D slices of 3D vector fields. In: Proceedings of the 2nd conference on Visualization 1991, pp. 148–153. IEEE Computer Society Press (1991)
15. Flury, B., Riedwyl, H.: Graphical representation of multivariate data by means of asymmetrical faces. J. Am. Stat. Assoc. **76**, 757–765 (1981)
16. Jacob, R.: Facial representation of multivariate data. In: Graphical Representation of Multivariate Data, pp. 143–168. Academic Press (1978)
17. Surtola, H.: The effect of data-relatedness in interactive glyphs. In: Proceedings of IV, pp. 869–876 (2005)
18. Stefaner, M.: The deleted (2014). http://notabilia.net/
19. Müller, B.: Poetry on the road (2014). http://www.esono.com/boris/projects/poetry05/
20. Kleiberg, E., van de Wetering, H., van Wijk, J.: Botanical visualization of huge hierarchies. In: IEEE Symposium on Information Visualization, 2001, INFOVIS 2001, pp. 87–94. IEEE (2001)
21. Sallaberry, A., Fu, Y.C., Ho, H.C., Ma, K.L.: Contacttrees: ego-centered visualization of social relations. Technical report (2012)

22. Stefaner, M.: Oecd better life index (2014). http://moritz.stefaner.eu/projects/oecd-better-life-index/

23. Beck, C.B.: An introduction to plant structure and development: plant anatomy for the twenty-first century. Cambridge University Press (2010)

24. Palmer, S.E.: Vision Science: Photons to Phenomenology, vol. 1. MIT press Cambridge, MA (1999)

25. Porter, T., Duff, T.: Compositing digital images. In: Proceedings of the 11th Annual Conference on Computer Graphics and Interactive Techniques. SIGGRAPH 1984, pp. 253–259. ACM, New York (1984)

26. Cortez, P., Morais, A.D.J.R.: A data mining approach to predict forest fires using meteorological data (2007)

27. Han, J., Kamber, M., Pei, J.: Data Mining: Concepts and Techniques, 3rd edn. Elsevier Ltd., Oxford (2011)

Compression and Heuristic Caching for GPU Particle Tracing in Turbulent Vector Fields

Marc Treib, Kai Bürger, Jun Wu$^{(\boxtimes)}$, and Rüdiger Westermann

Department of Computer Science, Technische Universität München,
Munich, Germany
{treib,buerger,jun.wu,westermann}@tum.de,
http://wwwcg.in.tum.de/

Abstract. Particle tracing in fully resolved turbulent vector fields is challenging due to their extreme resolution. Since particles can move along arbitrary paths through large parts of the domain, particle integration requires access to the entire field in an unpredictable order. Thus, techniques for particle tracing in such fields require a careful design to reduce performance constraints caused by memory and communication bandwidth. One possibility to achieve this is data compression, but so far it has been considered rather hesitantly due to supposed accuracy issues. We shed light on the use of data compression for turbulent vector fields, motivated by the observation that particle traces are always afflicted with inaccuracy. We quantitatively analyze the additional inaccuracies caused by lossy compression. We propose an adaptive data compression scheme using the discrete wavelet transform and integrate it into a block-based particle tracing approach. Furthermore, we present a priority-based GPU caching scheme to reduce memory access operations. In some experiments we confirm that the compression has only minor impact on the accuracy of the trajectories, and that on a desktop system our technique can achieve comparable performance to previous approaches on supercomputers.

Keywords: Vector fields · Turbulence · Particle tracing · Data compression · Data streaming

1 Introduction

One of the most intriguing and yet to be fully understood aspects in turbulence research is the statistics of Lagrangian fluid particles transported by a fully developed turbulent flow. Here, a fluid particle is considered a point moving with the local velocity of the fluid continuum. The analysis of Lagrangian statistics is usually performed numerically by following the time trajectories of fluid particles in numerically simulated turbulent fields. Let $\mathbf{x}(\mathbf{y}, t)$ and $\mathbf{u}(\mathbf{y}, t)$ denote the position and velocity at time t of a fluid particle originating at position $\mathbf{y}$ at time $t = 0$. The equation of motion of the particle is

$$\frac{\partial \mathbf{x}(\mathbf{y}, t)}{\partial t} = \mathbf{u}(\mathbf{y}, t),$$

© Springer International Publishing Switzerland 2016
J. Braz et al. (Eds.): VISIGRAPP 2015, CCIS 598, pp. 144–165, 2016.
DOI: 10.1007/978-3-319-29971-6_8

subject to the initial condition

$$\mathbf{x}(\mathbf{y}, 0) = \mathbf{y}.$$

The Lagrangian velocity $\mathbf{u}(\mathbf{y}, t)$ is related to the Eulerian velocity $\mathbf{u}^+(\mathbf{y}, t)$ via $\mathbf{u}(\mathbf{y}, t) = \mathbf{u}^+(\mathbf{x}(\mathbf{y}, t), t)$. By using a numerical integration scheme, the trajectory of a particle released into the flow can now be approximated.

Particle tracing in discrete velocity fields of a sufficient spatial and temporal resolution to resolve the higher wavenumber components in turbulent flows is nonetheless difficult. For reasonably-sized particle ensembles the performance is strongly limited by the available memory bandwidth capacities due to the massive amount of data to be accessed during particle tracing. Consequently, an effective performance improvement can be expected from data compression schemes which can read and decompress the data at significantly higher speed than reading the uncompressed data. We make use of a brick-based compression layer fulfilling this requirement [1].

Since in particle tracing the interpolation errors accumulate and are transmitted to the calculated trajectories, we analyze—compared to the established interpolation scheme on the uncompressed data—the inaccuracies in the computed trajectories which are caused by lossy compression.

Intuitively one might argue that lossy compression should not be considered, because it introduces an additional, non-acceptable error into particle tracing. On the other hand, in our application study the vector fields were simulated using a spectral method, meaning that the data values are a discrete sampling of a band-limited smooth function. Therefore, a ground truth interpolation exists—namely trigonometric interpolation—yet it is never used due to its high numerical complexity. Nevertheless it is clear that the established interpolation scheme already introduces an error, even though this error is generally accepted. As our major contribution we show that the additional inaccuracies caused by lossy data compression are in the same regions of variation in which the trajectories in the uncompressed field differ from the assumed ground truth trajectories. For the particular application this means that the trajectories extracted from the compressed data are as reliable as the trajectories usually used for analyzing the turbulence fields.

We focus on the analysis of the (spatial) interpolation error, because it is well known that interpolation is the major source of errors in numerical particle tracing in fully resolved turbulent flow fields. This is due to the fact that turbulent velocity fields are highly nonlinear. Since the time-step in turbulence simulations is commonly restricted to small values to enforce the Courant number stability condition, the time-stepping error in numerical integration is generally much less significant.

We use two vector-valued data sets describing turbulent flow fields to verify our approach. These data sets are the result of terascale turbulence simulations and originate from the JHU turbulence database cluster, which is publicly accessible at http://turbulence.pha.jhu.edu. Each is comprised of one thousand time steps of size 1024^3, making every time step as large as 12 GB (3 floating-point values per velocity sample). The data sets contain direct numerical simulations of magneto-hydrodynamic (MHD) turbulence and forced isotropic turbulence, and are called "MHD" and "Iso" in the following. For a description of the simulation methods used to compute these data sets let us refer to [2].

2   Related Work

We do not attempt here to survey the vast body of literature related to flow visualization approaches based on stream and path line integration because they are standard in flow visualization. For a thorough overview, however, let us refer to the reports by [3–5].

Teitzel, et al. [6] put special emphasis on the investigation of the numerical integration error and the error introduced by interpolation. They conclude that an RK3(2) integration scheme provides sufficient accuracy compared to linear interpolation, but they do not consider higher-order interpolation methods. There is also a number of works dealing especially with accuracy issues of particle tracing in turbulence fields [7–9]. One of the conclusions was that Lagrange interpolation of order 4 to 6 provides sufficient accuracy, and it is therefore the established scheme in practice (cf. [2]).

The use of graphics hardware is popular for interactive particle tracing approaches [10–14]. The fundamental problem in GPU-based approaches is the limited memory available on such architectures, allowing only data sets of rather moderate size to be handled efficiently. To the best of our knowledge, no previous approach has addressed the problem of GPU particle tracing when even a single time step does not fit into GPU memory.

An important topic related to our method is data compression using transform coding. For a general overview of data compression techniques we refer to the book by [15]. The recent survey by [16] provides a more focused treatment of techniques used in the context of volume visualization. Our GPU compression scheme builds upon previous work for performing Huffman and run-length decoding entirely on the GPU [1,17].

Precomputed Particle Traces: In a number of approaches it has been proposed to precompute and store particle trajectories for a number of prescribed seed points, and to restrict the visualization to subsets of these trajectories [18–20]. In this way, all computation is shifted to the preprocessing stage, and storage as well as bandwidth limitations at runtime can be overcome.

Conceptually, the approach to restrict the flow field analysis to a set of precomputed trajectories can be seen as a kind of lossy data compression, where the seeding positions are quantized rather than the flow data itself. However, since even very small perturbations of the seeding positions can lead to vastly different trajectories, the resulting visualizations might not contain all relevant structures that are present in the data.

Parallelization on Compute Clusters: Another possibility to address scalability issues in particle tracing is to employ parallel computing architectures such as tightly coupled CPU clusters or supercomputers. The larger memory capacities and I/O bandwidth on such systems make them attractive for handling large data sets. However, the highly data-dependent nature of particle tracing makes it difficult to effectively parallelize particle tracing on large distributed memory architectures.

The two basic parallelization strategies are *parallelize-over-seeds* (PoS) and *parallelize-over-blocks* (PoB) (cf. [21]). In both strategies, the data set is partitioned into blocks. In PoS, the seeding positions are distributed over the processors, and each processor dynamically loads those blocks required by its particles. This usually leads to fairly even load-balancing of computations, but it also results in the duplication of blocks in memory and increased I/O load since a block might be accessed by many processors. In PoB, the blocks are distributed across the processors, and each only handles particles within its assigned blocks. This avoids the duplication of blocks in memory, but causes severe load imbalance when many particles fall into the same processor's blocks while other processors remain idle. It also requires communication of particle positions between processors whenever a particle enters another processor's domain. A number of approaches have been presented to mitigate the drawbacks of PoS or PoB [21–25].

As reported e.g. in [22], particle tracing on compute clusters typically spends only a small fraction of the total time on the computation of particle traces. In many approaches, most of the time is spent on either node-to-node communication, I/O, or waiting due to load imbalances. It can be concluded that despite its embarrassingly parallel nature, particle tracing is not very well suited for computation on distributed memory clusters. The main benefit of such systems appears to be the large amount of aggregated memory, which can often prevent expensive trips to external memory such as hard disks.

3   Out-of-Core Particle Tracing

Our proposed system for out-of-core particle tracing takes as input a sequence of 3D velocity fields, each field representing the state of a flow field at a different time step. We assume that the values in each field are given on a Cartesian grid. In a preprocess, each grid is partitioned into a set of equally-sized bricks. A halo region is added around each brick to allow proper interpolation at brick boundaries. The bricks are compressed before being stored sequentially on disk. An index structure is stored along with the brick data to enable fast access to individual bricks at runtime. For one time step consisting of 1024^3 velocity values, this process takes about 5 min.

At runtime, the computation of particle trajectories is performed on the GPU. For that, bricks which are required to perform the numerical integration are requested from the CPU. The CPU decides based on a particular strategy (see Sect. 3.3) which bricks to upload to the GPU from main memory or disk. Compressed brick data is cached in CPU memory. The compression reduces disk bandwidth requirements and allows us to cache a large number of bricks.

For use on the GPU, the compressed brick data is uploaded into GPU memory and immediately decompressed. The decompressed brick is stored in a large 3D texture map, the so-called *brick atlas*. In this way, all GPU memory (apart from a small temporary buffer for the upload of compressed data) stores ready-to-use flow data. In the current implementation we use bricks of size 128^3 each (including a halo region of size 4). We have found that this size provides the best trade-off between locality of access and storage overhead for the halo voxels. The size of the *brick atlas* is chosen based on the amount of available GPU memory.

3.1 Particle Tracing in Rounds

Fluid particles are advected in parallel on the GPU to exploit memory bandwidth and computational capacities. We use the CUDA programming API and issue one thread per particle, grouped into thread blocks of size 128. Each thread advects the position of its particle while the required flow data is available in the *brick atlas*. Since the set of bricks which are required to perform the computation of all trajectories does not fit into GPU memory in general, only a subset can be made available at a time. An *index* buffer stores the mapping from spatial brick index to position in the brick atlas. It is indexed by the 3D brick index $\mathbf{b}$, which can be computed from a particle's position $\mathbf{p}$ and the brick size $\mathbf{s}$ as $\mathbf{b} = \lfloor \mathbf{p}/\mathbf{s} \rfloor$. If a brick is not currently available in the *brick atlas*, the corresponding *index* entry contains the value -1. Figure 1 illustrates the employed data structures and CPU-GPU interaction using a 2D example.

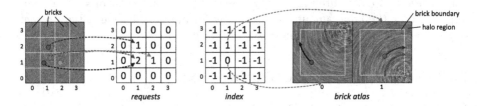

Fig. 1. Out-of-core particle tracing (2D example): (1) Particle positions are uploaded to the GPU where trajectories are computed in parallel. Requests of required bricks are issued (corresponding entries in the *requests* buffer are incremented). (2) The CPU downloads the *requests* buffer, uploads some requested bricks into the GPU *brick atlas*, and clears the *requests* buffer. The *index* buffer stores the index into the atlas of each uploaded brick and is also uploaded to the GPU. (3) The GPU advects each particle until it requires a brick not yet resident in GPU memory. The *requests* buffer is incremented again based on the new particle positions, and the process continues at step (2).

To start the computation, the user specifies the number of fluid particles to trace and the seed region in which they are spawned. Random positions inside the seed region are stored in the *particle* buffer on the GPU (not shown in the figure). Particle tracing then proceeds in a ping-pong fashion between the

GPU and the CPU: The GPU advances all particles in parallel. Whenever a particle enters a brick which is not stored in the *brick atlas* (indicated by a −1 entry in the *index*), the GPU requests this brick for the next round of tracing. This is realized by atomically incrementing the corresponding entry in a *requests* buffer. The particle's last position and any additional information (such as the current step size for adaptive integrators) are stored in the *particle* buffer. The GPU stops when all particles (a) must stop because they are waiting for a brick to be uploaded to the GPU, (b) have reached their maximum age, or (c) have been advanced by a fixed maximum number of steps (64 in the current implementation). The CPU then downloads the *requests* buffer and determines the bricks to be uploaded next into the atlas. With these bricks being available on the GPU, particle tracing is restarted. The process is finished once all particles have either reached their maximum age or left the domain.

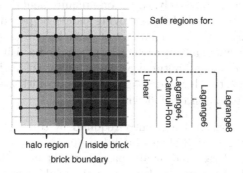

Fig. 2. Velocity sampling near a brick boundary: Texel boundaries are shown in gray; grid points and cell boundaries in black. A halo region is stored to allow sampling of velocity values in a small region outside of the brick. The size of the required halo region depends on the interpolation scheme.

3.2   Tracing Across Brick Boundaries

Special care has to be taken whenever a particle moves close to a brick boundary. In this case it has to be ensured that all velocity values required in the integration step are available in the current brick. The number of required values depends on the support of the interpolation kernel. Figure 2 depicts the admissible locations for velocity interpolation near a brick boundary for several interpolation schemes. For a multi-stage integration method, not only the initial particle location but also all intermediate stages of the integrator must lie within the admissible area. This can be guaranteed by limiting the maximum integration step size Δt appropriately: When the distance of the particle to the boundary of the admissible region in dimension i is b_i, then Δt must be limited to $\min_i(b_i/v_{i,\max})$, where $v_{i,\max}$ is the maximum absolute value of the i'th velocity component in the region. This value is computed in the preprocess and stored along with the brick data.

3.3   Heuristic Brick Selection and Paging

Since the full set of bricks required to trace a given set of particles cannot be stored in GPU memory, subsets of these bricks have to be paged in and out, and processed sequentially in a number of rounds. However, it can always happen that a brick which has been paged out is later visited by some particle and has to be paged in again. Thus, an appropriate paging strategy is required to reduce the number of bricks which are uploaded multiple times.

Besides multiple uploads of the same brick, the paging strategy also has to take into account the number of particles which can be advected using the currently available bricks. The massively parallel nature of GPUs can only be exploited to its full potential when many particles can be processed in parallel.

These two requirements, however, contradict each other: According to the first requirement, a brick should be kept in GPU memory as long as possible to avoid multiple uploads. Conversely, the second requirement demands that a brick through which few or no particles are moving should be paged out immediately, so that bricks required by a large number of particles can be paged in.

Since it is not known in advance which bricks are to be visited at which times, it is not possible to devise an optimal paging strategy. Instead, we have devised a simple yet very effective heuristic paging strategy which attempts to balance the two goals. It is based on the following observations:

- Paging out a brick which is currently required by some particles will always result in a repeated upload of this brick later on.
- A brick which is not required by any particle might be visited again later on, so it might still be beneficial to keep it on the GPU. However, since keeping such bricks at least currently wastes GPU memory, a balance must be found between minimizing premature swap-outs and maximizing GPU occupancy.
- Processing spatially close bricks at the same time tends to improve brick re-use and thus helps to avoid repeated uploads. It also increases the chances of particles moving between available bricks within one round, which increases the average number of active particles at any time and thus improves GPU utilization.

Based on these observations, our paging strategy operates as follows. Note that paging out a brick does not entail any data transfer, but simply means clearing the corresponding entry in the *index* buffer.

- We only ever page out bricks which are not required by any particle.
- A brick which is not required by any particle is kept on the GPU for a fixed number of rounds, r, before it is paged out. We have found that paging out such bricks after $r = 4$ rounds works well, with each round limited to a maximum of 64 integration steps per particle. Increasing the value of r could only reduce the number of brick uploads by up to 10 %, while increasing the particle advection time by an order of magnitude due to the less efficient GPU usage. Smaller values, on the other hand, significantly increased the number of brick uploads.

Whenever a slot in the brick atlas is available, the next brick to upload is selected according to the following priorities:

- Only bricks required by at least one particle are considered in order to avoid spurious uploads.
- Bricks which are required by a large number of particles are preferred.
- Bricks are favored or disfavored based on the availability of their neighbors in GPU memory, as well as the numbers of particles they contain. Particles expected to travel between the brick under consideration and an available neighbor result in a priority bonus in order to enhance locality and data re-use. Particles traveling in from an unavailable neighbor result in a penalty—in this case, it would be better to process the neighbor first, so these particles can coalesce with those in the current brick. Finally, particles traveling out into an unavailable neighbor carry neither bonus nor penalty.

Taking these rules into account, a heuristic load parameter, l, is computed for each brick b as follows:

$$l = \sqrt{c_b} + h \cdot \sum_{n \in N(b)} \begin{cases} \sqrt{c_b \cdot p_{b,n}} + \sqrt{c_n \cdot p_{n,b}}, & n \text{ on GPU} \\ -\sqrt{c_n \cdot p_{n,b}}, & \text{else} \end{cases}$$

This parameter is then used to assign priorities to the requested bricks. Here, c_b is the number of particles in brick b, and $N(b)$ is the set of neighbors of b. $p_{a,b}$ is the probability of a particle from brick a traveling into brick b (note that in general $p_{a,b} \neq p_{b,a}$). These probabilities can be precomputed by tracing a number of particles within each brick and storing in which direction they leave the brick, i.e. a flow graph (cf. [23]). Computing such a flow graph for a 1024^3 flow field takes around 1 min in our system (excluding disk I/O). However, we have found that when a flow graph is not available, simply substituting a constant value for the probabilities works surprisingly well, increasing the total computation time by less than 10 % is most cases.

The user-defined parameter $h \geq 0$ is used to weight the neighborhood-based bonus and penalty terms. Larger values correspond to a larger preference for locality. Figure 3 shows the time required to trace 4096 stream lines in two data sets for different values of h. Choosing an appropriate value for h reduces the total tracing time by about 30–40 % in both data sets. In the following experiments, we always use a value of $h = 10$.

Bricks are loaded from disk into CPU memory using the same priorities. However, bricks which currently do not contain any active particles are also pre-fetched into the cache if the disk would otherwise be idle.

3.4 Unsteady Flow

So far, we have addressed only the case of steady flow, i.e. stream line computation. Path lines are computed in much the same way, with some straightforward extensions to account for the time-dependent nature of the flow. Each slot in

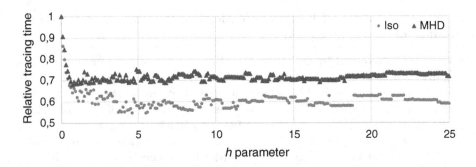

Fig. 3. Relative times for tracing 4096 stream lines in two data sets (including CPU-GPU data transfer and GPU advection), depending on the heuristic parameter h. The absolute times at $s = 0$ are 27.2 and 23.3 s for Iso and MHD, respectively.

the brick atlas now contains multiple time steps of the same spatial brick; the number of slots is reduced accordingly. The *index* records which time steps are currently available. Similarly, the *requests* buffer holds not only the number of requests to a spatial brick, but also the earliest time step that was requested. The CPU also tracks the earliest time step which was requested globally and pages out all brick time steps older than that, since they will not be visited again by the current particles. Finally, during the selection of bricks to upload, priority is given to older time steps, so that all path lines advance at roughly the same speed, and multiple uploads of the same data are avoided.

3.5 Interpolation Schemes

Our system supports a number of different interpolation schemes which are used in numerical particle integration. The simplest one is linear interpolation, which comes "for free" on the GPU. Lagrange interpolation of order n fits a polynomial of degree $n - 1$ through the n grid points centered around the interpolation point. Second-order Lagrange interpolation is thus equivalent to linear interpolation. We have implemented 4th, 6th, and 8th order Lagrange interpolation (called Lagrange4/6/8), corresponding to cubic, quintic, and septic polynomials, respectively. It is worth noting here that Lagrange6 can be considered one of the standard methods in turbulence research. The application of higher-order schemes can hardly be found in the literature. Additionally, we have implemented an interpolation scheme based on Catmull-Rom splines. This approach fits a cubic polynomial to the values and first derivatives, estimated via central differences, at two grid points. Compared to 4th order Lagrange interpolation, this has the advantage of creating a globally C1-continuous interpolant, while Lagrange4 is only C0. All described interpolation schemes are extended to multiple dimensions by a tensor product approach.

The minimum size required by the halo around each brick depends on the size of the chosen interpolation kernel, e.g. 2 voxels for Lagrange4 and 4 voxels for Lagrange8 (cf. Fig. 2). Thus, a selected halo size of 4 in our approach allows for any interpolation scheme with a support of up to 8^3 voxels. If only lower-order interpolation is required, a smaller halo size can be used which slightly reduces memory and bandwidth requirements.

4   Turbulent Vector Field Compression

In the absence of data compression, the performance of the proposed system for particle tracing in large turbulence fields is vastly restricted by bandwidth limitations when reading the data from disk. For instance, the computation of stream lines as shown in Fig. 4 in one single *uncompressed* time step involves a working set of almost 5 GB. The visualization takes roughly 45 s on our target architecture, of which over 98 % are spent waiting for data from disk. Thus, there is a dire need for compression in order to reduce the amount of data to be streamed.

Fig. 4. Stream lines in (*uncompressed*) MHD (1024^3).

When using data compression, the introduced compression error has to be carefully examined. Since no error is introduced by lossless compression schemes, they might be an attractive choice in particle tracing applications. However, for floating-point data the achieved compression rate is usually quite modest.

For instance, the lossless schemes proposed in [26, 27] can only compress the turbulence data to roughly $\frac{2}{3}$ of its original size. They achieve a decoding throughput of about 10 million floating point values per second, corresponding to over 600 ms for the decompression of a single 128^3 grid of 3D velocities. More sophisticated prediction schemes can slightly improve the compression rate [28], but they come at the expense of lower throughput.

In comparison, the lossy GPU compression scheme proposed in [17] provides high compression rate and decompression throughput. Even for vector data it was demonstrated that the throughput of a GPU decoder is significantly above disk speed [1]. The scheme is based on the discrete wavelet transform, followed by a quantization of wavelet coefficients and a final entropy coding of quantized coefficients. The process achieves a decoding throughput of over 650 million floating-point values per second, at 3 bits per velocity vector and a signal to noise ratio above 45 dB.

4.1   Interpolation Error Estimate

When a lossy scheme for vector field compression is used, it is clear that the reconstructed field is afflicted with some error compared to the initial field. At first, this seems to preclude lossy compression schemes in particle tracing, because the local reconstruction errors accumulate along the particle trajectories. On the other hand, this error has to be seen in relation to the error that is inherent to particle trajectories even when computed in the original data.

Even without compression the reconstructed samples are not exact in general, due to the interpolation which is used to reconstruct the data values from the initially given discrete set of samples. This interpolation makes assumptions on the continuous field which, in general, do not hold. As a consequence, it has to be accepted that the trajectories we compute numerically using interpolation diverge from those we would see in reality, even without compression.

It therefore makes sense to choose a compression quality so that the additional error introduced by the compression scheme is in the order of the error introduced by interpolation. It is worth noting, however, that without additional information about a data set it is impossible to accurately compute or even estimate the interpolation error. In some cases, theoretical error bounds depending on higher-order derivatives of the continuous function can be given; see, for instance, [29] for such a bound when linear interpolation is used. On the other hand, the derivatives of the continuous function are typically not known exactly. In that case, such bounds themselves come with some uncertainty. In addition, even with exact knowledge of the derivatives, they often overestimate the actual error significantly [30].

Therefore, we have adopted a different approach to estimate the interpolation error: We take the difference between interpolation results from a reference high-order interpolator and a lower-order interpolator as an estimate for the error in the low-order interpolator. For the two discrete turbulence data sets we analyze in this work (Iso and MHD), an exact interpolator is known. Due to the pseudo-spectral method that was used to simulate the turbulent motion [2], the velocity

Fig. 5. Path lines in the Iso and stream lines in the MHD data sets. Left: Particle trajectories using (ground truth) trigonometric (blue) and (established) Lagrange6 (red) interpolation in the original data for velocity sampling. Right: Particle trajectories using trigonometric interpolation (blue) and Lagrange6 interpolation (yellow) in the compressed data. The yellow traces appear to be of similar accuracy as the red lines (Color figure online).

field is guaranteed to be band-limited in the Fourier sense. As a consequence, Fourier or trigonometric interpolation using trigonometric polynomials of infinite support gives exact velocity values between grid points [9]. Due to efficiency reasons, however, what is used in practice for particle tracing is an interpolation scheme of "sufficiently high order" which resembles Fourier interpolation, e.g. Lagrange6. For instance, in Fig. 5 (left) the trajectories using trigonometric and Lagrange6 interpolation are compared. It is worth noting that even though in the turbulence community it is usually agreed that Lagrange6 is of sufficient accuracy for particle tracing, significant deviations from the ground truth can be observed.

The interpolation error over the whole volume for a given interpolator can now be computed: We upsample the volume to four times the original resolution using the interpolator under consideration as well as the reference interpolator. The RMS of the difference between the upsampled volumes then is a good approximation of the average error introduced by the interpolation. Since trigonometric interpolation has to be evaluated globally, to generate the interpolant in a computationally efficient way, we have adopted the following approach: First, we perform a fast Fourier transform (FFT) on the flow field using the FFTW library [31]. In the frequency domain, we then quadruple the data resolution in each dimension by zero padding. Finally, an inverse FFT is performed to generate a flow field of four times the original resolution. This field agrees with the original field at every fourth vertex, and the other vertices lie on the trigonometric interpolant between the original data samples. Figure 6 illustrates FFT-based upsampling in 1D, but only doubling the data resolution. Generating the 4096^3 trigonometric interpolant from a 1024^3 velocity field in this way takes about 1.5 h including disk I/O. Given the 4096^3 trigonometric interpolant, evaluating the interpolation errors in a 1024^3 velocity field for the listed interpolation schemes takes another 2 h including disk I/O.

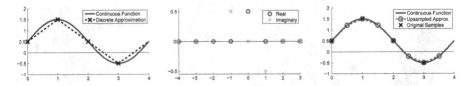

Fig. 6. FFT-based upsampling process. Left: A periodic band-limited function with a period of 4, and its discrete approximation sampled at a frequency of 1. Middle: FFT coefficients of the function in magenta and cyan. Because the input function was real, the coefficients have a Hermitian symmetry. The coefficients are padded with zeros to the left and right, corresponding to higher frequencies with an amplitude of zero. Right: The inverse FFT of the padded coefficients results in a higher-resolution approximation to the continuous function. Note that the even grid points of the upsampled approximation agree with the original grid points.

4.2   Error-Guided Data Compression

Equipped with the average interpolation error for a chosen interpolation scheme, we can choose a quantization step in the data compression scheme so that the compression error is equal to or falls below the interpolation error. In the wavelet-based compression scheme we use, the average error is roughly equal in magnitude to the quantization step and, thus, the acceptable error is a reasonable choice for the quantization step. Table 1 lists the RMS interpolation errors in both data sets for a number of different interpolation schemes. To verify that the lossy compression does not unduly affect the interpolant, we have computed the interpolation errors a second time after compression, comparing the reconstructed volumes to the original reference solution. It can be seen that by setting the quantization step equal to the RMS interpolation error, the error is increased by less than 50 % in all cases. It is worth mentioning that performing the same test with an upsampling factor of only 2 instead of 4 yields almost identical results (within 5 % of the listed numbers). This indicates that the discrete computation approximates the actual interpolation error very closely. This is expected, as the reference interpolant is by definition band-limited with respect to the original resolution, so no high-frequency deflections can occur between the original grid points.

It remains to show that the accumulation of the additional quantization errors does not introduce significantly larger regions of variation in the trajectories. A first experiment can be seen in Fig. 5 (right), where the trajectories computed on the compressed field using Lagrange6 interpolation are compared to the ground truth trajectories. Compared to Fig. 5 (left), the deviations seem to be in the same order of variation. A detailed quantitative accuracy analysis is given in the following section.

5   Evaluation

To evaluate the performance of our system as well as the accuracy of the resulting trajectories, we have conducted a number of experiments. In the first set of experiments we analyze the accuracy of trajectories when the proposed lossy compression scheme is used. In the scope of a second set of experiments we evaluate the performance of our system. In the following, we first introduce the error metrics we use to analyze the accuracy of the computed trajectories.

5.1   Error Metrics

Due to errors induced by the employed interpolation scheme and by lossy compression, a trajectory may gradually diverge from the ground truth over time. To evaluate the accuracy of computed trajectories, an error metric is required to quantitatively measure the difference between two trajectories starting at the same seed point.

One obvious metric is the maximum or average distance between trajectories $s_0(u)$, $s_1(u)$ along their parameter u. In addition, several metrics exist which measure some kind of distance between two curves, such as the (discrete) Fréchet distance [32] and the distance under dynamic time warping (DTW). While the Fréchet distance corresponds to a type of maximum distance, the DTW distance is akin to an average distance. Both disregard the u parametrization and instead are concerned only with the shape of the curves. All these metrics measure the distance along the complete trajectories. However, once two particles have diverged by some critical distance, their further behavior depends only on the characteristics of the flow field: They might diverge further or even converge again, but this provides no insight into the accuracy of the trajectory computation. Therefore, we introduce a new metric taking this into account, which we call the (clamped) divergence rate. Instead of measuring a distance between trajectories, it computes the rate at which they diverge. Given two trajectories $s_0(u)$, $s_1(u)$ over a parameter interval $[u_0, u_{\max}]$, we define their divergence rate as

Table 1. Root-mean-square interpolation errors for different interpolation schemes in two turbulent flow fields before (*orig*) and after compression (*comp*), compared to the reference trigonometric interpolant. The interpolation error has been evaluated in a grid of four times the original resolution.

Interpolation	Iso (range: 6.67)		MHD (range: 2.77)	
	orig	comp	orig	comp
Lagrange8	0.86E-3	1.26E-3	3.48E-4	4.97E-4
Lagrange6	1.10E-3	1.60E-3	4.52E-4	6.32E-4
Lagrange4	1.71E-3	2.41E-3	7.20E-4	9.63E-4
Linear	5.15E-3	6.65E-3	2.29E-3	2.81E-3

$$d_{s_0,s_1} := \frac{\text{dist}(u_{\text{div}})}{u_{\text{div}} - u_0}, \text{ where}$$

$$\text{dist}(u) := \|s_0(u) - s_1(u)\| \text{ and}$$

$$u_{\text{div}} := \max\left\{u \in [u_0, u_{\text{max}}] \mid \forall \tilde{u} \in [u_0, u] : \text{dist}(\tilde{u}) \leq \Delta s\right\}.$$

u_{div} is the last point along the trajectories at which they have not yet diverged by more than Δs. In the following experiments, we have set the critical distance Δs equal to the grid spacing.

Our definition of the trajectory divergence rate is similar in spirit to the idea of the finite-size Lyapunov exponent (FSLE) [33]. The FSLE measures the time it takes for two particles, initially separated only by an infinitesimal ϵ, to diverge by some given distance, usually specified as a multiple of ϵ. A fundamental difference is that in our case both trajectories start at exactly the same position, and we measure their divergence as an absolute distance rather than relative to their initial separation.

5.2 Accuracy Analysis

To compare the accuracy of particle trajectories computed in the original and compressed data sets, and via different interpolation schemes, a reference solution is required to which the trajectories can be compared. For the used turbulence data sets, trigonometric interpolation is known to be exact. Since evaluating the trigonometric interpolant during particle tracing is impracticable, we have upsampled the data sets to four times the original resolution (see Sect. 4.1) as the ground truth. Particle trajectories traced in the upsampled versions using 16th order Lagrange interpolation then act as the reference solution. While this is not equivalent to true trigonometric interpolation in the original data, the remaining error is expected to be negligible since the difference between the two times and four times upsampled versions is already very small (cf. Sect. 4.1).

For the accuracy analysis of computed trajectories, we have generated a set of 4096 seed points in each data set, randomly distributed over the entire

Table 2. File sizes and compression factors. For *Very Low*, *Low*, *Medium*, and *High*, the quantization step was chosen equal to the error in linear and Lagrange4/6/8 interpolation, resp. (cf. Table 1); for *Very High*, to half the error in Lagrange8.

Quality	Iso		MHD	
	size	factor	size	factor
Uncompressed	14.7 GB	–	14.7 GB	–
Very High	1.79 GB	8.21	1.55 GB	9.48
High	1.25 GB	11.8	1.06 GB	13.9
Medium	1.08 GB	13.6	942 MB	16.0
Low	843 MB	17.9	712 MB	21.1
Very Low	387 MB	38.9	331 MB	45.5

domain. Particles were traced from the seed points through different versions of the data sets: The upsampled reference version, the original uncompressed data, and compressed versions at different compression rates. The quantization steps for the compressed versions were chosen equal to the errors in linear and Lagrange4/6/8 interpolation as listed in Table 1. In addition, we generated one high-quality compressed version of each data set, where the quantization step was set to half the Lagrange8 interpolation error. The compressed file sizes and compression rates (as the ratio of original size to compressed size) are listed in Table 2.

To minimize the impact of inaccuracies due to numerical integration errors, in all of our experiments we used the Runge-Kutta method by [34]. The method provides a 5th order solution and a 4th order error estimate, which is used for adaptive step size control. The error tolerance for step size control was reduced until the accuracy of the results did not improve any further.

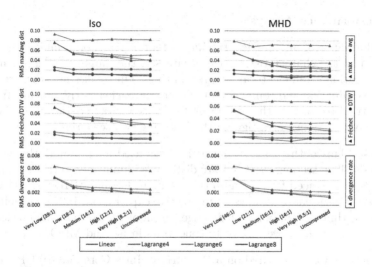

Fig. 7. Accuracy of stream lines vs. compression quality using different interpolation schemes. Accuracy is reported as the root mean square (RMS) of the individual trajectory distances (see Sect. 5.1), computed against trajectories traced using the (approximate) trigonometric reference interpolation.

Figure 7 provides the main results of our accuracy analysis. The graphs show the root mean square (RMS) of the average, maximum, Fréchet, and DTW distance, as well as the divergence rate, over all trajectories for different compression rates and interpolation schemes. For reference, the grid spacing is approximately 0.00614 in both data sets. The most prominent finding is that linear interpolation performs very poorly and eclipses the errors introduced at even the highest compression rates. The differences between the other interpolation schemes are comparatively small; as expected, with some advantage of the higher-order schemes. All distance metrics give qualitatively similar results. However, all metrics except for our novel divergence rate display a significant amount of noise in the results. This is caused by a few individual trajectories with a very large distance to their reference. These trajectories have a very large impact on the RMS distance, but actually carry little information on the accuracy of the results, as explained in Sect. 5.1. The divergence rate, on the other hand, handles such trajectories well.

The most important observation with regard to the lossy compression is that when the quantization step is chosen smaller than the interpolation error (e.g. Lagrange4 interpolation and a compression quality of "Medium" or higher), the additional error introduced by the compression is extremely small. For example, switching from Lagrange6 to Lagrange4 interpolation has a larger impact on the accuracy than switching from uncompressed data to the "High" compression quality in both data sets.

5.3   Performance Analysis

The performance of any particle tracing system depends on a multitude of factors, such as the characteristics of the data set, the number and placement of seeding locations, and the total integration time. This makes an exhaustive performance evaluation and comparison to other approaches fairly difficult. Instead, we have tried to capture the typical performance characteristics of our system. For both data sets, we investigated the following two scenarios:

1. **Sparse:** This is the same scenario that was used for pursuing the accuracy analysis. 4096 seeding locations are distributed uniformly in the domain, and particles are traced for 2.5 and 5 time units in Iso and MHD, respectively.
2. **Dense:** This scenario models an interactive exploration. 1024 seed points are placed within a small box with an edge length of 10 % of the domain size. The particles are traced over 5 and 10 time units in Iso and MHD, respectively.

All timings were measured on a PC with an Intel Core i5-3570 CPU (quad-core, 3.4 GHz) with 8 GB of DDR3-1600 RAM, equipped with an NVIDIA GeForce GTX 680 GPU with 4 GB of video memory. The size of the brick atlas was set to 64 bricks of size 128^3 each, corresponding to 2 GB of video memory (because CUDA does not support 3-channel textures, each velocity value had to be padded by an additional "w" component).

We have traced particles starting from the selected seed points in both the uncompressed and the compressed data sets to demonstrate the performance gains that can be achieved via compression. In all experiments, Lagrange6 interpolation was performed; the particle integration times are about 3× higher with Lagrange8, and about 4× lower with Lagrange4 or Catmull-Rom interpolation. When particle tracing was performed on the compressed data, the timings refer to the "High" compression quality. The decompression times for other compression rates differ only very slightly. To measure the impact of disk I/O, we have run every benchmark a second time, so that all required data was already cached in CPU memory. With uncompressed data, however, this was only possible for the dense scenario; in case of sparse particle seeding, the size of the working set exceeded the available CPU memory (even for stream lines extracted from a single time step). Table 3 lists the time required for running each scenario, and Table 4 lists the sizes of the corresponding working sets.

It can be seen that the use of compression allows us to trace thousands of characteristic lines within seconds in the dense seeding scenario. In the sparse seeding case, the required time is around an order of magnitude higher. The reason becomes clear when looking at the size of the working sets, which are larger by roughly the same factor in those cases.

Table 3. Times in seconds for computing stream lines, both for the cached case (C) and the un-cached case including disk access times (U). Individual times for uploading the data to the GPU (Upl, including decompression), particle integration (Int), and disk I/O (IO, overlapping Upl and Int) are listed separately.

Scenario	Quality	U	C	Upl	Int	IO
Iso dense	High	2.3	1.4	0.8	0.6	1.9
	Uncomp	18.6	1.3	0.7	0.6	17.8
Iso sparse	High	21.6	16.4	12.4	3.8	14.9
	Uncomp	156.9	n/a	10.7	3.8	156.4
MHD dense	High	3.4	2.3	1.4	0.8	2.8
	Uncomp	26.3	2.3	1.4	0.8	25.6
MHD sparse	High	19.9	16.2	11.8	4.2	13.6
	Uncomp	139.7	n/a	10.4	4.2	138.8

Table 4. Working set sizes in both compressed (*High*) and uncompressed (*Uncomp*) form. Also shown is the number of bricks in the working set ($\#B$) as well as the number of brick uploads during particle integration ($\#U$).

Scenario	High	Uncomp	#B	#U
Iso dense	165.9 MB	2155.5 MB	92	92
Iso sparse	1280.6 MB	15058.9 MB	728	1341
MHD dense	243.8 MB	3231.0 MB	138	155
MHD sparse	1095.5 MB	15066.5 MB	729	1298

Without compression, the overall system performance is clearly limited by disk bandwidth. In particular, in the sparse scenario, the working set was so much larger than main memory (cf. Table 4) that some bricks had to be loaded from disk multiple times. Even when all required data is already cached in CPU memory (which was only possible in the dense scenario), the performance of the compressed and uncompressed cases is very similar—the runtime overhead caused by the additional decompression step is very minor.

It is clear that when tracing path lines, the working sets are much larger because often many different time steps of the same spatial brick are required. In particular, the temporal distance between successive time steps is extremely small in both data sets (0.002 time units for Iso, 0.0025 for MHD). Because of this, the time required for path line computation is spent almost exclusively on disk-to-CPU data transfer and GPU decompression, and only a negligible amount of time is spent on the actual particle integration (less than 1 % in our tests). For example, tracing a set of path lines with the dense seeding configuration through Iso takes about 6 min, with a working set size of over 25 GB of compressed data. In the uncompressed setting, the working set comprises over 300 GB. Correspondingly, tracing these path lines in the uncompressed data set takes almost an hour, and most of that time is spent on disk I/O. In MHD, the time required for path line tracing is similar; in all cases, the time scales proportionally to the working set size.

5.4   Comparison to Previous Work

To the best of our knowledge, all previous techniques for particle tracing in very large flow fields have employed large compute clusters. Pugmire et al. [21] have used 512 CPUs to trace 10 K stream lines in two steady flow fields comprising 512 million grid cells each. They report wall times of 10 to 100 s. [22] later improved those timings to a few seconds for tracing thousands of stream lines on 128 cores. Nouanesengsy et al. [23] achieve timings between 10 and 100 s using 4096 cores for the computation of 256 K stream lines in regular grids of up to 1.67 billion grid points, but at the cost of an expensive preprocess. Peterka et al. [24] report computation times of about 20 s using 8192 cores for 128 K stream lines in a 1024^3 steady flow, and several minutes for 32 K lines in a $2304 \times 4096 \times 4096$ steady flow. In contrast to all other mentioned approaches, they have also addressed large *unsteady* flow fields. In a $1408 \times 1080 \times 1100 \times 32$ unsteady flow, the processing time is several minutes for 16 K path lines on 4096 cores.

While an exact performance comparison to our technique is not possible due to the different data sets and interpolation/integration schemes used, an order-of-magnitude comparison reveals that our method achieves competitive timings to the previous approaches in many cases, particularly in dense seeding scenarios, while making use of only a single desktop PC.

All in all it can be said that due to the use of an effective compression scheme, the performance of particle tracing in extremely large flow fields can be improved significantly. It is clear that due to the immense working set that is required when computing path lines, fully interactive rates cannot be achieved in this case. However, a simple preview mode which shows the already-computed parts of the current trajectories enables the interactive exploration of very large flow fields. For example, the preview allows the user to quickly discard trajectories originating from "uninteresting" seed points, instead guiding the process interactively towards more interesting regions.

6 Conclusion

In this paper we have presented an out-of-core system for particle tracing in very large and time-dependent flow fields. It does not require a high-performance computing architecture, but runs entirely on a desktop PC. Thus, the system can be used on demand by a turbulence researcher to explore data sets and validate hypotheses. We have employed lossy data compression to overcome bandwidth limitations due to the extreme data volumes that have to be processed.

In a number of experiments we have demonstrated that compared to interpolation errors in the reconstruction of the velocity field, the compression errors do not significantly affect the accuracy of the computed trajectories. In the statistical sense, the quality of the computed trajectories remains in the same order. A performance analysis indicates that such a system achieves a throughput that is comparable to that of previous systems running on high-performance architectures.

The most challenging future avenue of research will be the investigation of the effect of lossy data compression in scenarios other than turbulence research. The question will be whether lossy data compression can also be applied to other flow fields without unduly affecting the accuracy of the resulting trajectories. The main difficulty is that for most flow fields a "correct" interpolation scheme is not available, so the interpolation error can not be estimated accurately. However, different criteria might be found to steer the compression quality, e.g. given confidence intervals for the velocity values.

Acknowledgements. The work was partly funded by the European Union under the ERC Advanced Grant 291372: Safer-Vis - Uncertainty Visualization for Reliable Data Discovery. The authors want to thank Charles Meneveau from Johns Hopkins University and Tobias Pfaffelmoser from TUM for helpful discussions and constructive criticism.

References

1. Treib, M., Bürger, K., Reichl, F., Meneveau, C., Szalay, A., Westermann, R.: Turbulence visualization at the terascale on desktop PCs. IEEE Trans. Vis. Comput. Graphics **18**, 2169–2177 (2012)
2. Li, Y., Perlman, E., Wan, M., Yang, Y., Meneveau, C., Burns, R., Chen, S., Szalay, A., Eyink, G.: A public turbulence database cluster and applications to study Lagrangian evolution of velocity increments in turbulence. J. Turbul. **9**, N31 (2008)
3. Post, F.H., Vrolijk, B., Hauser, H., Laramee, R.S., Doleisch, H.: The state of the art in flow visualisation: feature extraction and tracking. Comput. Graph. Forum **22**, 775–792 (2003)

4. Laramee, R.S., Hauser, H., Doleisch, H., Vrolijk, B., Post, F.H., Weiskopf, D.: The state of the art in flow visualization: dense and texture-based techniques. Comput. Graph. Forum **23**, 203–221 (2004)
5. McLoughlin, T., Laramee, R.S., Peikert, R., Post, F.H., Chen, M.: Over two decades of integration-based, geometric flow visualization. Comput. Graph. Forum **29**, 1807–1829 (2010)
6. Teitzel, C., Grosso, R., Ertl, T.: Efficient and reliable integration methods for particle tracing in unsteady flows on discrete meshes. In: Lefer, W., Grave, M. (eds.) Visualization in Scientific Computing 1997. Eurographics, pp. 31–41. Springer, Vienna (1997)
7. Yeung, P.K., Pope, S.B.: An algorithm for tracking fluid particles in numerical simulations of homogeneous turbulence. J. Comput. Phys. **79**, 373–416 (1988)
8. Balachandar, S., Maxey, M.R.: Methods for evaluating fluid velocities in spectral simulations of turbulence. J. Comput. Phys. **83**, 96–125 (1989)
9. Rovelstad, A.L., Handler, R.A., Bernard, P.S.: The effect of interpolation errors on the Lagrangian analysis of simulated turbulent channel flow. J. Comput. Phys. **110**, 190–195 (1994)
10. Schirski, M., Gerndt, A., van Reimersdahl, T., Kuhlen, T., Adomeit, P., Lang, O., Pischinger, S., Bischof, C.H.: ViSTA FlowLib: a framework for interactive visualization and exploration of unsteady flows in virtual environments. In: 9th Eurographics Workshop on Virtual Enviroments, pp. 77–86 (2003)
11. Krüger, J., Kipfer, P., Kondratieva, P., Westermann, R.: A particle system for interactive visualization of 3D flows. IEEE Trans. Vis. Comput. Graph. **11**, 744–756 (2005)
12. Schirski, M., Bischof, C., Kuhlen, T.: Interactive particle tracing on tetrahedral grids using the GPU. In: Proceedings of Vision, Modeling, and Visualization (VMV), pp. 153–160 (2006)
13. Bürger, K., Schneider, J., Kondratieva, P., Krüger, J., Westermann, R.: Interactive visual exploration of unsteady 3D flows. In: Eurographics/IEEE VGTC Visualization (EuroVis) (2007)
14. Murray, L.: GPU acceleration of Runge-Kutta integrators. IEEE Trans. Parallel Distrib. Syst. **23**, 94–101 (2011)
15. Sayood, K.: Introduction to Data Compression, 3rd edn. Morgan Kaufmann Publishers Inc., San Francisco (2005)
16. Rodriguez, M.B., Gobbetti, E., Guitián, J.I., Makhinya, M., Marton, F., Pajarola, R., Suter, S.: A survey of compressed GPU-based direct volume rendering. In: Eurographics 2013 - STARs, pp. 117–136 (2013)
17. Treib, M., Reichl, F., Auer, S., Westermann, R.: Interactive editing of gigasample terrain fields. Comput. Graph. Forum **31**, 383–392 (2012)
18. Lane, D.A.: UFAT–a particle tracer for time-dependent flow fields. In: IEEE Visualization, pp. 257–264 (1994)
19. Bruckschen, R., Kuester, F., Hamann, B., Joy, K.I.: Real-time out-of-core visualization of particle traces. In: Proceedings of IEEE 2001 Symposium on Parallel and Large-Data Visualization and Graphics, pp. 45–50 (2001)
20. Ellsworth, D., Green, B., Moran, P.: Interactive terascale particle visualization. In: IEEE Visualization, pp. 353–360 (2004)
21. Pugmire, D., Childs, H., Garth, C., Ahern, S., Weber, G.H.: Scalable computation of streamlines on very large datasets. In: Proceedings of the Conference on High Performance Computing, Networking, Storage and Analysis, pp. 16: 1–16: 12 (2009)

22. Camp, D., Garth, C., Childs, H., Pugmire, D., Joy, K.: Streamline integration using MPI-hybrid parallelism on a large multicore architecture. IEEE Trans. Vis. Comput. Graph. **17**, 1702–1713 (2011)
23. Nouanesengsy, B., Lee, T.Y., Shen, H.W.: Load-balanced parallel streamline generation on large scale vector fields. IEEE Trans. Vis. Comput. Graph. **17**, 1785–1794 (2011)
24. Peterka, T., Ross, R., Nouanesengsy, B., Lee, T.Y., Shen, H.W., Kendall, W., Huang, J.: A study of parallel particle tracing for steady-state and time-varying flow fields. In: Parallel Distributed Processing Symposium (IPDPS), pp. 580–591 (2011)
25. Yu, H., Wang, C., Ma, K.L.: Parallel hierarchical visualization of large time-varying 3D vector fields. In: Proceedings of ACM/IEEE Conference on Supercomputing, pp. 24:1–24:12 (2007)
26. Isenburg, M., Lindstrom, P., Snoeyink, J.: Lossless compression of predicted floating-point geometry. Comput. Aided Des. **37**, 869–877 (2005)
27. Lindstrom, P., Isenburg, M.: Fast and efficient compression of floating-point data. IEEE Trans. Vis. Comput. Graph. **12**, 1245–1250 (2006)
28. Fout, N., Ma, K.L.: An adaptive prediction-based approach to lossless compression of floating-point volume data. IEEE Trans. Vis. Comput. Graph. **18**, 2295–2304 (2012)
29. Fout, N., Ma, K.L.: Fuzzy volume rendering. IEEE Trans. Vis. Comput. Graph. **18**, 2335–2344 (2013)
30. Zheng, Z., Xu, W., Mueller, K.: VDVR: Verifiable volume visualization of projection-based data. IEEE Trans. Vis. Comput. Graph. **16**, 1515–1524 (2010)
31. Frigo, M., Johnson, S.G.: The design and implementation of FFTW3. Proc. IEEE **93**, 216–231 (2005)
32. Eiter, T., Mannila, H.: Computing discrete Frèchet distance. Technical report CD-TR 94/64, Technische Universität Wien (1994)
33. Aurell, E., Boffetta, G., Crisanti, A., Paladin, G., Vulpiani, A.: Predictability in the large: an extension of the concept of Lyapunov exponent. J. Phys. A: Math. Gen. **30**, 1–26 (1997)
34. Dormand, J.R., Prince, P.J.: A family of embedded Runge-Kutta formulae. J. Comput. Appl. Math. **6**, 19–26 (1980)

Choosing Visualization Techniques for Multidimensional Data Projection Tasks: A Guideline with Examples

Ronak Etemadpour[1]([✉]), Lars Linsen[2], Jose Gustavo Paiva[3],
Christopher Crick[1], and Angus Graeme Forbes[4]

[1] Oklahoma State University, Stillwater, OK, USA
etemadp@okstate.edu
[2] Jacobs University, Bremen, Germany
[3] Federal University of Uberlandia, Uberlandia, MG, Brazil
[4] University of Illinois at Chicago, Chicago, IL, USA

Abstract. This paper presents a guideline for visualization designers who want to choose appropriate techniques for enhancing tasks involving multidimensional projection. Specifically, we adopt a user-centric approach in which we take user perception into consideration. Here, we focus on projection techniques that output 2D or 3D scatterplots that can then be used for a range of common data analysis tasks, which we categorize as pattern identification tasks, relation-seeking tasks, membership disambiguation tasks, or behavior comparison tasks. Our user-centric task categorization can be used to effectively guide the organization of multidimensional data projection layouts. Moreover, we present real-world examples that demonstrate effective choices made by visualization designers faced with complex datasets requiring dimensionality reduction.

Keywords: Multidimensional data analysis · Task taxonomy · Multi-dimensional data projection · User-centric evaluation

1 Introduction

Visualization is a crucial step in the process of data analysis. Often, when analyzing multidimensional data, dimensionality reduction (DR) techniques are displayed in form of 2D or 3D scatterplots that project the multidimensional points onto a lower-dimensional visual space. Methods using different algorithms to generate scatterplots with particular point placements are the most common visual encoding (VE) techniques for the resulting lower-dimensional data. DR techniques, coupled with appropriate VEs, enable an understanding of the relations that exist within the higher-dimensional data by displaying them in such a way that makes it easier for users to discover meaningful patterns [36].

Data analysis tasks are primarily concerned with the detection of structures such as patterns, groups, and outliers. Within a multidimensional data set, data

© Springer International Publishing Switzerland 2016
J. Braz et al. (Eds.): VISIGRAPP 2015, CCIS 598, pp. 166–186, 2016.
DOI: 10.1007/978-3-319-29971-6_9

points can be grouped manually into classes or automatically into clusters. For example, classes may be defined through manually labeling a collection of documents so that each document belongs to one topic within a set of topics, or by splitting an image collection into ten classes by assigning each image a particular theme from a set of ten themes. Clusters, on the other hand, are generated automatically using a clustering algorithm that may, for instance, identify groupings of similar points, or partition the data into dissimilar groups where each cluster contains similar items [25]. However, it may be difficult to see these clusters or classes when projected onto a lower-dimensional space. To make sense of this multidimensional data, it can be useful to know how the clusters or classes are defined and structured in the original multidimensional attribute space. However, multidimensional projection mappings are especially prone to distortion because projection methods may not necessarily preserve the spatial relations of the data. Thus, it is important to know how effective the scatterplots are at preserving segregation of the data [42]. Several studies evaluate the quality of projections with respect to preserving certain properties, thus guiding a user to select the most appropriate projection method for their task. Various numerical and visual methods have been introduced to quantify the accuracy of projection methods with respect to such properties [42, 46]. Recent studies [41] have shown that the quality of cluster separation by these measures was highly discrepant with user assessment of the cluster separation within the same data sets. Lewis et al. [24] believe that accurate evaluation of clustering quality is essential for data analysts, and they showed that such clustering evaluation skills are present in the general population. On the other hand, other studies have attempted to find a perception-based quality measure for scatterplots. They either evaluated users' performance on layouts generated by different projection techniques [14] and used eye-tracking while users are asked to perform typical analysis tasks for projected multidimensional data or allowed users to assess a series of scatterplots [2]. Other studies have investigated the perception of correlation in scatterplots from a psychological perspective; however these studies did not consider real-world data sets [34].

Because of the absence of a standard approach for evaluating multidimensional data projection, the results of these studies, and others like them, are difficult to compare. We present a taxonomy of visual analysis tasks for multidimensional data projection that we believe could be a useful means for evaluation. The idea of creating a task taxonomy has been recently explored by Brehmer and Munzner [7]. They contribute a multi-level typology of visualization tasks that augments existing taxonomies by filling a gap between low-level and high-level tasks. Specifically, they distinguish what the task inputs and outputs are, as well as why and how a visualization task is performed. In doing so, they more thoroughly organize the motivations for and methods of specific tasks for particular data analysis situations. Their task taxonomy is more general, and does not address multidimensional data projection in any detail. In this paper, we provide a taxonomy of visual analysis tasks related to multidimensional data projection. Our task taxonomy enables evaluation designers to investigate visualization performance effectively on both synthetic and real-world data sets. The main contributions of the paper are:

- We provide a systematic user-centric taxonomy of visual tasks related to projected multidimensional data.
- We divide the projection-related tasks into different categories based on their impact on the analysis of multidimensional data. The categories we identify are relation-seeking, behavior comparison, membership disambiguation, and pattern identification tasks.
- We enable, via our task taxonomy, visualization designers to improve visualization tasks related to the analysis of multidimensional data.
- We present our taxonomy as a guideline for researchers in choosing visualization techniques for these tasks, and provide explicit examples.
- We adapt multilevel typology of abstract visualizations to multidimensional data projection tasks [7].

In the next section, we provide a brief review of existing task taxonomies for DR and VE techniques. In Sect. 3, we introduce our task taxonomy for multidimensional data projection by describing new sets of tasks related to typical analysis tasks, including *pattern identification*, such as detecting clusters, *behavior comparison*, such as comparing characteristics of subsets, *membership disambiguation*, such as counting the number of objects in a cluster, and *relation seeking*, such as correlating subsets to each other. We discuss the effects of our proposed tasks on the evaluation of scatterplots by providing some examples of how different tasks support decision making respective to human perception over multidimensional data projections. We also characterize our proposed tasks using the multi-level typology of abstract visualization tasks [7]. We applied Brehmer and Munzner's multi-level topology concept for describing two tasks as guidelines, while the three questions (WHY, WHAT, HOW) can be used to structure the description of all tasks.

2   Related Work

Many projection methods exist to generate 2D similarity-based layouts from a higher-dimensional space. The design goals include maintaining pairwise distances between points [6] as implemented in multidimensional scaling (MDS), maintaining distances within a cluster, or maintaining distances between clusters [47]. Principal component analysis (PCA) generates similarity layouts by reducing data to lower dimensional visual spaces [22]. Some projection methods, such as isometric feature mapping (Isomap), favor maintaining distances between clusters instead. Isomap is an MDS approach that has been introduced as an alternative to classical scaling capable of handling non-linear data sets. It replaces the original distances by geodesic distances computed on a graph to obtain a globally optimal solution to the distance preservation problem [47]. Least-Square Projection (LSP) computes an approximation of the coordinates of a set of projected points based on the coordinates of some samples as control points. This subset of points is representative of the data distribution in the input space. LSP projects them to the target space with a precise MDS force-placement technique. It then builds a linear system from information given by the projected points and their neighborhoods [31]. The correlations of data points or clusters are not always known after they have been mapped from a higher-dimensional data space to 2D or 3D display space. Thus, several approaches evaluate the best views of multidimensional data sets. Sips et al. [42] provide measures for ranking scatterplots with classified and unclassified data. They propose two additional quantitative measures on class consistency: one based on the distance to the cluster centroids, and another based on the entropies of the spatial distributions of classes. They propose class consistency as a measure for choosing good views of a class structure in high-dimensional space. Tan et al. [44], Paulovich et al. [31], and Geng et al. [18] also evaluate the quality of layouts numerically. By ranking the perceptual complexity of the scatterplots, other studies investigate user perception by conducting user studies on scatterplots, finding that certain arrangements were more pleasing to most users [45]. However, these operational measures were not necessarily equivalent to the measures of user preference based on their qualitative perceptions. Sedlmair et al. [40] have discussed the influence of factors such as scale, point distance, shape, and position within and between clusters in qualitative evaluation of DR techniques. They examined over 800 plots in order to create a detailed taxonomy of factors to guide the design and the evaluation of cluster separation measures. They focused only on using scatterplot visualizations for cluster finding and verification. DimStiller [20] is a system to provide global guidance for navigating a data-table space through the process of choosing DR and VE techniques. This analysis tool captures useful analysis patterns for analysts who must deal with messy data sets. Rensink and Baldridge [34] explore the use of simple properties such as brightness to generate a set of scatterplots in order to test whether observers could discriminate pairs using these properties. They found that perception of correlations in a scatterplot is rapid, and that in order to limit visual attention to specific information it is more effective to group features together. Etemadpour et al. [17] postulate that cluster properties such as density, shape, orientation, and size influence perception when interpreting distances in scatterplots, and specifically, observe that the density of clusters is more influential than their size.

In general, little attention has been paid to providing details about low-level tasks that guide users to choose DR and VE techniques. However, both high-level goals and much more specific low-level tasks are important aspects of analytic activities. Amar et al. [3] presented a set of ten low-level analysis tasks that they found to be representative of questions that are needed to effectively facilitate analytic activity. Andrienko and Andrienko distinguish elementary tasks that address specific elements of a set and synoptic tasks that address entire sets or subsets, according to the level of analysis [4].

Brehmer and Munzer [7] emphasize three main questions, *why* the tasks are performed, *how* they are performed, and *what* are their inputs and outputs. These questions encompass their concept of multi-level typology. They believe that "low-level characterization does not describe the user's context or motivation; nor does it take into account prior experience and background knowledge." Their typology relies on a more abstract categorization based on concepts, rather than a taxonomy of pre-existing objects or tasks. In contrast, we attempt to specify tasks at the lowest level that can provide details about multidimensional data projection. However, the general approach of Brehmer and Munzner can be easily adopted as a tool to put these low-level tasks in context, facilitating the evaluation of user experiences by evaluation designers. This approach provides essential information, such as motivation and user expertise, for field studies that examine visualization usage. Therefore, we show how our defined tasks can be described according to a typology of abstract tasks relating intents and techniques (how) to modes of goals and tasks (why).

We (1) categorize possible tasks performed when analyzing a specific multidimensional data visualization, and (2) formulate guidelines for analysts to assist in selecting appropriate projection techniques for performing specific visualization tasks on data sets.

3   Task Taxonomy for Multidimensional Data Projection

We define a list of tasks from studies of different projection techniques and their 2D layouts such as PCA [22], Isomap [47], LSP [31], Glimmer [21], and NJ tree [29], as well as the applications behind the data (e.g. document and image data). We explain some of these tasks in detail and provide examples of effective data representations for relevant visual analysis tasks. As explained in Sect. 2, how well groups of points can be distinguished by users in scatterplots defines visual class separability. Our cluster-level tasks also focus on how easily a grouping of related points in multidimensional space (e.g., clusters) can be detected by users when projected into lower-dimensional space. However, rather than only looking at visual class separability, we consider how effective users are performing meaningful tasks related to the perceived clusters.

Although other researchers have explored some of these tasks, we systematically list the full range of analytic tasks for multidimensional projection techniques appropriate for large data sets. Additionally, our organization of these tasks takes into consideration user perception. We divided the tasks into four categories according to the typical visualizations required to support them:

Pattern Identification Tasks: We examine trends, which are more obvious for lower-dimensional data than for projected higher-dimensional ones. Relevant issues include cluster/class preservation and separation.

Relation-seeking Tasks: Relationships and similarities between different reference sets are considered.

Behavior Comparison Tasks: To compare characteristics of subsets (or clusters), we consider capturing different data behaviors (like asking the subjects to compare the point densities within clusters, where density is defined as the number of points per area).

Membership Disambiguation Tasks: Positional and distributional relation-ships within classes/clusters are particularly considered where objects occlude each other. Clutter and noise obscure the structure present in the data and make it hard for users to find patterns and relationships. Peng et al. [32] state that clutter reduction is a visualization-dependent task. Therefore, the DR and VE need to minimize the amount of confusing clutter. We believe that clutter can be measured by users using a wide variety of visualization techniques.

We now clarify these taxonomic categories by looking at common tasks found in the literature.

3.1 Pattern Identification Tasks

Multidimensional data sets may include hundreds or thousands of objects described by dozens or hundreds of attributes. Data characteristics regarding the distribution within multidimensional feature spaces vary for different appli-cation domains. For example, consider document data versus image data: text usually produces sparse spaces while imagery produces dense spaces. As Song et al. [43] state, traditional document representation like bag-of-words leads to sparse feature spaces with high dimensionality. This makes it difficult to achieve high classification accuracies. Figure 1 shows histograms of the distribution of the pairwise distances between four data objects after normalization to the interval [0,1]. The document data sets are referred to as CBR and KDViz[1]. The image data sets are referred to as Corel[2] and Medical[3]. The revealed histograms illus-trate different characteristics for document data sets and image data sets. Both image data sets exhibit lower mean distance values and much wider variance (representative of a denser feature space) than the document data sets.

Identifying patterns in high-dimensional spaces and representing them using dimensionality reduction techniques, in order to reveal trends, is a challenge in many scientific and commercial applications. To identify outliers, trends and interesting patterns in data, one of the many objectives of data exploration is to find correlations in the data, thus uncovering hidden relationships in the data distribution and providing additional insights about the high-dimensional data [53]. Therefore, a list of questions are suggested that can reveal user's perspective about local and global correlations with respect to features – for

---

[1] CBR comprises 680 documents, which include title, authors, abstract, and references from scientific papers in the four different subjects, leading to a data set with 680 objects and 1,423 dimensions. KDViz data has been generated from an Internet repository on the topics bibliographic coupling, co-citation analysis, milgrams, and information visualization, leading to 1,624 objects, 520 dimensions, and four highly unbalanced labels (http://vicg.icmc.usp.br/infovis2/DataSets).

[2] 1,000 photographs on ten different themes. Each image is represented by a 150-dimensional vector of SIFT descriptors (3UCI KDD Archive, http://kdd.ics.uci.edu).

[3] Each image is represented by 28 features, including Fourier descriptors and energies derived from histograms, as well as mean intensity and standard deviation com-puted from the images themselves. Hence, the data set contains 540 objects and 28 dimensions.

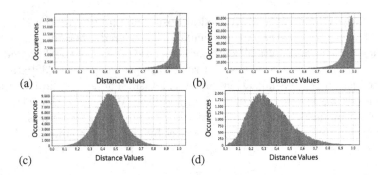

Fig. 1. Histograms of document data (top) and image data (bottom) exhibit characteristic distance distributions: (a) CBR. (b) KDViz. (c) Corel. (d) Medical.

instance, those subsets of data which form relevant patterns (e.g. subsets of data within dense feature groups): (1) Estimate the number of outliers in the given layout; (2) Estimate the number of observed clusters; (3) Find the number of clusters in a selected region; (4) Find the number of subclusters in a given cluster; (5) Find a cluster with a specific characteristic (e.g., longish); (6) Find the specific characteristics (e.g., sparsity) of a cluster; (7) Determine the number of outliers in a given cluster.

If researchers aim to find the user's performance on class segregation, it is important to draw the user's attention to global project views. Thus, we suggest asking *Estimate the number of clusters in the given layout* to identify the informative aspects of the data.

Pattern identification tasks often favor clear segregation by class, which means that techniques which incorporate cluster enclosing surfaces can be helpful. In some situations, the labeled classes in each data set can be considered as ground truth. For such cases, Poco et al. [33] developed a 3D projection method by generalizing the LSP technique from a 2D to a 3D scheme. A non-convex hull (of each cluster) that is computed from a 3D Voronoi diagram of the cluster points is illustrated in Fig. 4(a). This representation, when it is possible to construct, is both accurate and satisfying to users, compared to other techniques.

For situations in which a small set of representative instances from each class is available, or can be manually labeled from a large data set, Paiva et al. [30] proposed a semi-supervised dimensionality reduction approach that employs the Partial Least Squares (PLS) [52] technique, producing reduced spaces that favors class segregation. PLS models relations between sets of variables by estimating a low dimensional latent space, that maximizes the separation between instances with different characteristics, resulting in a low dimensional latent space in which instances from the same class are clustered. The proposed methodology employs visualization techniques to show the similarity structure of the collection, in order to guide the user in selecting representative instances to train the PLS model, that can then be applied to a much larger data set very effectively. Figure 2 shows the LSP projection of Corel data set, with the original dimensionality (a) and after a PLS reduction to 10 dimensions (b). One can notice that the groups are more dense on the reduced space, highlighting the class separability.

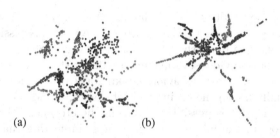

(a)   (b)

Fig. 2. Layouts for Corel data set obtained using LSP projection, using all 150 original attributes and using PLS reduced 10 attributes.

The methodology can also be used for situations in which the instances labels are not available. In this case, a clustering procedure is performed, and the cluster labels are then used to produce a PLS model. For data sets whose cluster structure reflects the class distribution, this methodology will create a reduced space that will favor class segregation.

Also for situations in which a labeled instances set is available, Paiva et al. [28] proposed a visual classification methodology (VCM) that integrates point-based visualization techniques and automatic classification procedures to support control over the whole classification process by users. It yields visual support to classify evolving data sets by allowing user interference, via similarity based visualizations, during supervised classification in an integrated form, promoting users control over model building, application, evaluation and evolution. User insertion is made by the selection of instances to create a classification model, and this selection is performed using the layout, whose structure and point organization is able to guide the user towards a relevant selection. The created classification model can then be employed in the classification of any collection bearing the same feature space. Similarity layouts may represent, in these scenarios, a potential tool to explore the structure and relationship among instances and thus identify the representative ones of each class. That can be achieved, e.g., by analyzing class segregation or by determining outliers that could distort the classifier behavior. Additionally, the methodology allows, in situations in which a ground truth exists, a visual inspection of the classification results using the same visual strategy, in a tool named *Class Matching*, which provides an understanding of the classifiers behavior, and how the data set structure influence this behavior. Finally, model updates can be performed by selecting additional instances from a visualization layout, that offers the possibility of several model updating strategies. Figure 3 shows three layouts, using a NJ tree, of a subset of the ETHZ[4] data set, containing 1,739 instances, with (a) representing the

---

[4] ETHZ represents a subset of the ETHZ dataset [13,38], with 2019 photographs of different people captured in uncontrolled conditions. It is divided into 28 unbalanced groups, and each image is represented by a vector of 3963 descriptors, combining Gabor filters, Histogram of Oriented Gradients (HOG), Local Binary Patterns (LBP) and mean intensity.

ground truth, (b) the result of a SVM classification on this data set, and (c) the corresponding class matching tree, exhibiting in red the misclassified instances. The training set used to build the SVM model contains 280 equally distributed instances. The layout provides several clues about the structure of this collection, as well as about the classifier behavior. Looking at (a), one can notice that the branches are usually homogeneous in terms of classes, as indicated by the circles colors. However, in (b) it is possible to see some heterogeneous branches, which coincide with most of the misclassified instances, indicating that the classifier is confuse about these instances. Moreover, it is possible to notice that class 6 instances are spread in four branches, which may indicate that this class is highly heterogeneous. The data set is originally unbalanced, and class 6 contains the highest number of instances, which may also cause instances from other classes to be classified as 6. By analyzing the confusion matrix, it is possible to notice that several instances from class 26 were classified as class 21 or 6. The layout shows that instances of these classes are positioned on the same branch, and it is possible that they share common attribute values, with similar content. The layout instances positions, allied with an adequate color coding, may facilitate the comprehension of the reasons by which the classifier took these decisions, as well as to indicate for which classes the classifier is deficient. Thus, users are capable to perform effective updates to refine the classification results.

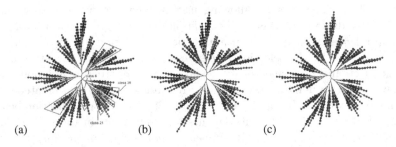

Fig. 3. NJ trees for ETHZ data set, showing (a) the ground truth, (b) the results of a SVM classification, and (c) the corresponding class matching layout.

While this projection works well when the data's pre-assigned class structure accurately models the data's inherent organization, this is often not feasible. In many situations, analysts want to leverage human perception to identify "visual groupings" of points, and in this case a point cloud representation produces favorable results. For example, when grouping information is not available, a point-based visualization as shown in Fig. 4(b) is still applicable. Also, Glimmer [21], as a technique representative of force-directed placement MDS, does not favor class segregation when employed on the KDViz data set. Thus, color coding to separate nodes of different classes can be useful as shown in Fig. 4(c). Therefore, if we have accurate class labels and good class separation, we suggest enclosing surfaces like nonconvex hulls. According to the eye-tracking study on Glimmer projection, the visual attention pattern is scattered and it is hard to identify any meaningful area of interest (AOIs) for Glimmer [17]. Hence, it is useful to differentiate classes when the projection doesn't reflect the class distribution at all.

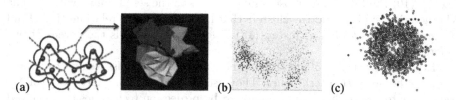

Fig. 4. Estimate the number of observed clusters: (a) Non-convex hulls computed from enclosing surfaces isodistant to cluster using LSP projection; (b) Point-based visualization using PCA projection taken from [37]; (c) The layout obtained with Glimmer projection on the KDViz data set. Circle color indicates instance class label (Color figure online).

3.2   Relation-Seeking Tasks

Relation-seeking tasks investigate the similarities and differences between subgroups which represent clusters or individual objects. Similarity layouts employ projection techniques to reducing data to lower-dimensional visual spaces, but in a different manner from that used in pattern identification. In this application, an analyst is interested in investigating whether a point (or object) is more similar to one cluster or to another, or whether a whole cluster is more similar to a second cluster or a third. We believe that relationship-seeking is a search task, Andrienko's visual task taxonomy model notwithstanding (in which search tasks are limited to lookup and comparison) [5]. In contrast, Zhang et al. [54] consider comparison and relationship-seeking to be compound tasks, containing at least two relationships, one being the data function and the other being relationships between values (or value sets) of a variable. Under this definition, we believe that finding similarities in projected high-dimensional data can be considered as a relation-seeking tasks. Users perform comparison tasks with respect to a given reference set, which can be a cluster or an individual object, and can undertake a similarity search by identifying a given cluster's neighbors. In such a search, the specified relationship is defined by a distance search within a high-dimensional data projection.

A list of potential tasks within the relation-seeking task category can be considered for multidimensional data visualization: (1) Identify the closest cluster to a given cluster; (2) Identify the most similar cluster to a given cluster; (3) Identify the closest cluster to a reference point; (4) Identify the most similar cluster to a given object; (5) Find k closest (most similar) objects to the given cluster; (6) Find k closest (most similar) objects to the reference object; (7) Find the closest (most similar) cluster to a cluster with a specific characteristic (e.g., Find the closest cluster to the longish cluster); (8) Identify the cluster to which the reference set/sets belong; (9) Find the closest (most similar) cluster to the set of points with specific characteristics (e.g., points that have identical movement); (10) Find k closest (most similar) points to the set of points with specific characteristics; (11) Find the clusters that have hierarchical relations; (12) Find k similar objects within a cluster; (13) Find a cluster that is the parent of two reference sets.

Etemadpour et al. [15] investigated how domain-specific issues affect the outcome of the projection techniques. They used a number of similarity interpretation tasks to assess the layouts generated by projection techniques as perceived by their users. To show that projection performance is task-dependent, they generated layouts of high-dimensional data with five techniques representative of different projection approaches. To find a perception-based quality measure, they asked individuals to identify the closest cluster to a given cluster and object. Users also ranked the k nearest objects to a given object. As shown in Fig. 5, the target cluster/object was shown in one color (red) and two other clusters in other colors (green and blue), from which the one closer to the target cluster/object should be identified.

Fig. 5. Task: determine whether green or blue cluster is closer to red object in order to investigate PCA projection performance (Color figure online).

Node-link diagrams have been studied in detail in many graph drawing topics or graph visualization approaches, where a node is representing an entity that is connected to other nodes through lines (i.e., links). Although the node-link diagram is an intuitive way to visually represent relationships between entities for relatively small data sets [19], there may be too many lines crossing with each other that obscure relationships among entities when dealing with larger data sets. In order to represent spatial distance visually in cases like these, a technique like the Force-Directed Placement approach [12] can be used to reveal connections and similarity magnitude between entities. This technique relies on iterative algorithms that model the data points as a system of particles attached to each other by springs. The length of the spring connecting two particles is given by the distance between their corresponding data points as shown in Fig. 6.

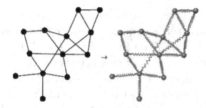

Fig. 6. The spring embedder model [11].

A spatial embedding is obtained with an iterative simulation of the spring forces acting on this hypothetical physical system, until it reaches an equilibrium state.

To *Find k closest objects to the reference object*, if the performance of a projection in terms of maintaining distances within a cluster is under investigation and the cluster structure is known, a combination of hull-based and point-based visualizations can be used. Schreck et al. [37] implemented an interactive system that combined these two visual presentations letting users choose the best visual representation of the projected data. They believed that such combined representations introduce visual redundancy; however, it can improve user's perception of the projection precision information depending on the application. Poco et al. [33] improved the performance of their 3D point representation when they combined standard point clouds with this user-guided process. Figure 7 demonstrates finding 3 closest objects to the red object within a cluster when the convex hull of the points is used.

Brehmer and Munzner's typology is intended to facilitate understanding of users' individual analytical strategies. We employ their multi-level code, used to label user behaviour, to enhance the evaluation of high-dimensional data projection. By utilizing the Brehmer and Munzner multi-level typology, we provide a systematic way of justifying the choice of a particular task through asking three main questions: Why, What and How. This multi-level typology of abstract visualization tasks fills the gap between low-level and high-level classification to describe user tasks in a useful way. This approach to analyzing visualization usage supports making precise comparisons of tasks between different visualization tools and across application domains [7]. For an effective design and evaluation of multidimensional data visualization tools, one should consider why and how our defined tasks should be conducted, and what are their potential inputs and outputs. Meanwhile, sequences of tasks can be linked, so that the output of one task may serve as input to a subsequent task. We focused on *Find k closest clusters to the given cluster* in the relation-seeking category. We did not consider any specific projection technique because it can be changed based on the evaluator's motivation.

Find k closest cluster to the given cluster: **WHY:** The goal is to *Discover* k groups that are closest to a given cluster. A known target (given cluster) and the whole projection visualization are provided. If the location of a given cluster was known (or given by the examiner), then participants perform a *Lookup*.

If the characteristic of the given cluster was given, the user can *Locate* the given cluster with specific characteristics (e.g., searching for a given cluster in which the elements are colored red). Then individuals search for k clusters that are in the neighborhood of the given cluster and list these groups. **WHAT:** The input for this task is a given cluster; this can be shown by the examiner or can be indicated by a particular characteristic like the color red. All other clusters in the entire visualization are also visible to the participants. The output is a list of k groups that are closest to the given cluster. **HOW:** Participants identify the k closest clusters to the given cluster. For example, they determine whether the green or blue cluster is closer to the red cluster. They provide a list of clusters that follow an ascending order, so that the distance of the first cluster in this list to the given cluster is shortest compared to the other clusters. *Select* refers to differentiating selected elements from the unselected remainder.

Fig. 7. Find 3 closest objects to the red object: Convex-hull of the point clusters (Color figure online).

Trees are a natural form for depicting hierarchical relations and can be used to *Find the clusters that have hierarchical relations*. A distinct category of 2D mapping employs tree layouts to convey similarity levels contained in a distance matrix. The algorithms to generate similarity layouts [9] are inspired by the well-known Neighbor-Joining (NJ) heuristic originally proposed to reconstruct phylogenetic trees. Similar points among members of the same subsets are placed at the ends of branches. The points nearer the root of the tree are less similar when compared with the points at the ends of branches. Similarity trees generate a hierarchy, creating a tree structure where interpretation is subject to organization of the branches; for example, mapping data setswith the NJ and LSP projections are compared in Fig. 8. In this example, the INFOVIS04 data set is composed of documents published in a conference on information visualization, and its content is homogeneous. Using NJ, documents with a high degree of similarity are placed along the same branch. The branches circled in the figure are examples of long branches without too many ramifications, and probably represent specific sub-topics inside the collection. LSP, on the other hand, has a tendency to create clusters in round clumps. This representation performs well for certain tasks, but is less useful for finding the closest clusters to selected objects [15].

Authors in [8] introduced BubbleSets as a visualization technique for data that makes explicit use of grouping and clustering information. Members of the same set are in continuous and concave isocontour, while a primary semantic

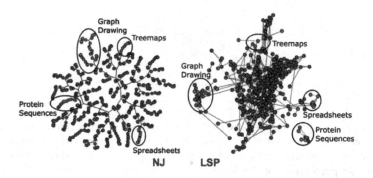

Fig. 8. Comparison of INFOVIS04 document data set map using Neighbor Joining and LSP projections: Four different topics of information visualization are identified by coloring points. Figure is taken from [9] (Color figure online).

data relation is maintained with spatial organization. These delineated contours do not disrupt the primary layout, so they avoid layout adjustment techniques. This visualization technique is designed in order to facilitate depicting more than one data relationship in data sets that contain multiple relationships. Using this concept, we suggest contours around nodes belonging to the same set to *Find k similar objects within a cluster* in a projection technique. Figure 9 shows an example that uses the BubbleSets concept for an NJ heuristic projection. The points that are sharing the same contour are members of the same set. These boundaries are used to indicate the grouping clearly.

Fig. 9. NJ projection: geometric relationships, hierarchy and cluster perimeter are all clearly defined using BubbleSets concept.

3.3   Behavior Comparison Tasks

A third way in which high-dimensional data projections can display data items in lower-dimensional subspaces can provide insight into important data dimensions and details. Our taxonomy distinguishes the subsets of tasks used for behavior comparison: (1) Find the cluster with the largest (smallest) occupied visual area;

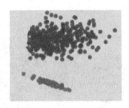

Fig. 10. Task: Compare the density of the longish cluster versus the roundish cluster. Scatter plots were generated with varying shapes, while holding density and size constant, in order to investigate the effect of cluster shape (in projected space) on a user's inferences and perceptions of the data.

(2) Find the cluster with the most (least) number of points or size; (3) Find densest (sparsest) cluster; (4) Given specific number of clusters (e.g. 5 clusters is given); (5) Rank the clusters by density; (6) Rank the clusters by their occupied visual area; (7) Rank the clusters by their size; (8) Compare density of two given clusters with different or similar characteristics (e.g., density of a longish cluster vs. a roundish cluster); (9) Compare the size of two given clusters with different or similar characteristics; (10) Compare the visual area of two given clusters with different or similar characteristics.

Density is an important metric because it indicates stronger relationships between points within a cluster. Moreover, many studies [1,39,49] have indicated that representations of density can play an important role in visualization. Further, studies in psychophysics have shown that visual search can be affected by the variance in the number of objects within groups [10,35,48]. Authors in [41] named density as one of the Within-Cluster factors, namely, the ratio between count and size. This can range from sparse, with few data points and a large spread, to dense, with many points and a small spread. If the task is to *Compare density of two given clusters with different or similar characteristics*(i.e. different shapes), we suggest a point-based visualization. This allows users to easily see the point distribution within a cluster and the occupied visual space. Moreover, as investigated in [17], according to the Gestalt principle [23], the shape and orientation of a cluster should also influence decisions during visual analysis. For example, when two stretched clusters are aligned, they may be perceived as a continuation of one cluster or in other words, characteristics of the clusters influence the visual analysis from a perceptual view. Following these ideas, continuity and closure create the perception of a whole cluster. Figure 10 illustrates the density of a longish cluster versus a cluster that looks more roundish. In this example, cluster shape (e.g., whether a cluster appears to be round or elongated) has been examined, while density and size of the clusters were the same. In addition, 2D scatter plots are manually generated using synthetic clusters [17]. Cluster shape (in projected space) influences users' performance on various inference tasks.

Again by utilizing the Brehmer and Munzner multi-level typology, we provide an example that shows how our defined tasks can be fitted to this multi-level typology of abstract visualization tasks, in order to concisely describe our pre-defined tasks. *Find the Cluster with the Highest Number of Sub-clusters* in the behavior comparison category has been considered. Additionally, we did not consider any specific projection technique because it can be changed based on the evaluator's motivation.

Find the Cluster with the Highest Number of Sub-clusters: **WHY**: The pur-pose is to *Discover* a cluster with the highest number of sub-clusters. The clus-ter characteristic is not provided; therefore, the search target is unknown and *Explore* entails searching for the cluster with the highest number of sub groups. Once the search process is done, *Identify* returns the desired reference. **WHAT**: The input for this task is the entire visualization, including all clusters and their sub-groups. The output is the identity of a cluster with the largest number of sub-clusters. **HOW**: Individuals need to estimate the number of sub-clusters of each cluster. This involves counting sub-groups within successive clusters until the largest number is found. Therefore, they must *Derive* new data elements, then *Select* the desired cluster.

3.4 Membership Disambiguation

It is desirable for the visual representation to avoid clutter, resolve ambiguity and handle noise. At times, "identifying overlaps" may indicate that the classes are not clearly separable, which suggests that the overriding task is one of pattern identification. However, too much data on too small an area of the display, such as a dense region of entangled clusters, diminishes the potential usefulness of the projections even if the projection consists of some clearly separated clusters. This is especially true when the user is exploring the data to: (1) Estimate the number of objects in a selection; (2) Find an object with specific characteristic (e.g. labeled point) within a cluster; (3) Count the number of objects in a given cluster; (4) Identify the objects that overlap in a selected area.

When *Finding an Object with a Specific Characteristic within a Cluster*, a visualization can favor good performance in preserving distances and relation-ships, but only at the expense of producing visual clutter. As an example, the PCA scatterplot of KDViz is too cluttered and distinguishing a specific object within a cluster is not an easy task (Fig. 11).

To *Estimate the number of objects in a selection*, a target cluster/selection can be highlighted with a different color as shown in Fig. 12.

A recent study [16] showed that a density-based motion can enhance pattern detection and cluster ranking tasks for multidimensional data projections and also uncover hidden relationships in scatterplots.

3.5 Meta-Projection

The tasks that are explained above can be used as given, or can be combined into multi-step macrotasks. We note that the tasks that we have provided may not cover all possible tasks of a given type, but they can be used as exemplars when defining new tasks. Sub-clusters of a given cluster or group of points can

Fig. 11. Find a purple object within the green cluster. Using a PCA projection employed on the KDViz data set, it is hard to distinguish the purple point (Color figure online).

Fig. 12. Estimate the number of objects in a selection in LSP projection.

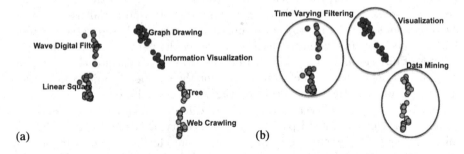

Fig. 13. A meta-projection: (a) sub-clusters; (b) clusters (meta-objects).

be considered as a meta-object. Meta-objects can create a meta-projection, and new tasks can be executed on this projection based on this process. In Fig. 13(a), the task is: "*Find the closest cluster to the given cluster*". For instance, as apparent "Linear Square" is the closest sub-cluster to the "Information Visualization" sub-cluster and "Tree" is the closest sub-cluster to "Graph Drawing". Therefore, as shown in Fig. 13(b) we can analyze the meta-projection to see that "Time Varying Filtering" is the closest cluster to the "Visualization" cluster and similarly "Visualization" is the closest cluster to "Data Mining". Using this meta-projection, we can get more insight into our data.

Thus, in Sect. 3, we saw examples of how appropriate visualization methods could be determined for specific tasks.

4 Conclusion

Our user-centric guideline supports precise comparisons across different multi-dimensional data projection techniques. However, it could be further extended by considering a wider range of application domains that could introduce new visualization scenarios, such as volumetric data sets with continuous scatter-plots. The tasks we have defined are specific neither to a particular projection algorithm nor dataset. Although we delineate a number of example tasks within each of our taxonomic task classifications, they are not intended to be exhaustive. We believe that our guideline could easily incorporate additional tasks; in future work we plan to extend it with further user-centric tasks. We argue that pro-jection methods are distinct in their characteristics in terms of both sparseness and distance distribution, and that the nature of the task (in taxonomic terms) should guide the visualization design. Our taxonomy can be used for examining projection layouts and scatterplots in order to analyze how users perceive mul-tidimensional data in a variety of situations. We also incorporate recent findings about perception rules and cognitive processes as a valuable source of informa-tion for such analyses; our guideline can help in categorizing possible tasks when analyzing multidimensional data visualizations. These user-centric tasks could be used as a guideline for assessing when other scatterplot visualization tech-niques are appropriate, such as Star Coordinates [50], StretchPlots [26,27], or even animations based on point cloud datasets [51]; future work will explore the application of our guideline to a wider range of existing techniques.

References

1. Ahuja, N., Tuceryan, M.: Extraction of early perceptual structure in dot patterns: integrating region, boundary, and component gestalt. Comput. Vision Graph. Image Process. **48**(3), 304–356 (1989)
2. Albuquerque, G., Eisemann, M., Magnor, M.: Perception-based visual quality mea-sures. In Proceedings of IEEE Symposium on Visual Analytics Science and Tech-nology (VAST), pp. 13–20 (2011)
3. Amar, R., Eagan, J., Stasko, J.: Low-level components of analytic activity in infor-mation visualization. In: Proceedings of the 2005 IEEE Symposium on Informa-tion Visualization, INFOVIS 2005, p. 15. IEEE Computer Society, Washington, DC (2005)
4. Andrienko, G., Andrienko, N., Bak, P., Keim, D., Kisilevich, S., Wrobel, S.: A conceptual framework and taxonomy of techniques for analyzing movement. J. Vis. Lang. Comput. **22**(3), 213–232 (2011)
5. Andrienko, N.V., Andrienko, G.L., Gatalsky, P.: Visualization of spatio-temporal information in the internet. In: 11th International Workshop on Database and Expert Systems Applications (DEXA 2000), 6–8 September 2000, Greenwich, London, UK, pp. 577–585 (2000)

6. Borg, I., Groenen, P.J.F.: Modern Multidimensional Scaling Theory and Applications. Springer Series in Statistics, 2nd edn. Springer, New York (2010)
7. Brehmer, M., Munzner, T.: A multi-level typology of abstract visualization tasks. IEEE Trans. Visual. Comput. Graphics (TVCG) **19**(12), 2376–2385 (2013). (Proc.InfoVis)
8. Collins, C., Penn, G., Carpendale, S.: Bubble sets: revealing set relations with isocontours over existing visualizations. IEEE Trans. Visual Comput. Graphics **15**(6), 1009–1016 (2009)
9. Cuadros, A. M., Paulovich, F. V., Minghim, R., Telles, G. P.: Point placement by phylogenetic trees and its application to visual analysis of document collections. In: Proceedings of the 2007 IEEE Symposium on Visual Analytics Science and Technology, pp. 99–106. IEEE Computer Society (2007)
10. Duncan, J., Humphreys, G.: Visual search and stimulus similarity. Psychol. Rev. **96**, 433–458 (1989)
11. Eades, P., Huang, W., Hong, S.: A force-directed method for large crossing angle graph drawing. CoRR, abs/1012.4559 (2010)
12. Eades, P.A.: A heuristic for graph drawing. Congressus Numerantium **42**, 149–160 (1984)
13. Ess, A., Leibe, B., Schindler, K., van Gool, L.: A mobile vision system for robust multi-person tracking, pp. 1–8, Anchorage, AK, USA (2008)
14. Etemadpour, R., da Motta, R.C., de Souza P., Gustavo, J., Minghim, R., Ferreira, M.C., Linsen, L.: Role of human perception in cluster-based visual analysis of multidimensional data projections. In: 5th International Conference on Information Visualization Theory and Applications (IVAPP), pp. 107–113, Lisbon, Portugal (2014a)
15. Etemadpour, R., Motta, R., de Souza Paiva, J.G., Minghim, R., de Oliveira, M.C.F., Linsen, L.: Perception-based evaluation of projection methods for multidimensional data visualization. IEEE Trans. Visual. Comput. Graphics **21**(1), 81–94 (2014b)
16. Etemadpour, R., Murray, P., Forbes, A.G.: Evaluating density-based motion for big data visual analytics. In: IEEE International Conference on Big Data, pp. 451–460, Washington, DC (2014c)
17. Etemadpour, R., Olk, B., Linsen, L.: Eye-tracking investigation during visual analysis of projected multidimensional data with 2D scatterplots. In: 5th International Conference on Information Visualization Theory and Applications (IVAPP), pp. 233–246, Lisbon, Portugal (2014d)
18. Geng, X., Zhan, D.-C., Zhou, Z.-H.: Supervised nonlinear dimensionality reduction for visualization and classification. IEEE Trans. Syst. Man Cybern. Part B **35**(6), 1098–1107 (2005)
19. Henry, N., Fekete, J.: Matrixexplorer: a dual-representation system to explore social networks. IEEE Trans. Visual. Comput. Graphics **12**, 677–684 (2006)
20. Ingram, S., Munzner, T., Irvine, V., Tory, M., Bergner, S., Mller, T.: Workflows for dimensional analysis and reduction. In: IEEE VAST, pp. 3–10. IEEE (2010)
21. Ingram, S., Munzner, T., Olano, M.: Glimmer: multilevel MDS on the GPU. IEEE Trans. Visual. Comput. Graphics **15**(2), 249–261 (2009)
22. Jolliffe, I.T.: Pincipal Component Analysis. Springer-Verlag, New York (1986)
23. Koffka, K.: Principles of Gestalt Psychology. Lund Humphries, London (1935)
24. Lewis, J.M., Ackerman, M.: A comparative study. In: 34th Annual Conference of the Cognitive Science Society, pp. 1870–1875 (2012)
25. Müller, E., Günnemann, S., Assent, I., Seidl, T.: Evaluating clustering in subspace projections of high dimensional data. PVLDB **2**(1), 1270–1281 (2009)

26. Murray, P., Forbes, A.G.: Interactive visualization of multi-dimensional trajectory data. In: Proceedings of IEEE Visual Analytics Science and Technology (VAST), pp. 261–262, Paris, France (2014a)
27. Murray, P., Forbes, A.G.: Interactively exploring geotemporal relationships in demographic data via stretch projections. In: Proceedings of the ACM SIGSPA-TIAL International Workshop on Interacting with Maps (MapInteract), pp. 29–35, Dallas, Texas (2014b)
28. Paiva, J., Schwartz, W., Pedrini, H., Minghim, R.: An approach to supporting incremental visual data classification. IEEE Trans. Visual. Comput. Graphics **21**(1), 4–17 (2015)
29. Paiva, J.G.S., Florian, L., Pedrini, H., Telles, G.P., Minghim, R.: Improved similarity trees and their application to visual data classification. IEEE Trans. Visual. Comput. Graphics **17**(12), 2459–2468 (2011)
30. Paiva, J.G.S., Schwartz, W.R., Pedrini, H., Minghim, R.: Semi-supervised dimensionality reduction based on partial leastsquares for visual analysis of high dimensional data. Comput. Graph. Forum **31**(3pt4), 1345–1354 (2012)
31. Paulovich, F.V., Nonato, L.G., Minghim, R., Levkowitz, H.: Least square projection: a fast high-precision multidimensional projection technique and its application to document mapping. IEEE Trans. Visual. Comput. Graphics **14**(3), 564–575 (2008)
32. Peng, W., Ward, M.O., Rundensteiner, E.A.: Clutter reduction in multidimensional data visualization using dimension reordering. In: Ward, M.O., Munzner, T. (eds.) INFOVIS, pp. 89–96. IEEE Computer Society (2004)
33. Poco, J., Etemadpour, R., Paulovich, F.V., Long, T.V., Rosenthal, P., de Oliveira, M.C.F., Linsen, L., Minghim, R.: A framework for exploring multidimensional data with 3D projections. Comput. Graph. Forum **30**(3), 1111–1120 (2011)
34. Rensink, R.A., Baldridge, G.: The perception of correlation in scatterplots. Comput. Graph. Forum **29**(3), 1203–1210 (2010)
35. Rosenholtz, R., Twarog, N.R., Schinkel-Bielefeld, N., Wattenberg, M.: An intuitive model of perceptual grouping for HCI design. In: Proceedings of the SIGCHI Conference on Human Factors in Computing Systems, CHI 2009, pp. 1331–1340. ACM, New York (2009)
36. Samet, H.: Foundations of Multidimensional and Metric Data Structures (The Morgan Kaufmann Series in Computer Graphics and Geometric Modeling). Morgan Kaufmann Publishers Inc., San Francisco (2005)
37. Schreck, T., von Landesberger, T., Bremm, S.: Techniques for precision-based visual analysis of projected data. In: Park, J., Hao, M.C., Wong, P.C., Chen, C. (eds.) VDA. SPIE Proceedings, vol. 7530, p. 75300. SPIE (2010)
38. Schwartz, W.R., Davis, L.S.: Learning discriminative appearance-based models using partial least squares. Rio de Janeiro, Brazil (2009)
39. Sears, A.: Aide: a step toward metric-based interface development tools. In: Proceedings of the 8th Annual ACM Symposium on User Interface and Software Technology, UIST 1995, pp. 101–110. ACM, New York (1995)
40. Sedlmair, M., Brehmer, M., Ingram, S., Munzner, T.: Gaps and guidance - UBC computer science technical report tr-2012-03. Technical report, The University of British Columbia (2012a)
41. Sedlmair, M., Tatu, A., Munzner, T., Tory, M.: A taxonomy of visual cluster separation factors. Comp. Graph. Forum **31**(3pt4), 1335–1344 (2012b)
42. Sips, M., Neubert, B., Lewis, J.P., Hanrahan, P.: Selecting good views of high-dimensional data using class consistency. Comput. Graph. Forum **28**(3), 831–838 (2009). (Proc. EuroVis 2009)

43. Song, Y., Zhou, D., Huang, J., Councill, I.G., Zha, H., Giles, C.L.: Text categorization for unstructured data on the web. In: the Sixth IEEE international Conference on Data Mining, (ICDM 2006). IEEE (2006)
44. Tan, P.-N., Steinbach, M., Kumar, V.: Introduction to Data Mining. Addison-Wesley Longman, Boston (2005)
45. Tatu, A., Bak, P., Bertini, E., Keim, D.A., Schneidewind, J.: Visual quality metrics and human perception: an initial study on 2D projections of large multidimensional data. In: Proceedings of the Working Conference on Advanced Visual Interfaces (AVI 2010), pp. 49–56 (2010)
46. Tatu, A., Theisel, H., Magnor, M., Eisemann, M., Keim, D., Schneidewind, J., et al.: Combining automated analysis and visualization techniques for effective exploration of high-dimensional data (2009)
47. Tenembaum, J.B., de Silva, V., Langford, J.C.: A global geometric faramework for nonlinear dimensionality reduction. Science 290, 2319–2323 (2000)
48. Treisman, A.: Perceptual grouping and attention in visual search for features and for objects. Exper. Psychol. Hum. Percept. Perform. 8(2), 194–214 (1982)
49. Tullis, T.S.: A system for evaluating screen formats: research and application. In: Hartson, H.R., Deborah, H. (eds.) Advances in Human-Computer Interaction, vol. 2, pp. 214–286 (1988)
50. Van Long, T., Linsen, L.: Visualizing high density clusters in multidimensional data using optimized star coordinates. Comput. Stat. 26(4), 655–678 (2011)
51. Villegas, J., Etemadpour, R., Forbes, A.G.: Evaluating the perception of different matching strategies for time-coherent animations. In: Proceedings of SPIE-IS&T Electronic Imaging Human Vision and Electronic Imaging XX 939412 (HVEI), San Francisco, California, vol. 9394, pp. 1–13 (2015)
52. Wold, H.: Partial Least Squares. Wiley, New York (2004)
53. Zhang, X., Pan, F., Wang, W.: Finding local linear correlations in high dimensional data. In: IEEE 30th International Conference on Data Engineering, pp. 130–139 (2014)
54. Zhang, Y., Passmore, P.J., Bayford, R.H.: Visualization of multidimensional and multimodal tomographic medical imaging data, a case study. Philos. Trans. R. Soc. A: Math. Phys. Eng. Sci. 367(1900), 3121–3148 (2009)

Computer Vision Theory
and Applications

A Hybrid Approach for Individual and Group Activity Analysis in Crowded Scene

K.N. Tran, Xu Yan, I.A. Kakadiaris, and S.K. Shah[✉]

Quantitative Imaging Laboratory, Department of Computer Science,
University of Houston, Houston, TX 77204-3010, USA
khaitran@cs.uh.edu, xyan5@uh.edu, {ikakadia,sshah}@central.uh.edu
http://qil.uh.edu/qil/

Abstract. This paper presents an efficient hybrid (top-down and bottom-up) framework for activity recognition based on analyzing group context in crowded scenes. The approach presented starts by discovering interacting groups of people using a graph based clustering algorithm. Given the interacting groups, a novel group context activity descriptor is computed that captures not only the focal person's activity but also the behaviors of neighbors in the group. Finally, for a high-level of understanding of human activities, we propose a bottom-up approach using a random field model to encode activity relationships between people in the scene. We evaluate our approach on two public benchmark datasets and compare the utility of our proposed descriptor with other descriptors using the same baseline recognition framework. The results of both the steps show that our approach with the proposed descriptor achieves recognition rates comparable to state-of-the-art methods for activity recognition in crowded scenes.

1 Introduction

The goal of human activity recognition is to automatically recognize ongoing activities from an unknown video (i.e. a sequence of image frames). Recent approaches have demonstrated great success in recognizing action performed by one individual. However, a vast number of activities involve multiple people and their interactions. This poses a rather challenging problem in activity recognition due to variations in the number of people involved, and more specifically the different human actions and social interactions exhibited within people and groups [1–4].

Understanding groups and their activities is not limited to only analyzing movements of individuals in group. The environment in which these groups exist provide important contextual information that can be invaluable in recognizing activities in crowded scenes. Perspectives from sociology, psychology, and computer vision suggest that human activities can be understood by investigating a subject in the context of social signaling constraints [5–8]. Exploring the spatial and directional relationships between people can facilitate the detection of social interactions in a group. Thus, activity analysis in crowded scenes can

© Springer International Publishing Switzerland 2016
J. Braz et al. (Eds.): VISIGRAPP 2015, CCIS 598, pp. 189–204, 2016.
DOI: 10.1007/978-3-319-29971-6_10

often be considered a multi-step process, one that involves individual person activity, individuals forming meaningful groups, interaction between individuals and interactions between groups. In general, the approaches to group activity analysis can be classified into two categories: bottom-up and top-down. The bottom-up approaches rely on recognizing activity of each individual in a group. Vice versa, top-down approaches model the entire group as a whole rather than each individual separately.

By looking at group of people in a scene, top-down approaches quickly give an answer of question: *'what are people doing in a particular group?'* rather than first answer to more detail questions: *'what is each individual doing?'* and then quickly find out the number of people groups and the dominant activity of people in each group. In other words, top-down approaches roughly recognize activity of people in a particular group at a glance. However, there are limitation in using this approach. First, it is lack of understanding individual activity in a group. Second, from previous top-down approaches, there are limited number of group activity classes which can be recognized.

Bottom-up approaches show the understanding of activities at the individual level. By looking at each person in group context, bottom-up approaches will answer questions: *'what is each individual doing?'* in with or without consideration of group context. This type of approaches not only summarize the dominant activities in a group but also give detail understanding of each individual's activity. However, this is an exhausted approach for analyzing a large number of people in a crowded scene where we only want to roughly estimate the dominant activities of people.

There are advantages and limitations of using top-down and bottom-up approaches in analyzing people activities in a crowded scene as illustrated in Fig. 1. With different approaches for analyzing activity in a crowded scenes, there are different methodologies of designing activity descriptors for achieving better recognition. Intuitively, the bottom-up approach relying on recognizing at individual level will need more granularity descriptor on each subject than top-down approach which need more concentrate on interactions of people in a group. An ideal robust activity descriptor should encode most of the descriptive and discriminative information of individual level while being sufficient rich at capturing group interaction level in a crowded scene.

In this paper, we present a social context framework for recognizing human activities in crowded scenes by taking advantage of both top-down and bottom-up approaches. Our hybrid framework localizes groups through social interaction analysis using a top-down approach and analyzes individual activity based on social context within the group using a bottom-up mechanism. We propose a novel group context activity descriptor capturing characteristics of individual activity with respect to the behavior of its neighbors along with an efficient conditional random field model to learn and classify human activities in crowded scenes.

The main contributions of our work are:

1. *A group context activity descriptor.* We use a top-down approach to dynamically localize interacting groups capturing behaviors of individuals. We form

a group context activity descriptor that is a combination of individual activity and its neighbor's behavior, represented using the Bag-of-Words (BoW) representation.

2. *An efficient conditional random field framework to learn and classify human activities in context.* We present a recognition framework that jointly captures the individual activity and its activity relationships with its neighbors.

This paper is an extension of our work in [9] and the extensions are: (1) introducing more comprehensive analysis perspectives (top-down and bottom-up) of individual and group activities in a crowded scene. (2) presenting an baseline model to evaluate the performance of using our group context activity descriptor in comparison of using other well-known contextual activity descriptors. The rest of the paper is organized as follows. We review related work on activity analysis in crowded scenes in Sect. 2. Section 3 describes the human activity descriptor in group context along with the conditional random field model used to address the activity recognition task. Experimental results and evaluations are presented in Sect. 4. Finally, Sect. 5 concludes the paper.

Fig. 1. By using top-down approach, we can quickly summary that there are only two main activities *'Running'* and *'Talking'* in video. However, two *'unknown'* activities are not recognized. By using bottom-up approach, we slowly scan through each person in the scene and recognize what is each person doing? However, we might end up with assigning wrong *'Talking'* action for person number 8 who is mingling with 6 and 7.

2   Related Work

In this section, we review related work on human activity analysis in crowded scenes that use a top-down or bottom-up approach. In bottom-up approaches, group context is used to differentiate ambiguous activities e.g. *'Standing'* and *'Talking'*, which are normally represented by the same local descriptors. Most approaches integrate contextual information by proposing a new feature descriptor extracted from an individual and its surrounding area. Lan et al. [10] propose an Action Context (AC) descriptor capturing the activity of the focal person and the behavior of other people nearby. AC descriptor is computed by concatenating the focal person's action probability vector (computed using Bag-of-Words approach with SVM classifier), and the context action probability vectors capturing the activities of other neighborhood people. However, this AC descriptor only can capture spatial proximity information by using 'near by' context. Considering a more sophisticated contextual descriptor, Choi et al. [11] propose Spatio-Temporal Volume (STV) descriptor, which captures spatial distribution of pose and motion of individuals in a scene to analyze group activity. STV descriptor centered on a person of interest or an anchor is used for classification of the group activity. The descriptor is a histogram of people and their poses in different spatial bins around the anchor. These histograms are concatenated over the video to capture the temporal nature of the activities. SVM using pyramid kernels is used for classification. The similar descriptor named as Randomized Spatio-Temporal Volume (RSTV) is leveraged in [12] but Random Forest classification is used for group activity analysis. In addition, random forest structure is used to randomly sample the spatio-temporal regions to pick most discriminative features. Recently, Amer et al. [13] introduced Bags-of-Right-Detections (BORD) seeking to remove noisy people detection in groups. BORD is a histogram of human poses detected in a spatio-temporal neighborhood centered at a point in the video volume. The BORD is not computed from all neighborhood people, but only from those detections that are considered to take part in the target activity. A two-tier MAP inference algorithm is proposed for the final recognition step.

In contrast to bottom-up approaches, top-down methods model the entire group as a whole rather than each individual separately. Khan and Shah [14] use rigidity formulation to represent parade activities. They modeled group shape as a 3D polygon with each corner representing a participating person. The tracks from person in group are treated as tracks of feature points in a 3D polygon. Using rank of track matrix, activities are classified as parade or just random crowds. Vaswani et al. [15] model an activity using a polygon and its deformation over time. Each person in the group is treated as a point on the polygon. The model is applied to abnormality detection in a crowded scene. Multi-camera multi-target tracks are used to generate dissimilarity measure between people, which in turn are used to cluster them into groups in [16]. Group activities are recognized by treating the group as an entity and analyzing the behavior of the group over time. Mehran et al. [17] built a 'Bag-of-Forces' model of the movements of people using social force model in a video frame to detect abnormal

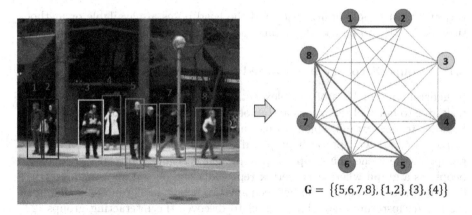

$\mathbf{G} = \big\{\{5,6,7,8\}, \{1,2\}, \{3\}, \{4\}\big\}$

Fig. 2. Illustration of group discovery. Human interactions in a group is represented as an undirected edge-weighted graph. Dominant set based clustering algorithm is used to localize interacting groups. There are four discovered groups from the scene: {5,6,7,8}, {1,2}, {3} and {4}.

crowd behavior. Close to top-down approach, Ryoo et al. [1] present an approach that splits group activity into sub-events like person activity and person to person interactions. Each portion is represented using context free grammar and the probability of their occurrence given a group activity or time periods. A hierarchical recognition algorithm based on Markov Chain Monte Carlo density sampling technique is developed. The technique identifies the groups and group activity simultaneously.

Recently, several approaches that leverage social signaling cues for analyzing crowded scenes have been proposed. Group activities can be better inferred from valuable social interactions cues between people present in the scene. Several approaches are proposed to identify meaningful group from the videos using spatial and orientational arrangement of people in the scene as a cue based on social signaling principles [5,18,19]. Lan et al. [20] present a bottom-up approach integrating social role analysis to understand activities in crowd scene. Different from above approaches, our approach takes advantage of both bottom-up and top-down mechanisms by designing a group context activity descriptor capturing individual activity and behavior of its neighbors within its groups. Once meaningful groups are identified from the videos by using top-down approach, group context activity descriptor is built for each individual in discovered group. Using this descriptor a random field model is built to recognize individual activities using a bottom-up approach.

3   Approach

In this paper, we mainly focus on recognizing human activities in crowded scenes. Thus, we assume that people in a crowded scene have been detected and the trajectories of people in 3D space and the head poses are available or methods such as [11,21] can be used to obtain the same.

3.1 Group Discovery in Crowded Scene

In general, the analysis of complex activity in crowded scene is a challenging task, due to noisy observations and unobserved communication between people. In order to understand which people in the scene form meaningful groups, we employ a top-down approach proposed in [5] to discover socially interacting groups in the scene. This top-down approach basically represents all detected people as a graph where each vertex represents one person and weighted edges describe the social interaction between any two people in a group. The dominant set based clustering algorithm is used to discover the interacting groups [22]. Figure 2 depicts the overview of group discovery in the crowded scene.

3.2 Model Formulation

Given a set $\mathbf{N} = \{1, ..., n\}$ of all the people detected in the scene, let $\mathbf{x} = \{x_1, x_2, ..., x_n\}$ be the set of people activity descriptors; $\mathbf{a} = \{a_1, a_2, ..., a_n\}$ be the set of individual activity labels, where x_i is feature vector and $a_i \in \mathbf{A}$ is activity label associated with person $i \in \mathbf{N}$ ($\mathbf{A}$ is set of all possible activity labels). As a result of clustering people in the scene to different interacting groups, let us define $\mathbf{G} = \{\mathbf{G}_1, \mathbf{G}_2, ..., \mathbf{G}_m\}$ as the set of discovered groups where $\mathbf{G}_c$ is set of people clustered in group c and $\cup_{c=1}^{m} \mathbf{G}_c = \mathbf{N}$. We introduce a standard conditional random field model to learn the strength of the interactions between activities in discovered groups. The activity interaction is conditioned on image evidence, so that the model not only takes into account which activity each person is engaged in, but also higher-order dependencies between activities. Our model is represented as:

$$\Psi(\mathbf{a}, \mathbf{x}) = \sum_{i \in \mathbf{N}} \phi(a_i, x_i) + \sum_{c=1}^{m} \sum_{(i,j) \in \mathbf{G}_c} \phi(a_i, a_j) \tag{1}$$

where $\phi(a_i, x_i)$ is a singleton factor that models the probability of person i's activity label $a_i \in \mathbf{A}$ given its feature vector x_i. $\phi(a_i, a_j)$ is the pairwise factor that models the probability between pair of individual activities a_i and a_j, where (i,j) belong to the same group $\mathbf{G}_c$ discovered by using top-down approach described in Sect. 3.1. A graphical illustration of our model discovering meaningful groups and formulation of conditional random field model is shown in Fig. 3.

The model described in Eq. 1 only captures high-level activity-to-activity relationships of people in discovered groups. This limit us from analyzing the low-level interaction detail between pair of individuals and the effect caused by their neighbors' behavior. Thus we introduce a low-level group context activity descriptor that encodes detailed individual activity interactions within a group to overcome the limitation.

Fig. 3. Illustration of our conditional random field model for each discovered interacting group. The activity-to-activity relationships in each group are represented by dashed lines.

3.3 Group Context Activity Descriptor

An ideal context activity descriptor can efficiently incorporate the focal person's activity in spatio-temporal relationship with activities in its spatial proximity. Lan et al. [10] propose an Action Context (AC) descriptor capturing the activity of the focal person and the behavior of other people nearby. AC descriptor uses spatial proximity as an indicator of context. They do not consider whether the people near-by are engaged in meaningful interactions or not, effectively leading to a semantically noisy descriptor. This limitation is clearly showed in Fig. 4 where person 1 is not involving in *'Talking'* with other people $\{2, 3, 4\}$. However, spatial proximity as indicator of context will consider him as a part of *'Talking'* group. This leads to a misunderstanding of his activity as well as other people activities. Moreover, we argue that AC is represented by concatenating a set of probability vectors computed using Bag-of-Words approach with SVM classifier that adds to ambiguity already existent in the representation (BoW) for each person. Choi et al. [11,12] employs well-known shape context idea [23] to propose Spatio-Temporal Volume (STV) and its enhanced version Randomized Spatio-Temporal Volume (RSTV) descriptors, which capture spatial distribution of pose and motion of individuals in a scene. The descriptor centered on a person of interest or an anchor is represented as histograms of people and their poses in different spatial bins around the anchor. Both STV and RSTV descriptors can effectively capture higher-level spatial relationship of individual interactions.

Nonetheless, because of only capturing spatial distribution of pose and positions of people so they are too coarse to capture finer semantically driven contextual relationship of individual activities in detail.

We develop a novel group context activity (GCA) descriptor that exploits strategies from above approaches. Our descriptor is centered on a person (the focal person), and describes the behaviors of focal person and its semantic neighbors represented by arranging individual activity descriptors in polar view. Let $\mathbf{f} = \{f_1, f_2, ..., f_n\}$ be the set of local activity descriptors formed using Bag-of-Words representation for all people in the scene, where f_i is K-dimensional vector representing person i's activity (K is number of visual codewords). Dense trajectory based descriptors have shown to be efficient for representing actions in video, thus we employ approach proposed in [24] to extract motion boundary histogram (MBH) as local activity descriptors. Given the i-th person in discovered group $\mathbf{G}_c$ as the focal person, we divide its context region into P sub-polar context regions characterized by number of orientation bins and radial bins [23, 25]. Using spatial relationship between people in discovered group $\mathbf{G}_c$, we extract descriptors in each sub-polar context region around the focal person. As a result, the group context activity descriptor x_i for person i is represented as a $(P+1) \times K$ dimensional vector including focal activity descriptor computed as follows:

$$x_i = [f_i, \sum_{j \in S_1(i)} f_j, \sum_{j \in S_2(i)} f_j, ..., \sum_{j \in S_P(i)} f_j] \tag{2}$$

where $S_p(i)$ is set of people in the p-th sub-polar context region of person i. Figure 4 shows the extraction of group context activity descriptor for a selected person in a discovered group.

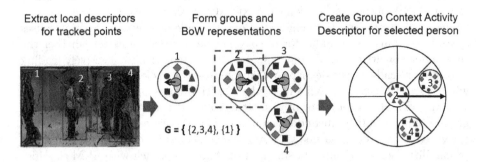

Extract local descriptors for tracked points

Form groups and BoW representations

Create Group Context Activity Descriptor for selected person

$G = \{\ \{2,3,4\}, \{1\}\ \}$

Fig. 4. Depiction of Group Context Activity (GCA) descriptor extraction. From left to right, people are localized in different groups using a top-down approach from [5]; local descriptors are extracted from dense trajectories [24]; local BoW is computed for each person's activity; GCA descriptor is extracted for a selected person in a discovered group by computing descriptor for each sub-polar context bin.

3.4   Inference and Learning

Our model is a standard Conditional Random Field (CRF) with no hidden variables. We train a multi-class SVM classifier based on GCA descriptors and their associated labels to learn and compute singleton factor $\phi(a_i, x_i)$. Given an observation x_i, we use SVM parameters to compute probabilities for all possible activity labels. From training data, we use top-down approach to discover interacting groups in the scene. All pairs of activity labels in discovered groups are counted to compute pairwise factor $\phi(a_i, a_j)$.

Given a new testing scene, our inference task is to find best activity label assignments for all people detected in the scene. The prediction assignment $\mathbf{a}^*$ is computed by running MAP inference on the network as:

$$\mathbf{a}^* = \arg\max_{\mathbf{a}} \Psi(\mathbf{a}, \mathbf{x}) \tag{3}$$

where $\Psi(\mathbf{a}, \mathbf{x})$ is specified in Eq. 1.

4   Experiment and Results

In this section, we describe the experiments designed to evaluate the performance of the proposed group context activity (GCA) descriptor and framework for human activity recognition in crowded scenes.

4.1   Datasets

In this work, we choose to use two challenging benchmark datasets to evalute our proposed approach in recognizing human activities in a crowded scene. The first benchmark dataset is Collective Activity dataset [11]. The old version of dataset contains 5 activities in group (*Crossing, Waiting, Queuing, Walking* and *Talking*) and recently, the authors presented a new version of dataset including two additional activities (*Dancing* and *Jogging*). HOG based human detection and head pose estimation along with a probabilistic model is used to estimate camera parameters [11]. Extended Kalman filtering is employed to extract 3D trajectories of people in the scene. These automatically extracted 3D trajectories and head pose estimates are provided as a part of the dataset. Thus, the dataset represents real world, noisy observations with occlusions and automatic person detection and trajectory generation.

The second benchmark dataset is UCLA Courtyard dataset recently introduced by Amer et. al [26]. This dataset contains 106 min of high resolution videos at 30 fps from a bird-eye view of a courtyard at the UCLA campus. The annotations in term of bounding boxes, poses, and activity labels are provided for each frames in video. The dataset contains 10 primitive human activities which are *Riding Skateboard, Riding Bike, Riding Scooter, Driving Car, Walking, Talking, Waiting, Reading, Eating,* and *Sitting*.

4.2   Model Parameters

In using the group discovery algorithm [5], we set parameters that maintain the ratio proposed in [27] and the social distance function is modeled as the power function $F_s(r) = (1 - r)^n, n > 1$. We define 56 activity labels (8 head poses × 7 activity labels) for new version of Collective Activity dataset and 40 activity labels (8 head poses × 5 activity labels) for UCLA Courtyard dataset by combining the head poses and activity labels. We train a multi-class SVM classifier which is used to compute singleton factors by utilizing the libSVM library [28] with linear kernel on GCA descriptor. Using discovered groups from top-down approach, respectively, matrices of size 56 × 56 and 40 × 40 are used to learn and look up pairwise factors for Collective Activity and UCLA Courtyard datasets. For recognition, we use libDAI [29] to perform inference in our conditional random field model. To compute the MBH descriptors, we set the neighborhood size $N = 32$ pixels, the spatial cell $n_\sigma = 2$, the temporal cells $n_\tau = 3$, trajectory length $L = 10$, and dense sampling step size $W = 5$ for dense tracking. This setting claims to empirically give good results for a wide range of datasets (see [24] for parameter details). In designing GCA descriptor, we select codebook size of $K = 200$ by clustering a subset of $100,000$ randomly selected training features using k-means. In addition, we evaluate our proposed model in different settings of P, which is number of sub-polar context regions around a focal person. Basically $P = R \times O$ where R is number of radial bins and O is number of orientation bins. However, given a focal person within his discovered group, context activity descriptor differs from others by discriminating in orientation distribution rather than radial distribution. Thus in our case, R is set to 1 and our GCA descriptor is controlled by $P = O$ number of orientation bins. For the special case when $P = 0$, GCA descriptor amounts to the focal person local activity descriptor without using context ($x_i = f_i$). This is the same for a non-group person in the scene who does not belong to any discovered groups. Experiments show that $P = 4$ and $P = 16$ achieves the best performances in Collective Activity and UCLA Courtyard datasets, respectively.

Table 1. Recognition rates of various proposed methods on Collective Activity dataset [11].

| Accuracy (%) | | | | | | | | | |
Approach	Year	Walk	Cross	Queue	Wait	Talk	Jog	Dance	Avg.
Choi [11]	2009	57.9	55.4	63.3	64.6	83.6	N/A	N/A	65.9
Choi [12]	2011	N/A	76.5	78.5	78.5	84.1	94.1	80.5	82.0
Amer [13]	2011	72.2	69.9	96.8	74.1	99.8	87.6	70.2	81.5
Amer [26]	2012	74.7	77.2	95.4	78.3	98.4	89.4	72.3	83.6
Lan [10]	2012	68.0	65.0	96.0	68.0	99.0	N/A	N/A	79.1
Our Method		60.4	60.6	89.1	80.9	93.1	93.4	95.4	82.9

Table 2. The recognition rates of using the baseline recognition method with different context activity descriptors.

Accuracy (%)		
Approach	5 Activities	6 Activities
AC [10]	70.9	N/A
STV [11]	64.3	N/A
RSTV [12]	67.2	71.7
GCA	**72.6**	**74.2**

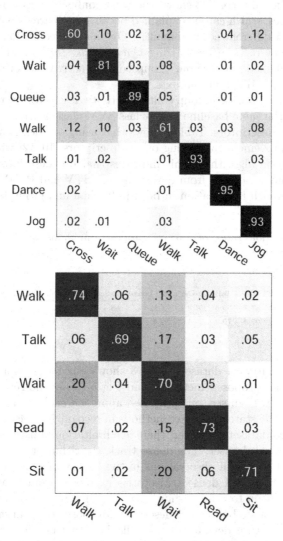

Fig. 5. Confusion matrices for Collective Activity dataset (Top) and UCLA Courtyard dataset (Bottom).

4.3 Human Activity Recognition Evaluation

We summarize the recognition results obtained using our method and other approaches in Table 1 for Collective Activity dataset using standard 4-folds cross-validation scheme. As we can see, our proposed approach achieves recognition rates comparable to state-of-the-art methods in the new version of Collective Activity dataset. Figure 5(Top) shows the confusion matrices obtained on Collective Activity dataset. It lists the recognition accuracy for each activity individually. The low values of the non-diagonal elements imply that the descriptor is highly discriminative with very low decision ambiguity between different activities. The confusion matrix also shows the most confusion between *Walking* and *Crossing* activities, which can be explained because both are essentially *Walking* activity but with different scene semantic. Overall, the confusion matrix shows very high accuracy rates in recognizing *Queue, Talk, Dance* and *Jog* activities. This can be explained because our group context activity descriptor efficiently encodes activities in different contexts.

To evaluate our proposed group context activity descriptor we report the performance using the same baseline multi-class SVM classification algorithm with different activity descriptors. Table 2 shows that using our descriptor we achieve a significant improvement over using other descriptors [10–12] with the same a baseline recognition algorithm. This indicates that our descriptor took advantages of exploiting strategies from designing AC, STV and RSTV descriptor to provide sufficient rich information in both individual and group levels.

Table 3. Recognition rates of various proposed methods on UCLA Courtyard dataset [26].

Accuracy (%)										
Approach	Walk	Wait	Talk	D-Car	R-S-board	R-Scooter	R-Bike	Read	Eat	Sit	Avg.
Amer [26]	69.1	67.7	69.6	70.2	71.3	68.4	61.4	67.3	71.3	64.2	68.1
Our Method	74.3	69.9	70.0	N/A	N/A	N/A	N/A	72.8	N/A	70.8	71.4

For UCLA Courtyard dataset, Table 3 shows our recognition rate in comparison with other proposed methods, and Fig. 5 shows the recognition confusion matrices. As we can see, our proposed approach achieves recognition rates that outperform the state-of-the-art methods in recognizing selected activities in UCLA Courtyard dataset. However, there is a limitation in using our framework to UCLA Courtyard dataset. The dense track algorithm proposed in [24] does not perform well across all observations in the UCLA Courtyard videos. There are very small number of dense trajectories extracted from people in shadow regions in comparison to other regions (see Fig. 6). Thus, there are not enough extracted descriptors to build GCA descriptor for those people in shadow regions. Using alternate feature detectors could alleviate this problem and hence the limitation towards computing the local activity descriptor for UCLA Courtyard videos. Due to this limitation, not all activities are included in our evaluation.

Fig. 6. Depiction of computed dense tracks in UCLA Courtyard dataset. Due to low resolution, there are few tracked trajectories extracted for people in shadow regions. Thus, there are very few local activity descriptors extracted for people in those regions.

Some activities such as *Riding Skateboard, Riding Bike, Riding Scooter, Driving Car,* and *Eating* are limited and hence do not provide sufficient exemplars for learning.

5   Conclusion

In this paper, we have proposed an efficient hybrid (top-down and bottom-up) framework for activity recognition based on analyzing group context in crowded scenes. A novel group context activity descriptor is introduced that is designed to capture the focal person's activity along with the behavior of the neighbors. We have also proposed a high-level recognition framework that jointly utilizes the individual person's activity and the relationships with the activity of the neighbors. The experimental results show that a baseline recognition using our proposed descriptor achieved a significant improvement over using other descriptors. We have also evaluated our hybrid framework on two public benchmark datasets and the results demonstrate that our approach obtains results comparable to state-of-the-art approaches in recognizing human activities in crowded scenes.

Acknowledgements. This work was supported in part by the US Department of Justice 2009-MU-MU-K004. Any opinions, findings, conclusions or recommendations expressed in this paper are those of the authors and do not necessarily reflect the views of our sponsors.

References

1. Ryoo, M., Aggarwal, J.: Stochastic representation and recognition of high-level group activities. Int. J. Comput. Vis. **93**, 183–200 (2011)
2. Tran, K.N.: Contextual descriptors for human activity recognition, Ph.D. Thesis, University of Houston (2013)
3. Zhou, B., Tang, X., Wang, X.: Learning collective crowd behaviors with dynamic pedestrian-agents. Int. J. Comput. Vis. **111**(1), 50–68 (2015). doi:10.1007/s11263-014-0735-3
4. Yi, S., Wang, X., Lu, C., Jia, J.: L0 regularized stationary time estimation for crowd group analysis. In: IEEE Conference on Computer Vision and Pattern Recognition (CVPR), 2014, pp. 2219–2226 (2014). doi:10.1109/CVPR.2014.284
5. Tran, K., Gala, A., Kakadiaris, I., Shah, S.: Activity analysis in crowded environments using social cues for group discovery and human interaction modeling. Pattern Recogn. Lett. **44**, 49–57 (2014). Pattern Recognition and Crowd Analysis
6. Smith, K., Ba, S., Odobez, J.-M., Gatica-Perez, D.: Tracking the visual focus of attention for a varying number of wandering people. IEEE Trans. Pattern Anal. Mach. Intell. **30**, 1212–1229 (2008)
7. Helbing, D., Molnár, P.: Social force model for pedestrian dynamics. Phy. Rev. E **51**(5), 4282–4286 (1995)
8. Cristani, M., Bazzani, L., Paggetti, G., Fossati, A., Tosato, D., Bue, A.D., Menegaz, G., Murino, V.: Social interaction discovery by statistical analysis of f-formations. In: Proceedings of the British Machine Vision Conference, pp. 23.1–23.12 (2011)

9. Tran, K., Yan, X., Kakadiaris, I., Shah, S.: A group contextual model for activity recognition in crowded scenes. In: Proceedings of the International Conference on Computer Vision Theory and Applications (2015)
10. Lan, T., Wang, Y., Yang, W., Robinovitch, S., Mori, G.: Discriminative latent models for recognizing contextual group activities. IEEE Trans. Pattern Anal. Mach. Intell. **34**, 1549 (2012)
11. Choi, W., Shahid, K., Savarese, S.: What are they ng? : collective activity classification using spatio-temporal relationship among people. In: Proceedings Visual Surveillance Workshop, ICCV, pp. 1282–1289 (2009)
12. Choi, W., Shahid, K., Savarese, S.: Learning context for collective activity recognition. In: Proceedings of the Computer Vision and Pattern Recognition, Spring CO, USA, pp. 3273–3280 (2011)
13. Amer, M.R., Todorovic, S.: A chains model for localizing participants of group activities in videos. In: Proceedings of the IEEE International Conference on Computer Vision (2011)
14. Khan, S.M., Shah, M.: Detecting group activities using rigidity of formation. In: Proceedings of the ACM International Conference on Multimedia, MULTIMEDIA 2005, NY, USA, pp. 403–406. ACM, New York (2005). http://doi.acm.org/10.1145/1101149.1101237. doi:10.1145/1101149.1101237
15. Vaswani, N., Roy Chowdhury, A., Chellappa, R.: Activity recognition using the dynamics of the configuration of interacting objects. In: Proceedings of the Computer Vision and Pattern Recognition, vol. 2, pp. II-633–II-40 (2003). doi:10.1109/CVPR.2003.1211526
16. Chang, M.-C., Krahnstoever, N., Lim, S., Yu, T.: Group level activity recognition in crowded environments across multiple cameras. In: Proceedings of the IEEE International Conference on Advanced Video and Signal Based Surveillance, DC, USA, pp. 56–63 (2010)
17. Mehran, R., Oyama, A., Shah, M.: Abnormal crowd behavior detection using social force model. In: Proceedings of the Computer Vision and Pattern Recognition, pp. 935–942 (2009). doi:10.1109/CVPR.2009.5206641
18. Farenzena, M., Tavano, A., Bazzani, L., Tosato, D., Pagetti, G., Menegaz, G., Murino, V., Cristani, M.: Social interaction by visual focus of attention in a three-dimensional environment. In: Proceedings of the Workshop on Pattern Recognition and Artificial Intelligence for Human Behavior Analysis at AI*IA (2009)
19. Farenzena, M., Bazzani, L., Murino, V., Cristani, M.: Towards a subject-centered analysis for automated video surveillance. In: Foggia, P., Sansone, C., Vento, M. (eds.) ICIAP 2009. LNCS, vol. 5716, pp. 481–489. Springer, Heidelberg (2009)
20. Lan, T., Sigal, L., Mori, G.: Social roles in hierarchical models for human activity recognition. In: Proceedings of the Computer Vision and Pattern Recognition, pp. 1354–1361 (2012). doi:10.1109/CVPR.2012.6247821
21. Hoiem, D., Efros, A., Hebert, M.: Putting objects in perspective. In: Proceedings of the Computer Vision and Pattern Recognition, vol. 2, pp. 2137–2144 (2006) doi:10.1109/CVPR.2006.232
22. Pavan, M., Pelillo, M.: Dominant sets and pairwise clustering. IEEE Trans. Pattern Anal. Mach. Intell. **29**, 167–172 (2007)
23. Belongie, S., Malik, J., Puzicha, J.: Shape matching and object recognition using shape contexts. IEEE Trans. Pattern Anal. Mach. Intell. **24**(4), 509–522 (2002)
24. Wang, H., Klaser, A., Schmid, C., Liu, C.-L.: Action recognition by dense trajectories. In: Proceedings of the Computer Vision and Pattern Recognition, pp. 3169–3176 (2011). doi:10.1109/CVPR.2011.5995407

25. Tran, K., Kakadiaris, I., Shah, S.: Part-based motion descriptor image for human action recognition. Pattern Recogn. **45**(7), 2562–2572 (2012)
26. Amer, M.R., Xie, D., Zhao, M., Todorovic, S., Zhu, S.-C.: Cost-sensitive top-down/bottom-up inference for multiscale activity recognition. In: Proceedings of the European Conference on Computer Vision, pp. 187–200 (2012)
27. Was, J., Gudowski, B., Matuszyk, P.J.: Social distances model of pedestrian dynamics. In: El Yacoubi, S., Chopard, B., Bandini, S. (eds.) ACRI 2006. LNCS, vol. 4173, pp. 492–501. Springer, Heidelberg (2006)
28. Chang, C.-C., Lin, C.-J.: LIBSVM: a library for support vector machines. ACM Trans. Intell. Syst. Technol. **2**, 27:1–27:27 (2011)
29. Mooij, J.M.: libDAI: a free and open source C++ library for discrete approximate inference in graphical models. J. Mach. Learn. Res. **11**, 2169–2173 (2010)

Fusing Intertial Data with Vision for Enhanced Image Understanding

Osian Haines[✉], David R. Bull, and J.F. Burn

University of Bristol, Bristol, UK
research@osianh.com, {dave.bull,j.f.burn}@bristol.ac.uk

Abstract. In this paper we show that combining knowledge of the orientation of a camera with visual information can be used to improve the performance of semantic image segmentation. This is based on the assumption that the direction in which a camera is facing acts as a prior on the content of the images it creates. We gathered egocentric video with a camera attached to a head-mounted display, and recorded its orientation using an inertial sensor. By combining orientation information with typical image descriptors, we show that segmentation of individual images improves in accuracy compared with vision alone, from 61 % to 71 % over six classes. We also show that this method can be applied to both point and line based features from the image, and that these can be combined together for further benefits. Our resulting system would have applications in autonomous robot locomotion and guiding visually impaired humans.

Keywords: Vision guided locomotion · Segmentation · Image interpretation · Scene understanding · Inertial sensors · Oculus Rift · Mobile robotics

1 Introduction

The ability to safely traverse rough terrain is crucial to the survival of almost all land animals, and is a crucial requirement in order to hunt prey, forage for food, escape predators, find mates, migrate, and so on. Vision is a very important sense for this, and can provide a rich depiction of the surrounding world; but vision is rarely used in isolation, and sound, scent and touch all provide important information too. Another very important source of information is the vestibular system, allowing accelerations and absolute orientations to be perceived independently of visual or other cues [1], and is crucial for balance and normal locomotion [29]. Vestibular information becomes even more important when vision is impaired [7], and its absence can lead to problems in interpreting visual information [30]. There is also some evidence that the central nervous system dynamically controls the relative importance of visual and vestibular signals [7], and that reciprocal inhibition of visual and vestibular signals allows perception of self-motion in situations with conflicting stimuli [4], showing that there is significant and important interaction between the two senses.

© Springer International Publishing Switzerland 2016
J. Braz et al. (Eds.): VISIGRAPP 2015, CCIS 598, pp. 205–226, 2016.
DOI: 10.1007/978-3-319-29971-6_11

There are a large number of applications in computer vision and robotics where visual information is used to guide wheeled or legged vehicles over unknown terrain (see for example [8,20]). However, the use of orientation information alongside visual information has been less well studied, despite the apparent biological motivation for doing so. Fusing information from these sensing modalities offers great potential for assisting in tasks relevant to locomotion, for example the structural interpretation of image content which we consider in this work. This is based on the observation that the content of an image typically changes in relation to its real-world orientation – for example a downward pointing camera tends to be looking at the ground, while a sideways facing camera can expect to see a combination of walkable terrain and obstacles. This information alone is not sufficient for predicting image content of course, since it disregards any specific information about the current scenario; but it can serve as a useful prior for the kinds of structures to be expected, when used in combination with visual information.

(a)   (b)

Fig. 1. Typical results of our algorithm, showing how segmentation results using only vision (left) can be improved by taking into account the camera orientation (right). In both examples knowledge of the camera orientation avoids misclassifying vertical walls as ground (yellow). See Fig. 7 for full color legend. All images are best viewed in color (Color figure online).

Our work develops these ideas and presents the first method, to our knowledge, for combining visual and vestibular information for semantic image segmentation. Using this we show that by combining camera orientation, measured with an inertial sensor, with visual features extracted from images from the camera, we can achieve improved performance in an image segmentation task. The ultimate aim of this work is to build a system enabling autonomous locomotion by legged robots. In order to work towards this goal we focus on developing a method for guiding humans through urban landscapes. Not only is this a convenient test-bed for evaluating vision guidance algorithms without the constraints of robot locomotor capability, but it also demonstrates a potential application in guiding visually impaired humans, where knowledge of the scene structure is of great importance [28]. With this in mind we developed an algorithm which segments images into relevant structural regions, such as the walkable ground region, impassable obstacles and intermediate surfaces such as stairs, and displays the result through a head-mounted display unit. Examples outputs of the algorithm are shown in Fig. 1.

The next section discusses related work in the field. A brief overview of our method is given in Sect. 3, followed by Sect. 4 which describes the data acquisition process. Section 5 gives full details of how our algorithm works. We then present extensive results and examples in Sect. 6, before concluding in Sect. 7. Please note that this paper is an extended version of our earlier work [13], incorporating new results and examples.

2   Related Work

Using inertial sensors has been known to improve performance in a variety of computer vision tasks. One of these is visual simultaneous localisation and mapping, in which the pose of a camera with respect to a map, and the unknown map itself, must be recovered. The pose estimate given by an inertial sensor can be fused with that derived from vision to improve robustness [22] and help to mitigate scale drift [24]. A rather different example from [15] uses inertial information for blur reduction, by using estimates of the camera's motion derived from an inertial sensor during an exposure to guide deconvolution.

The work of Hoiem at el. [14] is more closely related to ours, in that images are segmented into geometrically consistent regions. As well as visual features, this uses the position of a segment within the image as a feature during classification, to learn that sky occurs toward the top of the image, for example – although in this work the camera is assumed to be in an upright position with no roll. Visual segmentation can be enhanced using other 3D information – for example using features extracted from a point cloud to help classify objects in road scenes [26]; or by jointly segmenting individual video frames and labelling structures in a 3D reconstruction from those frames [18]. While the orientation of the camera may be implicitly included in these methods via the 3D map, this is not directly investigated, and furthermore is estimated from the image stream itself.

The use of inertial data for terrain classification was investigated by [25], where the inertial data themselves are used as features to encode the vehicle vibration and accelerations for different terrains, in order to predict the terrain type over which the vehicle traverses. This bears some similarity to our approach, in that inertial data is being used for classification, but it does not attempt to make use of the relationship between class and pose.

While these show interesting uses of information not directly present in the image to aid labelling, they are not making use of the information potentially provided by the camera orientation itself. Similarly, while some of the above mentioned works use inertial data to aid vision tasks, this is generally in a purely geometric sense, and they have not exploited the relevance to semantic attributes in the image. We investigate ways to do this in the following sections.

3   Overview

In this paper we present an algorithm which takes as input a single image and its associated 3D orientation, measured with an inertial measurement unit (IMU), and produces a segmentation of the image into distinct regions, corresponding to classes relevant to the task of locomotion. The classes used are: ground (walkable), plane (non-walkable, usually vertical), obstacle (non-walkable and not planar), stairs (walkable with caution), foliage (possibly traversable, maybe with a different gait), and sky (neither traversable nor obstacular). These are colored yellow, red, magenta, cyan, green and blue respectively in all examples. This particular choice of classes is somewhat arbitrary – and our algorithm is not specific to this choice of course – but we believe this represents a reasonably minimal set of necessary classes to facilitate locomotion through different environments.

To demonstrate the use of orientation in enhancing segmentation, we developed a relatively simple means of classifying and segmenting images. We segment an image by describing a grid of points with a collection of feature vectors, consisting of visual and pose information. These are used to predict the most likely class for each point with a pre-trained classifier. Since each point is classified independently, this initial segmentation exhibits much noise. To mitigate this, we experiment with a Markov random field (MRF) algorithm to enforce a smoothness constraint across the grid of points; or alternatively a conditional random field (CRF) to create more detailed segmentations. Using any of these approaches we show that fusing visual and orientation information can substantially improve segmentation accuracy over using either alone; and crucially, that orientation information enhances performance beyond using position within the image as a feature.

We also show that classification of lines detected in the image can be enhanced by adding location and orientation features, as well as features encoding properties of the lines themselves (non-visual features are collectively referred to as pose features). Finally, we show that combining the results from point and line classification can improve performance over either in isolation.

The result of our method is a segmentation of the image, comprised of sets of contiguous points with the same classification which, as Fig. 1 shows, divides the image into regions appropriate for a navigation task (here showing MRF segmentation). The basic algorithm does not produce a per-pixel segmentation, due to the resolution of the grid we use, but every pixel in the image is covered, and every pixel is used for the description; the CRF segmentation goes beyond this by giving an individual label to each pixel (see Fig. 12).

4   Data Acquisition

To develop and evaluate the algorithms in this paper, we gathered long video sequences (totalling around 90 min of footage) using an IDS uEye USB 2.0

camera[1] fitted with a wide-angle lens (approximately 80° field of view). This provides images at a resolution of 640 × 480, at a rate of 30 Hz.

Our aim is to use this method to guide humans through outdoor environments. Therefore, all our data were gathered from a camera mounted on the front of a virtual reality headset, worn by a person traversing various urban environments. While walking, the subject saw only the view through the camera. This was done to make the data as close as possible to what would be encountered in a real application, both in terms of the camera being mounted in the same way, and the movements of the head being typical of a human with limited visibility.

The hardware we used for this was the Oculus Rift[2] (Dev. Kit 1), which has a large field of view and sufficiently high framerate (up to 60 Hz). The camera was mounted sideways, so that the images have a portrait orientation – this is because the view for each eye is higher than it is wide. We correct for barrel distortion introduced by the lens to produce an image approximating a pinhole camera, using camera parameters obtained with the OpenCV calibration tool.[3]

To gather orientation information we used the inertial sensor built into the Oculus Rift. This comprises a three-axis accelerometer, gyroscope and magnetometer, which are combined with a sensor fusion algorithm to give estimates of orientation in a world coordinate frame at 1000 Hz. We retrieve the orientation as three Euler angles, and discard the yaw angle (rotation about the vertical axis), since in general this will not have any relationship to semantic aspects of the world. Conversely, pitch and roll are important since they encode the camera pose with respect to the horizon line, and thus whether the camera is looking up/down or is tilted. This has an influence on the likelihood of different classes being observed.

From these videos, a subset of frames are hand picked for labelling, for training or testing. They are manually segmented into disjoint regions, built from straight-line segments. Each region is assigned a ground truth label from our

Fig. 2. Example ground truth – the manual labelling of regions (left) and ground truth segmentation (right).

---

[1] en.ids-imaging.com.

[2] www.oculus.com.

[3] www.docs.opencv.org.

set of classes. This is by nature a subjective task, since image content is often ambiguous, but the labelling is as consistent as possible. Some truly ambiguous regions are not labelled, which are omitted from all training and testing.

Examples of ground truth data can be seen in Fig. 2. The labelling is independent of the points and lines which are later created in the image. We also show ground truth segmentations derived from these, in which a grid of points has been assigned labels according to the underlying ground truth (where the blockiness due to the grid is clearly visible). This is the best possible segmentation, against which we evaluate our algorithms in Sect. 6.

5   Classification and Segmentation

In this section we describe the process by which an image is segmented, according to either the visual features, pose features, or combinations thereof; and how these features are created in regions surrounding grid points, detected lines, or both. An overview of the whole system is presented in Fig. 3.

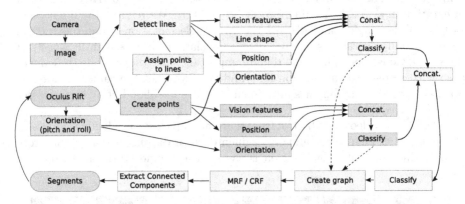

Fig. 3. This block diagram showing how the system as a whole works. The dotted lines show what happens when points or lines are used alone; if both are used, their outputs are concatenated and passed to the meta-classifier.

5.1   Structures

While applying classification at the level of individual pixels is a valid option, this would be very computationally expensive, and the information at one pixel (e.g. its color and location) is unlikely to be sufficiently discriminative. Instead, we use a combination of point and line structures. The points, organized in a grid, are described by features created from their surrounding pixels. Using a regular grid of points also makes segmentation with graphical models more straightforward than using only salient points for example [12]. Lines are detected in the image use the LSD line segment detector[4] [31] (we discard lines under 6 pixels long and 3 pixels wide (LSD gives a width value for each line) since these are likely to be noise). These are used in order to represent high-frequency image content, and distinctive appearance changes over discontinuities, which would be missed by the smaller and more localized point features. Feature vectors for lines are created from surrounding pixels, extending along their length and covering a region of fixed width on either side.

In order to combine lines with the point-based segmentation, we assign points to lines if they lie within the region enclosed by the line feature (a point may be assigned to multiple lines). It is this assignment of points to lines which later allows line classifications to be transferred to points for segmentation; similarly, the ground truth label of a line is obtained via the points, whose label in turn comes from the marked ground truth regions (thus a line's label vector is the mean label vector of all points inside the area used to describe it).

5.2 Features

The features with which we classify structures in the image are divided into two broad categories: visual features and pose features. The former uses information derived from the image pixels to describe local regions of the image; the latter comprise other information not directly present in the image, but pertaining to properties of image structures or the image as a whole.

Visual Features. The visual features we use to describe points are histograms encoding the distribution of gradients and colors in a surrounding square patch. Histograms of gradients describe the local texture, and are built by first convolving the image with gradient filters in the x and y directions, to obtain at each pixel gradient responses g_x and g_y. For each pixel we calculate the gradient angle $\theta = \tan^{-1} \frac{g_y}{g_x}$ and gradient magnitude $m = \sqrt{g_x^2 + g_y^2}$. These are used to build the gradient histogram for a patch by quantising the angles into bins, and weighting the contribution to each bin by their magnitudes. To encode richer structure information we create a separate histogram for each quadrant of the patch and concatenate them together, in the manner of [12].

In addition, color descriptors are created for these patches. These are histograms built in HSV space, which combine a histogram of quantized hue values, weighted by the saturation (since the saturation represents the degree to which the hue is relevant), and an intensity histogram. These are included alongside texture information since color is beneficial when classifying and segmenting images [14,17].

As mentioned above, we also perform classification on lines detected in the image. In order to create a description better suited to lines, for both gradient and color we create pairs of histograms from the pixels in rectangular regions either side of the line, and concatenate them. Thus, the gradient descriptor has

---

[4] Code available at www.ipol.im/pub/art/2012/gjmr-lsd.

half the dimensionality compared to the point case (which had four quadrants), while the color descriptor is twice as long.

Pose Features. The most basic of our pose features is simply the position (of the point, or the line's midpoint) in the image, where we use the x and y coordinates (normalized by image size) directly. This is to represent any dependence on image location which may be exhibited by different classes. Note that the use of location without orientation was not investigated in the original version of this work [13].

The main contribution of this paper is the use of the orientation of the camera as a feature. To obtain this, we use the pitch and roll values from the Oculus Rift IMU, each normalized to the range $[0, 1]$. The orientation features are always combined with image location, since otherwise the orientation feature would be the same for all points in the image: it is the interaction between image position and camera orientation which gives rise to cues of different types of structure at different locations in space.

For the line regions only, we also use a shape descriptor, which comprises simply a line's length, width, and orientation in the image, each appropriately normalized. This is to add extra information – usually at a larger scale – about the scene structure which may be ignored by both the visual and location/orientation features.

5.3   Classification

After extracting features for all points and lines in our training set, each being paired with a ground truth label, we train a set of classifiers. The classifier we use for this work is multivariate Bayesian linear regression [3], chosen because it is both fast to train, and very fast to evaluate for a new input. It is similar to standard regularized linear regression, except that the optimal value for the regularisation parameter can be chosen directly, under the assumption that the data have a Gaussian distribution.

To use it, each class label is represented as a 1-of-K vector (for the $K = 6$ classes), where dimension k is 1 for class k, and zero otherwise. The classifier outputs a K-dimensional vector, which after normalization to sum to 1, is treated as the estimated probability for each class. Prediction is simply a matter of multiplying the feature vector by the $M \times K$ weight matrix (for features of dimensionality M). Rather than the raw feature vector – which would allow for learning only linear combinations of the inputs – we use fourth order polynomial basis functions.

5.4   Combining Information

Different combinations of structures and features lead to different versions of our algorithm. The most basic is using points only (**P**) with visual features (**V**), an algorithm which we will denote **P-V**. Similarly, we can experiment using only location information (which we denote with an 'X'), orientation information (O), and combinations thereof. For algorithms using line structures (L), we can also add the shape feature, denoted by 'S'.

In order to combine different features together, we simply concatenate them to create one long feature (an alternative method was not found to improve accuracy, c.f. [13]). These combinations will be expressed as **P-VX** for example (points with visual and location features concatenated).

This concatenation is also done for line regions' features (e.g. **L-XOS**, which combines location, orientation and shape features). However, we cannot combine points with lines by simply concatenating their features, because their features are created over different image regions. Instead, we retain separate classifiers for both structures, and combine their outputs by a process known as 'meta-learning' (or sometimes 'stacked generalisation') [2]. The K-dimensional predicted label vectors from the two classifiers are concatenated, and treated as a new feature vector. This is input to a second round of classification, the output of which is another K-d label vector, representing the final probability estimate for each class, thus combining information from the points and lines. We run the classifiers whose outputs we wish to combine (e.g. **P-V** and **L-V**) on the training data to gather example outputs. These predicted label vectors are concatenated and paired with the known ground truth label for each point, so that the meta-classifier can be trained (e.g. resulting in **PL-VO**). Note that this concatenation is done at the points, where the points receive labels from the lines in which they lie. Points not within any line regions simply keep their own predicted label.

5.5 Segmentation

The result of any of the above algorithms is a set of points in the image, each having a predicted label vector, from which we choose the most likely class assignment as the dimension with the highest value. Since each point is classified individually, there is no guarantee that neighbouring points will have similar labels, even if they belong to perceptually similar regions of the image; this is especially true when using line regions, as adjacent points may be assigned to different lines.

Markov Random Field. To address this we formulate the problem as a Markov random field (MRF). This allows us to choose the best label for each point according to its observation (i.e. classification result), while also incorporating a smoothness constraint imposed by its neighbours.

We create a grid graph to represent all the points in the image, by connecting each point to its 4-neighbours. The aim when optimising a MRF is to maximize the probability of the configuration of the field (i.e. an assignment of labels to points); this is equivalent to minimising an energy function over all cliques in the graph [19] (we use up to second-order cliques, i.e. unary and pairwise terms). A configuration of the MRF is represented as $p = (p_1...p_N)$, where $p_i \in \mathcal{L}$ is the class assigned to point i of N and $\mathcal{L}$ is the set of possible labels. The goal is to find the optimal configuration p^*, such that $p^* = \text{argmin}_p E(p)$, where $E(p)$ is the posterior energy of the MRF. We define this as:

$$E(p) = \alpha \sum_{i=1}^{N} \psi_d(p_i) + \sum_{i=1}^{N} \sum_{j \in \mathcal{N}_i} \psi_s(p_i, p_j) \tag{1}$$

where the first term sums over all points in the graph, and the second sums over all neighbours $\mathcal{N}_i$ for each point i. α is a weight parameter, balancing the effects of the data and the smoothness terms. The unary and pairwise potentials are:

$$\begin{aligned} \psi_d(p_i) &= \|\mathbf{p}_i - \mathbf{c}_i\| \\ \psi_s(p_i, p_j) &= V_{ij} T(p_i \neq p_j) \end{aligned} \tag{2}$$

$\mathbf{p}_i$ denotes the label p_i represented as a 1-of-K vector, and $\mathbf{c}_i$ is the K-d output of the classifier (thus taking into account the predicted probability for all the classes). $T(.)$ is an indicator function, returning 1 iff its argument is true, and V_{ij} is a pairwise interaction term, controlling the degree to which label dissimilarity is penalized at sites i and j. This is set to $V_{ij} = \beta - \min(\beta, |m_i - m_j|)$, where m_i is the median intensity over the patch at point i. This penalizes differences in label more strongly between points with similar appearance, in order to adapt the segmentation to the underlying image contours. We set the parameters to $\alpha = 60$ and $\beta = 90$ empirically based on observations on the training set (note the pixel intensities are in the range $[0, 255]$). We optimize the MRF using graph cuts with alpha-expansion[5] [6]. After optimising the MRF we perform connected-component analysis to recover the segments. Examples of results before and after MRF segmentation can be seen in Fig. 12.

Conditional Random Field. We also describe an alternative way to segment the image, using a conditional random field (CRF). CRFs have an advantage over MRFs in that they model the conditional distribution of the labels with respect to the features, rather than the full joint distribution. This means an accurate conditional model can have a much simpler structure than a fully generative joint model [27]. Recent advances allow for extremely efficient CRF optimisation, so much so that it is now possible to use a fully-connected graph, as opposed to the grid-structured graph described above, connecting every pixel to every other pixel. To do this we use the algorithm of Krähenbühl and and Koltun[6] [17]. For the unary potential at pixel i and label k we use:

$$U_{ik} = \begin{cases} -ln(\mathbf{p}_{ki}^+) & \text{if point } i \text{ has a label} \\ -ln(\frac{1}{K}) & \text{otherwise} \end{cases} \tag{3}$$

where x^+ is the value of x if it is greater than zero (zero otherwise), and $\mathbf{p}_{ik}$ is the kth element of the K-d probability vector predicted at point i. Since

---

[5] Using the 'gco-v3.0' code at vision.csd.uwo.ca/code.

[6] Using code available at www.philkr.net/home/densecrf.

the CRF is defined over every pixel, most nodes will not have an initial label. The inference algorithm is otherwise used unchanged, except we doubled the standard deviation of the color-independent term, as we found this to improve performance.

6   Results

To evaluate our algorithms, we gathered two datasets as described in Sect. 4. All data were obtained from the same camera, having a (rotated) resolution of 480×640, and were corrected for barrel distortion due to a wide-angle lens.

The first dataset was designated the training set, and contained 178 manually labelled images. This set was used for cross-validation experiments, to demonstrate the claims made above. The second set of 156 images was the test set, which came from different video sequences recorded in physically distinct locations to the training set. This was done to verify that the algorithms generalize beyond the training set, and to show example images (all examples in the paper come from this set). All our labelled data are available online.[7]

Our algorithm has a large number of parameters which will effect its operation. The most important ones are described here, with typical values given. The grid density (distance between points) was set to a value of 15 pixels (making a grid of approximately 30×40 points), to give a compromise between an overly coarse representation/segmentation, and the quadratic increase in computational time for denser grids. The patches around every pixel, from which visual features are built, were squares of side 20 pixels. The width of line regions was set to 30. The basic gradient histogram was 12-d, making the concatenated quadrant feature on points 48-d; and color histograms had 20 dimensions each for the hue and intensity parts. As described earlier, location and orientation features had two dimensions each, while line shape features have three.

These parameters were set to values which appear sensible or are supported by related literature. However, we make no claim that these were the optimal parameters, and much further tuning could be done, although the best settings would depend on the dataset used. We emphasize that this does not alter the central claims of this work, i.e. that making use of orientation information, using either points or lines, can improve segmentation. All parameters were kept constant across evaluations, so all results are relative.

6.1   Cross-Validation

We begin with results obtained through cross-validation on our training set. This was done by running five independent runs of five-fold cross-validation on the data (to mitigate artefacts due to particular choices of training/test splits). For comparison we use classification accuracy, i.e. the average number of times a point was assigned the correct class. Segmentation was evaluated point-wise,

---

[7] Our dataset can be found at www.osianh.com/inertial.

i.e. looking at every point individually, since for the time being we are not concerned with the issue of true segments being wrongly split or merged. We first analyse the performance of the algorithms without the benefits of segmentation: the MRF and CRF are not used here, and we directly used the labels assigned to points by the classifiers.

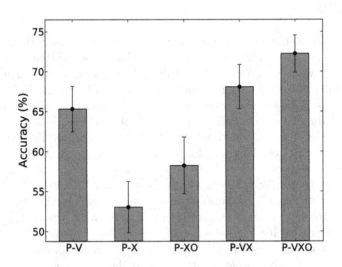

Fig. 4. Adding orientation features to vision features for points. All error bars show a 95 % confidence interval.

First, we ran an experiment to evaluate classification using only points, with various combinations of vision and pose features. The results are shown in Fig. 4. The bars indicate the average accuracy over all the runs of cross-validation, and the error bars are drawn to show a 95 % confidence interval, based on the average standard error over all runs of cross-validation.

Using vision features alone (**P-V**) provides a reasonable baseline performance. We ran experiments using only the location feature (**P-X**) or location plus orientation (**P-XO**) – as one might expect, these perform much worse than using only vision, since no image information is actually used. Nevertheless, it is encouraging to see that adding orientation already improves the accuracy, and it can be surprising what orientation alone can tell us about what an image is expected to contain, as we will show in the next section.

One of the key results of this paper is that combining vision with orientation information (**P-VXO**) is significantly better than using vision alone. Crucially, we also show that while adding image position as a feature (**P-VX**) does give some improvement (as per [14]), it is the combination of inertial information with image location which gives the largest increase. This was not shown in the original paper [13], but is important in demonstrating that the prior introduced by where the camera is pointing is relevant to classification.

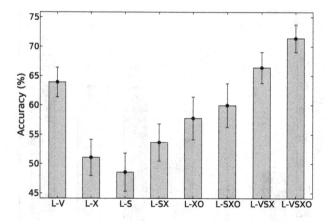

Fig. 5. Using orientation with vision features for lines.

The next experiment was the same as the above, but for line regions instead. Note that these evaluations were done using the points which were assigned labels from classified lines, not on the lines themselves (points not in lines were excluded from the evaluation). As shown in Fig. 5, we tested the use of line location (**L-X**) and shape on their own (**L-S**), in combination (**L-SX**), and combined with orientation (**L-SXO**), again showing the increased performance when using orientation. When combining with visual information (**L-VSXO**), we show that this is superior to using visual information alone, or even visual information with shape and location (**L-VSX**), once again demonstrating that the addition of orientation information is of primary importance, and is what allows us to obtain significantly better classification rates.

Finally, we investigated the effect of combining point and line features. Figure 6 first shows both point and lines separately, using only vision features (the same results from the two previous graphs), then the result obtained when combining both point and line classification (**PL-V**). It can be seen that this improves performance compared to using either structure in isolation. We then see the same trend when adding location, orientation and shape information: the graph shows the results of points and lines individually with the full set of features (once again the same as the previous graphs), and finally the result using both structures and all features (**PL-VSXO**). This results in an improved accuracy, suggesting that combining information from multiple types of patch/region is indeed beneficial, albeit by a smaller margin than the above experiments.

In Fig. 7 we show a confusion matrix, obtained as the mean confusion matrix over all runs of cross-validation, using the full algorithm **PL-VSXO**. The diagonal is pleasingly prominent, though there is significant confusion between stairs and ground (when the true class is stairs), which is somewhat unfortunate from a safety point of view. Vertical surfaces also tend to be confused with other obstacles and foliage, which is less of a concern. For our task, ground identification is perhaps the most important criterion, which appears to be the strongest result.

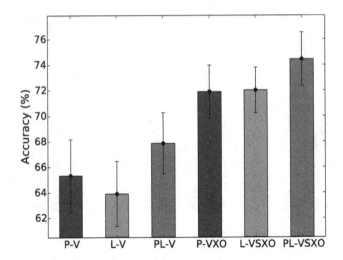

Fig. 6. Combining predictions from both points and lines.

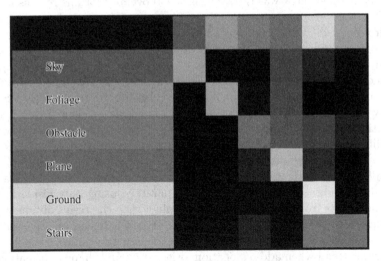

Fig. 7. Confusion matrix over all runs of cross-validation, for complete **PL-VSXO** algorithm. Rows correspond to the true classes, while columns represent the predicted classes. Colors correspond to those used through all segmentation examples.

6.2   Independent Data and Examples

After the cross-validation experiments, we trained sets of classifiers corresponding to different variants of the algorithm, using the training set above (plus copies obtained by reflecting across the vertical image axis). We used these to evaluate performance on the independent test set. Results are shown in Table 1. This confirms the important result of the paper: that combining orientation information

Table 1. Comparison of the different algorithms on independent test data. Using a MRF to smooth away spurious local detections increases accuracy slightly in all cases; a fully connected CRF does not give better performance than the MRF as measured here, but gives more detailed segmentation.

Algorithm	Accuracy	MRF	CRF
P-V	61.0%	64.7%	63.1%
P-VXO	71.1%	73.8%	72.2%
PL-VSXO	71.5%	74.6%	72.5%

is beneficial, exhibiting around 10% increase in overall accuracy. Adding line information did confer a further improvement, although this was only slight. We also show results after applying MRF and CRF segmentation, both of which increased accuracy by a few percent.

We now show example results taken from the test set, showcasing the differences between the algorithms presented above and demonstrating what they are capable of. In all example images in the paper (except Fig. 12), the MRF segmentation has been run, to remove noise and give a tidier segmentation.

(a)   (b)

Fig. 8. Example results, showing segmentation using only vision features (left) and combined with orientation features (right). See color legend in Fig. 7.

First, Fig. 8 shows side by side examples of the basic vision version (**P-V**), and the effect of adding pose information (**P-VXO**). In (a), the building façade is partly mistaken for the ground by the visual features, whereas knowing the camera is pointing upwards corrects this. In (b) the miss-classification of the road as stairs is also corrected.

In the next example (Fig. 9) we show the effect of adding line classifications to the points-only segmentation, in both cases using all visual and pose features (**P-VXO, PL-VSXO**, respectively). These examples show how the information gleaned from the lines can aid segmentation, for example by disambiguating stairs and planes, or finding non-planar objects. However, as our results below will show, lines can sometimes be detrimental.

Fig. 9. Examples showing how adding line classifications (centre) in conjunction with point features (**P-VXO**, left) can help improve segmentation (**PL-VSXO**, right).

It is interesting to see what effect the orientation features have, independently of the vision features, so in Fig. 10 we show results generated using **P-XO** and **PL-XOS**, i.e. there are no visual features at all being used in these segmentations (image information is being used only for line detection). Figure 10(a) appears to be correctly segmented, but only because this is a common and rather empty configuration of ground and walls; whereas the cars in (b) are obviously ignored. (c) is interesting since it shows that with the camera looking down at a certain angle, stairs are predicted – in this case correctly. This raises the interesting issue that stairs are predicted here not just because they are likely to be below the viewer, but because the viewer is likely to look downward when walking up stairs. In Fig. 10(d) and (e) the use of lines has altered the segmentation, to give the impression it is seeing the bollards and the sky (the points assume there is sky above, but lines even at such a height are rarely labelled as sky in the training set). In (f) the lines themselves are shown, and it can be seen how their orientation in the image has an effect, since the bollards and paving stones are classified differently, despite being at around the same image height.

More examples are shown in Fig. 11. Here, we show the input image for clarity, plus the ground truth segmentation. The contributions from vision (**P-V**), orientation (**P-VXO**), and lines (**L-VSXO**) to the final segmentation (**PL-VSXO**) are shown. Figure 11(a) and (b) again show orientation information being used to improve classifications, the latter being an interesting example where adding lines improves segmentation even in the presence of motion blur. Note that the different orientations of the camera, such as in (c) and (d), illustrates why using only position in the image as a prior is inferior.

(a)   (b)   (c)   (d)   (e)   (f)

Fig. 10. Example segmentations using only orientation information features – points only (a–c) and points with lines (d–f). (f) shows the lines themselves, showing the effect of the lines' orientations within the image, aiding detection of vertical posts.

The segmentation in Fig. 11(e) also benefits from classification of lines along the steps. Similarly in Fig. 11(c) lines help to correctly identify the step-edges, but the step faces are classified as ground. In a way this is correct, since stairs are made up of periodic walkable regions, but this result would be marked mostly incorrect compared to our ground truth, which is labelled at a coarser resolution. This echoes our comment in Sect. 4 about the world being ambiguous; but also that some regions may belong to multiple classes simultaneously at different scales.

The example in Fig. 11(g) also shows orientation information being used to correctly identify the non-ground surface; however, the addition of lines in this case degrades the result. The final two rows show examples where our augmented algorithms fail to provide any benefit. In Fig. 11(h), the initial **P-V** segmentation is correct, and is unchanged by the addition of orientation or lines (of course, if we could achieve perfect segmentation, no amount of prior knowledge would help). On the other hand, this illustrates why it is so important that orientation does not impose a hard constraint on surface identity: even when orientation features are added, the grass (foliage class) remains. In Fig. 11(i), none of the versions of the algorithm are able to detect either the ground plane or the foliage, perhaps due to the lower level of illumination.

Our implementation, consisting of single-threaded C++ code running on a desktop PC (Intel i5, 2.40 GHz), processes one image in around 0.3 seconds on average (including MRF optimisation). This is below the camera rate, but fast enough for real-time use when run in a separate thread; further improvements could be made by parallelising the code or using a GPU. Videos of our code running can be seen on our website.[8]

6.3   Detailed Segmentation

Finally, we show results of the algorithm when segmenting using the CRF (see Sect. 5.5). This is much slower than the MRF (taking over 1 s per image), but since it uses a fully-connected graph of every pixel in the image it results in

---

[8] Videos available at www.osianh.com/inertial.

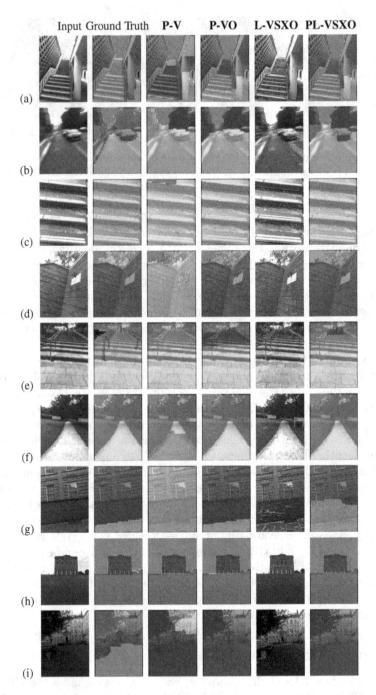

Fig. 11. Example results of the various algorithms. After the input and ground truth, we show the baseline result, of points with only vision features (**P-V**), followed by adding orientation information (**P-VXO**). Detected and vision-classified lines are shown, before the final result, combining everything (**PL-VSXO**).

Fig. 12. This shows the result of the pointwise classification before any segmentation or smoothing (second column). The MRF reduces noise by imposing a smoothness constraint, and groups together points with the same class (third column). We also show results using a dense fully-connected CRF (fourth column), which assigns a label to every pixel using the pointwise result as input.

much more detailed segmentations. The fully-connected nature of the graph means that relationships between distant regions of the image can be taken into account, which can help improve the robustness of the segmentation; however, this also means that classifications in one region can be influenced by those in another, so that a region's label may change as different parts of the scene come into view. As Table 1 shows, the CRF gives some improvement on the raw output, but does not (in its current configuration) out-perform the MRF (note that we are only evaluating using the points, as before, so the evaluation cannot benefit from the improved resolution). Nevertheless, as the examples in Fig. 12 show, the dense CRF can give a significantly more precise and detailed segmentation of the image. It helps to more clearly delineate small objects such as bollards (Fig. 12(a)) and complex boundaries like trees (c,d), although it can also introduce some misclassifications (e). Videos of this being run in a threaded real-time system are also available on our website.

7   Conclusion

We have presented a way of combining information about the real-world orientation of a camera, obtained through inertial measurements, with more traditional vision features, for an image segmentation algorithm. This focused on our example application of scene segmentation for locomotion in outdoor environments, but we would expect the results to be applicable to other types of classification, segmentation, scene understanding and image parsing tasks where the orientation of the camera is likely to effect the image content. We have also shown that adding orientation information is beneficial for line regions; and that combining points and lines in a similar manner can lead to some further improvement.

Our experiments used a comparatively basic design of segmentation algorithm to highlight the effect of using extra prior information. While we have also experimented with using more advanced segmentation techniques, the CRF segmentation took as input our classified image. An interesting avenue of further research would be to combine the orientation prior with the visual features within the graphical model framework [27], to make use of the graph structure at the classification stage too.

Other future work will look at ways of using this method with other sources of information, for example making use of temporal information to enforce consistency across frames, or to combine with depth and 3D data.

Acknowledgments. This work was funded by the UK Engineering and Physical Sciences Research Council (EP/J012025/1). The authors would like to thank Austin Gregg-Smith and Geoffrey Daniels for help with hardware and data, and Adeline Paiement for all the enlightening discussions.

References

1. Angelaki, D.E., Cullen, K.E.: Vestibular system: the many facets of a multimodal sense. Ann. Rev. Neurosci. **31**, 125–150 (2008)
2. Bi, Y., Guan, J., Bell, D.: The combination of multiple classifiers using an evidential reasoning approach. Artif. Intell. **172**(15), 1731–1751 (2008)
3. Bishop, C.M.: Pattern Recognition and Machine Learning. Springer, New York (2006)
4. Brandt, T., Bartenstein, P., Janek, A., Dieterich, M.: Reciprocal inhibitory visual-vestibular interaction. Brain **121**(9), 1749–1758 (1998)
5. Dahlkamp, H., Kaehler, A., Stavens, D., Thrun, S., Bradski, G.R.: Self-supervised monocular road detection in desert terrain. In: Robotics Science and Systems (2006)
6. Delong, A., Osokin, A., Isack, H.N., Boykov, Y.: Fast approximate energy minimization with label costs. Int. J. Comput. Vis. **96**(1), 1–27 (2012)
7. Deshpande, N., Patla, A.E.: Visual-vestibular interaction during goal directed locomotion: effects of aging and blurring vision. Exp. Brain Res. **176**(1), 43–53 (2007)
8. De Souza, G.N., Kak, A.C.: IVision for mobile robot navigation: a survey. IEEE Trans. Pattern Anal. Mach. Intell. **24**(2), 237–267 (2002)
9. Domke, J.: Learning graphical model parameters with approximate marginal inference. IEEE Trans. Pattern Anal. Mach. Intell. **35**(10), 2454 (2013)
10. Gould, S., Fulton, R., Koller, D.: Decomposing a scene into geometric and semantically consistent regions. In: IEEE International Conference on Computer Vision (2009)
11. Gupta, S., Arbeláez, P., Girshick, R., Malik, J.: Indoor scene understanding with RGB-D images: bottom-up segmentation, object detection and semantic segmentation. Int. J. Comput. Vis. **112**, 1–17 (2014)
12. Haines, O., Calway, A.: Recognising planes in a single image. IEEE Trans. Pattern Anal. Mach. Intell. **37**(9), 1849–1861 (2014)
13. Haines, O., Bull, D., Burn, J.F.: Using inertial data to enhance image segmentation. In: International Conference on Computer Vision Theory and Applications (2015)
14. Hoiem, D., Efros, A.A., Hebert, M.: Recovering surface layout from an image. Int. J. Comput. Vis. **1**(75), 151–172 (2007)
15. Joshi, N., Kang, S.B., Zitnick, C.L., Szeliski, R.: Image deblurring using inertial measurement sensors. ACM Trans. Graph. **29**(4), 30 (2010)
16. Kleiner, A., Dornhege, C.: Real-time localization and elevation mapping within urban search and rescue scenarios. J. Field Robot. **24**(8–9), 723–745 (2007)
17. Krähenbühl, P., Koltun, V.: Efficient inference in fully connected CRFs with gaussian edge potentials. In: Advances in Neural Information Processing Systems (2011)
18. Kundu, A., Li, Y., Dellaert, F., Li, F., Rehg, J.M.: Joint semantic segmentation and 3D reconstruction from monocular video. In: Fleet, D., Pajdla, T., Schiele, B., Tuytelaars, T. (eds.) ECCV 2014, Part VI. LNCS, vol. 8694, pp. 703–718. Springer, Heidelberg (2014)
19. Li, S.Z.: Markov Random Field Modeling in Image Analysis. Springer, New York (2009)
20. Lorch, O., Albert, A., Denk, J., Gerecke, M., Cupec, R., Seara, J.F., Gerth, W., Schmidt, G.: Experiments in vision-guided biped walking. In: IEEE International Conference on Intelligent Robots and Systems (2002)

21. Maimone, M., Cheng, Y., Matthies, L.: Two years of visual odometry on the mars exploration rovers. J. Field Robot. **24**(3), 169–186 (2007)
22. Nützi, G., Weiss, S., Scaramuzza, D., Siegwart, R.: Fusion of IMU and vision for absolute scale estimation in monocular SLAM. J. Intell. Robot. Syst. **61**, 287–299 (2011)
23. Patla, A.E.: Understanding the roles of vision in the control of human locomotion. Gait Posture **1**(5), 54–69 (1997)
24. Piniés, P., Lupton, T., Sukkarieh, S., Tardós, J.D.: Inertial aiding of inverse depth SLAM using a monocular camera. In: International Conference on Robotics and Automation (2007)
25. Sadhukhan, D., Moore, C., Collins, E.: Terrain estimation using internal sensors. In: International Conference on Robotics and Applications (2004)
26. Gould, S., Fulton, R., Koller, D.: Combining appearance and structure from motion features for road scene understanding. In: British Machine Vision Conference (2009)
27. Sutton, C., McCallum, A.: An introduction to conditional random fields for relational learning. In: Introduction to Statistical Relational Learning, pp. 93–128 (2006)
28. Tapu, R., Mocanu, B., Zaharia, T.: A computer vision system that ensure the autonomous navigation of blind people. In: Conference on E-Health and Bioengineerin (2013)
29. Vidal, P.P., Degallaix, L., Josset, P., Gasc, J.P., Cullen, K.E.: Postural and locomotor control in normal and vestibularly deficient mice. J. Physiol. **559**(2), 625638 (2004)
30. Virre, E.: Virtual reality and the vestibular apparatus. Eng. Med. Biol. Mag. **15**(2), 41–43 (1996)
31. Von Gioi, R.G., Jakubowicz, J., Morel, J., Randall, G.: LSD: a fast line segment detector with a false detection control. IEEE Trans. Pattern Anal. Mach. Intell. **32**(4), 722–732 (2010)

Multiple View 3D Reconstruction with Rolling Shutter Cameras

Gaspard Duchamp[✉], Omar Ait-Aider, Eric Royer, and Jean-Marc Lavest

Institut Pascal, Aubiere, France
gaspard.duchamp@gmail.com

Abstract. Nowadays Rolling shutter CMOS cameras are embedded on a lot of devices. This type of cameras does not have its retina exposed simultaneously but line by line. The resulting distortions affect structure from motion methods developed for global shutter, like CCD cameras. The bundle adjustment method presented in this paper deals with rolling shutter cameras. We use a projection model which considers pose and velocity and needs 6 more parameters for one view in comparison to the global shutter model. We propose a simplified model which only considers distortions due to rotational speed. We compare it to the global shutter model and the full rolling shutter one. The model does not need any condition on the inter-frame motion so it can be applied to fully independent views, even with global shutter images equivalent to a null velocity. We also propose a way to handle epipolar geometry for rolling shutter. It is shown that constraint using essential matrix becomes non linear, and we show how to use it to recover poses and speeds from matched points. Results with both synthetic and real images shows that the simplified model can be considered as a good compromise between a correct geometrical modelling of rolling shutter effects and the reduction of the number of extra parameters.

1 Introduction

Origin. The rolling shutter is a way to acquire an image which is a consequence of the electronic of the retina. In argentic photography, the chemical sensor used to record the image has a sensibility to light. According to some condition like ambient lightning, it needs a precise duration of exposition to achieve a correct exposition, and this time is managed by a mechanical obturator which block the light. With a numerical sensor, there is no need of physical shutter, the electronic can command the start and the stop of the light capture.

A numerical sensor is an array of photo-sensible cells (pixels) which will individually catch rays of light and convert it in a digital signal. First, the cells unload their electrical charges, the reset, then begin the exposition where the capted photons are converted to electric charges. At the end of the exposition, the charge is proportional to the quantity of light got. Then the charges are measured and so is obtained the image.

As the number of pixels is high, the electronic system in charge of the collect of the charges has to sequence the transfer to the system to form the image.

© Springer International Publishing Switzerland 2016
J. Braz et al. (Eds.): VISIGRAPP 2015, CCIS 598, pp. 227–239, 2016.
DOI: 10.1007/978-3-319-29971-6_12

The two main technologies of digital camera sensor are Charge Coupled Device (CCD) and Complementary Metal Oxyde Semiconductor (CMOS). The main difference between these two technologies is the sequencing of charge transfer.

CCD cameras have been mainly used for years, the charges can be stored in its cells so in general there is only one circuit to convert them. The acquisition is done for the whole image and treatment time is quite long. The CMOS technology is more recent in image sensing, nowadays it is quasi omnipresent in embedded devices such as phones, notepads and even in cameras. It allows lighter electronic systems making cameras cheaper and less power-hungry. Basic CMOS can't store its charges, so there is an amplifier for each cell, the read out is done sequentially line by line and so is the acquisition. This introduce a delay between each line and acquiring a dynamic scene will produce some distortion in the image.

Overview. There are different way to correct distortion introduced by the rolling shutter or to take advantage from them in the literature. One way is to undistort the entire image [1–3]. This kind of methods gives correct visual results but is not satisfying because it does not deal with 3D structure of the scene. [4] and [5] solved the PNP problem of a moving object of known geometry by taking advantage of image distortion to get the speed of the target simultaneously with the pose. [6] gives a method to get a temporal calibration of the rolling shutter, and a correct model for small rotational speed and fronto-parallel motion. [7] propose a polynomial projection model and a constrained global optimization technique in order to solve the minimal PNP problem without any initial guess of the solution making the method more suitable for automatic 3D-2D matching in a RANSAC framework. [8] proposes a unifying model for both motion blur and rolling shutter distortion for dense registration.

Recently, few works adressed the problem of structure from motion using rolling shutter image sequences. [9] studied 3D reconstruction and egomotion recovering using a calibrated stereo rig. [10] presents a bundle adjustment method which computes structure and motion from a rolling shutter video exploiting the continuity of the motion across a video sequence. [11] consider the stereo in the case of a fast moving vehicle where rotational speed is neglected and where Rolling Shutter effects are supposed to be affected principally by the depth of the scene. A recent way to handle reconstruction is not to consider discrete poses of a camera along a trajectory, but a continuous time motion in space as do [12] non sequential captor and [13] for cameras including rolling shutter ones. Finally, [14,15] proposes an approach to correct the reconstruction based on Kalman filter using inertial sensors which are more and more embedded on devices like smartphones or notepads.

2   Formalism

We consider 3D object expressed as a set of point $\mathbf{Q}$, a camera has a pose defined by an orientation and a position $[\mathbf{R}\ \mathbf{T}]$. The amount of rotation completed by the camera under a rotational motion during a scan time from the origin of time of an image to some line is $\delta\mathbf{R}$. For a translational motion it is $\delta\mathbf{T}$. For a global shutter camera, the equation of projection is

$$\tilde{\mathbf{m}} = \mathbf{K}[\mathbf{R}^{\mathbf{T}} \quad - \mathbf{R}^{\mathbf{T}}\mathbf{T}]\tilde{\mathbf{Q}} \qquad (1)$$

where $\mathbf{K}$ is the matrix the instrinsic parameters of the camera, $\mathbf{m} = [\mathbf{u}, \mathbf{v}]^T$ the perspective projection of $\mathbf{Q}$ noting $\tilde{\mathbf{m}}, \tilde{\mathbf{Q}}$ the homogeneous coordinates of $\mathbf{m}, \mathbf{Q}$.

The projection equation of this point considering a uniform motion during the time of one image scanning with a rolling shutter camera is:

$$\tilde{\mathbf{m}}_{\mathbf{i}} = \mathbf{K}[\delta\mathbf{R}_{\mathbf{i}}\mathbf{R}^{\mathbf{T}} \quad - \delta\mathbf{R}_{\mathbf{i}}\mathbf{R}^{\mathbf{T}}(\mathbf{T} + \delta\mathbf{T}_{\mathbf{i}})]\tilde{\mathbf{Q}}_{\mathbf{i}} \qquad (2)$$

where τ is the delay from a line to the next one, $\delta\mathbf{R}_i$ is the amount of rotation due to rotational velocity at the time corresponding to the line $\tau \cdot v_i$, and $\delta\mathbf{T}_i$ the amount of translation due to translational velocity at this time. The index i is for the ith 3D point, v_i its corresponding line on the sensor and $t_i = \tau \cdot v_i$ the delay in acquisition from the first line to the line of the current 3D point i.

3   Related Work

Bundle Adjustment. The closest work to ours, is the one presented in [10,16]. Rolling shutter bundle adjustment is achieved by exploiting the continuity of the motion across a video sequence. A key rotation and translation is associated to the first row of each frame as in classical bundle adjustment. In addition, the poses attached to the rest of image rows are interpolated from each pair of successive key pose parameters. The basic idea is to assume that the trajectory and pose variation between frames are smooth.

$$\tilde{\mathbf{m}}_i = \mathbf{K}\left[\mathbf{R}_{j,j+1}(\mathbf{v})^T \quad -\mathbf{R}_{j,j+1}(\mathbf{v})^T\mathbf{T}_{j,j+1}(\mathbf{v})\right]\mathbf{Q}_i \qquad (3)$$

where $\mathbf{R}_{j,j+1}(v)$ corresponds to the interpolated rotation between the jth and the $j^{th} + 1$ image at the line v, The interpolation method used by authors is SLERP [17]. And $\mathbf{T}_{j,j+1}(v)$ a linear interpolation of the positions between the jth and $j^{th} + 1$.

The advantage of this approach is that only six extra parameters are used for the entire sequence in comparison to classical bundle adjustment. Nevertheless, it seems evident that the inter-frame motion should be very small to ensure that interpolated rotations and translation fit the actual values. As result, the approach requires a high frame rate as in the experiments presented in the paper. This increases the risk of data bottleneck and/or limits the dynamic performances in real time applications such as SLAM with mobile robots. in addition, the quality of motion estimation and triangulation is always better when the inter-frame motion is significant. Therefore, it seems to us that a method which estimates independent cameras without any assumption about motion parameters during inter-frame intervals is more pertinent.

Stereo Rig. A triangulation method for a stereo rig which at least one rolling shutter camera is presented in [9]. Considering the first camera as reference frame, the projection equation for each are

$$\mathbf{m}_i = \mathbf{K} \begin{bmatrix} \delta\mathbf{R}_i & \delta\mathbf{T}_i \end{bmatrix} \mathbf{Q}_i \tag{4}$$

$$\mathbf{m}'_i = \mathbf{K}' \begin{bmatrix} \mathbf{R}\delta\mathbf{R}'_i & \mathbf{T} + \delta\mathbf{T}'_i \end{bmatrix} \mathbf{Q}_i \tag{5}$$

Here rotational speed is presented as an instantaneous axis of rotation $\mathbf{a}$ and an angular speed ω and et a vector $\mathbf{V}$ pour la translation. Matrix are obtained thanks to Rodrigues formula

$$\delta\mathbf{R}_i = \mathbf{a}\mathbf{a}^T \left(1 - \cos(t_i\omega)\right) + I\cos(t_i\omega) + \hat{\mathbf{a}}\sin(t_i\omega) \tag{6}$$

$$\delta\mathbf{T}_i = t_i\omega\mathbf{V} \tag{7}$$

$$\delta\mathbf{R}'_i = \mathbf{b}\mathbf{b}^T \left(1 - \cos(t'_i\omega)\right) + I\cos(t'_i\omega) + \hat{\mathbf{b}}\sin(t'_i\omega) \tag{8}$$

$$\delta\mathbf{T}'_i = t'_i\omega\mathbf{V}' \tag{9}$$

Parameters of speed in the second camera $\mathbf{b}$ and $\mathbf{V}'$ are obtained from those of the first one $\mathbf{a}$ and $\mathbf{V}$ by application of the kinematic twist transformation.

$$\begin{bmatrix} \mathbf{V}' \\ \mathbf{b} \end{bmatrix} = \begin{bmatrix} \mathbf{R} & [\mathbf{T}]_\times\mathbf{R} \\ \mathbf{0}_3 & \mathbf{R} \end{bmatrix} \begin{bmatrix} \mathbf{V} \\ \mathbf{a}\omega \end{bmatrix} \tag{10}$$

It is shown some ambiguities appear for a motion following the epipolar line.

Multi-view Stereo. In the work presented in [11] the authors presented a dense multi-view stereo algorithms that solve for time of exposure and depth, even in the presence of lens distortion. The camera is supposed to be embedded on vehicle and rotations are neglected so that Rolling Shutter effects are supposed to be affected principally by the depth of the scene. The projection equation becomes linear.

$$\tilde{\mathbf{m}}_i = \mathbf{K} \begin{bmatrix} \mathbf{I} - (\mathbf{T} + \delta\mathbf{T}_i) \end{bmatrix} \mathbf{Q}_i \tag{11}$$

Unfortunately, as it stated in [8,18], in practice the lateral rotational movements are the most significant image deformation components.

4   Impact of Speeds on Optical Flow and Distortion

The third model tested here is made with the assumption that the rolling shutter effect due to translation is negligible compared to the one due to rotation. The effect induced by a linear motion parallel to the retina is slightly the same as a rotational motion of the camera according to an axe perpendicular to the linear displacement. In the case of a frontal motion (camera placed at the front or the rear of a vehicle), no rotational motion can have the same effect and the model with only rotational speed cannot compensate, but in this case the optical flow consecutive to the linear motion is reduced. Figure 1 shows the optical flow induced by the motion of the camera. Translations have a very lower effect and become quickly indistinguishable with depth of the view.

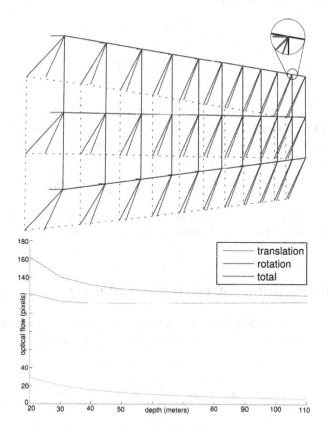

Fig. 1. Simulated image of an object and the optical flow inherent to the camera motions, green: translation (10 m/s), red: rotation (2 rad/s), blue: both (a), optical flow expressed in pixels according to depth of the scene for each type of motion (Color figure online).

5   Epipolar Geometry with Rolling Shutter

Let's consider the fundamental matrix:

$$\mathbf{F} \sim \mathbf{K}_2^{-T} \mathbf{E} \mathbf{K}_1^{-1} \tag{12}$$

and the projection equation of a point in an image:

$$\tilde{\mathbf{m}} \sim \mathbf{K} \left[\mathbf{R}_i^T \; -\mathbf{R}_i^T \mathbf{T}_i \right] \mathbf{Q}_i \tag{13}$$

The epipolar constraint led us to get

$$0 = \tilde{\mathbf{m}}_2 \mathbf{K}^{-T} \mathbf{T}_\times \mathbf{R} \mathbf{K}^{-1} \tilde{\mathbf{m}}_1 \tag{14}$$

with

$$\mathbf{R} = \mathbf{R}_{i1}^T \mathbf{R}_{i2} \tag{15}$$

$$\mathbf{T} = \mathbf{R}_{i1}^T(\mathbf{T}_{i2} - \mathbf{T}_{i1}) \tag{16}$$

for

$$\mathbf{R}_{i1} = \left[\mathbf{R}_1\delta\mathbf{R}_{i1}\right], \mathbf{R}_{i2} = \left[\mathbf{R}_2\delta\mathbf{R}_{i2}\right] \tag{17}$$

and

$$\mathbf{T}_{i1} = \left[\mathbf{T}_1 + \delta\mathbf{T}_{i1}\right], \mathbf{T}_{i2} = \left[\mathbf{T}_2 + \delta\mathbf{T}_{i2}\right] \tag{18}$$

The essential matrix is different for each point due to the scan-line time different for each one. It is possible to recover the parameters of pose and speed by minimizing an epipolar error

$$\epsilon(\mathbf{R}_1, \mathbf{R}_2, \mathbf{T}_1, \mathbf{T}_2, \mathbf{W}_1, \mathbf{W}_2, \mathbf{V}_1, \mathbf{V}_2) = \tilde{\mathbf{m}}_2\mathbf{F}\tilde{\mathbf{m}}_1 \tag{19}$$

where $\mathbf{R}, \mathbf{T}$ stand for camera pose and $\mathbf{W}, \mathbf{V}$ for camera speeds (rotational and translational). For k 3D points detected in n views, we obtain $k(n-1)$ residuals to minimize.

6 Bundle Adjustment for Rolling Shutter

This leads us to minimize the following cost function with respect to pose $(\mathbf{R}_j, \mathbf{T}_j)$, rotational speed $(\mathbf{W}_j)$ and translational speed $(\mathbf{V}_j)$ parameters of jth view:

$$\epsilon(\mathbf{R}, \mathbf{T}, \mathbf{W}, \mathbf{V}, \mathbf{Q}) = \sum_{j=1}^{k} \sum_{i=1}^{l} [\mathbf{m_{ij}} - \mathbf{p_{ij}}]^2 \tag{20}$$

$\mathbf{p}_{ij}$ being the detected points in j^{th} image and $\mathbf{m}_{ij}$, the projection in j^{th} image of the 3D point $\mathbf{Q}_i$ associated to $\mathbf{p}_i$.

7 Results

7.1 Synthetic Data Experiment

To test the reconstruction, a first stage was to use synthetic data. An object was created as a 3D point cloud. Cameras with their own speed and pose were placed around watching towards it. there is no correlation between the orientation of the speed and the displacement between each view. Each image of the scene is considered as totally independent. Images were obtained by application of the rolling shutter projection equation. Resolution of the problem was then done for each model, global shutter, rolling shutter and simplified rolling shutter. The virtual object was included in a cube of 35 m edge placed at 60 m to 70 m from the cameras.

Table 1. List of parameters used for the simulation.

	Views	3d pts	Angular speed	Linear speed	Noise
Min	2	100	0	0	0
Max	12	500	1.5	20	1
Step	2	200	0.25	5	0.1
Steps	6	3	7	5	11

Bundle Adjustment. *Speeds.* Simulated cameras resolution was 1600 per 1200 pixels, the focal was 6.5 mm, τ was set to $25\,\mu s$. Cameras speed magnitude was in range $[0, 20]\,m/s$ and $[0, 1.5]rad/s$ for linear and angular speed. Those parameters were chosen to keep the object in the vision field and the speeds according to the ones available for a pedestrian or the autonomous vehicle VIPA (http://www.ligier.fr/ligier-vipa). A noise on measures was applied in range $[0, 1]\,pixel$, the number of 3D points in range $[100, 500]points$ and the number of views in range $[2, 12]images$. All of those parameters are shown in the Table 1. For each set, 100 simulation were done and solved with each model. The number of simulations is 2673000.

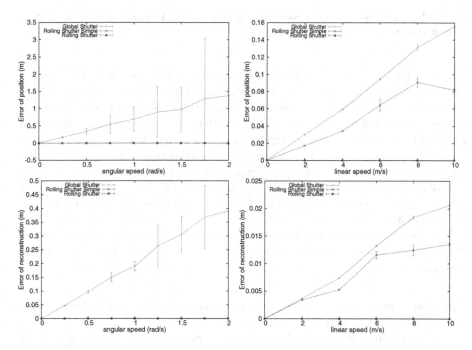

Fig. 2. Errors of position according to speeds, Errors of reconstruction according to speeds.

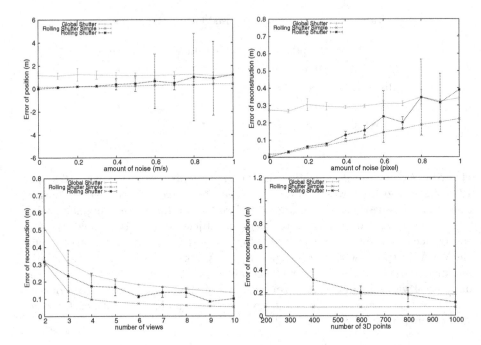

Fig. 3. Errors of position and reconstruction according to noise, Errors of position according to noise.

Noise. In this section we study robustness of the model to noise. We added a random geometric noise following a uniform distribution from 0 to 1 pixel every 0.1 pixel. Each measure is a mean of 100 simulations following the same scheme as previously for a speed corresponding to $[10\,m/s, 2\,rad/s]$. As one can see on Fig. 3, the addition of degrees of freedom to the system makes it less robust to noise. It needs more views of the scene and more 3D points to get the system constrained enough and the reconstruction robust to a high noise level when using the complete rolling shutter model. The simplified rolling shutter model is more robust. Figure 3 shows the precision of reconstruction according to the number of cameras for a speed and noise corresponding to $[10\,m/s, 2\,rad/s, 0.5\,pixel]$ and 600 3D points, and the precision of reconstruction according to the number of 3D points for a speed and noise corresponding to $[10\,m/s, 2\,rad/s, 0.5\,pixel]$ and 4 views.

Epipolar Constraint. *Speeds.* The same approach was used to test motion reconstruction with epipolar constraint.

This constraint seems to be less sensitive to rolling shutter effects with exacts data, but it is when there is noise cf Fig. 4.

Noise. The addition of noise, even a small one has a big impact on the global convergence for global shutter and full rolling shutter methods. As we can see on Fig. 5.

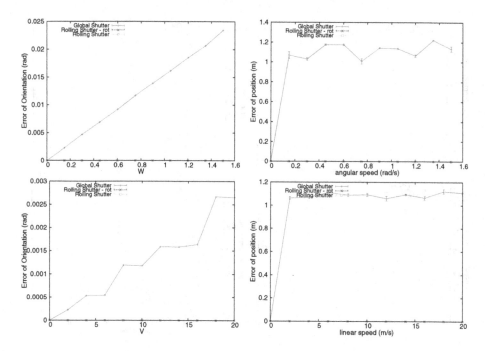

Fig. 4. Errors of orientation and position according to rotationnal speed, Errors of orientation and position according to translationnal speed.

Table 2. Results of the reconstruction.

	3dpts	Observation	rms
Global shutter	2537	10615	0.86
Rolling shutter	3273	13204	0.84
Rolling shutter simple	3446	13639	0.77

7.2 Real Data Experiment

To illustrate the relative robustness of the simplified rolling shutter model beside the complete one, and the gain of precision from the global shutter one, it was tested to reconstruct a 3D structure. The chosen structure is a corner to easily visually check if any deformations occurs during the reconstruction. We can see on Fig. 6 that the reconstruction presents no apparent distortions on the global structure, this is more evident to see on the attached video. The camera used is a webcam logitech C310, used with a resolution of 640 by 480 pixels and a focal distance of 4 mm. The detector used is a Harris corner detector and the outlier rejection is done using Least Square Median. Feature are matched from frame to frame and images used for reconstruction are selected using Mouragnon's method [19].

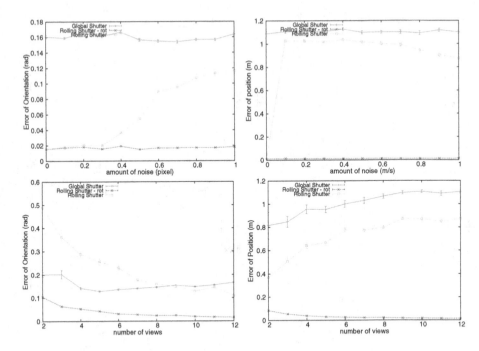

Fig. 5. Errors of orientation and position according to noise, and according to the number of views.

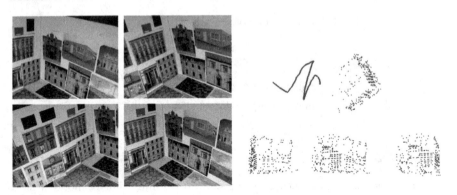

Fig. 6. Motion and reconstruction of a trajectory, (top) some images used for the reconstruction, (bottom) in blue the motion of the camera and in black the 3D reconstruction from different point of view (Color figure online).

The final parameters of reconstruction with the number of inlier 3D points, inlier observations, and final reprojection error RMS are shown in Table 2.

Another test was done with a gopro camera at the resolution of 1080 p and 60 fps. The camera was mount on a race bycicle whose structure is very rigid and in spite of the speed of the shutter of the camera, the effect is sensible due to the roughness of the ground. It is hard to see because of the radial distortion but it makes the reconstruction with the global shutter model fails. The result is shown on Fig. 7.

Fig. 7. Motion and reconstruction of a trajectory, (a) some image of the scene, (b) in blue the motion of the camera and in black the 3D reconstruction from different point of view (Color figure online).

8 Discussion

In addition to the presentation of the simplified rolling shutter model, the results in Figs. 2 and 4 show that the impact from linear speed on the quality of the reconstruction is less than the one from the rotational speed in the same way of the optical flow previously studied. As well the addition of variables to the system makes it less constrained and so cause a decay in its robustness to noise.

According to Figs. 3 and 5 the simplified rolling shutter model is more robust than the complete one. In addition, it is faster to solve (less parameters to opti- mise, less derivation a fortiori numerical ones, smaller jacobians). Less variables reduces too the probability to have local minima.

A system which doesn't need successive sequences of near images allow to work with spatially and time spaced images (leading better triangulation due to a more pronounced stereo), the inclusion of images taken out of the sequence both rolling and global shutter. It results a lighter application with less processor charge and less data transfer via the bus. Currently the methods in reconstruc- tion are not in using all the images from the camera but selecting them, as seen in [19]. The presented method is suitable in the actual state of art SLAM by its spatially and temporally spaced acquisition robustness.

An embedded camera suffers the motion of its shelf. The global shutter cam- eras are suitable to get its global motion whereas rolling shutter cameras are sensitive to the micromotion due to the fast but non instantaneous acquisition of one frame. Some methods tries to interpolate lines of the image between frames, or following splines, this is suitable for electrical vehicles with a suspen- sion system and fat tyres at low speed level on smooth roads. On rough ground, for hand held cameras, or diesel engine, the vibration induced by the shelf are fast and the use of splines for its modeling would be very expensive.

9   Conclusion

We presented a method to deal with rolling shutter distortion for SFM applica- tions relevant in the current state of art. The method is accurate thanks to the modelling of the motion; generic, it can deal with both rolling shutter and global shutter images; robust thanks to the use of only very useful parameters; usable with very low frame rate video. We think that this method can be very useful in many applications in robotics, or in augmented reality applications with the use of devices such as phones or notepads whose embedded cameras are rolling shutter. We envisage to use the effect of rolling shutter on primitives to get a priori on motion and robustify matching.

References

1. Liang, C.K., Chang, L.W., Chen, H.H.: Analysis and compensation of rolling shut- ter effect. IEEE Trans. Image Process. **17**, 1323–1330 (2008)
2. Baker, S., Bennett, E., Kang, S.B., Szeliski, R.: Removing rolling shutter wobble. In: 2010 IEEE Conference on Computer Vision and Pattern Recognition (CVPR), pp. 2392–2399. IEEE (2010)
3. Bradley, D., Atcheson, B., Ihrke, I., Heidrich, W.: Synchronization and rolling shut- ter compensation for consumer video camera arrays. In: IEEE Computer Society Conference on Computer Vision and Pattern Recognition Workshops, 2009, CVPR Workshopps 2009, pp. 1–8. IEEE (2009)

4. Ait-Aider, O., Andreff, N., Lavest, J.-M., Martinet, P.: Simultaneous object pose and velocity computation using a single view from a rolling shutter camera. In: Leonardis, A., Bischof, H., Pinz, A. (eds.) ECCV 2006. LNCS, vol. 3952, pp. 56–68. Springer, Heidelberg (2006)
5. Ait-Aider, O., Bartoli, A., Andreff, N.: Kinematics from lines in a single rolling shutter image. In: IEEE Conference on Computer Vision and Pattern Recognition, CVPR 2007, pp. 1–6. IEEE (2007)
6. Meingast, M., Geyer, C., Sastry, S.: Geometric models of rolling-shutter cameras. arXiv preprint cs/0503076 (2005)
7. Magerand, L., Bartoli, A., Ait-Aider, O., Pizarro, D.: Global optimization of object pose and motion from a single rolling shutter image with automatic 2D-3D matching. In: Fitzgibbon, A., Lazebnik, S., Perona, P., Sato, Y., Schmid, C. (eds.) ECCV 2012, Part I. LNCS, vol. 7572, pp. 456–469. Springer, Heidelberg (2012)
8. Meilland, M., Drummond, T., Comport, A.I.: A unified rolling shutter and motion blur model for 3d visual registration. In: 2013 IEEE International Conference on Computer Vision (ICCV), pp. 2016–2023. IEEE (2013)
9. Ait-Aider, O., Berry, F.: Structure and kinematics triangulation with a rolling shutter stereo rig. In: 2009 IEEE 12th International Conference on Computer Vision, pp. 1835–1840. IEEE (2009)
10. Hedborg, J., Forssen, P.E., Felsberg, M., Ringaby, E.: Rolling shutter bundle adjustment. In: 2012 IEEE Conference on Computer Vision and Pattern Recognition (CVPR), pp. 1434–1441. IEEE (2012)
11. Saurer, O., Koser, K., Bouguet, J.Y., Pollefeys, M.: Rolling shutter stereo. In: 2013 IEEE International Conference on Computer Vision (ICCV), pp. 465–472. IEEE (2013)
12. Anderson, S., Barfoot, T.D.: Towards relative continuous-time slam. In: 2013 IEEE International Conference on Robotics and Automation (ICRA), pp. 1033–1040. IEEE (2013)
13. Lovegrove, S., Patron-Perez, A., Sibley, G.: Spline fusion: a continuous-time representation for visual-inertial fusion with application to rolling shutter cameras. In: British Machine Vision Conference (2013)
14. Li, M., Kim, B.H., Mourikis, A.I.: Real-time motion tracking on a cellphone using inertial sensing and a rolling-shutter camera. In: 2013 IEEE International Conference on Robotics and Automation (ICRA), pp. 4712–4719. IEEE (2013)
15. Li, M., Mourikis, A.I.: Vision-aided inertial navigation with rolling-shutter cameras. Int. J. Robot. Res. 33, 1490–1507 (2014)
16. Hedborg, J., Ringaby, E., Forssén, P.E., Felsberg, M.: Structure and motion estimation from rolling shutter video. In: 2011 IEEE International Conference on Computer Vision Workshops (ICCV Workshops), pp. 17–23. IEEE (2011)
17. Shoemake, K.: Animating rotation with quaternion curves. In: ACM SIGGRAPH computer graphics, vol. 19, pp. 245–254. ACM (1985)
18. Ringaby, E., Forssén, P.E.: Efficient video rectification and stabilisation for cellphones. Int. J. Comput. Vis. 96, 335–352 (2012)
19. Mouragnon, E., Lhuillier, M., Dhome, M., Dekeyser, F., Sayd, P.: Generic and realtime structure from motion using local bundle adjustment. Image Vis. Comput. 27, 1178–1193 (2009)

Fully-Automatic Target Detection and Tracking for Real-Time, Airborne Imaging Applications

Tunç Alkanat[1,2]([✉]), Emre Tunali[1,2], and Sinan Öz[1,2]

[1] Microelectronics, Guidance and Electro-Optics Division,
ASELSAN Inc., Ankara, Turkey
{talkanat,etunali,soz}@aselsan.com.tr
[2] Department of Electrical and Electronics Engineering,
Middle East Technical University, Ankara, Turkey

Abstract. In this study, an efficient, robust algorithm for automatic target detection and tracking is introduced. Procedure starts with a detection phase. Proposed method uses two alternatives for the detection phase, namely maximally stable extremal regions detector and Canny edge detector. After detection, regions of interest are evaluated and eliminated according to their compactness and effective saliency. The detection process is repeated for a predetermined number of pyramid levels where each level processes a downsampled version of input image to achieve scale invariance. Then, temporal consistency for detections from all scales is evaluated and target likelihood map is constructed using kernel density estimation in order to merge all target hypotheses. Finally, outstanding targets are selected from target likelihood map and tracking is achieved by minimizing spatial distance between the selected targets in consecutive frames.

Keywords: Real-time target detection · Multiple target tracking · Temporal consistency · Data association · Target probability density estimation · Adaptive target selection

1 Introduction

Multiple target detection and tracking plays an important role in many applications such as reconnaissance and surveillance where the main purpose is to describe trajectories of the targets throughout the scenario. Many of recent electro-optical systems have a requirement to achieve this task in a fully automated manner. Many multi-tracking algorithms have two fundamental stages; the time independent automatic multi-target detection and association of the detections in temporal space. In spite of many research [1–3] on the subject in recent years, the problem remains to be challenging mainly due to unknown and changing number of targets; noisy and missing observations; interaction of multiple targets. As for real-time applications, all these challenges are to be addressed in a time efficient way.

© Springer International Publishing Switzerland 2016
J. Braz et al. (Eds.): VISIGRAPP 2015, CCIS 598, pp. 240–255, 2016.
DOI: 10.1007/978-3-319-29971-6_13

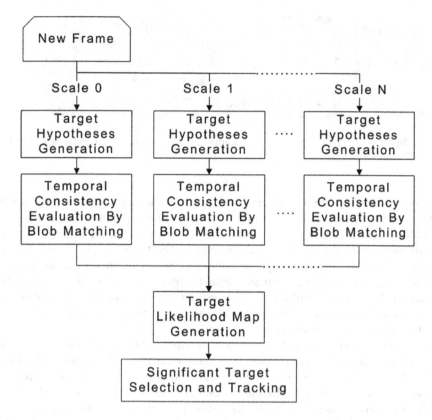

Fig. 1. General overview of the proposed solution.

The outstanding target detection concept can be interpreted differently depending on the task in hand. As a starting point, interest point detection techniques can be utilized to detect such objects on an image. Many versions of interest point detection methodologies were carried out in the literature. Some major examples are based on blob detection [4–6], corner detection [7,8] and edge detection [9–11]. Defining the outstanding object using contrast information is preferred in our application since we are interested in object as a whole, rather than some parts only. This requirement can be handled by blob or edge detection methodologies. Predictably, usage of blob or edge detectors has advantages and disadvantages compared to each other. Given the fact that the detection phase is intended to be used to detect regions contrasting their vicinity, among all blob detection techniques, maximally stable extremal regions(MSER) detection [6] can be considered as a good candidate for our purpose. On the other hand, among edge detection methods, Canny edge detection reveals its superiority due to its ability of generating closed contours by merging weak edges with the strong edges around their vicinity. Moreover, both methods are appropriate for real-time processing because of their low computational cost which is one of the major goals.

Fig. 2. Sample output of the proposed solution demonstrating successful tracking for variable number of targets.

Determining the number of targets dynamically is another important aspect of the detection phase as the selection of predetermined number of targets would be problematic. In plain words, using static number of targets may compel an algorithm to introduce insignificant targets to the track list if the number of targets is smaller than the expected number. Similarly, in the scenes having higher number of significant targets than the expected, some of them will be discarded. To deal with unknown and varying number of targets and to develop an unsupervised approach, the proposed method is designed to determine the number of targets without any supervision.

One of the most important issues in multi-target tracking is temporal association of detections. Despite the fact that there exist many detection methodologies, robustness cannot be guaranteed for different scenes. Consequently, missing and noisy observations occur frequently and the outlier data are left to be tackled in the data association stage. In order to address this problem, Kalman filtering [12] is often utilized. Despite the effective usage of this methodology in many applications [13,14], it needs significant restrictions on the predetermined motion model. The particle filters [15] provide relaxation on the constraints of the Kalman filters by exploring multiple hypotheses. Yet, particle filters have a high computational cost that may not be suitable for real-time processing. On the other hand, there exist some other widely used techniques such as joint probability density association filters (JPDAF [16]) and multiple hypothesis tracking (MHT [17]). JPDAF uses soft data assignment by considering the probability of a measurement belonging to more than one track which results in a single hypothesis summarizing all previous measurements. Although JPDAF is a useful tool, it has some assumptions that may decrease its range of applications. For example, JPDAF assumes that the number of targets is fixed. In other words, JPDAF is not able to deal with targets entering/leaving the scene. In MHT, this problem is suppressed by using integrated track initiation. Association algorithm of MHT is a hypothesis based brute force implementation to generate all possible hypotheses. For that reason, it requires high computational cost. Furthermore, MHT necessitates a large memory space to be used since hypotheses from previous frames have to be stored. Due to the disadvantages of mentioned techniques, we have used a different approach for data association: The proposed method acquires measurements using a pyramid structure and profits from motion heuristics together with a probability density estimation methodology which is designed for merging measurements from different levels of the observation pyramid. The density estimation method is based on Parzen windowing [18], and benefits from a weighting scheme to eliminate noisy and missing observations with low computational cost.

The rest of the paper is organized as follows: The proposed target detection and tracking methods are described in Sect. 2. In Sect. 3, the experimental results are presented. Finally, the study is concluded and the contributions of the paper are reviewed in Sect. 4.

2   Proposed Method

The proposed multiple target detection and tracking method consists of four fundamental stages. During the first stage, target hypotheses are produced for different scales by taking distinctiveness and compactness assumptions of target model into consideration. In the second stage, in order not only to reject outliers and but also to compensate missing detections, temporal consistency of each target hypothesis is evaluated. In the third stage, a target likelihood map representing the target existence likelihood at each pixel, is generated from each scale of the observation pyramid by using consistent target hypotheses. In the fourth and the last stage, outstanding (relevant) targets are chosen from the likelihood map by using an adaptive thresholding algorithm. General overview of the proposed methodology is depicted in Fig. 1. Also, sample results of the proposed method demonstrating successful tracking for variable number of targets are given in Fig. 2.

2.1   Target Hypotheses Generation

To achieve automatic target detection, each target candidate fulfilling some preliminary requirements should be further analyzed to decide whether it is a relevant target or not. The target candidates are referred as target hypotheses and generated at each scale of the observation pyramid, where each level of the pyramid structure processes a downsampled version of the original frame. Therefore, for both hypothesis generation and selection, some assumptions are made to describe the target model.

The first assumption is the distinctiveness assumption stating that target candidates should be distinctive from their surroundings. Actually, this assumption is made based on human visual attentional system in which robust saliency detection mechanisms provide focus of attention to the salient regions pre-attentively for further processing. Again similar to human visual system, the distinctiveness is measured by the intensity difference. Most of the saliency detection methods are founded on the same principle; however saliency detection in global scale (by considering the whole scene) would generally require high

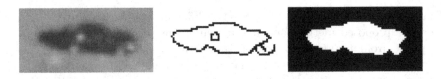

Fig. 3. Effect of filling on a target with layered contours.

processing time which may not be suitable for real-time applications. Considering the fact that the computational complexity is one of the key issues, we propose two starting points for the target hypothesis generation in this paper: Canny edge detection and maximally stable extremal regions(MSER) detection.

Canny edge detector is a useful tool for proposed method because of its suitable computational load for real-time applications and its capability of uniting weak edges with other edges in their neighborhood. However, due to noise in the scene, Canny edge detector may fail to generate closed contours. This problem can be avoided by applying morphological closure on edge map. After the morphological operation, the resulting edge map is then applied filling operation to generate blobs. A useful outcome of the filling operation appears in the case of targets with layered structure, as illustrated in Fig. 3 where the reflected daylight generates a closed contour inside the object. In such a scenario, selection of the outermost closed contour yields better localization.

Fig. 4. From left to right: Original image, Canny edge detection results, MSER detection results. Figure illustrates an important advantage of MSER detection over Canny edge detection that is its ability to separate regions by their intensity difference.

Another alternative method of generating blobs is to use maximally stable extremal regions(MSER) detection. Similar to the Canny edge detector, this algorithm is also suitable to be used in a real-time system because of its low computational demands. Moreover, using MSER detection has several advantages compared to Canny edge detection. Firstly, given that MSER detection returns regions on an image that are nearly uniform inside and are contrasting their vicinity, this method tends to respond more likely to the human attentional system than the Canny edge detector. More formally, MSER detection forms regions by not only considering distinctiveness, but also variance of intensity values of the pixels that are located inside a blob candidate. This property of MSER detection results in a better separation as demonstrated in Fig. 4. Secondly, MSER detection does not require any morphological operations which reduces the computational cost.

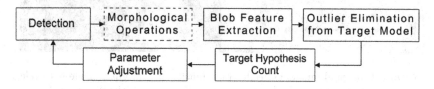

Fig. 5. Flowchart for target hypotheses generation.

Despite the fact that both Canny edge detection and MSER detection methods are appropriate considering the requirements of the proposed algorithm, performance of both detection methodologies rely greatly on the choice of parameters due to different contrast spans of scenes. Improperly adjusted parameters may cause the algorithms not detecting any edges/regions especially in low contrast scenes where outstanding edges/regions are present and perceptible by human visual system. Conversely, in high contrast case, the detectors may return false edges/regions. To overcome this problem, proposed method uses a dynamic parameter selection scheme with a feedback loop (See Fig. 5) that uses the number of target hypotheses from the previous frame as input. The utilized feedback loop changes the high threshold of the Canny edge detector or the threshold step parameter of MSER detector by their corresponding predetermined factor when the number of hypotheses from the previous frame is lesser/higher than the desired number of target hypotheses. In addition, the proposed dynamic parameter selection scheme also indirectly upper limits the number of generated target hypotheses and thus the number of declared targets.

In spite of the fact that the detection process reveals regions contrasting their surroundings, proposed detection methodologies do not provide a measure of distinctiveness for the detected blobs. To evaluate the distinctiveness, we have used a modified version of the metric referred as effective saliency proposed in [19]. However, [20] stated that the proposed saliency calculation method [21] of the metric has crucial drawbacks. For instance, [21] treats all image boundaries as background. This assumption does not hold even when a small portion of the salient object hits the boundary. To overcome this problem, we have used the proposed effective saliency metric along with the saliency calculation method of [20].

Since both of the detection methods reveal regions considering the pixel intensity values only, outlier observations may be present in the detection results. To avoid noisy observations, compactness of a region is also evaluated to eliminate some of the detected blobs. The used compactness criterion for a detected blob

Fig. 6. On left, original image. On right, blob mask of the original image with relevant targets, inconsistent targets, blobs violating compactness, blobs violating distinctiveness marked with green, white, red and blue, respectively (Color figure online).

is defined as the scalar that is the proportion of area of a blob to the area of its minimum sized bounding box. This metric introduces an assumption on target shape: Blobs degraded from rectangular shape are eliminated as illustrated in Fig. 6. After the elimination process of detected blobs by compactness and distinctiveness metrics, the remaining detected blobs are referred as target hypotheses and are further processed to evaluate their temporal consistency.

2.2 Temporal Consistency Evaluation by Blob Matching

As discussed, Canny edge detector and MSER detector fulfill the requirements of target hypothesis generation stage. However, both detection methods are vulnerable to noise. For example, Canny edge detection may fail to provide closed contours, yielding missing observations in some frames or can produce artificial closed contours due to noise. As for MSER detector, highly textured objects in the scene may only partially be detected or may not be detected at all. To avoid this problem, the temporal consistency of hypotheses should be evaluated to compensate erroneous detections.

Generation of faulty observations is a common problem. The solutions for this problem are based on probabilistic model on target behavior in some well-known techniques [12,16]. However, for the scenes where the imaging device is moving with a complex motion pattern, probabilistic motion modeling approach may be over-constraining. Hence, in order to reject outliers and handle missing data, proposed method specifies an observation point as a target hypothesis if and only if the observation point preserves its presence for multiple frames. Therefore, temporal consistency of a target is ensured. Proposed method achieves this by using a scoring scheme in which higher score of a target hypothesis represents higher reliability.

The proposed scoring scheme is applied at each frame and starts with associating newly generated and existing target hypotheses. At first, for each new target hypothesis, existing hypotheses are searched in a neighborhood to satisfy the

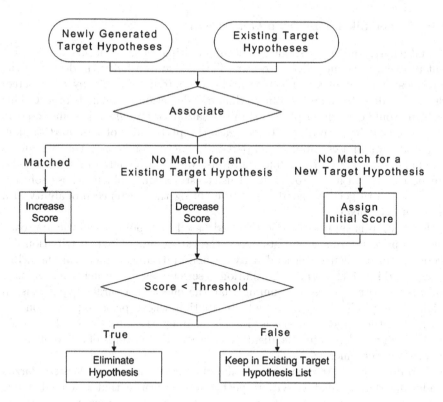

Fig. 7. Proposed scoring scheme.

motion heuristic known as maximum velocity as described in [22]. Usage of such a simple model is both less restricting and requires much less computational cost compared to other motion models. Existence of a match is decided by minimizing the norm of vectors that contain spatial distance and mean intensity difference of a new hypothesis to all existing target hypotheses within the neighborhood. If a match is found, the score of the matched existing target is increased. After matching all new target hypotheses, the score of the remaining (unmatched) existing target hypotheses are decreased. Then, unmatched new target hypotheses are considered as possible new targets entering the scene and initial scores are assigned according to their similarity to the target model description. After adding new target hypotheses to the existing target list and adjusting their scores, target hypotheses list is reconstructed by eliminating hypotheses that have a score below the minimum score threshold. Hence, the missing observations for a target hypothesis would be tolerated for a limited number of frames. In a similar fashion, the observations that are generated due to noise will also be eliminated within a limited number of frames since they are not persistent. The proposed scoring scheme is summarized in Fig. 7.

2.3   Target Likelihood Map Generation

The false partitioning of a single object into multiple closed contours is an important problem in Canny edge detection. This is due to a failure in detecting the outermost contour of an object as a closed contour. Like Canny edge detector, MSER detector may yield more than one detection for a single object. This problem would cause multiple target initialization for a single object and appears more often for larger objects. To decrease the probability of false partitioning, proposed method uses a multi-level pyramid structure and consequently requires merging of information from different levels. Merging the data of different scales can be considered as a probability density estimation problem whose solution identifies the target likelihood map representing the existence probability of a target at each pixel.

Estimation is supposed to be achieved based on a non-parametric approach since no prior information exists about the target probability distribution. To overcome this challenge, kernel density estimation (Parzen window method [18]) is employed in which normal distribution is selected as the kernel function. Normal distribution is chosen assuming that the effect of a target hypothesis on neighboring pixels yields a normal distribution whose peak is placed on the centroid of the target hypothesis. In this manner, the variance of the normal distribution will determine the distance between the centroids of different target hypotheses to be merged.

In order to generate the target likelihood map, different from classical Parzen windowing, data is weighted with respect to its significance that is defined by two scalars; temporal consistency and scale weights. Since the importance of a target increases with its temporal consistency, consistency weight (w_c) is obtained by

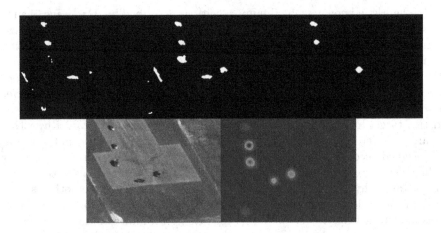

Fig. 8. From top to bottom and left to right: Target hypotheses at scale 1 (original image), target hypotheses at scale 2 (2x downsampled), target hypotheses at scale 3(3x downsampled), original image, and generated target likelihood map. Masks for each scale are resized for visualization.

the score whose calculation is explained in Sect. 2.2. Thus, while decreasing the effect of mis-detected hypotheses from one scale of the pyramid, the weights of the corresponding target hypotheses are increased at the relevant scale yielding better localization. The second scalar, scale weight (w_s) is designed to select the importance of different scales of the pyramid. Since the partitioning occurs generally for the large objects; to compensate the erroneous data, detections acquired from lower resolutions (downsampled by a higher factor) are weighted proportional to the downsampling factor. The formal definition of the target likelihood for each pixel (x, y) is given in Eq. 1,

$$P(x, y) = \frac{\sum\limits_{j \in H} w_c \cdot w_s \cdot exp\left(-\frac{(x-x_j)^2 + (y-y_j)^2}{2\sigma^2}\right)}{\sum\limits_{\forall pixels} \sum\limits_{j \in H} w_c \cdot w_s \cdot exp\left(-\frac{(x-x_j)^2 + (y-y_j)^2}{2\sigma^2}\right)}, \tag{1}$$

where (x_j, y_j) is the locations from the set of target hypotheses H. Target likelihood map generation process is illustrated in Fig. 8.

2.4   Target Selection and Tracking

After acquiring target likelihood map, target selection problem becomes nothing but a threshold selection problem that determines the lowest probability in the target likelihood map that will be considered as a target. The simplest solution to problem is to utilize a static threshold. Nevertheless, dynamic thresholding is preferable because of the scoring scheme applied to the target hypotheses. For this purpose, the dynamic thresholding methodology proposed in [23] is followed to disclose distinctive intensity falls on a given image. Analyzing the relationship between the local maxima of input image, the method computes the threshold using weighted average of local maxima. Obviously, the crucial part is to acquire the proper weights. For the calculation of weights, first, the local maxima are detected. Then, they are sorted in descending order to form a vector ($LocalMax_{sorted}$). As higher laplacian represents distinctive falls, the weights are calculated by computing the normalized laplacian of this vector. This approach fits well to our problem since distinctive falls indicate splits between different target hypothesis groups having similar likelihood values; so it achieves successful separation of distinctively more remarkable target hypotheses. The formal definition of the weighting procedure is shown in Eqs. 2 and 3.

$$Thr = LocalMax_{sorted}^T \cdot \nabla_{norm}^2 (LocalMax_{sorted}), \tag{2}$$

$$\nabla^2_{norm}(f) = \frac{\nabla^2(f) - min(\nabla^2(f))}{\sum_i \nabla^2(f)|_i - min(\nabla^2(f))}. \tag{3}$$

Following the process of choosing the relevant target hypotheses, the tracking is simply accomplished by matching the relevant targets from successive frames just by minimizing spatial distance.

3   Experiments

The testing procedure for the proposed algorithm is designed to test both detection and tracking capabilities. For the detection phase, two metrics are used to determine the performance of the proposed algorithm: False discovery rate (Eq. 4) and true positive (Eq. 5) rate.

$$FDR = \frac{FP}{FP + TP} \ , \tag{4}$$

$$TPR = \frac{TP}{TP + FN} \ . \tag{5}$$

Besides detection, another important task that should be achieved is tracking of the detected targets. Despite existence of multiple targets in each scenario of the VIVID dataset [24], the ground truth is only provided for the primary target. Due to lack of ground truth data for secondary targets, we followed the same procedure used in [25]. Thus, the tracking performance of the proposed method was evaluated by manually labeling the results as good tracking; tracking had drifted off center, or lost. A track is described as good track when the track center is within the object; labeled as drifted track when the track center is located outside of the object boundary and a track stated to be lost whenever track gate ceases its existence in the presence of the target. One exemplary illustration is given in Fig. 9 for good and drifted tracks respectively.

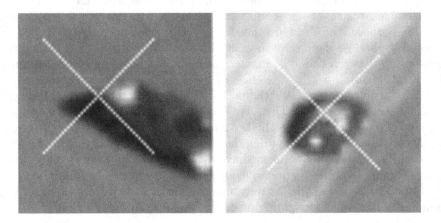

Fig. 9. Examplary outputs showing the drifted track, on left, and successful track, on right.

To evaluate the results, VIVID dataset is preferred since it is a widely-used, public dataset that contains many challenges such as: Out of plane rotation, pose variation, occlusion, low contrast, existence of similar targets in the vicinity and defocusing. The experiments on VIVID dataset are conducted after the conversion of the dataset from three-band to single-band. Yet, note that, the algorithm could also be modified to work using three-band images by changing the detection phase only.

For each scenario, effective saliency and compactness thresholds were set to 0.7 and 0.45, respectively. The variance of the normal distribution that was used to generate target likelihood maps was set to 0.15 and a three-level pyramid structure was used: 1st level processing original image, 2nd level processing original image downsampled by 2 and 3rd level by 3. Since optimum number of scales depends on the span of expected target size, minimum number of scales should manually be selected considering the application. Also, initial score, increase score step, decrease score step, maximum euclidean distance for inter-frame target matching, saliency evaluation patch size parameters are selected as; 200, 20, 10, 21, 41, respectively. Canny edge detector's low threshold and MSER detector's maximum area variation parameter were chosen to be static: 0.1 and 10. For the parts of the procedure that uses Canny edge detector, morphological operation structuring element was selected to be circular with a diameter of 3 pixels.

Fig. 10. Sample result on VIVID dataset. Top row, columns 1–2: Scale changes. Top row, columns 2–4: Defocusing. Bottom row, columns 1–2: Different motion patterns, changing number of targets. Bottom row, columns 3–4: Occlusion

In Fig. 10, some of the important findings of the experimental results are demonstrated. The first two images of the first row illustrates the success of the algorithm against scale changes which is achieved with the usage of pyramid structure. Remaining images of the first row demonstrates the behavior of the proposed method in case of missing observations. In this scenario, the target detection fails for a while due to defocusing of the camera. Despite the missing observations, tracks are continued without breaking and the targets are again well localized after refocusing of the camera. The importance of the selection of a simple motion model (maximum velocity) is illustrated on the first two images of the 2nd row. If a restrictive probabilistic motion model was used, some of the targets having different turning angles would be lost. Moreover, these sub-figures also visualize the success in handling varying number of targets. Finally, last two images of the second row visualizes the major weakness of the proposed algorithm which is the incapability of occlusion handling resulting in track losses.

Table 1. Performance results of proposed method for detection and tracking on VIVID dataset (in percentage %).

Method	Name of Sequence	# of Frames	TPR	FDR	TQ
Edge Det.	egTest01	1820	92.7	24.6	90.6
Edge Det.	egTest02	1300	83.8	31.8	84.4
Edge Det.	egTest03	2571	45.6	38.8	83.2
Edge Det.	egTest04	1833	95.2	54.3	81.0
Subtotal		7524	75.7	37.9	84.6
Edge Det. + Saliency	egTest01	1820	94.1	18.6	98.7
Edge Det. + Saliency	egTest02	1300	77.2	20.8	99.4
Edge Det. + Saliency	egTest03	2571	74.1	32.8	88.6
Edge Det. + Saliency	egTest04	1833	83.2	34.4	81.5
Subtotal		7524	0.822	27.7	92.1
MSER Det.	egTest01	1820	99.7	04.5	90.9
MSER Det.	egTest02	1300	85.4	12.0	87.2
MSER Det.	egTest03	2571	68.3	05.0	86.1
MSER Det.	egTest04	1833	90.8	49.1	75.0
Subtotal		7524	84.3	16.8	84.7
MSER Det. + Saliency	egTest01	1820	97.4	03.4	99.1
MSER Det. + Saliency	egTest02	1300	88.4	12.6	96.2
MSER Det. + Saliency	egTest03	2571	79.1	02.7	93.2
MSER Det. + Saliency	egTest04	1833	93.6	41.8	92.3
Subtotal		7524	88.7	14.1	94.9
Grand Total		30096	82.7	24.1	89.1

The quantitative results of experimental procedure are summarized in Table 1. In addition, sample qualitative results are demostrated in Fig. 11. According to the results, algorithm with MSER detector gives better results as implied by both higher true positive rate and lower false discovery rate. Moreover, increase in performance when using MSER detector for egTest03 sequence implies that, MSER detector outperforms the Canny edge detector in low contrast scenes. Another interesting conclusion of the experimental results is the increase in performance when saliency calculation is used. Since saliency calculation is not only intended for elimination, but also target center correction, track quality of the algorithm with saliency calculation is significantly higher for both detection techniques. Likewise, true positive rate is increased when saliency is used since shifting to target center using saliency map provides robustness to the detection phase. In addition to the detection and tracking performance of the proposed method, another important aspect is the computational load. The proposed solution was tested using un-optimized C++ code running on a single core of an Intel i5-3470 3.2GHz CPU. For the variant that uses MSER detection and saliency calculation, the algorithm was able to run at a minimum frame rate

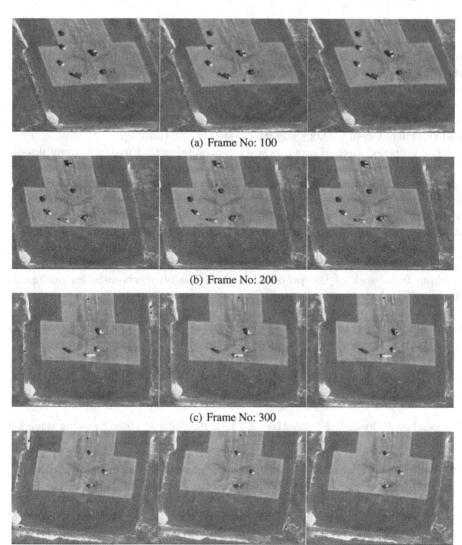

(a) Frame No: 100

(b) Frame No: 200

(c) Frame No: 300

(d) Frame No: 400

Fig. 11. Sample results of the proposed method using three alternatives for the detection phase. Detection methods used from left to right: Edge Det. + Saliency, MSER Det., MSER Det. + Saliency.

of 25.33 fps and an average rate of 26.9 fps for maximum 256 target hypotheses at each scale of the pyramid. Note that, the frame-rate can further be improved by using parallel processing or advanced optimization techniques.

4 Conclusions

In this study, a method designed for multiple target detection and tracking in real-time airborne imagery is introduced. Also, performance of the algorithm is quantitatively evaluated, presented and compared for several underlying detection methodologies. Results imply that, even with a simple tracking approach, it is possible to achieve a high true positive, low false discovery rate detection and tracking system without any user assistance or training.

The proposed method provides several advantages that makes it a useful tool for airborne reconnaissance and electro-optical surveillance systems. Firstly, low computational cost and complexity of proposed method open up new possibilities for airborne applications where real-time processing is preferred. Secondly, the algorithm requires no user assistance or training which provides practicality. Another advantage arises from the fact that proposed method yields invariance to scale, number of targets, out of plane rotation and contrast span of the scene. On the other hand, proposed algorithm has certain drawbacks. For instance, simple tracking framework of the proposed method has no mechanism for occlusion handling which decreases the performance.

As a future work, we plan to employ tracklet concept to increase the performance of the proposed method on the scenes where frequent occlusions are present. Also, we plan to work on three-band detection to extend the possible areas of use for the algorithm.

References

1. Berclaz, J., Fleuret, F., Turetken, E., Fua, P.: Multiple object tracking using k-shortest paths optimization. IEEE Trans. Pattern Anal. Mach. Intell. **33**, 1806–1819 (2011)
2. Niedfeldt, P.C., Beard, R.W.: Multiple target tracking using recursive ransac. In: American Control Conference (ACC), pp. 3393–3398. IEEE (2014)
3. Andriyenko, A., Schindler, K.: Multi-target tracking by continuous energy minimization. In: 2011 IEEE Conference on Computer Vision and Pattern Recognition (CVPR), pp. 1265–1272. IEEE (2011)
4. Lowe, D.G.: Distinctive image features from scale-invariant keypoints. Int. J. Comput. Vis. **60**, 91–110 (2004)
5. Bay, H., Ess, A., Tuytelaars, T., Van Gool, L.: Speeded-up robust features (surf). Comput. Vis. Image Underst. **110**, 346–359 (2008)
6. Matas, J., Chum, O., Urban, M., Pajdla, T.: Robust wide-baseline stereo from maximally stable extremal regions. Image Vis. Comput. **22**, 761–767 (2004)
7. Harris, C., Stephens, M.: A combined corner and edge detector. In: Alvey Vision Conference, UK, Manchesterm, vol. 15, p. 50 (1988)
8. Rosten, E., Drummond, T.W.: Machine learning for high-speed corner detection. In: Leonardis, A., Bischof, H., Pinz, A. (eds.) ECCV 2006, Part I. LNCS, vol. 3951, pp. 430–443. Springer, Heidelberg (2006)
9. Canny, J.: A computational approach to edge detection. IEEE Trans. Pattern Anal. Mach. Intell. **6**, 679–698 (1986)
10. Prewitt, J.M.: Object enhancement and extraction. Picture Process. Psychopictorics **10**, 15–19 (1970)

11. Sobel, I., Feldman, G.: A 3x3 isotropic gradient operator for image processing. a talk at the Stanford Artificial Project, pp. 271–272 (1968)
12. Kalman, R.E.: A new approach to linear filtering and prediction problems. J. Fluids Eng. **82**, 35–45 (1960)
13. Tsai, C., Dutoit, X., Song, K., Van Brussel, H., Nuttin, M.: Robust face tracking control of a mobile robot using self-tuning kalman filter and echo state network. Asian J. Control **12**, 488–509 (2010)
14. Ramakoti, N., Vinay, A., Jatoth, R.K.: Particle swarm optimization aided kalman filter for object tracking. In: International Conference on Advances in Computing, Control, & Telecommunication Technologies, ACT 2009, pp. 531–533. IEEE (2009)
15. Ristic, B., Arulampalam, S., Gordon, N.: Beyond the Kalman Filter: Particle Filters for Tracking Applications, vol. 685. Artech House, Boston (2004)
16. Fortmann, T.E., Bar-Shalom, Y., Scheffe, M.: Multi-target tracking using joint probabilistic data association. In: 19th IEEE Conference on Decision and Control including the Symposium on Adaptive Processes, vol. 19, pp. 807–812. IEEE (1980)
17. Reid, D.B.: An algorithm for tracking multiple targets. IEEE Trans. Autom. Contr. **24**, 843–854 (1979)
18. Parzen, E.: On estimation of a probability density function and mode. Ann. Math. Stat. **33**, 1065–1076 (1962)
19. Alkanat, T., Tunalı, E., Öz, S.: A real-time, automatic target detection and tracking method for variable number of targets in airborne imagery. In: Proceedings of the 10th International Conference on Computer Vision Theory and Applications, pp. 61–69 (2015)
20. Zhu, W., Liang, S., Wei, Y., Sun, J.: Saliency optimization from robust background detection. In: 2014 IEEE Conference on Computer Vision and Pattern Recognition (CVPR), pp. 2814–2821. IEEE (2014)
21. Wei, Y., Wen, F., Zhu, W., Sun, J.: Geodesic saliency using background priors. In: Fitzgibbon, A., Lazebnik, S., Perona, P., Sato, Y., Schmid, C. (eds.) ECCV 2012, Part III. LNCS, vol. 7574, pp. 29–42. Springer, Heidelberg (2012)
22. Yilmaz, A., Javed, O., Shah, M.: Object tracking: a survey. ACM Comput. Surv. (CSUR) **38**, 13 (2006)
23. Aytekin, C., Tunalı, E., Öz, S.: Fast semi-automatic target initialization based on visual saliency for airborne thermal imagery. In: Proceedings of the 9th International Conference on Computer Vision Theory and Applications, Visapp 2014, pp. 490–497 (2014)
24. VIVID (2005). http://vision.cse.psu.edu/data/vivideval/datasets/datasets.html
25. Bolme, D.S., Beveridge, J.R., Draper, B.A., Lui, Y.M.: Visual object tracking using adaptive correlation filters. In: IEEE Conference on Computer Vision and Pattern Recognition (CVPR), pp. 2544–2550 (2010)

A Generalized Structure from Motion Framework for Central Projection Cameras

Christiano Couto Gava[1,2]([✉]) and Didier Stricker[1,2]

[1] Department Augmented Vision, German Research Center for Artificial Intelligence,
Trippstadterstr. 122, 67663 Kaiserslautern, Germany
{Christiano.Gava,Didier.Stricker}@dfki.de
[2] Department of Computer Science, Kaiserslautern University of Technology,
Gottlieb-Daimlerstr., 67663 Kaiserslautern, Germany
http://av.dfki.de/en/, http://www.informatik.uni-kl.de/en/

Abstract. This paper presents a novel Structure from Motion (SfM) framework designed for central projection cameras. The goal is to support future large scale multi-view 3D reconstruction algorithms. We believe that these algorithms will be able to benefit from several different sources of visual information. Accordingly, SfM approaches will need to handle this variety of image sources, such as perspective, wide-angle and spherical images. However, this issue has not yet been addressed. Current state of the art techniques are not able to handle heterogeneous images simultaneously. Therefore, we introduce SPHERA, a generalized SfM framework designed for central projection cameras. By adopting the unit sphere as underlying model it is possible to treat single effective viewpoint cameras in a unified way. We validate our framework on synthetic and real datasets. Results show that SPHERA is a powerful framework to support upcoming algorithms and applications on large scale 3D reconstruction.

Keywords: Structure from Motion · Spherical images · Multi-view 3D reconstruction · Large scale

1 Introduction

The popularity of full panoramic images has significantly increased during the past few years. This is confirmed by the growing variety of spherical image acquisition hardware and software packages available nowadays [1–5]. Mobile devices such as smartphones and tablets feature easy-to-use Apps that allow the user to capture panoramas within seconds. Additionally, panoramic images offer the possibility to create immersive environments where the user experiences a first-person view, such as Google Street View [6]. Immersive visualization systems find appliance in a number of applications, e.g. documentation, education, preservation of cultural heritage, gaming, city planing, etc. Clearly, these applications can further benefit from 3D information. This makes full spherical images specially attractive for immersive visualization as well as 3D reconstruction.

© Springer International Publishing Switzerland 2016
J. Braz et al. (Eds.): VISIGRAPP 2015, CCIS 598, pp. 256–273, 2016.
DOI: 10.1007/978-3-319-29971-6_14

There are several ways to classify multi-view 3D reconstruction algorithms. One of them concerns the distance between the images relative to the scene, the so called baseline. Recent *narrow* baseline approaches are capable of simultaneously recovering camera poses and 3D geometry from a video sequence [7,8]. However, these approaches are normally restricted to indoor, office-like, environments. *Wide* baseline techniques, on the other hand, are better suited for large scale reconstruction, but assume camera poses have been previously determined [9,10]. In other words, they implicitly demand Structure from Motion (SfM) to recover the camera poses before the 3D model can be computed.

To perform SfM, spherical images are more suitable than standard perspective images. Due to their wide field of view, scene features are observed in more images, thus increasing the number of constraints on camera poses. Consequently, methods have been derived to perform SfM on wide field of view cameras. More specifically, [11–13] address SfM on omnidirectional images, while [14–16] deal with full spherical images. Not surprisingly, perspective SfM has been extensively studied e.g. by [17–23]. Although these approaches have shown to work well for the specific image type they were designed for, up to the authors knowledge they are unable to handle images of any other type.

Another relevant aspect of SfM algorithms is whether the camera poses are estimated globaly or incrementaly. Usually, global methods split the camera pose estimation into two parts. The first part aims at recovering the rotation matrices of all cameras. The second part uses the global rotations obtained in the first part to determine the translation of all cameras. The later may be performed independently of the scene structure [20] or along with it [21]. The main reason for this splitting is that the estimation of relative translations is inaccurate in case of narrow baseline, whereas relative rotations can be precisely recovered regardless of the baseline, provided enough point correspondences. Global methods have the advantages of evenly distributing errors among all cameras and being independent of an initial pair of cameras. They traditionally solve a linear system of equations (which minimize an *algebraic* error), combined or followed by bundle adjustment (BA) [24] to refine camera poses. However, if a new camera is added afterwards, the entire pipeline has to be executed again.

Incremental methods are initialized by computing the poses of a selected camera pair. Then, point correspondences are triangulated and the resulting 3D points are used to select the next camera. Once the pose of the new camera is determined, new 3D points are created and the procedure is repeated. In other words, the poses of all cameras along with a sparse representation of the scene structure are recovered by alternating between triangulation and resectioning. An advantage of these methods is the possibility to obtain the optimal pose every time a new camera is added. This happens because each pose is estimated using BA which minimizes the reprojection error, carrying along a meaningful *geometric* interpretation, instead of an algebraic error. Another advantage of incremental methods is the ability to later include new cameras without necessarily rerunning the entire pipeline. Incremental pipelines may suffer from drift caused by accumulated errors. Consequently, loop closure may become an issue.

Nevertheless, it has been shown in [22] that re-triangulation of existing point correspondences is able to redistribute accumulated errors as well as deal with loop closures. Moreover, one of the most successful SfM algorithms is Bundler [17], which implements an incremental pipeline.

Given the current effort to reconstruct ever growing environments [10,25,26], every source of visual information shall be taken into account, regardless of the shape of image surface. This is an issue that has not yet been addressed. Apart from performance and accuracy, another highly desirable feature of 3D reconstruction algorithms is to update and improve the scene model whenever new images are available. Here again, the ability to deal with different camera types is essential. Therefore, we present SPHERA, a novel Structure from Motion framework to bridge the gaps between current SfM methods for central projection cameras. We build on the model proposed in [27] and adopt the unit sphere to represent images and to treat heterogeneous camera types in a unified way. Our approach dynamically selects the best information available to recover camera poses and scene structure, allowing new images to be integrated efficiently. Experiments on synthetic and real image sequences validate our framework as a valuable contribution to support large scale 3D reconstruction algorithms.

1.1   Related Work

The work presented in [11] uses epipolar geometry to compute scene structure from an omnidirectional vision system mounted on a robot platform. However, the camera pose problem is not addressed. In [12], Micusik and Pajdla focus on omnidirectional images with a field of view larger than 180° and devise a camera model specific for that type of image. Although scene structure can be recovered, the technique is limited to the two-view geometry problem. Consequently, the proposed camera model can hardly be used in a more generic SfM approach. Bagnato et al. present in [13] a variational approach to achieve ego-motion estimation and 3D reconstruction from omnidirectional image sequences. Nonetheless, the environment must be densely sampled so that the relationship between image derivatives and 3D motion parameters is still valid. Thus, this approach can not be used in a more general, sparse SfM pipeline.

A method to recover camera poses from a set of spherical images on a sparsely sampled environment is presented in [14]. However, SfM is performed based on panoramic cubes computed for each spherical image. The camera poses are recovered by casting the spherical problem back to the standard perspective problem. In [16], spherical images are used to estimate the relative camera poses and to build a map of the environment. To simplify the problem, Aly and Bouguet assume planar motion, i.e. all camera frames must lie on the same plane. This assumption strongly limits the applicability of the proposed technique. Our approach is closely related to [15], as both exploit full spherical images to deliver a sparse representation of the scene along with recovered camera poses. Nevertheless, the method presented by Pagani and Stricker was designed exclusively for spherical cameras, whereas our framework naturally handles any kind of central projection camera. Additionally, SPHERA allows to dynamically select a subset of the cameras to optimize and speed up BA with little to no loss of accuracy, as detailed in Sect. 3.

Not surprisingly, our pipeline has similarities with some SfM methods derived exclusively for perspective images. For instance, Wu proposes an incremental SfM where loop closure does not need to be explicitly detected [22]. His algorithm tracks *under-reconstructed* camera pairs, i.e. pairs with low ratio between their common 3D points and number of point correspondences. Then, based on a geometric sequence, re-triangulation is performed for all under-reconstructed camera pairs. Wu shows that this re-triangulation is able to reduce drift errors without explicitly detecting loops even for long image sequences. Our framework incorporates this idea. However, as we aim at high accuracy, re-triangulation is performed during every step of BA, instead of following a predefined sequence. Another incremental method has been proposed in [23]. The authors introduce an *algebraic* cost function formulated on pairwise epipolar constraints as a more efficient alternative to the traditional reprojection error. Their algorithm eliminates structure from BA aiming at speeding up convergence. Nevertheless, their final solution lacks the accuracy of geometric based cost functions. Therefore, their pipeline requires two or three additional iterations of the classical BA (which takes the structure back into account) at the end to improve precision. As described in Sect. 3, we also consider only the camera parameters for BA to reduce the dimension of the parameter search space. Contrary to [23], we implicitly model scene structure through the re-triangulation mentioned above. Moreover, instead of an algebraic error, SPHERA minimizes a reprojection error defined directly on the surface of the unit sphere.

2   Background

2.1   Spherical Images

A spherical image is a $180^{\circ} \times 360^{\circ}$ environment mapping that allows an entire scene to be captured from a single point in space. Consequently, every visible 3D point P_W given in world coordinate system can be mapped onto the image surface. This is done by a two-step process. First, analogue to the perspective case, P_W is represented in the camera coordinate system as $P_C = RP_W + t$, with R and t representing the camera rotation matrix and translation vector. Second, and different from the perspective projection, P_C is projected onto the image surface by scaling its coordinates, as shown in Fig. 1-(a). Without loss of generality, we assume a unit sphere. Thus, the scaling becomes a normalization and $p = P_C / \|P_C\|$.

Spherical images are stored as a 2D pixel-map as depicted in Fig. 1-(b). This map is obtained using a latitude-longitude transformation, with $0 \le \phi \le \pi$ and $0 \le \theta \le 2\pi$.

2.2   Sphere as Unifying Model

Our approach is grounded on the seminal work developed in [27], where the authors proposed a unifying model for the projective geometry of vision

(a)   (b)

Fig. 1. (a) Spherical coordinates and illustration of the spherical projection. (b) Pixel-map of a spherical image.

systems having a single effective viewpoint. These vision systems are commonly referred to as central projection cameras and include catadioptric sensors featuring conic mirrors of different shapes, such as parabolic, hyperbolic or elliptic. Geyer and Daniilidis showed that any central catadioptric projection is equivalent to a two-step mapping via the sphere. It is well known from the pinhole model that standard perspective imaging characterizes a single viewpoint system. Nonetheless, perspective images are also central catadioptric systems with a virtual planar mirror and are, therefore, covered by the aforementioned model. In practice, that means it is possible to treat these central projection systems as spherical cameras, provided the mapping from the original image surface to the sphere is known. This mapping may be seen as a warping transformation from the original image to the unit sphere. As an example, Fig. 2 shows the result of warping a perspective image onto the sphere.

(a)   (b)

Fig. 2. Example of (a) an original perspective image [28] and (b) its warped version. The warped image appears mirrored due to the viewpoint ("outside" the unit sphere).

2.3   Spherical Camera Pose Estimation

Epipolar Geometry. The epipolar geometry for full spherical cameras has already been presented in [29]. Thus, here we provide a short overview. Consider a pair of spherical cameras C_0 and C_1. Let R and t be the associated rotation matrix and translation vector. A point p_0 on the surface of C_0, along with the centers of the cameras, define a plane Π that may be expressed by its normal vector $n_\Pi = Rp_0 \times t = [t]_\times Rp_0$, where $[t]_\times$ is the skew-symmetric matrix representing the cross-product. For any point p_1 on C_1 belonging to Π the condition $p_1^T n_\Pi = 0$ holds, which is equivalent to $p_1^T [t]_\times Rp_0 = 0$, where $E = [t]_\times R$ is the essential matrix [18]. The condition $p_1^T E p_0 = 0$ is known as the epipolar constraint and is the same result obtained in the perspective case. This shows that the epipolar constraint is independent of the shape of the image surface (Fig. 3).

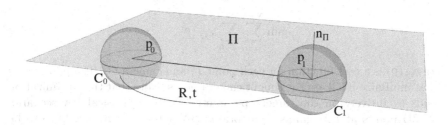

Fig. 3. Epipolar geometry for two spherical images.

Pose Estimation. There are mainly two techniques for computing camera poses. The first is useful for *relative* pairwise pose estimation, typically when only 2D image correspondences (2D-2D correspondences) are available. Without loss of generality, one of the cameras is assumed as reference and R and t represent the pose of the second camera. In this case, R and t may be determined with e.g. the 5-point algorithm [30]. The second technique is normally used when a number of 3D scene points and their respective projections onto an image are known, i.e. a set of 2D-3D correspondences is available. This configures a *Perspective-n-point* (PnP) problem, which can be solved with a minimum of 6 correspondences [31].

3   The Proposed Approach

Given a set of images of a scene, our goal is to accurately estimate the pose of all cameras as well as to recover a sparse 3D point cloud of the underlying scene representing its geometry. The set of central projection cameras is defined as

$$C = \left\{ C_j = \left[\hat{R}_j | \hat{t}_j \right] \mid \hat{R}_j \in SO(3), \; \hat{t}_j \in \mathbb{R}^3 \right\}, \tag{1}$$

where $j = 0, .., M-1$, M is the total number of cameras and $\hat{R}_j$ and $\hat{t}_j$ are the rotation matrix and translation vector representing the estimated pose of camera C_j.

To aid the non-linear optimization, we adopt an axis-angle parameterization for the rotation matrix and C_j is then parameterized by a vector $\rho_j \in \mathbb{R}^6$. All together, the cameras are parameterized by a vector $\rho \in \mathbb{R}^m$, with $m = 6M$.

Likewise, we denote the set of sparse 3D points reconstructed along with the camera poses as

$$\mathcal{P} = \left\{ \hat{P}_i \in \mathbb{R}^3 \right\}, \tag{2}$$

where $i = 0, .., N - 1$, N is the number of points and $\hat{P}_i$ holds the estimated coordinates of a scene point P_i.

We then formulate the problem of recovering all camera poses along with a sparse point representation of the scene as a non-linear optimization problem. The parameter vector ρ is optimized in order to minimize

$$\min_{\rho} \sum_{i=0}^{N-1} \sum_{j=0}^{M-1} e_{ij}(\rho), \tag{3}$$

where $e_{ij}(\rho)$ is a cost function for each point $\hat{P}_i$ and camera C_j. The parameters ρ^+ that minimize Eq. 3 are the sought camera poses. Note that this optimization depends exclusively on the camera parameters ρ. In a classical BA scenario, for N 3D points and M cameras, a total of $3N + 6M$ parameters have to be optimized. Different from the classical BA, we reduce the complexity of the problem by dropping the structure and considering only the camera parameters. This leads to an important advantage: the dimension of the parameter search space is at most $6M$, a significant reduction compared to $3N + 6M$, what is particularly convenient in case of large scale scenes. Nonetheless, structure is jointly estimated. Inspired by [22], point correspondences are re-triangulated at every step of our BA, updating the structure with the most recent camera parameters and reducing drift due to accumulated errors.

3.1   Reprojection Error and Visibility Map Models

The cost function $e_{ij}(\rho)$ in Eq. 3 represents the reprojection error of a point $\hat{P}_i$ on camera C_j and is defined as

$$e_{ij}(\rho) = cos^{-1}(p_{ij}\hat{p}_{ij}), \tag{4}$$

where $p_{ij}\hat{p}_{ij}$ is the scalar product between the expected projection p_{ij} and the measured projection $\hat{p}_{ij}$ obtained with $\hat{P}_i$, $\hat{R}_j$ and $\hat{t}_j$. The expected projection p_{ij} is determined by the keypoint location corresponding to P_i. Note that as $-1 \leq p_{ij}\hat{p}_{ij} \leq 1$, we have $0 \leq e_{ij}(\rho) \leq \pi$ and it is not necessary to take the absolute value in Eq. 4. Furthermore, we do not use any approximation of the reprojection error as in [15]. As we aim at high accuracy, the error defined in Eq. 4 is the exact geodesic distance, i.e. the exact angular deviation, between p_{ij} and $\hat{p}_{ij}$. Additionally, to each point P_i we associate a visibility map

$$\mathcal{V}_i = \left\{ (C_j, p_{ij}) \mid C_j \in \mathcal{C}, \ p_{ij} \in \mathcal{S}^2 \right\}, \tag{5}$$

where $\mathcal{S}^2$ represents the unit sphere. We denote the pair (C_j, p_{ij}) as the *observation* of a scene point P_i on camera C_j.

3.2 Sub-set Constraints

SPHERA implements an incremental pipeline, that is, starting from an initial pair, cameras are sequentially added until all poses have been estimated. More specifically, we draw from C an initial pair of cameras and once their poses are determined they are used to initialize a set $\acute{C}$ representing the current set of calibrated cameras. Then, one by one, cameras are added to $\acute{C}$ until $|\acute{C}| = |C|$. After adding a camera to $\acute{C}$, BA is performed to refine the poses of all calibrated cameras. However, this is not always necessary. As calibration progresses, previously added cameras become more stable, i.e. their poses no longer change significantly. After some time, refining their poses brings no improvement. This is often true for large image datasets. The exceptions to this are loop closures and later addition of new images to the dataset.

To address this issue, we introduce a sub-set $C^* \subset \acute{C}$ to hold the cameras for which pose refinement is unnecessary. Cameras belonging to C^* will be regarded as fixed and their poses will not be updated during BA. A camera $C_j \in \acute{C}$ is added to C^* when the update on its pose is no longer significant. This is achieved with the introduction of two measurements in the following way. After BA, we measure the update on its rotation matrix δ_{R_j} and translation vector δ_{t_j} computed as

$$\delta_{R_j} = \|log\left(R_j^{k-1}\left(R_j^k\right)^T\right)\|, \tag{6}$$

$$\delta_{t_j} = \frac{\|t_j^k - t_j^{k-1}\|}{\|t_j^{k-1}\|}, \tag{7}$$

where k stands for calibration step, i.e. it is incremented after each BA. The right-hand side of Eq. 6 is a metric in $SO(3)$ and can be efficiently computed with quaternions [32]. Then, if $\delta_{R_j} < \tau_r$ and $\delta_{t_j} < \tau_t$, with $\tau_r \geqslant 0$ and $\tau_t \geqslant 0$, C_j is added to C^*. Clearly, once C_j is added to C^*, $\delta_{R_j} = 0$ and $\delta_{t_j} = 0$ in the subsequent calibration steps. Therefore, to correctly handle loop closures and to locally update camera poses whenever new images are included in the dataset, a third measurement is required. This measurement allows to *remove* cameras from C^* so that they may be optimized once again. It is based on the visibility of scene points and works as follows. Assume $C_j \in C^*$ and C_j observes N_j 3D points. The visibility measurement δ_{v_j} of a camera C_j is then defined as

$$\delta_{v_j} = \|\nu_j^k - \nu_j^{k-1}\|, \quad with \tag{8}$$

$$\nu_j = \sum_{n=0}^{N_j-1} \eta_n, \quad \eta_n = \begin{cases} 1, & if\ \mathcal{V}_n\ increased \\ 0, & otherwise \end{cases}. \tag{9}$$

Note that δ_{v_j} is independent of the camera pose. Also, it does not measure how many new 3D points C_j observes. Instead, it measures, among all 3D points visible in C_j, how many had their visibility maps updated, i.e. are now visible in at least one new camera. Then, if $\delta_{v_j} > \tau_v$, C_j is removed from C^* and will be taken into account in the next calibration step. In our implementation, we also use the visibility measurement along with the first two to decide whether a camera should be added to C^*. Together, τ_r, τ_t and τ_v form the sub-set constraints.

Remark 1. The re-triangulation of point correspondences is beneficial as it reduces drift due to accumulated errors. However, it increases the overall computational cost. The sub-set constraints prevent the re-triangulation of points $\hat{P}_i$ that are seen exclusively by cameras in C^*, thus further improving performance.

3.3   Minimizing the Reprojection Error

As discussed above, recovering the camera poses and scene structure can be achieved by solving a bundle adjustment problem [24]. SPHERA minimizes a reprojection error formulated directly on the surface of the unit sphere (see Eqs. 3 and 4). This is interpreted as finding the camera poses that maximize the alignment between the rays defined by all predicted and measured projections. This is true for any central projection camera.

After the introduction of the visibility map in Sect. 3.1, we may now rewrite Eq. 3 in the form shown in Eq. 10. We adopted the framework available in [33] as the core non-linear solver upon which SPHERA is built.

$$\min_{\rho} \sum_{i=0}^{N-1} \sum_{j=0}^{M-1} \gamma_{ij} e_{ij}(\rho), \ \gamma_{ij} = \begin{cases} 1, & if \ C_j \in \mathcal{V}_i \\ 0, & otherwise \end{cases} \tag{10}$$

In practice, we solve a modified version of Eq. 10, where only cameras $C_j \in \mathcal{C} \setminus C^*$ and the most reliable points are used. These points are defined as

$$\mathcal{P}^* = \left\{ \hat{P}_i \in \mathcal{P} \mid e_{ij}(\rho) < \tau_e, \ \forall (C_j, p_{ij}) \in \mathcal{V}_i \right\}, \tag{11}$$

where τ_e is a threshold imposed to all individual reprojection errors $e_{ij}(\rho)$.

4   Evaluation

4.1   Preliminaries

Keypoints are detected and matched using the method proposed in [34], where a multi-scale keypoint detector and matcher was developed for high resolution spherical images. Nonetheless, it is worth mentioning that SPHERA is completely independent of how keypoints are detected, described and matched. Consequently, any other keypoint detector and matcher may be adopted (see Sect. 4.3).

We validate our framework using synthetic spherical as well as real perspective and spherical images. The resolution of all spherical images presented below is 14142×7071 (100 Mega-pixels). Experiments are divided into four categories: The first category consists of a set of synthetic spherical cameras where the goal is to validate our framework on spherical images using groundtruth. The second is composed exclusively of real perspective images. Here, the idea is to show that our framework is suitable for standard SfM, i.e. it may be used even when no spherical image is available. The third category consists of spherical images only, where we compared SPHERA to the work presented in [15] in two different real world scenarios. The fourth and last category is a hybrid dataset where real perspective and spherical images are used simultaneously. The aim is to demonstrate SPHERA's ability to improve scene geometry estimation whenever more images are available, independent of their types[1]. Whenever available, groundtruth data is used for evaluation. Otherwise, we rely on the global mean reprojection error computed taking all images and all reconstructed points into account.

4.2 Synthetic Dataset

An artificial room with dimensions $6 \times 6 \times 3\,\mathrm{m}$ was created using [35] and 72 spherical images were rendered (see Fig. 4-(a)). The poses of these artificially generated cameras were used as groundtruth. Additionally, the depth map shown in Fig. 4-(b) was stored and serves to measure the accuracy of the recovered scene geometry.

(a)   (b)

Fig. 4. (a) Sample image of the synthetic dataset. (b) Groundtruth depth map used to evaluate the accuracy of scene geometry estimation (contrast enhanced to improve visualization).

After detecting and matching keypoints with Gava's approach, camera poses and scene structure were recovered with SPHERA. Residual errors were computed in the following way. The position error is the Euclidean distance between the groundtruth and estimated camera positions. To measure the orientation error, we chose again the function presented in [32], which in this context may be written as $\|log\left(R\hat{R}^T\right)\|$, with R the desired rotation and $\hat{R}$ the estimated rotation matrix. For details we refer to [32]. The residual error of a reconstructed point $\hat{P}_i$ is computed as $\|\hat{P}_i - P_i\|$, where the coordinates of P_i are obtained as

---

[1] Assuming central projection cameras.

Table 1. Errors in camera poses and sparse scene reconstruction for the synthetic dataset. Mean and standard deviation are identified by μ and σ, respectively.

	orient. error [degree]	pos. error [mm]	recon. error [mm]
μ	0.009	0.68	0.482
σ	0.03	2.8	0.9

follows. A virtual spherical camera is located at the origin of the global coordinate system. The projection of $\hat{P}_i$ onto this virtual camera delivers p_i'. Then $P_i = I_{dm}\left(p_i'\right)p_i'$, where $I_{dm}\left(p_i'\right)$ is the groundtruth depth retrieved from the stored depth map.

We first ran our pipeline ignoring the sub-set constraints and with τ_e equivalent to 5 pixels. Although τ_e is an angular deviation, for convenience we converted and presented it in pixels. Table 1 presents the resulting errors in camera poses and scene reconstruction.

Table 2. Average errors in camera poses and sparse scene reconstruction for the synthetic dataset with sub-set constraints. The last line shows the running time relative to the total time needed when no sub-set constraints are used.

τ_v [# points]	100		1000	
τ_t [%]	1	5	1	5
orient. error [degree]	0.0091	0.0091	0.0094	0.0095
pos. error [mm]	0.76	0.78	0.76	0.79
recon. error [mm]	0.487	0.494	0.581	0.604
time [%]	49.7	48.2	15.4	14.9

Figure 5 shows the reconstructed point cloud, with approximately 156K points. The rendered spheres and their corresponding coordinate frames reflect the recovered camera poses. We adopted the Odysseus Studio [36] to visualize and present our results.

A second experiment aimed at evaluating the impact of the sub-set constraints on camera pose estimation, the sparse reconstruction of the scene and the overall performance gain. We ran our pipeline varying the sub-set constraints within the ranges $\tau_r = [0.25°, 2°]$ in steps of 0.25°, $\tau_t = [0.01, 0.05]$ and $\tau_v = [100, 1000]$. We noticed that, for this experiment, varying τ_r had little impact on the final results. Thus, Table 2 summarizes the average values. Time values are relative to the total time required when no sub-set constraints are used. The standard deviations for the rotation, position and reconstruction errors were below 0.04°, 3.25 mm and 1.2 mm, respectively. On the other hand, the standard deviations for the performance gain were approximately 10 % for

Fig. 5. Reconstructed point cloud and recovered camera poses obtained with SPHERA. Details on the floor and walls can be easily seen.

$\tau_v = 100$ and 7% for $\tau_v = 1000$. This is probably due to different gradient descent paths chosen by the non-linear optimizer [33].

4.3 Perspective Datasets

To validate our approach on perspective images, we compared it to Bundler [17], a popular software developed for SfM on standard perspective images. Bundler is the camera calibration tool currently used in [10,26,37], and is publicly available.

The experiments presented in this section were carried out on the datasets published in [28]. For each dataset, we ran Bundler on the original images and SPHERA on the corresponding warped images as shown in Fig. 2. To ensure a fair comparison, we ran our pipeline using the same keypoints detected by Bundler [38] after warping their coordinates to the unit sphere. This eliminates the influence of image feature location on the evaluation. Moreover, it shows SPHERA's independence of keypoint detectors as pointed out in Sect. 4.1. Results on camera pose estimation are summarized in Fig. 6. Orientation errors were obtained as in the previous section. Position errors, however, were computed after preprocessing the estimated camera positions. To account for the differences in scale, the baseline between the closest camera pair was normalized and the remaining camera positions were scaled accordingly. After that, the Euclidean distance was measured as in Sect. 4.2.

As can be seen, Bundler performs slightly better and the reason is as follows. Bundler works exclusively on perspective images and optimizes the camera poses

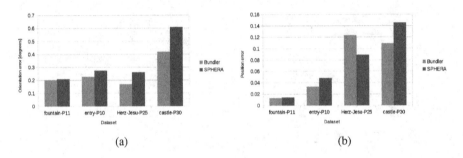

(a)   (b)

Fig. 6. (a) Orientation error and (b) position error on perspective image datasets obtained with Bundler and SPHERA. See text for details.

along with their individual intrinsic parameters such as focal length and lens distortion. In contrast, SPHERA has been designed to operate on any kind of central projection camera, but the optimization of intrinsic parameters has not been integrated yet. Therefore, for the experiments presented in this section, we used a constant focal length in our pipeline and a variable focal length for Bundler. In fact, the differences observed in Fig. 6 are proportional to the variance of the focal length within each dataset, see Table 3. The exception is Herz-Jesu-P25, where Bundler delivers smaller orientation error whereas SPHERA provides better camera positions.

Table 3. Variation of focal lengths estimated with Bundler. The second column shows the standard deviation and the third column the difference between maximum and minimum values. Note that, except for the Herz-Jesu-P25 dataset, the differences in Fig. 6 are proportional to the variation of the focal length.

dataset	σ_f [pixel]	range [pixel]
fountain-P11	8.49	23.02
entry-P10	10.97	28.41
Herz-Jesu-P25	4.01	16.15
castle-P30	20.44	118.86

4.4   Spherical Datasets

In this section we compare SPHERA and the approach presented in [15]. We ran both pipelines on two datasets. The first dataset consists of 9 spherical images captured inside one of the Mogao Caves, in China. The second dataset contains 35 spherical images taken at the Saint Martin Square in Kaiserslautern, Germany, and represents outdoors, more challenging, environments. Due to the lack of groundtruth data for these datasets, we based our evaluation on the global mean reprojection error. The assumption is that the correlation observed in Sect. 4.2 can be used to infer the relative accuracy of the estimated scene geometry.

As can be seen in Fig. 7, SPHERA improves the reprojection error on both datasets, specially on the St. Martin Square. In the case of the Mogao Cave, due to its simple geometry and rich texture (Fig. 8-(a)), only few points are discarded based on Eq. 11, what explains the small difference in the reprojection error for this dataset. The St. Martin Square dataset is more challenging (Fig. 8-(b)). It contains many low textured regions, depth discontinuites, occlusions as well as repetitive patterns. Therefore, several points are inconsistent and discarding them from the camera pose estimation leads to the difference observed in Fig. 7. These results suggest that SPHERA delivers more accurate scene structures. Figure 8 displays the sparse point clouds yielded by our framework, where details of the surroundings are accurately reconstructed.

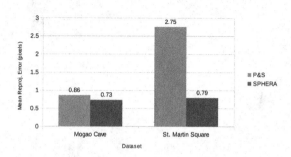

Fig. 7. Global mean reprojection error on spherical image datasets obtained with [15] and SPHERA. See text for details.

4.5   Hybrid Dataset

In this section we evaluate the SPHERA framework on a hybrid dataset composed of perspective and spherical images. The idea is to show that our framework naturally handles different central projection cameras simultaneously. This dataset is composed of the same 35 spherical images used in the previous experiment and additional 11 perspective images of resolution 3888×2592 pixels. As shown in Fig. 9, the reprojection error obtained with spherical images (same as previous experiment) is better than the error for perspective images.

The main reason spherical camera pose estimation is better than its perspective counterpart is due to their wide field of view. As can be seen in Fig. 10, matches between spherical images cover the entire scene and thus impose more constraints on cameras' poses. As expected, the reprojection error decreases when perspective and spherical images are used simultaneously.

Fig. 8. First row: Sample images of the Mogao Cave and St. Martin Square datasets. Second to fourth rows: reconstructed point clouds delivered by SPHERA, containing approximately 106 K and 197K 3D points for the Mogao Cave and St. Martin Square, respectively.

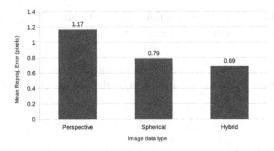

Fig. 9. Global mean reprojection error for the hybrid St. Martin Square experiment. Note how it decreases when perspective and spherical images are used together.

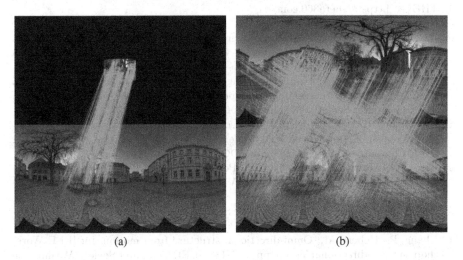

(a)   (b)

Fig. 10. (a) Symmetric matches between a warped perspective image and a spherical image. (b) Symmetric matches between two full spherical images.

5   Conclusions

This paper presents SPHERA, a novel unifying Structure from Motion framework designed for central projection cameras. The goal is to cover the gaps between pipelines developed for perspective, spherical and catadioptric images and to support future large scale 3D reconstruction algorithms. Through extensive quantitative evaluation on synthetic and real image sequences, we showed that our approach delivers high quality camera pose as well as scene geometry estimations when compared to state of the art approaches optimized for specific camera types.

Future work aims at integrating the optimization of intrinsic parameters to increase the accuracy of pose estimation of perspective cameras. Additionally, we plan to validate our framework on larger, hybrid image datasets, supported by groundtruth data. Finally, SPHERA will be the underlying SfM mechanism in our upcoming dense multi-view reconstruction approach.

Acknowledgments. The authors would like to thank Richard Schulz for the creation of the synthetic dataset. This work was funded by the project DENSITY (01IW12001).

References

1. Civetta 360° Digital Imaging. http://www.weiss-ag.org/solutions/civetta/
2. LizardQ. http://www.lizardq.com
3. Seitz Roundshot. http://www.roundshot.ch
4. THETA. https://theta360.com/en/
5. New House Internet Services BV. http://www.ptgui.com
6. Anguelov, D., Dulong, C., Filip, D., Frueh, C., Lafon, S., Lyon, R., Ogale, A., Vincent, L., Weaver, J.: Google street view: capturing the world at street level. Computer **43**, 32–38 (2010)
7. Klein, G., Murray, D.: Parallel tracking and mapping for small AR workspaces. In: IEEE and ACM International Symposium on Mixed and Augmented Reality. IEEE Computer Society, Washington DC (2007)
8. Newcombe, R.A., Lovegrove, S.J., Davison, A.J.: DTAM: dense tracking and mapping in real-time. In: International Conference on Computer Vision, pp. 2320–2327. IEEE Computer Society, Washington DC (2011)
9. Hiep, V.H., Keriven, R., Labatut, P., Pons, J.-P.: Towards high-resolution large-scale multi-view stereo. In: Computer Vision and Pattern Recognition, pp. 1430–1437. IEEE Computer Society (2009)
10. Furukawa, Y., Curless, B., Seitz, S.M., Szeliski, R.: Towards internet-scale multi-view stereo. In: Computer Vision and Pattern Recognition (2010)
11. Chang, P., Hebert, M.: Omni-directional structure from motion. In: IEEE Workshop on Omnidirectional Vision, pp. 127–133. IEEE Computer Society, Washington (2000)
12. Micusik, B., Pajdla, T.: Structure from motion with wide circular field of view cameras. Trans. Pattern Anal. Mach. Intell. **28**, 1135–1149 (2006)
13. Bagnato, L., Frossard, P., Vandergheynst, P.: A variational framework for structure from motion in omnidirectional image sequences. J. Math. Imag. Vis. **41**, 182–193 (2011)
14. Kangni, F., Laganiere, R.: Orientation and pose recovery from spherical panoramas. In: IEEE International Conference on Computer Vision, pp. 1–8. IEEE Computer Society, Los Alamitos (2007)
15. Pagani, A., Stricker, D.: Structure from Motion using full spherical panoramic cameras. In: OMNIVIS (2011)
16. Aly, M., Bouguet, J.-Y.: Street view goes indoors: Automatic pose estimation from uncalibrated unordered spherical panoramas. In: Winter Conference on Applications of Computer Vision (2012)
17. Snavely, N., Seitz, S.M., Szeliski, R.: Photo tourism: exploring photo collections in 3D. In: SIGGRAPH, pp. 835–846. ACM, New York (2006)
18. Hartley, R.I., Zisserman, A.: Multiple View Geometry in Computer Vision. Cambridge University Press, Cambridge (2004)
19. Ma, Y., Soatto, S., Kosecka, J., Sastry, S.: An Invitation to 3D Vision, from Images to Models. Springer Verlag, New York (2003)

20. Arie-Nachimson, M., Kovalsky, S.Z., Kemelmacher-Shlizerman, I., Singer, A., Basri, R.: Global motion estimation from point matches. In: 3DIMPVT (2012)
21. Olsson, C., Enqvist, O.: Stable structure from motion for unordered image collections. In: Heyden, A., Kahl, F. (eds.) SCIA 2011. LNCS, vol. 6688, pp. 524–535. Springer, Heidelberg (2011)
22. Wu, C.: Towards linear-time incremental structure from motion. In: 3DV (2013)
23. Rodríguez, A.L., López-de-Teruel, P.E., Ruiz, A.: Reduced epipolar cost for accelerated incremental SfM. In: Computer Vision and Pattern Recognition (2011)
24. Triggs, B., Mclauchlan, P.F., Hartley, R.I., Fitzgibbon, A.W.: Bundle adjustment: a modern synthesis. In: International Workshop on Vision Algorithms: Theory and Practice, pp. 298–372 (1999)
25. Frahm, J.-M., et al.: Building rome on a cloudless day. In: Daniilidis, K., Maragos, P., Paragios, N. (eds.) ECCV 2010, Part IV. LNCS, vol. 6314, pp. 368–381. Springer, Heidelberg (2010)
26. Agarwal, S., Snavely, N., Simon, I., Seitz, S.M., Szeliski, R.: Building Rome in a day. In: International Conference on Computer Vision, Kyoto, pp. 72–79 (2009)
27. Geyer, C., Daniilidis, K.: Catadioptric projective geometry. Int. J. Comput. Vis. **43**, 223–243 (2001)
28. Strecha, C., Hansen, W., Van-Gool, L., Fua, P., Thoennessen, U.: On benchmarking camera calibration and multi-view stereo for high resolution imagery. In: Computer Vision and Pattern Recognition (2008)
29. Torii, A., Imiya, A., Ohnishi, N.: Two- and three- view geometry for spherical cameras. In: OMNIVIS (2005)
30. Stewénius, H., Engels, C., Nistér, D.: Recent developments on direct relative orientation. J. Photogram. Remote Sens. **60**, 284–294 (2006)
31. Quan, L., Lan, Z.: Linear N-point camera pose determination. Trans. Pattern Anal. Mach. Intell. **21**, 774–780 (1999)
32. Huynh, D.Q.: Metrics for 3D rotations: comparison and analysis. J. Math. Imag. Vis. **35**, 155–164 (2009)
33. Agarwal, S., Mierle, K.: Others: Ceres Solver. https://code.google.com/p/ceres-solver/
34. Gava, C.C., Hengen, J.M., Tätz, B., Stricker, D.: Keypoint detection and matching on high resolution spherical images. In: International Symposium on Visual Computing, Rethymnon, pp. 363–372 (2013)
35. Blender. http://www.blender.org/
36. Odysseus Studio. http://av.dfki.de/odysseus-studio
37. Furukawa, Y., Ponce, J.: Accurate, dense, and robust multi-view stereopsis. Trans. Pattern Anal. Mach. Intell. **32**, 1362–1376 (2010)
38. Lowe, D.G.: Distinctive image features from scale-invariant keypoints. Int. J. Comput. Vis. **60**, 91–110 (2004)

TVL₁ Planarity Regularization for 3D Shape Approximation

Eugen Funk[1,2]([⊠]), Laurence S. Dooley[1], and Anko Börner[2]

[1] Department of Computing and Communications, The Open University,
Milton Keynes, UK
eugen.funk@dlr.de, l.s.dooley@open.ac.uk
[2] Department of Information Processing for Optical Systems,
Institute of Optical Sensor Systems, German Aerospace Center (DLR), Berlin,
Germany
anko.boerner@dlr.de

Abstract. The modern emergence of automation in many industries has given impetus to extensive research into mobile robotics. Novel perception technologies now enable cars to drive autonomously, tractors to till a field automatically and underwater robots to construct pipelines. An essential requirement to facilitate both perception and autonomous navigation is the analysis of the 3D environment using sensors like laser scanners or stereo cameras. 3D sensors generate a very large number of 3D data points when sampling object shapes within an environment, but crucially do not provide any intrinsic information about the environment which the robots operate within.

This work focuses on the fundamental task of 3D shape reconstruction and modelling from 3D point clouds. The novelty lies in the representation of surfaces by algebraic functions having limited support, which enables the extraction of smooth consistent implicit shapes from noisy samples with a heterogeneous density. The minimization of total variation of second differential degree makes it possible to enforce planar surfaces which often occur in man-made environments. Applying the new technique means that less accurate, low-cost 3D sensors can be employed without sacrificing the 3D shape reconstruction accuracy.

1 Introduction

The analysis and perception of environments from static or mobile 3D sensors is widely envisioned as a major technological breakthrough and is expected to herald a significant impact upon both society and the economy in the future. As identified by the German Federal Ministry of Education and Research [25], spatial perception plays a pivotal role in robotics, having an impact onmany vital technologies in the fields of navigation, automotive, safety, security and human-robot-interaction. The key task in spatial perception is the reconstruction of the shape of the observed environment. Improvements in shape reconstruction have direct impact on three fundamental research disciplines: *self localization* from camera images [13], *inspection* in remote sensing [26] and *object recognition* [12].

© Springer International Publishing Switzerland 2016
J. Braz et al. (Eds.): VISIGRAPP 2015, CCIS 598, pp. 274–294, 2016.
DOI: 10.1007/978-3-319-29971-6_15

Applying 3D sensors in uncontrolled practical environments, however, leads to strong noise and many data outliers. Homogeneous surface colours and dynamic illumination conditions lead to outliers and reduce drastically the quality of computed 3D samples. Figure 1 shows an example 3D point cloud obtained from a stereo camera traversing a building. Many 3D points such as marked by ① suffer from strong noise. Occlusions or over-exposure frequently occur in realistic scenes ② and make automated shape reconstruction even more challenging.

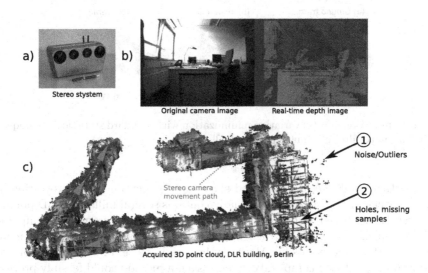

Fig. 1. (a) Stereo System, (b) sample RGB and depth image, (c) acquired 3D point cloud.

Dealing with noise and outliers inevitably involves applying statistical techniques. In the last decade, so-called kernel-based methods have become well-accepted in statistical processing. Successful techniques like *deep learning* or *support vector machines* exploit kernel-based methods in the fields of machine learning and robotics for interpolation and extrapolation [36]. Since shape interpolation and extrapolation are required when dealing with error-prone 3D samples, the application of kernel-based techniques for shape approximation is especially relevant to this domain. The initial aim was the investigation and development of a suitable kernel for geometrical shape modelling from noisy 3D samples.

Many indoor and urban outdoor environments can be represented by a small set of planar shapes. This information can be exploited to help to achieve higher approximation accuracy. Integrating piecewise smoothness into the approximation task has attracted a lot of interest in the image processing community. Several research groups applied a regularization technique, also known as *Total Variation* (TV) minimization, to penalize strong variations in the colour values [21,34]. Bredies [9] extended the traditional TV approach to second derivatives of the filtered image pixels. Figure 2 shows the comparison of Bredies's TVL₁ approach with state-of-the-art statistical filtering techniques. The extension of the

TV technique to 3D shapes is still a fertile area of research which is considered as the second aim in this work.

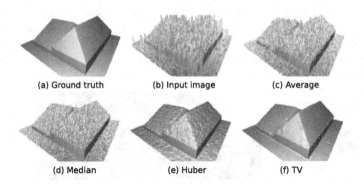

Fig. 2. Comparison of total variation minimization with standard statistical techniques for height maps filtering. Image courtesy: [9].

A further challenge in automated shape approximation is the processing of large datasets. A realistic dataset usually contains several millions of 3D points. However, kernel-based and total variation techniques suffer from high computational complexity prohibiting their application to datasets which contain more than a few thousand points [5]. Methods that require a set of linear equations to be solved incur $\mathcal{O}(N^3)$ complexity. Even if such a method could feasibly process $N = 1,000$ points in 10 ms, it would still take 115 days to process $1,000,000$ points. This major complexity issue motivated the third aim of this work which is to develop efficient strategies for handling non-smooth (L_1) total variation regularization on large datasets.

The remainder of this paper is organized as follows: A short overview of 3D shape reconstruction approaches is provided in Sect. 2, including issues such as approximation quality and stability. Section 3 discusses the three main contributions of this work: (i) application of smooth kernels for implicit 3D shape modelling, (ii) integration of non-smooth TVL_1 regularization for noise suppression, and (iii) efficient optimization reducing the computation complexity from $\mathcal{O}(N^3)$ to $\mathcal{O}(N)$. A critical quantitative analysis is presented in Sect. 4, and concluding comments are provided in Sect. 5.

2 Literature Review

The problem of reconstructing a surface of an object from a set of scattered 3D points attracted a lot of attention [2,23,29]. This section will review existing techniques relating to the aims of this paper, namely: shape representation using radial basis functions, statistical planarity-aware regularization model, and efficient optimization.

2.1   Shape Reconstruction

Two general shape representation approaches for 3D data currently exist: explicit and implicit representations.

Explicit models are polygon meshes, *non-uniform B-Splines* (NURBS) or Bezier curves [27]. Research in computer graphics leads to a large number of software frameworks such as OpenGL [42] that enables the visualization of parametric polygon meshes with the help of parallel graphics hardware. For this reason initial research on automated shape reconstruction from 3D scattered points focused on the direct construction of triangle meshes, also known as Delaunay-Triangulation. Methods such as α shapes [6,7,18] aim at creating a polygonal mesh by connecting the input samples with triangle edges. This, however, leads to inaccurate results when error-prone samples are provided. Another family of parametric shapes are NURBS [32,33] and Bezier curves [1], which are commonly used in 3D modelling. These methods are able to create smooth surfaces for non-uniform control point sets. In order to apply these methods to automated shape reconstruction from scattered 3D points, the surface is defined as a graph in the parameter space. This makes the problem *non-polynomial* (NP) hard so its application to larger datasets is prohibited [44].

Implicit Models: Several state-of-the-art techniques represent a shape *implicitly* by an indicator function $f(\mathbf{x})$ to indicate *inside* $f(\mathbf{x}) < 0$ or *outside* $f(\mathbf{x}) > 0$ of the object with $\mathbf{x} \in \mathbb{R}^3$ as the location in the 3D space. The surface of the object is the set of all $\mathbf{x}$ where f gives zero. Figure 3 illustrates an implicit shape where the dots indicate the samples on the surface ($f(\mathbf{x}) = 0$) and the point orientations the normal of the shape ($\nabla f(\mathbf{x}_i) = \mathbf{n}_i$). This representation allows to extract smooth surfaces from irregularly sampled, noisy and incomplete datasets [23].

Facing the noise sensibility issues of Delaunay-Triangulation techniques, Alexa et al. proposed to apply *moving least squares* (MLS) for smoothing (averaging) the point samples prior to reconstructing a mesh via a Delaunay-Triangulation technique [2]. A simple implicit shape is for instance a plane defined by its four parameters $\mathbf{n}^T\mathbf{x} + d = 0$ with $\mathbf{n} \in \mathbb{R}^3$ as the plane normal vector and d as offset to the origin along $\mathbf{n}$. Defining a shape function as $f(\mathbf{x}) = \mathbf{b}(\mathbf{x})^T\mathbf{u}$ with $b(\mathbf{x}) = (x_1, x_2, x_3, 1)$ and $\mathbf{u}$ as the plane

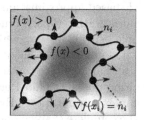

Fig. 3. Smooth shape representation from scattered points and surface orientations (arrows) via an implicit function $f(\mathbf{x})$.

coefficients $\mathbf{u} = (n_1, n_2, n_3, d)$ allows to find $\mathbf{u}$ via a regression task [3]. Similarly, Guennebaud extended the shape model to spheres and proposed the popular *Algebraic Point Set Surfaces* (APSS) method [24]. Ohtake et al. and Oztireli et al. addressed the over-smoothing issues by applying non-linear regression for shape approximation [29,31]. The MLS techniques are well capable of filtering datasets with moderate or small noise. However, it is still not feasible for realistic datasets as introduced in Fig. 1.

Implicit Models with Basis Functions: Motivated by the drawbacks of MLS approaches, Calakli and Taubin proposed applying a global optimization process [11]. Acquired 3D samples are structured with an octree and the implicit values of $f(x)$ are distributed on the corners of the octree nodes (voxels). This approach enables large holes to be closed and allows to handle sparse spatial samples which lead to isolated fragments when MLS is applied. A similar approach is proposed by Kazhdan and Hoppe, where the voxel corners are the B-Splines control points [28]. Both approaches suffer from the fundamental drawback that a priori information is required from an expert user to define the depth of the octree structure, which makes using it in automated applications very difficult.

Another family of implicit surface reconstruction algorithms uses smooth *radial basis functions* (RBF). The main difference between RBF-based approximation and discrete octree models [11,28] is that RBFs are not necessarily centred on the octree leaves but directly on the samples. This reduces the risk of applying inappropriate discretization and to lose shape details [14,23].

Novel approaches [43] propose creating a dense grid of a user-specified resolution and to use the L_1 norm to penalize the changes between the implicit grid corner values. Accurate results are achieved when a fine grid is applied, although the approach does not consider the smoothness of the second derivative of the shape leading to non-smooth reconstruction. Another drawback of the method is that it is restricted to small and compact objects since the computation time and memory consumption for the dense grid quickly become prohibitive.

Bredies et al. proposed to apply so-called generalized total variation minimization on depth images to penalize the variance in the second derivatives leading to piecewise smooth shapes (Fig. 2). The accuracy of the method motivates its extension to 3D shapes, which has not been reported in the literature. Bredies et al. state that the stability of the approach heavily depends on the smoothness of the data, which is feasible when smooth RBFs are applied [9]. Thus, when developing an RBF-based approximation model with a TV regularization, the choice of an appropriate RBF type is crucial.

With a popular RBF example being Gaussian, which is of infinite differential degree but tends to smooth out fine detail, Wahba studied the application of Duchon's *Thin Plate Splines* [16] that facilitate control of the smoothness degree [39]. Due to their global definition domain, Thin Plate Splines do not result in sparse systems and lead to complex computations. Even more adverse, a change of a single RBF centre affects the complete shape model in the full approximation domain, which is not the case for RBF using compact support such as Gaussians. Later, Wendland proposed several RBF types with compact support of minimal smoothness degree [40]. Wendland's RBFs also control the smoothness of the approximated function and still lead to sparse and efficient linear regression systems. Moreover, as presented in Sect. 3, the smaller the smoothness degree the more stable is the regression process. The Thin Plate RBFs, however, have been shown to achieve superior approximation quality in the presence of noise [37]. Important aspects when selecting an appropriate RBF type are presented in Sect. 3.1.

Table 1. Shape approximation comparison. Here + indicates that a method is moderately successful in a particular aspect, and ++ indicates that a method is very successful.

Method	Missing data	Noise	Outliers	Comput. speed	Sharp edges
α shapes [18]				++	+
Adaptive α shapes [6]	+			++	+
APSS [24]		+	+	+	+
SSD [11]	++	+	+	+	
Poisson [28]	++	+	+	+	
TVL$_1$ depth fusion [9]	++	++	++		++

2.2 Efficient L_1 Optimization

Extending the shape approximation with a L_1 penalty requires more advanced techniques to solve the optimization task. This issue has been discussed for some time in the statistics and numerical optimization community. However, efficient techniques being capable of dealing with thousands or millions of data samples are focussed in current research.

Tibshirani proposed the *Least Absolute Shrinkage and Selection* (Lasso) technique to minimize cost functions such as

$$\| \mathbf{y} - K\boldsymbol{\alpha} \|_2^2 + \| \boldsymbol{\alpha} \|_1 \tag{1}$$

with $\| \boldsymbol{\alpha} \|_1 = \sum_j^N |\alpha_i|$ enabling its application on images with several hundreds of thousands of entries in $\boldsymbol{\alpha}$ [38]. This form is common for regression problems where the signal $\mathbf{y}$ is approximated linearly by the model matrix K. The additional $\| \cdot \|_1$ penalty term enforces only a small amount of non-zeros entries in $\boldsymbol{\alpha}$. This behaviour is suitable for problems where the vector $\boldsymbol{\alpha}$ is expected to have many zero entries. A common application is for example signal approximation by only a small set of frequencies represented by $\boldsymbol{\alpha}$.

When representing a shape with N RBFs

$$f(\mathbf{x}) = \sum_i^N \varphi(\mathbf{x}, \mathbf{x}_i)\alpha_i = k^T \boldsymbol{\alpha} \tag{2}$$

with $k_i = \varphi(\mathbf{x}, \mathbf{x}_i)$, its second derivatives are penalized by $\| D\boldsymbol{\alpha} \|_1$ with $D_{j,i} = \partial_{xx}^2 \varphi(\mathbf{x}_j, \mathbf{x}_i)$. This way $\| D\boldsymbol{\alpha} \|_1$ penalizes the amount of non-smooth regions in the extracted model. However, since the entries in $D\boldsymbol{\alpha}$ are not separated as it is the case in (1), such problems are solved with more difficulty and using the Lasso technique is not possible. Initially, interior active sets methods have been applied to solve the TVL$_1$ objective [4]. Chen et al. additionally demonstrated that the efficiency of primal-dual methods is of magnitudes higher than that of the interior methods [15]. Also Goldstein et al. proposed a primal-dual approach known as the *Bregman Split* [10] to separate the smooth data term $f_d = \| \mathbf{y} - K\boldsymbol{\alpha} \|_2^2$ from the non-smooth regularization term $f_r = \| D\boldsymbol{\alpha} \|_1$ and to optimize each of them independently [22]. Boyd et al. extended the Bregman Split approach by Dykstra's alternating projections technique [17] and proposed the *Alternating Direction Method of Multipliers* (ADMM) [8], which further improves the convergence. Discussions related to applications of ADMM are reported by Parikh and Boyd (2014).

The bottleneck of ADMM is the minimization of the smooth part $f_d = \| \mathbf{y} - K\boldsymbol{\alpha} \|_2^2$. Solving this for $\boldsymbol{\alpha}$ with efficient Cholesky factorization suffers from a complexity of $\mathcal{O}(N^3)$. However, an iterative linear solver such as Jacobian or Gauss-Seidel may reduce the complexity to $\mathcal{O}(N)$ as discussed by Saad or Friedman et al. relating to L_1 regularization [20,35]. Nevertheless, further investigations on the applicability of iterative linear solvers and ADMM on 3D shape modelling do not exist.

2.3   Summary

The presented state of the art in robust shape approximation and optimization methods covers several appropriate options for investigation. Table 1 shows the seminal methods summarizing the benefits and drawbacks of each technique. The TVL$_1$ approach [9] delivers high quality with artefacts such as missing data, noise, outliers, or sharp edges in the image domain. This technique, however, suffers from high computational complexity and needs to be extended to 3D shape approximation. Section 2.2 states that the ADMM technique is expected to outperform the efficiency of existing TVL$_1$ algorithms when extended with an iterative solver.

The next section investigates the impact of different RBFs applied for signal and shape approximation from scattered 3D points before the new ADMM technique for TVL$_1$ optimization on large datasets is presented.

3   The Method

The first part of this section pursues the first research objective and discusses the fundamentals of RBF-based approximation and compares different types of RBFs with respect to quality and stability when least squares optimization is performed. Section 3.2 applies the proposed RBFs and defines the convex optimization task to perform shape reconstruction from scattered 3D samples augmented with a TV regularization term. The last part of this section presents the developed optimization technique that allows to reduce the computational complexity while still being able to solve TVL$_1$ regularized approximation tasks.

3.1 ˙ Interpolation with Radial Basis Functions

When approximating any signal from a set of measurements, the general aim is to determine a function $f : \mathbb{R}^d \mapsto \mathbb{R}$ from a set of N sample values at $x_i \in \mathbb{R}^d$. The core idea of RBF-based approximation is that the function $f(x)$ may be represented by a linear combination of M weighted functions such as

$$f(x) = \sum_j^M \varphi(x, x_j)\alpha_j. \tag{3}$$

Each of the basis functions $\varphi(x, x_j)$ is centered at each measurement x_j, and basically computes the similarity between x and $x_j \in \mathbb{R}^d$. One possible form for φ is a Gaussian $\varphi(x, x_j) = e^{-\|x-x_j\|/\sigma}$ with σ influencing the width of the support.

The underlying idea of RBF approximation is illustrated for a one-dimensional signal in Fig. 4, where f is defined as a sum of all given Gaussians with their weights α_j respectively. Usually it is assumed that the widths σ of the basis functions are known a priori so only the weighting factors α_j are to be found, leading to $f(x)$.

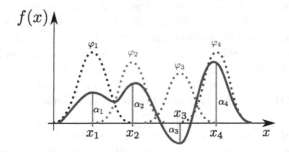

Fig. 4. Illustrative example of smooth $f(x)$ (red line) constructed by a weighted linear combination of Gaussian radial basis functions φ_i (Color figure online).

The task is therefore to perform regression over N samples and to find M weights via minimization of

$$\min_{\alpha} \sum_i^N (y_i - \sum_j^M \alpha_j \varphi(x_i, x_j))^2,$$

where y_i is the i-th measured sample at position x_i. This can also be rewritten in matrix-vector form as:

$$\min_{\alpha} \| \mathbf{y} - K\boldsymbol{\alpha} \|_2^2 \tag{4}$$

where K is often referred to as the design matrix or the *kernel* matrix with $K_{i,j} = \varphi(x_i, x_j)$. The solution is obtained via

$$\boldsymbol{\alpha} = A^{-1}K^T\mathbf{y} \tag{5}$$

with $A = (K^T K)$. This is the well known linear *least squares regression*. Note that the function $f(x)$ itself is not restricted to be linear.

In the last two decades several types of RBFs have been proposed for different applications. For the application on shape approximation three RBF types are investigated. (i) The Gaussian which is the state of the art, (ii) Thin-plate splines [16] with global smooth properties and the (iii) compactly supported RBFs (CSRBFS) [40] which enable sparse regression systems to be created and to control the smoothness of the solution. Table 2 shows the three types of the investigated RBFs with the corresponding explicit forms for data dimension $d = 3$. Note that the scaling of each RBF type is achieved by scaling the argument

$$\varphi_s(r) = \varphi(\frac{r}{s}) \text{ with } r = \| x_i - x_j \|_2 . \tag{6}$$

In order to make a systematic decision which RBF type is best suited for the underlying application, the stability and the approximation quality is considered. When solving for $\boldsymbol{\alpha}$ in (5), the condition of K cond $K = |\frac{\lambda_{max}}{\lambda_{min}}|$ plays an important role. λ_{min} and λ_{max} are the minimal and maximal eigenvalues of K respectively. In practice, it is not feasible to evaluate the condition number on large systems since the computation of the eigenvalues has a complexity of $\mathcal{O}(N^3)$. Therefore, a generalized approach to assess the stability a priori is proposed.

Considering the minimal distance between two samples as $q_x := \frac{1}{2}\min_{j \neq i}$ $\| x_i - x_j \|_2$ and interpreting $f(x) = \sum_j^M \varphi(x)\alpha_j$ as a transfer function, it is proposed to analyse the system behaviour in the frequency domain. The key to this is the Fourier-Bessel transform of $\varphi(r)$ [41]. Interpreting the frequency ω as

Table 2. Investigated radial basis functions for data dimension $d = 3$.

Type	$\varphi(r)$	Cont. m
Gaussian	e^{-r^2}	C^∞
CSRBF	$(1-r)_+^2$	C^0
	$(1-r)_+^4(4r+1)$	C^2
	$(1-r)_+^6(35r^2+18r+1)$	C^4
Thin-plate	r^{2m-3}	C^m

the minimal distance q_x between the approximated samples provides the best-case stability of the regression model without having to perform experiments on data. More practically, the boundaries for the lowest eigenvalue are discussed and put in relation to the expected sample radius q_x. This enables qualitative assessment of a basis function without performing any numerical experiments. Regarding the stability evaluation presented in Fig. 5a, the Thin-Plate RBF is the only type, which remains stable for all q_x. Depending on the smoothness, CSRBF with C^0 and C^2 follow. The Gaussian RBF is the least stable basis function with λ_{min} slightly below zero.

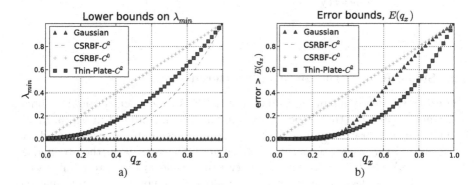

Fig. 5. (a) Lower bounds for λ_{min} (higher is better), (b) lower bounds (lower is better) for the approximation error of each RBF type.

Another important aspect when selecting an RBF type is the approximation quality. Similarly to the case of stability assessment, numerical experiments often only indicate the behaviour of the RBFs restricted to the given dataset. Thus, it is proposed to appraise the theoretical error bounds in a similar way, as has been shown with the generalized stability. The diagram in Fig. 5b presents the best achievable error up to a positive scale factor for each RBF type given the sampling density q_x. It is clear that the higher the sampling density, the better the approximation quality. Notably, the CSRBF-C^2 achieves higher quality than other compact RBFs with lower sampling density q_x and is very similar (overlays) with the global Thin-Plate RBF.

This evaluation indicates the superior performance of the Thin-Plate RBF, though this is not applicable in most realistic applications because the support is not restricted to the neighbouring domain. Furthermore, the presented evaluations claim that applying the compactly supported RBFs with C^2 or C^0 achieves comparable properties. Table 3 shows the summarized investigation results, where a higher number of plus signs reflect better performance. According to the evaluation it is clear that CSRBF is more stable and more accurate than the Gaussian RBFs and provides comparable performance to the Thin-Plate splines without requiring global support. These key observations imply that using CSRBF for 3D data approximation is an attractive option which will now be examined in the next Section.

Table 3. Comparative overview of the RBF models.

	Gaussian	Thin plate	CS-RBF
Stability		$+++$	$+++$
Approximation	$++$	$+++$	$+++$
Smoothness	$+++$	$+++$	$++$
Efficiency	$+$		$++$

3.2 Shape Reconstruction from Scattered Points

The principal idea of shape modelling with RBF is to extract an implicit function which represents the shape by its zero value as introduced in Fig. 3. More formally, an algebraic function $f(\mathbf{x}), f : \mathbb{R}^3 \mapsto \mathbb{R}$ needs to be constructed by regression. Given a set of measured 3D points, the task is further to find a function $f(\mathbf{x})$ which returns zero on every i-th sample $\mathbf{x}_i$ and interpolates well between the samples. Since the zero level alone does not provide information about the orientation of the surface, the surface normals $\mathbf{n}_i$ at every sample are used as constraints for the gradient $\nabla f(\mathbf{x})$ wrt. $\mathbf{x}$. The task is now to find $f(\mathbf{x}_i) = 0$ giving zero at every sample position *and* $\nabla f(\mathbf{x}_i) = \mathbf{n}_i$. Integrating all this information, a convex cost functional is defined.

$$\min_{\alpha} \sum_i^N \parallel f(\mathbf{x}_i) \parallel_2^2 + \parallel \mathbf{n}_i - \nabla f(\mathbf{x}_i) \parallel_2^2 \tag{7}$$

To simplify the optimization problem the normalization term $\parallel \nabla f(\mathbf{x}_i) \parallel_2 = 1$ is omitted. In order to obtain the gradient ∇f, only the gradient of φ needs to be computed, which is precomputed analytically. Putting (7) into matrix notation leads to the short form of the cost function

$$\min_{\alpha} \parallel K\alpha \parallel_2^2 + \parallel \mathbf{n} - K_{\nabla}\alpha \parallel_2^2 . \tag{8}$$

The matrix K contains the values of the RBFs $K_{n,m} = \varphi(\mathbf{x}_n, \mathbf{x}_m) \in \mathbb{R}$ and $K_{\nabla n,m} = \nabla\varphi(\mathbf{x}_n, \mathbf{x}_m) \in \mathbb{R}^3$ represents the derivatives of φ wrt. $\mathbf{x}_n$. The matrices are of sizes $K \in \mathbb{R}^{N \times M}$ and $K_{\nabla} \in \mathbb{R}^{3N \times M}$. At this point it becomes clear that radial basis functions with local support return for distant points $\mathbf{x}_n, \mathbf{x}_m$ zeros which leads to sparse matrices K and K_{∇} improving the storage and the computation efficiency. Figure 6 shows example matrices K and K_{∇} when a RBF with compact support is applied. Black dots illustrate the non-zero matrix values. Since most of the entries in K and K_{∇} are zero they can be dismissed in the computation process.

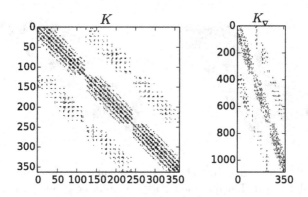

Fig. 6. Example of sparse matrices K_∇ and K for Eq. (10) when CSRBF is applied.

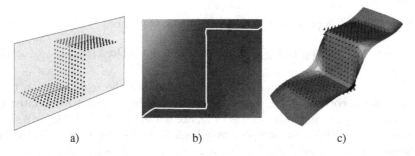

a)   b)   c)

Fig. 7. (a) Synthetic input points, (b) the cut plane visualizing $f(\mathbf{x})$ as red ($f < 0$) green ($f > 0$), (c) reconstructed shape from (a).

Figure 7 shows an example of applying CSRBF-C^2 on a synthetic point set. The red line in Fig. 7(a) indicates the cut-plane at which the Fig. 7(b) has been rendered, while Fig. 7(c) shows the 3D shape reconstruction.

Next, it is proposed to extend the cost term with an additional total variation regularization term enforcing piecewise smoothness. In computer vision it is accepted practise [21] to measure the total variation by computing the Frobenius norm of a Hessian matrix. In contrast, it is proposed to compute the second derivatives with respect to the radius r of the RBF $\varphi(r)$. Comparing the single computation $\partial^2_{rr}\varphi(r)$ to the evaluation of the 3×3 Hessian matrix with nine elements, this reduces the computational cost by a factor of nine and is easier to compute analytically. Similar to the case when computing the gradients of f, the second derivative is also a sum of derived RBFs

$$TV(\mathbf{x}) = \sum_{m}^{M} \partial^2_{rr}\varphi(r)\alpha_m \tag{9}$$

with $r = \| \mathbf{x}_m - \mathbf{x} \|_2$.

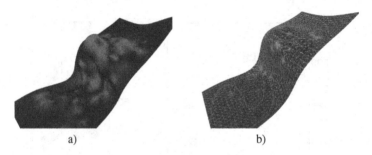

a)   b)

Fig. 8. (a) The TV cost (red) overlaid with the unregularized shape obtained via LSQ. (b) The reduced TV cost (less red colour) after performing regularized approximation following (10).

Applying the TV regularization, the cost function becomes

$$\min_{\alpha} \| K\alpha \|_2^2 + \| \mathbf{n} - K_\nabla \alpha \|_2^2 + \lambda \| D\alpha \|_1 \qquad (10)$$

with $D_{n,m} = \partial_{rr}^2 \varphi(\mathbf{x}_n, \mathbf{x}_m)$ and λ as the weighting of the regularization term. The factors α_m corresponding to the largest eigenvalue of D are attenuated the most while weights lying in the kernel of D, are not affected at all. Figure 8 shows an example when the input samples have been perturbed by noise (Fig. 9a) and the shape is reconstructed via (a) simple least squares (LSQ) and (b) TVL_1. In both images the red colour corresponds to the TV cost intensity (9). Clearly, when applying TV minimization the shape accuracy of the reconstruction is improved and the red TV intensity is reduced significantly.

Increasing the normally distributed sample noise up to $\sigma \approx 30\%$ of the bounding box, the effect of the regularization is demonstrated in Fig. 9. While the simple LSQ model does not achieve a smooth shape (Fig. 9b) the new regularized approach in Fig. 9(c) shows considerable perceptual improvement in terms of the quality of the shape reconstruction.

In the next section, the proposed numerical technique to efficiently solve the TVL_1 task is presented.

3.3 TVL$_1$ Solver

To minimize the task (10) it is proposed to apply the Lagrangian approach from the *Alternating Direction Method of Multipliers* (ADMM) [8]. Formally, (10) is restated to

$$\min_{\alpha, z} L(\alpha, \mathbf{z}) = f_1(\alpha) + f_2(\mathbf{z}) + \mathbf{b}^T (D\alpha - \mathbf{z}) + \frac{\rho}{2} \| D\alpha - \mathbf{z} \|_2^2 \qquad (11)$$

where $f_1(\alpha)$ is the data part from (10) depending on α, and $f_2(z) = \lambda \| \mathbf{z} \|_1$ is the non-smooth regularization part weighted by λ. The basic approach is to minimize for α, then for $\mathbf{z}$ iteratively. The terms $\mathbf{b}^T (D\alpha - \mathbf{z})$ and $\| D\alpha - \mathbf{z} \|_2$

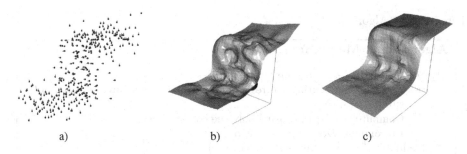

Fig. 9. (a) Noisy 3D samples of the step function. (b) Direct LSQ. (c) TVL_1 regularized approximation.

make sure that $D\boldsymbol{\alpha}$ is close to $\mathbf{z}$ after an iteration finishes reducing the duality gap. This restriction is controlled by ρ which is usually a large scalar. The iterative optimization process between $\boldsymbol{\alpha}$ and $\mathbf{z}$ is summarized in Algorithm 1. The minimization for $\mathbf{z}$ involves a sub gradient over $\|\cdot\|_1$ and its solution is known as the shrinkage operator [19] being applied on each element z_i independently:

$$z_i = \text{shrink}(a, b)$$
$$= a - b \cdot \text{sign}(b - a)_+ \tag{12}$$
$$\text{where } a = \frac{b_i}{\rho} + (D\boldsymbol{\alpha})_i, \ b = \frac{\lambda}{\rho}$$

with $(D\boldsymbol{\alpha})_i$ as the i-th element of the vector $D\boldsymbol{\alpha}$ and $\text{sign}(b-a)_+$ gives 1 if $b > a$ and zero otherwise.

Algorithm 1. ADMM for L_1 approximation

1. Solve for $\boldsymbol{\alpha}$: $(K_\nabla^T K_\nabla + K^T K + \rho D^T D)\boldsymbol{\alpha} = K_\nabla^T \mathbf{n} + D^T(\rho \mathbf{z} - \mathbf{b})$.
2. Evaluate: $z_i^{k+1} = \text{shrink}(\frac{b_i}{\rho} + (D\boldsymbol{\alpha})_i, \lambda/\rho)$
3. Evaluate: $\mathbf{b}^{k+1} := \mathbf{b}^k + (D\boldsymbol{\alpha}^{k+1} - \mathbf{z}^{k+1})\rho$

While steps 2 and 3 are direct evaluations and can be performed in parallel after $D\boldsymbol{\alpha}^{k+1}$ has been precomputed, step 1 incurs high computational complexity. It is proposed to solve $\boldsymbol{\alpha}^{k+1}$ via Gauss-Seidel iterations which are well known from large scale linear system optimization [35]. However, the standard Gauss-Seidel process suffers from difficult convergence conditions. Thus, *successive over relaxation* (SOR) is applied with a weight factor ω. By applying SOR in step 1, Algorithm 1 changes to

$$\alpha_i^{k+1} = \alpha_i^k + \omega \frac{y_i - (K_{\nabla_i}^T K_\nabla + K_i^T K + \rho D_i^T D)\boldsymbol{\alpha}^k}{K_{\nabla_i}^T K_{\nabla i} + K_i^T K_i + \rho D_i^T D_i} \tag{13}$$

with $y_i = K_{\nabla_i}^T \mathbf{n} + D_i^T(\mathbf{z}\rho - \mathbf{b})$ and i-th columns of a matrix respectively. Considering that K_∇, K and D are sparse when CSRBF is applied, the computation is reduced to Algorithm 2.

Algorithm 2. Matrix free TVL$_1$

1. For each RBF centre α_i compute:
   (a) Find all neighbouring centres and all neighbouring samples located in the support of α_i.
   (b) Compute the Eq. (13) using only the collected neighbours.
   (c) Precompute $D\alpha$ with the new α_i^{k+1}.
2. Evaluate: $z_i^{k+1} = \mathrm{shrink}(\frac{b_i}{\rho} + (D\alpha)_i, \lambda/\rho)$
3. Evaluate: $\mathbf{b}^{k+1} := \mathbf{b}^k + (D\alpha^{k+1} - \mathbf{z}^{k+1})\rho$

The optimization is controlled by two important parameters: ω for the successive over relaxation and the RBF scaling s as introduced in Table 2 and (6). Figure 10 shows the effect of these parameters on the approximation quality and the achieved convergence rates. The experiments have been performed on the synthetic dataset from Fig. 7. Figure 10a illustrates the approximation quality over the scaling s. The quality attains its optimum when $s = 10$ is reached. This observation corresponds to the generalized investigations from Fig. 5, where $s = 10$ is $q_x = 0.1$. Furthermore, the empirical impact analysis of the over relaxation parameter ω on the convergence concludes that $\omega \leq 0.15$ allows to remove the instability issues for CSRBF-C^2 and CSRBF-C^4 when SOR is applied. Note that when applying the Gaussian RBF, ω is required to be very small ($\omega \approx 1e - 3$), leading to an impractically high number of iterations. This fact is a consequence of the stability properties of the Gaussian investigated in Sect. 3.1.

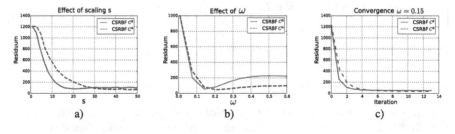

a)   b)   c)

Fig. 10. (a) The impact of the scaling parameter s, (b) the over relaxation weighting ω, (c) convergence behaviour for $\omega = 0.15$ when CSRBF-C^2 or CSRBF-C^4 are applied.

The next section evaluates the proposed TVL$_1$ shape approximation framework with respect to existing methods by applying the algorithms on a large dataset with an existing ground truth.

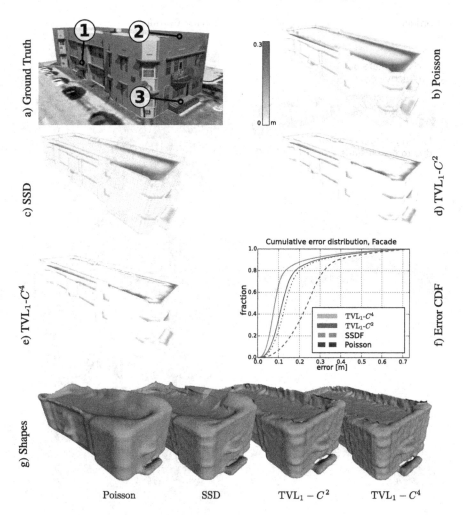

Fig. 11. Evaluation results for the proposed TVL_1 and the compared techniques. See text for details (Color figure online).

4   Evaluation

This section evaluates the proposed shape reconstruction framework on different datasets and compares it with two successful surface reconstruction techniques: the *Poisson* approximation [28] and the *Smooth-Signed-Distance* (SSD) algorithm [11]. The selected methods have been identified as successful techniques for shape reconstruction under strong noise. Both use the implicit model to represent the shapes. However, in contrast to the presented work, the compared methods structure the data via an octree of predefined depth and apply discrete optimization via finite differences to extract the zero level of the surface.

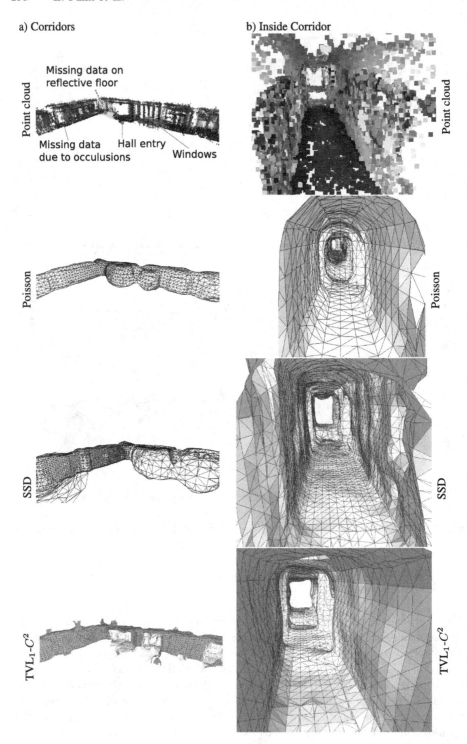

Fig. 12. Shape reconstruction from real time point clouds. See text for details.

The presented analysis uses a 3D point cloud as input. A virtual camera flight was simulated in order to generate error prone data with an established ground truth. The simulation of the moving camera and noisy 3D measurements were achieved by extending the CAD software Blender [30]. This enabled the control of the noise level on the samples and an accurate model of the observed object. An outdoor scene was selected since similar environments are used in many robotic applications. Figure 11a shows the ground truth model used for the simulated measurements in assessing the quality of reconstructed 3D shapes in Fig. 11g. Figure 11b–e show the model coloured according to the local reconstruction error. Red areas indicate larger errors. The facade consisted of large planar areas with a number of sharp edges as identified in point ①. The proposed TVL$_1$ technique performs significantly better than the Poisson approach and similarly well compared to SSD. During the simulation several areas ② have been occluded by the railing, thus have not been sampled. This increases the difficulty of the reconstruction task. At these locations TVL$_1$ interpolates a shape which is more similar to the ground truth than other techniques. The area marked by ③ is the balcony, where only a small part of the floor has been sampled. In such areas, both TVL$_1$ and SSD perform well, significantly outperforming the Poisson approach. The diagram in Fig. 11f shows the cumulative error distribution of the reconstructed shapes. It states for example, that only 77 % of all samples have a smaller error than $0.3m$ when Poisson is applied. The diagram is produced by re-sampling the ground truth model and the approximated shapes with $5M$ points and by measuring the distance between a reconstructed sample and its nearest neighbour from the ground truth set. The point-to-point (ptp) error is shown on the horizontal axis. The increased accuracy of TVL$_1$ techniques can also be observed in the coloured error models in Fig. 11b–e.

Another evaluation scenario has been considered where real point clouds from a mobile stereo system have been processed to shapes. As previously illustrated in Fig. 1, the small stereo system has been carried inside of a building at high speed, computing the 3D point clouds in real time. This data has been processed by the Poisson, SSD and TVL$_1$-C^2 techniques, and is shown in Fig. 12. Figure 12a and below show larger overview, which has been selected because of the difficult conditions. Floor reflections, occlusions and blending from ceiling illumination lead to error prone data. Poisson and SSD are designed to reconstruct closed surfaces and thus generate wrong surfaces even in the open entries. TVL$_1$ however, extrapolates the measurements to some extent but does not close the entries correctly indicating open and traversable space.

Figure 12b show a part from the DLR building. It can be observed, that Poisson provides over-smoothed surfaces, SSD tends to interpolate noisy samples and TVL$_1$ manages smooth surface but also representing the edges between the walls and the floor.

5   Conclusion

This paper has presented a new 3D shape modelling strategy for noisy error prone 3D data samples. Modelling 3D shapes with radial basis functions has been proposed with the choice of the most appropriate RBF corroborated using generalized stability and approximation quality assessments. The shape regression model has been extended by non-smooth L_1 regularization assuming planar areas to improve the accuracy of the reconstructed shape in indoor and urban environments. Since the TVL_1 optimization task is computationally expensive, a low complexity optimization technique has been developed. The optimization process exploits the Lagrangian form of the optimization task with an iterative over relaxation technique. This enables realistic datasets containing several million points to be effectively processed. Quantitative analysis confirms that the proposed method achieves superior accuracy on the synthetic objects.

For future research, the presented solution will be adapted and extended to recursive, real-time 3D mapping applications where environment measurements are received as a data stream. The corresponding 3D shape approximation model then will be able to recursively modify its shape as new measurements become available.

References

1. Agoston, M.: Computer Graphics and Geometric Modelling: Implementation & Algorithms, 1st edn. Springer, London (2005)
2. Alexa, M., Behr, J., Cohen-Or, D., Fleishman, S., Levin, D., Silva, C.T.: Point set surfaces. In: Proceedings of the Conference on Visualization 2001, VIS 2001, pp. 21–28. IEEE Computer Society, Washington, DC, USA (2001)
3. Alexa, M., Behr, J., Cohen-Or, D., Fleishman, S., Levin, D., Silva, C.T.: Computing and rendering point set surfaces. IEEE Trans. Vis. Comput. Graph. 9(1), 3–15 (2003)
4. Alizadeh, F., Alizadeh, F., Goldfarb, D., Goldfarb, D.: Second-order cone programming. Math. Program. **95**, 3–51 (2003)
5. Bach, F.R., Jenatton, R., Mairal, J., Obozinski, G.: Optimization with sparsity-inducing penalties. Found. Trends Mach. Learn. 4(1), 1–106 (2012)
6. Bernardini, F., Mittleman, J., Rushmeier, H., Silva, C., Taubin, G.: The ball-pivoting algorithm for surface reconstruction. IEEE Trans. Vis. Comput. Graph. **5**(4), 349–359 (1999)
7. Bodenmüller, T.: Streaming surface reconstruction from real time 3D-measurements. Ph.D. thesis, Technical University Munich (2009)
8. Boyd, S., Parikh, N., Chu, E., Peleato, B., Eckstein, J.: Distributed optimization and statistical learning via the alternating direction method of multipliers. Found. Trends Mach. Learn. 3(1), 1–122 (2011)
9. Bredies, K., Kunisch, K., Pock, T.: Total generalized variation. SIAM J. Img. Sci. **3**(3), 492–526 (2010)
10. Bregman, L.: The relaxation method of finding the common point of convex sets and its application to the solution of problems in convex programming. USSR Comput. Math. Math. Phys. **7**(3), 200–217 (1967)
11. Calakli, F., Taubin, G.: SSD: smooth signed distance surface reconstruction. Comput. Graph. Forum **30**(7), 1993–2002 (2011)
12. Canelhas, D.R.: Scene representation, registration and object detection in a truncated signed distance function representation of 3D space. Ph.D. thesis, Örebro University (2012)

13. Canelhas, D.R., Stoyanov, T., Lilienthal, A.J.: SDF tracker: a parallel algorithm for on-line pose estimation and scene reconstruction from depth images. In: IEEE/RSJ International Conference on Intelligent Robots and Systems (IROS), pp. 3671–3676. IEEE (2013)

14. Carr, J.C., Beatson, R.K., Cherrie, J.B., Mitchell, T.J., Fright, W.R., McCallum, B.C., Evans, T.R.: Reconstruction and representation of 3D objects with radial basis functions. In: Proceedings of the 28th Annual Conference on Computer Graphics and Interactive Techniques, SIGGRAPH 2001, pp. 67–76. ACM, New York, NY, USA (2001)

15. Chen, X., Lin, Q., Kim, S., Peña, J., Carbonell, J.G., Xing, E.P.: An efficient proximal-gradient method for single and multi-task regression with structured sparsity. CoRR, abs/1005.4717 (2010)

16. Duchon, J.: Splines minimizing rotation-invariant semi-norms in sobolev spaces. In: Schempp, W., Zeller, K. (eds.) Constructive Theory of Functions of Several Variables. Lecture Notes in Mathematics, vol. 571, pp. 85–100. Springer, Berlin (1977)

17. Dykstra, R.: An algorithm for restricted least squares regression. Technical report, Mathematical Sciences. University of Missouri-Columbia, Department of Statistics (1982)

18. Edelsbrunner, H., Mücke, E.P.: Three-dimensional alpha shapes. ACM Trans. Graph. **13**(1), 43–72 (1994)

19. Efron, B., Hastie, T., Johnstone, I., Tibshirani, R.: Least angle regression. Ann. Stat. **32**, 407–499 (2004)

20. Friedman, J.H., Hastie, T., Tibshirani, R.: Regularization paths for generalized linear models via coordinate descent. J. Stat. Softw. **33**(1), 1–22 (2010)

21. Getreuer, P.: Rudin-Osher-Fatemi total variation denoising using split bregman. Image Process. On Line **2**, 74–95 (2012)

22. Goldstein, T., Osher, S.: The split bregman method for l1-regularized problems. SIAM J. Img. Sci. **2**(2), 323–343 (2009)

23. Gomes, A., Voiculescu, I., Jorge, J., Wyvill, B., Galbraith, C.: Implicit Curves and Surfaces: Mathematics, Data Structures and Algorithms, 1st edn. Springer Publishing Company, London (2009)

24. Guennebaud, G., Gross, M.: Algebraic point set surfaces. ACM Trans. Graph. **26**(3), July 2007

25. Hägele, M.: Wirtschaftlichkeitsanalysen neuartiger Servicerobotik-Anwendungen und ihre Bedeutung für die Robotik-Entwicklung (2011)

26. Hirschmüller, H.: Semi-global matching - motivation, developments and applications. In: Fritsch, D. (ed.) Photogrammetric Week, pp. 173–184. Wichmann, Heidelberg (2011)

27. Hughes, J., Foley, J., van Dam, A., Feiner, S.: Computer Graphics: Principles and Practice. The Systems Programming Series. Addison-Wesley, Boston (2014)

28. Kazhdan, M., Hoppe, H.: Screened poisson surface reconstruction. ACM Trans. Graph. **32**(3), 29:1–29:13 (2013)

29. Ohtake, Y., Belyaev, A., Alexa, M., Turk, G., Seidel, H.-P.: Multi-level partition of unity implicits. ACM Trans. Graph. **22**(3), 463–470 (2003)

30. Open Source Community. Blender, open source film production software (2014). http://blender.org. Accessed on 6 June 2014

31. Oztireli, C., Guennebaud, G., Gross, M.: Feature preserving point set surfaces based on non-linear kernel regression. Comput. Graph. Forum **28**(2), 493–501 (2009)

32. Piegl, L., Tiller, W.: The NURBS Book. Monographs in Visual Communication. Springer, Heidelberg (1997)
33. Rogers, D.F.: Preface. In: Rogers, D.F. (ed.) An Introduction to NURBS. The Morgan Kaufmann Series in Computer Graphics, pp. 105–107. Morgan Kaufmann, San Francisco (2001)
34. Rudin, L.I., Osher, S., Fatemi, E.: Nonlinear total variation based noise removal algorithms. Phys. D **60**(1–4), 259–268 (1992)
35. Saad, Y.: Iterative Methods for Sparse Linear Systems, 2nd edn. Society for Industrial and Applied Mathematics, Philadelphia (2003)
36. Schölkopf, B., Smola, A.J.: Learning with Kernels: Support Vector Machines, Regularization, Optimization, and Beyond. MIT Press, Cambridge (2001)
37. Tennakoon, R., Bab-Hadiashar, A., Suter, D., Cao, Z.: Robust data modelling using thin plate splines. In: 2013 International Conference on Digital Image Computing, Techniques and Applications (DICTA), pp. 1–8. November 2013
38. Tibshirani, R.: Regression shrinkage and selection via the lasso. J. Roy. Stat. Soc. Ser. B **58**, 267–288 (1994)
39. Wahba, G.: Spline models for observational data, vol. 59 of CBMS-NSF Regional Conference Series in Applied Mathematics. Society for Industrial and Applied Mathematics (SIAM), Philadelphia (1990)
40. Wendland, H.: Piecewise polynomial, positive definite and compactly supported radial functions of minimal degree. Adv. Comput. Math. **4**(1), 389–396 (1995)
41. Wendland, H.: Scattered Data Approximation. Cambridge University Press, Cambridge (2004)
42. Wolff, D.: OpenGL 4 Shading Language Cookbook, 2nd edn. Packt Publishing, Birmingham (2013)
43. Zach, C., Pock, T., Bischof, H.: A globally optimal algorithm for robust TV-L1 range image integration. In: IEEE 11th International Conference on Computer Vision, 2007, ICCV 2007, pp. 1–8, October 2007
44. Zhao, H., Oshery, S., Fedkiwz, R.: Fast surface reconstruction using the level set method. In: Proceedings of the IEEE Workshop on Variational and Level Set Methods, VLSM 2001 (2001)

Traffic Sign Recognition Using Visual Attributes and Bayesian Network

Hamed Habibi Aghdam[✉], Elnaz Jahani Heravi, and Domenec Puig

Department of Computer Engineering and Mathematics,
University Rovira i Virgili, Tarragona, Spain
{hamed.habibi,elnaz.jahani,domenec.puig}@urv.cat

Abstract. Recognizing traffic signs is a crucial task in Advanced Driver Assistant Systems. Current methods for solving this problem are mainly divided into traditional classification approach based on hand-crafted features such as HOG and end-to-end learning approaches based on Convolutional Neural Networks (ConvNets). Despite a high accuracy achieved by ConvNets, they suffer from high computational complexity which restricts their application only on GPU enabled devices. In contrast, traditional classification approaches can be executed on CPU based devices in real-time. However, the main issue with traditional classification approaches is that hand-crafted features have a limited representation power. For this reason, they are not able to discriminate a large number of traffic signs. Consequently, they are less accurate than ConvNets. Regardless, both approaches do not scale well. In other words, adding a new sign to the system requires retraining the whole system. In addition, they are not able to deal with novel inputs such as the false-positive results produced by the detection module. In other words, if the input of these methods is a non-traffic sign image, they will classify it into one of the traffic sign classes. In this paper, we propose a coarse-to-fine method using visual attributes that is easily scalable and, importantly, it is able to detect the novel inputs and transfer its knowledge to a newly observed sample. To correct the misclassified attributes, we build a Bayesian network considering the dependency between the attributes and find their most probable explanation using the observations. Experimental results on a benchmark dataset indicates that our method is able to outperform the state-of-art methods and it also possesses three important properties of novelty detection, scalability and providing semantic information.

Keywords: Traffic sign recognition · Visual attributes · Bayesian network · Most probable explanation · Sparse coding

1 Introduction

Traffic sign detection and recognition is one of the major tasks in advanced driver assistant systems and intelligent cars. A traffic sign detection and recognition

© Springer International Publishing Switzerland 2016
J. Braz et al. (Eds.): VISIGRAPP 2015, CCIS 598, pp. 295–315, 2016.
DOI: 10.1007/978-3-319-29971-6_16

system is composed of two modules namely *detection* and *recognition*. The input of the detection module is the image of a scene and its output is the areas of the image where they contain a traffic sign. Then, the recognition module analyses these areas and recognizes the traffic sign.

One of the important characteristics of traffic signs is their design simplicity which facilitates their detection and recognition for a human driver. First, they have a simple geometric shape such as circle, triangle, polygon or rectangle. Second, they are distinguishable from *most* of objects in the scene using their color. To be more specific, traffic signs are usually composed of some basic colors such as red, green, blue, black, white and yellow. Finally, the meaning of traffic sign is acquired using the pictograph in the center. Even though the design is clear and discriminative for a human, but there are challenging problems in real world applications such as shadow, camera distance, weather condition, perspective and age of the sign that need to be addressed in the traffic sign detection and recognition systems.

Moreover, there are two difficulties that must be coped by the recognition module in real-world applications. First, traffic sign recognition is a multi-category classification problem that may have hundreds of classes. Second, assuming the fact that it is probable to have some false-positive outputs from the detection module, the recognition module must discard these false-positive inputs. In other words, the recognition module must deal with the *novel* inputs that have not been observed during the training stage.

To the best of our knowledge, most of works in the recognition module have only focused on increasing the performance under more realistic conditions and on a limited number of classes. Further, none of the methods in the literature have been tried to recognize traffic signs in a *coarse-to-fine* fashion. Despite the impressive results obtained by different groups in the German traffic sign benchmark competition [1], all of these methods suffer from some common problems.

First, none of the methods in the literature are able to deal with novel inputs. For example, given the image of a non-traffic sign object (*e.g.* false-positive results of the detection module), the state-of-art methods classify the novel input into one of traffic sing classes. Second, they are not easily scalable. On the one hand, adding a new class to the recognition module might require to re-train the whole system. On the other hand, they use the conventional classification method in which we consider that all classes are well separated in the same feature space and, using this assumption, a single model is trained for whole classes. While this assumption can be true for a few number of classes but if the number of classes increases and we map all images onto the same feature space, it is probable that there will be an overlap between classes. Third, they do not take into account the *visual attributes* of traffic signs.

Attributes are high level concepts that provide useful information about the objects. For example, if we observe that the input image "has a red margin" and "is triangle" and its pictograph depicts an object that "is pointing to the left", the input image is high probable to be the "dangerous curve to the right" traffic sign. In this case, we could recognize the traffic sign using three attributes. As the second example, assume the attributes "has a red rim", "is circle" and "contains a two-digit number" have been observed. These attributes reveal that the input image indicates a "speed limit" traffic sign. Considering that there are at most 10 speed limit traffic signs, we only need to do a 10-class classification instead of hundreds-class classification[1] if we observe the mentioned attributes before the final classification. In sum, we believe a successful traffic sign recognizer must have the following characteristics: (1) The cost of adding a new class to the model should be low (scalability). (2) Novel inputs must be rejected (novelty detection) and (3) it should follow a coarse-to-fine classification approach.

In this paper, we propose a *coarse-to-fine* method for recognizing the *large number of traffic signs* with *ability to identify the novel inputs*. In addition, adding a new class to the system requires to update a few models instead of the whole system. It should be noted that our goal is not to notably improve the numerical results of the state-of-art methods since the current performance is $\sim 99\%$ but to propose a more scalable and applicable method with better performance which is also able to detect the novel inputs and provide some high level information about the any inputs.

To achieve this goal, we first perform a *coarse* classification on the input image using semantic visual attributes and classify it into one of the abstract traffic sign categories. An abstract category contains the traffic sign with similar attributes. Then, a fine-grained classification is done on the signs of the detected category. However, because the attributes of a sign are detected using a *one-versus-all* classifier, it is possible that some attributes of the object are not detected and some irrelevant attributes are detected for the same sign. To deal with this problem, we take into account the correlation between the attributes as well as the uncertainty in the observations and build a Bayesian network. Next, we enter our observation to a Bayesian network and select the **most probable explanation** of the attributes. Finally, the refined attributes are used to find the category of the traffic sign or ascertain if it is a novel input.

Contribution: One of the important aspects of the proposed method is that all signs in the same category share the same attributes. For example, all speed limit traffic signs are triangle, have a red rim and contain a two-digit or three-digit number. In our proposed method, the input image is in the category of the speed limit traffic signs if it possesses all these three attributes. Otherwise, it does not belong to this category. Using this property, we are able to identify the novel inputs. More precisely, if the input image does not belong to any of the abstract categories, it is classified as a novel input. Our second contribution is proposing a scalable method. That said, the proposed framework can be effectively extended to hundreds of classes. Our third contribution is dividing the hundreds of classes into fewer categories and building separate fine-grained classifiers for every category. For instance, the category "speed limit" may contain 10 classes including 8 signs with different two-digit numbers and 2 signs with different three-digit numbers. Clearly, there are subtle differences between these signs. For example, the traffic sign "speed limit: 70 Km/h" is visually very similar to the "speed limit: 20 Km/h" sign. As a result, the classification approach

---

[1] We consider that there are at most 100 traffic signs to be recognized.

must take into account the subtle differences rather than more abstract characteristics. Another advantage of dividing the problem into smaller problems is that in the case of adding a new sign to the model, we only need to find its relevant category and update the classification model of this category. Last but not the least, in the case that our system cannot find the category of the object or it is not confident about the classification result, it provides more abstract information which can be fused with the context and temporal information for inference.

The rest of this paper is organized as follows: Sect. 2 reviews the state-of-art methods for recognizing traffic signs as well as the methods for detecting the visual attribute of objects. Then, the proposed method is described in Sect. 3 where we mention the feature extraction method and the Bayesian network model. Next, we show the experimental results in Sect. 4 and finally, the paper concludes in Sect. 5.

2   Related Work

Traffic sign recognition has been extensively studied and some impressive results on uncontrolled environments have been reported. In general, the methods for recognizing the traffic signs can be divided into three different categories namely *template matching*, *classification* and *deep networks*.

Template Matching: In early works, a traffic sign was considered as a rigid and well-defined object and their image were stored in the database. Then, the new input image was compared with the all templates in the database to find the best matching. The methods based on template matching usually differ in terms of similarity measure or template selection. In general, these methods are not stable and accurate in uncontrolled environments. For more detail the reader can refer to [2,3].

Classification: Recently, classification approaches have achieved accurate results on more realistic databases. These approaches consist of two major stages. First, features of the image are extracted and, then, they are classified using machine learning approaches. Stallkamp *et al.* [1] achieved 95 % classification accuracy on German traffic sign benchmark database [4] by extracting the HOG features and classifying the images into 43 classes using the linear discriminant analysis. Zaklouta and Stanciulescu [5,6] extracted the same HOG features on the same database and classified them using the random forest model. They could increase the performance up to 97.2 %. Similarly, Sun *et al.* [7] utilized extreme learning machine method for classification of the HOG features and achieved 97.19 % accuracy on the same database.

Maldonado *et al.* [8,9] recognized the traffic signs by recognizing the pictographs using support vector machine. Most recently, Liu *et al.* [10] extracted the SIFT features of the image after transforming it to the log-polar coordinate system and found the visual words using k-means clustering. Then, the feature vectors were obtained using a novel sparse coding method and, finally, the traffic signs were recognized using support vector machines.

Different from the previous approaches, Wang *et al.* [11] employed a two step classification. In the first step, the input image is classified into 5 super-classes using HOG features and support vector machine. In the second stage, the final classification is done using HOG and support vector machine after doing perspective adjustment on the image taking into account the information from the super class. For more detailed information about classification based methods the reader can refer to [12].

Deep Network: recently, CNN could beat the human performance by correctly classifying more than 99 % of the images on a practical dataset called German Traffic Sign Benchmark [1]. To be more specific, the winner algorithm proposed by Cireşan *et al.* [13] computes the average score of 25 CNNs with the same architecture trained on 5 variations of the original dataset. In addition, the second place was also awarded to another CNN proposed by Sermanet and Lecun [14]. In contrast to the winner, the second algorithm uses only one CNN to recognize the traffic signs. In fact, the data-augmentation procedure utilized by both teams is almost identical and the only difference is the architecture of the networks. Each CNN in the winner algorithm requires optimizing $1,543,443$ parameters. However, the second algorithm trains a network with $1,437,791$ parameters. Last not the least, the first algorithm uses hyperbolic tangent as the activation function and the second algorithm utilizes the rectified sigmoid activation function.

Discussion: Despite the impressive results achieved by both deep networks and classification methods, but they are still far from the real applications. First, a deep network is slow and it cannot currently be used in real-time applications. Second, finding the optimal structure of the deep network is a time consuming task which depends on the number of the classes. In the other words, if the number of the classes changes, the whole network need to be trained again. Third, neither deep network nor the above classification methods are not able to deal with the novel inputs and they will classify every input image into one of the traffic sign classes. To address all these problems, in this paper, we have formulated the traffic sign recognition problem in terms of visual attributes and fine-grained classification.

Visual attributes was first proposed by Ferrari and Zisserman [15] and, later, it has been successfully used for defining the objects [16]. Cheng and Tan [17] classified the flowers by learning attributes using sparse representation. Farhadi *et al.* [18] described the objects using semantic and discriminative attributes. Semantic attributes are more comprehensive and they are the ones that human use to describe the objects. They can include **shape**, **material** and **parts**. In contrast, discriminative attributes are the ones that does not have a specific meaning for human but they are utilized for better separating the objects. One important advantage of visual attributes is their ability to transfer the knowledge to the newly observed classes of objects and learn them without examples. This is called *zero shot learning* and it is illustrated in Fig. 1.

Here, 7 different attributes are learned and they can be identified in the input images. As it is shown in this figure, by detecting the correct attributes of the

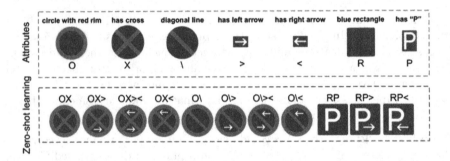

Fig. 1. Zero-shot learning using a set of attributes.

input image we are able to recognize 11 signs without observing them during the training phase. This is an important property which can help us to extend our models with a few efforts through transferring the knowledge from observed classes to the new classes [19,20].

3 Proposed Method

Traffic sign recognition is a multi-category classification problem with hundreds of classes. Also, it is not trivial to collect a large number of real-world images of every sign. Further, some signs are observed more frequently than other signs. For example, it is more probable to see the "curve" signs instead of the "be aware of snow" sign. For this reason, the collected database might be highly unbalanced. Consequently, the trained model for the signs with fewer data can be less accurate than the ones with more data. One feasible remedy to this problem is to update the models through time. However, if we build a single model for classification of all signs, it will be a time consuming task to re-train this model. But, if we group the N traffic signs into $M < N$ categories[2], then, we can train a different model for each category and in the case of adding a new sign, we only need to find its relevant category and re-train the model of this category.

From another perspective, temporal information plays an important role in human inference system. For example, if we observe the "no passing" sign at time t_1 we expect to see the "end of no passing zone", at time t_2. Assume the sign "end of no passing zone" is impaired because of its age and it is hard to see its pictograph and the stripped crossing. In this case, if we follow the classification approaches that we mentioned in the previous section, the "end of no passing zone" sign can be incorrectly classified. However, if we provide some more abstract information such as "the input image has a circular shape and a black-white color," the traffic sign recognition system can infer that the image is related to the previously observed "no passing" sign. Hence, it probably indicates the "end of no passing zone" traffic sign.

---

[2] A category may contain more than one traffic sign.

In this paper, we propose a coarse-to-fine classification approach using the semantic attributes of the object. Figure 2 shows the overview of the proposed algorithm. In the first stage, the image is divided into several regions and each region is coded using a sparse coding method. Then, the feature vector is obtained by concatenating the locally pooled coded vectors (Sect. 3.1). Next, the feature vector is individually applied on the attribute classifiers and the classification score of each attribute is computed (Sect. 3.2). Finally, the certain state (*i.e.* absence or presence) of each attribute is estimated by plugging the scores into a Bayesian network and calculating the *most probable explanation* of the attributes (Sect. 3.3). In the next step, the category of the image is found using the attribute configuration (Sect. 3.4). Having the sign category found, the fine-grained classifier of this category is used to do the final classification (Sect. 3.5).

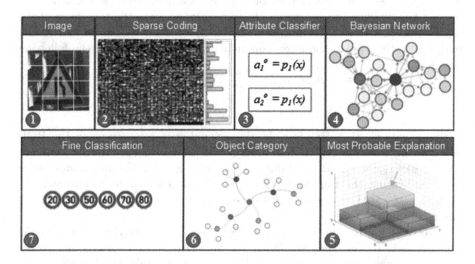

Fig. 2. Overview of the proposed method (best viewed in color).

3.1 Feature Extraction

In order to train the attribute classifiers, we first need to extract the features of the traffic sign. The extracted feature must be able to encode the color, the shape and the content of the traffic sign in the same vector. One of the characteristics of the traffic sign is that they are rigid and their geometrical features (*e.g.* shape, size and orientation) as well as their appearance (*e.g.* color and content) remain relatively unchanged. From this point of view, a simple template matching approach can be useful for the recognition task. However, some important issues such as motion blur, weather condition and occlusion cause the template matching approach to fail.

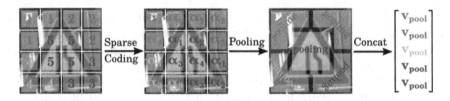

Fig. 3. Feature extraction scheme.

Nonetheless, it is possible to divide the image of the traffic signs into smaller blocks and *learn* the most dominant exemplars of each block, independently. Then, we can reconstruct the original block by linearly combining the exemplars. This is the idea behind *sparse coding* approach [21]. More specifically, as it is shown in Fig. 3, we divide the input image into 5 different regions and each region is divided into a few smaller blocks. For example, the region indicated by number 1 is divided into 3 blocks. Then, in order to learn the templates of the region r, we first collect the images of the blocks of this region from all training images and, then, learn the most dominant exemplars by solving the following equation:

$$minimize_{D^r, \alpha^r} \frac{1}{n} \sum_{i=1}^{n} \frac{1}{2} \|x_i^r - D^r \alpha_i^r\|_2^2$$

$$subject \ to \qquad \|\alpha_i^r\|_1 <= \lambda$$

(1)

In this equation, $x_i^r \in \mathbb{R}^M$ is a M-dimensional vector representing the RGB values of the blocks in region r, D^r is a $\mathbb{R}^{M \times K}$ matrix storing the K dominant templates of region r in the training images, $\alpha_i^r \in \mathbb{R}^K$ is a K-dimensional sparse vector indicating the templates which have been selected to reconstruct the block x_i^r and λ is the value to control the sparsity. The value of λ is determined empirically by the user.

After training the matrices D^r, we use them to extract the features of the input images. To this end, we divide the input image into regions and blocks in the same way that it is shown in Fig. 3. Then, we take the blocks of each region r, separately, and minimize (1) assuming that the values of D^r are fixed in order to compute the vector α_i^r of each block. At this step, we have a few K-dimensional vectors for each region. For example, we will obtain four vectors from region 5. Then, the feature vector of region r is computed by pooling the vectors in that region:

$$f_r = \sum_{i=1}^{n_r} \alpha_i^r$$

(2)

In this equation, n_r is the number of the blocks in region r. Finally, the feature vector of the image is obtained by concatenating the vectors $f_r, r = 1 \ldots 5$ into a single vector and normalizing it using L_1 norm.

Computing the sparse vectors $\alpha_i^r \in \mathbb{R}^K$ is called *inference* and it is a computational task since it requires solving an optimization problem. To alleviate this problem, we can train a non-linear regressor that maps the input x_i^r to sparse vector α_i^r. Specifically, we can train the following non-linear regressor:

$$H(x_i^r) = A \times tanh\,(Wx_i^r + B) \tag{3}$$

where $W \in \mathbb{R}^{K \times M}$ is a weight matrix, $B \in \mathbb{R}^K$ is a vector of biases, $tanh(.)$ is the hyperbolic tangent function and $A \in \mathbb{R}^{K \times K}$ is a diagonal matrix for scaling the result of $tanh$ function. In order to train the above regressor, we only need to compute the α_i^r of each training sample and minimize the mean square error function:

$$E_r = \frac{1}{n} \sum_{i=1}^{n} \|\alpha_i^r - H(x_i^r)\| \tag{4}$$

By training the regressor $H(x_i^r)$ we will be able to estimate the α_i^r of the input image patches using a few matrix operations instead of solving an optimization problem.

Convolutional Neural Network. It is worth mentioning that our proposed framework does not depend on the image representation method. In fact, the feature extraction method mentioned in the previous section can be replaced with any other methods such as HOG. However, it is shown that Convolutional Neural Networks (CNNs) provide a richer representation on challenging datasets such as CIFAR [22] and ImageNet [23–25] compared with conventional methods such as HOG. As the result, it is possible to design a new CNN and train it on the dataset of traffic sign images. Then, the trained CNN can be used to extract feature vector of the input image.

3.2 Attribute Classifier

A traffic sign can be defined using three sets of visual attributes. These are illustrated in Fig. 4. Dashed arrows show a soft dependency relation and we will discuss about them in the next section. In fact, there is a causal relationship between these attributes and the traffic signs. In other words, we can verify the validity of this relationship using the concept of *ancestral sampling*. Given the color, shape and content attributes, we can randomly generate new traffic signs using the probability distribution function $p(traffic\ sign|color, shape, content)$. For instance, while $p(traffic\ sign\ =\ curve\ left|color\ =\ red, shape\ =\ circle, content\ =\ has\ number)$ might be close to zero but $p(traffic\ sign\ =\ speed\ limit\ 60|color = red, shape = circle, content = has\ number)$ is high.

Taking this causal relationship into account, we have defined three sets of attributes including color (4 attributes), shape (3 attributes) and content (12 attributes). These attributes are listed in Table 1. Each traffic sign in our experiments is described using these attributes. However, they can be easily extended to more attributes without affecting the general model we have proposed in this paper.

Detecting the attributes of the input image is done through the attribute classifiers. For this reason, we need to train 19 binary classifiers as follows. For each attribute, we select the images having that attribute as the positive samples and the rest of the images as the negative samples. Then, we train a random

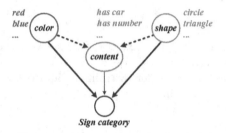

Fig. 4. Causal relationship between the attributes and the traffic signs (color figure online).

Table 1. Sets of attributes for describing the traffic signs.

Content		
has human(a_1)	danger road(a_2)	pointing up(a_3)
end of(a_4)	2-digit number(a_5)	pointing right(a_6)
has car(a_7)	3-digit number(a_8)	pointing left(a_9)
has truck(a_{10})	irregular object(a_{11})	is blank(a_{12})

Color		
red(a_{13})	blue(a_{14})	yellow(a_{15})
	black-white(a_{16})	

Shape		
circle(a_{17})	triangle(a_{18})	polygon(a_{19})

forest model on the collected data. At the end, we will have 19 random forest models for finding the attributes of the input image.

3.3 Bayesian Network Model

Figure 5 shows the general model for the classification of the images using attributes where x indicates the feature vector, $a_i, i = 1 \ldots N$ is a binary value indicating the presence or absence of the ith attribute and $y_k, k = 1 \ldots K$ is the class label.

Based on this model, it is easy to show that the classification will be done by finding the maximum a posteriori of the class labels:

$$y^* = \arg\max_{k=1\ldots K} \sum_{a=0,1} p(a|x)p(y_k|a) \tag{5}$$

where $a = a_i | i = 1 \ldots N$ is a binary vector. There are two important issues with this model. First, it does not take into account the causal relationship between the attributes and it considers them completely independent. This means, using

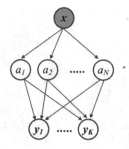

Fig. 5. General classification model using attributes.

this model, the attribute "danger in road" does not longer depend on the shape attributes. But, all traffic signs indicating the danger will be only shown in the *red* and *triangle* signs. Suppose that we observe the attributes "is blue", "is triangle" and "pointing left". Obviously, there is no traffic sign with this configuration. However, if the shape had been detected as "is circle" or the color had been detected as "is red", the configuration was valid. But, with the model of Fig. 5 it is difficult to find which attribute has been falsely classified. The reason is it does not take into account the dependency between attributes and the uncertainty of the observations. The second issue is that using this model, detecting the novel inputs is not a trivial task. In order to detect the novel inputs, we need to define a threshold which can be compared with the maximum a posteriori value for this purpose. However, determining the value of the threshold is an empirical task and it highly depends on the conditional distribution model of each attribute. On the other hand, if one of the models changes, we need to find the threshold value, again.

As we mentioned in Fig. 4, the image of the traffic sign can be described in terms of color, shape and content (pictograph). However, there is also a soft dependency between the content and other attributes (dashed lines). This is because some attributes can happen regardless of the shape and color. For example, the content attribute "is blank" can happen on every possible combination of the color and the shape attributes. In other words, the attribute "is blank" **can be** independent of the other attributes. In addition, there is also an intra-dependency between the content attributes. For example, if we observe "has truck" attribute, it is probable to observe "has car" attribute, as well (*e.g.* "no passing" traffic sign). To find the dependencies between the all attributes in Table 1, we calculated the co-occurrence matrix of the attributes. This is illustrated in Fig. 6.

The co-occurrence matrix is a 19×19 matrix where the element (i, j) in this matrix indicates the probability of observing i^{th} and j^{th} attributes at the same time among the whole classes of traffic signs. Using the co-occurrence matrix, we create our Bayesian network by discarding the relations where their probability in the co-occurrence matrix is less than the threshold T. Figure 7 shows the obtained Bayesian network. The nodes in this Bayesian network depicts the state of an attribute. In other words, all of the nodes in this network are binary

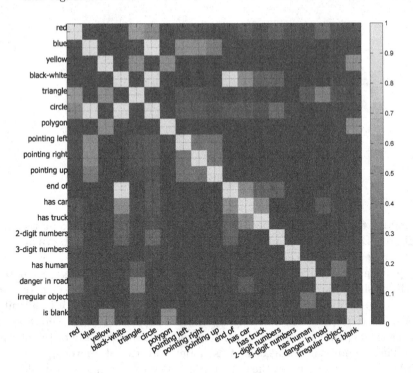

Fig. 6. Co-occurrence matrix of the attributes.

nodes and they represent the conditional probabilities of the child attributes given their parents that are acquired using the training data. Our goal is to find the optimal state of the attributes using the identified evidences from the image. For this reason, we add another 19 observation nodes to the network. This is illustrated in Fig. 8 where the solid gray circles are the hidden nodes and the white circles are the evidence nodes. Our observations are the scores of the attribute classifiers (random forests) that is a number between 0 and 100. Given the evidence from the attribute classifiers, our goal is to maximize following function:

$$a_1^* \ldots a_{19}^* = \underset{a_1 \ldots a_{19} \in [0\ 1]^{19}}{\arg\max} \ p(a_1 \ldots a_{19}, O_{a_1} \ldots O_{a_{19}}) \tag{6}$$

where $O_{a_i}, i = 1 \ldots 19$ is the score of the ith attribute obtained from the ith random forest model. According to the equation, we are looking for the state of the hidden variables (actual state of the attributes) such that the joint probability of the hidden variables and the observed attributes are maximum. This is called *most probable explanation* problem.

3.4 Category Finding

Given the set of 19 attributes, we can cluster the traffic signs into smaller categories. To this end, we manually specify which attributes are active for each

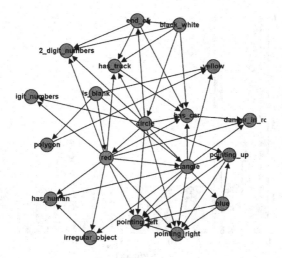

Fig. 7. Bayesian network without observation nodes.

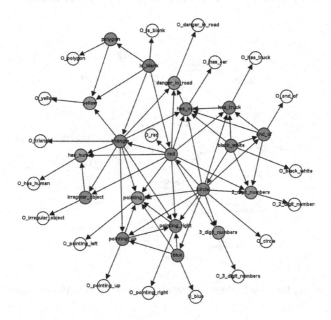

Fig. 8. Bayesian network after adding observation nodes.

class of traffic signs. For example, only attributes *red, circle, 2-digit number* are active on traffic sign "speed limit 60". Then, we cluster the classes with exactly the same active attributes into one category. We applied this procedure on the German traffic sign database and reduced the number of classes from 43 classes into 29 categories. Figure 9 shows the statistic of different categories as well as

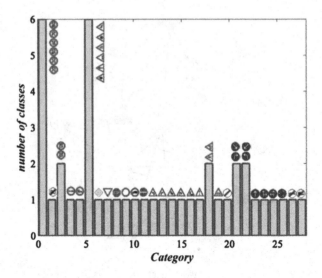

Fig. 9. Clustering the traffic signs using visual attributes (Color figure online).

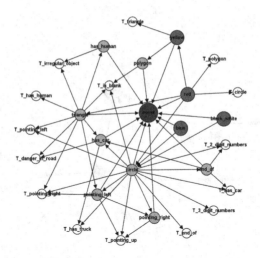

Fig. 10. Parsing tree for finding the category of the image (Color figure online).

the traffic signs inside each category. As it is clear from the figure, there are 6 categories with more than one traffic sign and other categories contain only one traffic sign. This means that 23 traffic signs can be recognized using their visual attributes and they do not need a finer classification. In contrast, the traffic signs inside the other 6 categories can be recognized by the fine classification models.

Having the optimal state of the attributes estimated using (6), we can find the category of the new image using the parsing tree illustrated in Fig. 10. In this tree, orange nodes are the starting points and the white nodes are the leaf nodes.

Novelty Detection: One interesting property of the parsing tree in Fig. 10 is that we can use it for finding the novel inputs. To achieve this, we start by the comparing the optimal state of the attributes obtained from (6) with the starting nodes. For example, assume the state of the attribute *blue* is active and state of the other color attributes are inactive. According to the parsing tree, there is only one outgoing path from the node *blue* that is the *circle* attribute. If it is active, then we keep do parsing, otherwise the input is novel because the node *blue* is not a leaf node and the parsing is not successful. In sum, an input image is novel if there is no active path from the starting points to the leaf nodes.

3.5 Fine Classification

As it is shown in Fig. 10, there are a few categories containing more than one traffic sign. If the input image belongs to one of these categories, then, the actual class of the image is found using a fine classification model which is trained on the images of the category. In other words, we create an individual model for every object category with more than one class inside the category. We follow the same method as in [5] for classifying the objects within the same category. Moreover, using our method, we reduce the number of the classes from 43 to 6 in German traffic sign benchmark database which is about 7 times reduction in the number of classes. Therefore, we only need to do a 6-class classification instead of 43-class classification that can be more accurate and flexible.

4   Experiments

We have applied our proposed method on the German traffic sign benchmark database [1]. This database consists of 43 classes. It also includes two different sets for training and testing. We have resized the all images into 40×40 pixels before applying any feature extraction method. We have two sets of feature vectors. The first set which is obtained by sparse coding method mentioned in this paper is for recognizing the attributes of the image and the second set is the HOG features for fine-classification. For sparse coding approach we applied our proposed method on both RGB and the distance transform of the edge image. Then, we concatenated the pooled vectors to build the final feature vector. For HOG features, we utilized the same configuration in [5]. Next, we trained two sets of random forest model one for the attribute classification (19 classifiers) and one for the fine classification of the categories with more than one traffic sign (6 classifiers) using only the training set.

It is worth mentioning that the conditional probabilities of the hidden variables of the Bayesian network are modelled using the *conditional probability tables* and the conditional probability of the observations are modelled using Gaussian distribution of the attribute scores.

Table 2. Precision, Recall and F_1 measure of the attribute classifiers.

Attribute	precision	recall	F_1 measure
red	0.993	0.990	0.992
blue	0.988	0.992	0.990
yellow	0.971	0.948	0.959
black-white	1.0	0.992	0.996
triangle	0.985	0.995	0.990
circle	0.997	0.987	0.992
polygon	0.991	0.988	0.990
pointing left	0.967	0.947	0.975
pointing right	0.982	0.935	0.957
pointing up	0.994	0.954	0.973
end of	1.0	0.992	0.996
has car	0.993	0.983	0.988
has truck	1.0	0.977	0.988
2-digit number	0.983	0.973	0.978
3-digit number	0.959	0.965	0.962
has human	0.988	0.896	0.940
danger in road	0.932	0.960	0.946
irregular object	0.977	0.913	0.944
is blank	0.991	0.993	0.992

Table 2 shows the results of the attribute classification on the test dataset. Apparently, the attribute classifiers have achieved high accuracy in detecting the attributes of the input images. Next, we tried to find the category of the test images using the attribute classification model depicted in Fig. 5 and our proposed method. Tables 3 and 4 show the results of the category classification using the proposed method and the general model, respectively. Clearly, our method has outperformed the general attribute classification model. The reason is that, using our method, we are able to model the uncertainties of the observations and correct the mistakes. Consequently, the number of the samples that pass the tests in the parsing tree increases.

In addition, Fig. 9 revealed that some categories contain only one class. However to compare our results with other methods, we also applied the fine classification model on the categories with more than one class inside. Then, to be consistent with the state-of-art results, we computed the mean accuracy of the classifications. Table 5 shows the results. As it is clear, there is a significant improvement in recognition of de-restriction signs (you can refer to [1] for definitions of different signs). This is because the shape of de-restrictions signs is very similar to some of the signs in other classes. For example, "end of no-passing" sign has a very similar edge features to the "no passing" signs. On the other hand, two other methods represented in this paper have utilized HOG features for classification. Obviously, because of shape similarity of the other signs with the de-restrictions signs, their feature vector will be similar, as well. For this reason, there is an overlap between the feature vector of this signs with other signs in the feature space which causes the misclassification.

However, because our method utilizes the attributes of the image, it is able to model the color, shape and content of each sign explicitly. For this reason, when an image from de-restrictions group is given to our method, it is able to distinguish between them with other signs simply using the color attributes. As the result, it is able to improve the accuracy of the classification.

Table 3. The result of category classification using the proposed method. The categories are indexed according to Fig. 9.

Category	C_0	C_1	C_2	C_3	C_4	C_5	C_6	C_7	C_8	C_9	C_{10}	C_{11}	C_{12}	C_{13}
precision	0.998	1.0	0.995	0.993	0.996	0.992	1.0	0.996	0.995	0.993	1.0	0.967	1.0	0.960
recall	0.995	1.0	0.987	1.0	0.996	0.992	1.0	0.998	1.0	1.0	0.980	0.991	0.936	0.941
accuracy	0.994	1.0	0.983	0.993	0.993	0.984	1.0	0.994	0.995	0.993	0.980	0.960	0.936	0.905

Category	C_{14}	C_{15}	C_{16}	C_{17}	C_{18}	C_{19}	C_{20}	C_{21}	C_{22}	C_{23}	C_{24}	C_{25}	C_{26}	C_{27}	C_{28}
precision	1.0	0.976	0.985	1.0	0.988	0.961	1.0	0.998	0.992	0.991	0.965	0.965	0.961	1.0	1.0
recall	0.985	0.984	0.987	0.969	0.988	1.0	1.0	0.992	0.977	0.986	0.988	0.965	1.0	1.0	1.0
accuracys	0.985	0.960	0.973	0.969	0.977	0.961	1.0	0.990	0.970	0.978	0.955	0.933	0.961	1.0	1.0

Table 4. The result of category classification using the general model in Fig. 5. The categories are indexed according to Fig. 9.

Category	C_0	C_1	C_2	C_3	C_4	C_5	C_6	C_7	C_8	C_9	C_{10}	C_{11}	C_{12}	C_{13}
precision	0.998	1.0	0.981	0.996	1.0	0.972	1.0	0.998	1.0	0.993	1.0	0.725	0.943	1.0
recall	0.958	0.895	0.896	0.972	0.946	0.925	0.899	0.966	0.979	0.979	0.981	0.952	0.688	0.365
accuracy	0.956	0.895	0.881	0.969	0.946	0.901	0.899	0.964	0.979	0.973	0.981	0.699	0.660	0.365

Category	C_{14}	C_{15}	C_{16}	C_{17}	C_{18}	C_{19}	C_{20}	C_{21}	C_{22}	C_{23}	C_{24}	C_{25}	C_{26}	C_{27}	C_{28}
precision	1.0	1.0	0.991	1.0	0.929	0.814	1.0	0.996	0.962	0.996	0.977	1.0	0.886	1.0	1.0
recall	0.718	0.944	0.844	0.715	0.824	0.946	0.778	0.930	0.933	0.982	0.977	0.966	0.984	0.840	0.947
accuracy	0.718	0.944	0.838	0.715	0.775	0.778	0.778	0.927	0.899	0.978	0.955	0.966	0.873	0.840	0.947

Table 5. The result of category classification using the proposed method. The categories are indexed according to Fig. 9.

	Speed limits	Other prohibitions	De-restriction	Mandatory	Danger	Unique
Our method	97.01	99.25	100	97.09	96.31	98.76
Random forests	95.95	99.13	87.50	99.27	92.08	98.73
LDA	95.37	96.80	85.83	97.18	93.73	98.63

4.1 Novelty Detection

Assume a traffic sign recognition system where the inputs are provided by a traffic sign detection module. Clearly, training a traffic sign detector with *false-positive rate* equal to zero is not trivial. Consequently, some false-positive (*i.e.* non-traffic sign) images might be fetched into the recognition module by the detection module. If the recognition model is not equipped with a novelty detection module, it will classify the non-traffic sign image into one of traffic sign classes. This might produce some fatal mistakes in the case of driver-less cars. We showed in Fig. 10 that our attribute based method can be used to identify the novel inputs.

To assess this property, we added some non-traffic sign images into the validation set as the novel data. Then, we trained state-of-art novelty detectors such as K-NN, K-means, Guassian Mixture Model, One-class SVM, Support Vector Data Descriptor (SVDD) using PRTools[3], data description toolbox[4] and LibSVM. Table 6 shows the results.

Table 6. Result novelty detection using the proposed method and state-or-art methods.

Dataset		Our	GMM	K-means	SVM	SVDD	KNN
GTSR	AUC	**0.68**	0.502	0.49	0.55	0.58	0.51
	TPR	0.96	0.64	0.22	0.32	0.90	0.37
	FPR	0.61	0.63	0.24	0.21	0.73	0.35

We observe that our method has produced the best result compared with the other methods applied on traffic sign recognition problem. The feature vectors extracted from the GTSRB dataset are sparse vectors. Consequently, the minimum enclosing ball found by SVM and SVDD methods will include a large area of novel vectors. Therefore, the false-positive rate of these methods is high on this dataset compared with their true-positive rate. Further, GMM method is not able to model the distribution of the vectors accurately. This is due to the fact that the number of data compared with the dimensions of feature vectors is not enough to partition the dataset into many clusters and find their covariance matrix. Hence, GMM partitions the dataset into very small number of groups (3 clusters in this experiment). In addiction, because the vectors inside each cluster are sparse, they cannot be modelled using a Gaussian density function accurately. As the results, the overall accuracy of GMM drops. Similarly, KNN and K-means methods suffer from the fact that the feature vectors are sparse and they do not select a proper threshold values in their novelty score function.

5 Conclusion

In this paper, we proposed a probabilistic framework for recognizing traffic signs using visual attributes and Bayesian networks. Specifically, we define each traffic

---

[3] http://prtools.org/.
[4] http://prlab.tudelft.nl/david-tax/dd_tools.html.

sign in terms of color, shape and pictograph attributes which are collectively represented using 19 binary attributes. Thus, 19 random forests are trained in order to recognize the attributes of each traffic sign. Taking into account the fact that attribute classifiers are very likely to have some false-positive and false-negative results, we design a Bayesian network including observed nodes and hidden nodes to refine the observed attributes. To this, we enter the classification score of the attribute classifiers to the Bayesian network and calculate the most probable explanation of the hidden nodes. The hidden nodes are binary valued nodes indicating presence and absence of each attribute. The most probable explanation of these nodes illustrates the most probable attributes of the input image. Finally, the refined attributes are used to classify the traffic signs. In this paper, we used sparse coding approach for extracting features of the input image and training the attribute classifiers. However, more richer representation can be obtained using convolutional neural networks.

One of the important characteristics of our framework is its ability to identify novel inputs. Concretely, there might be some false-positive results from the traffic sign detection stage which are fetched into the recognition module. If the recognition module is not equipped with a novelty detection model, it will classify the non-traffic sign images into one of traffic sign classes. However, because we represent each traffic sign using visual attributes, it is possible to identify the novel input using their attribute pattern. In other words, visual attributes pattern of a non-traffic sign image might be different from traffic signs. Consequently, using our proposed parse tree we can determine if the input image follows the visual attributes of the traffic signs in the dataset. If it does not follow the learnt attributes, then, it is recognized as a novel input.

Last but not the least, our framework is more scalable and in some cases it is possible to learn the new classes without any training samples (zero-shot learning). Further, in the case that zero shot learning is not applicable, the system only requires to update the models locally instead of the whole models. In addition, this framework is easily expendable to hundreds of classes of traffic signs since it breaks the hundred classes to the categories with much less traffic signs which make them more tractable to classify without affecting the accuracy of the system. Our experiments on the German traffic sign benchmark dataset indicates that in addition to improving in the results compared with state-of-art methods, our method is also able to identify novel inputs more accurately than state-of-art novelty detection methods.

References

1. Stallkamp, J., Schlipsing, M., Salmen, J., Igel, C.: Man vs. computer: benchmarking machine learning algorithms for traffic sign recognition. Neural Netw. **32**, 323–332 (2012). Selected Papers from IJCNN 2011
2. Piccioli, G., De Micheli, E., Parodi, P., Campani, M.: A robust method for road sign detection and recognition (1996)
3. Paclik, P., Novovicova, J., Duin, R.P.W.: Building road sign classifiers using trainable similarity measure. IEEE Trans. Intell. Transp. Syst. **7**, 309–321 (2006)

4. Houben, S., Stallkamp, J., Salmen, J., Schlipsing, M., Igel, C.: Detection of traffic signs in real-world images: the German Traffic sign detection benchmark. In: International Joint Conference on Neural Networks (2013). Number 1288

5. Zaklouta, F., Stanciulescu, B.: Warning traffic sign recognition using a HOG-based K-d tree. In: 2011 IEEE Intelligent Vehicles Symposium (IV), pp. 1019–1024 (2011)

6. Zaklouta, F., Stanciulescu, B.: Real-time traffic sign recognition in three stages. Robot. Auton. Syst. **62**, 16–24 (2014). New Boundaries of Robotics

7. Sun, Z.L., Wang, H., Lau, W.S., Seet, G., Wang, D.: Application of BW-ELM model on traffic sign recognition. Neurocomputing **128**, 153–159 (2014)

8. Maldonado-Bascon, S., Lafuente-Arroyo, S., Gil-Jimenez, P., Gomez-Moreno, H., Lopez-Ferreras, F.: Road-sign detection and recognition based on support vector machines. IEEE Trans. Intell. Transp. Syst. **8**, 264–278 (2007)

9. Bascón, S.M., Rodríguez, J.A., Arroyo, S.L., Caballero, A.F., López-Ferreras, F.: An optimization on pictogram identification for the road-sign recognition task using SVMs. Comput. Vis. Image Und. **114**, 373–383 (2010)

10. Liu, H., Liu, Y., Sun, F.: Traffic sign recognition using group sparse coding. Inf. Sci. **266**, 75–89 (2014)

11. Wang, G., Ren, G., Wu, Z., Zhao, Y., Jiang, L.: A hierarchical method for traffic sign classification with support vector machines. In: 2013 International Joint Conference on Neural Networks (IJCNN), pp. 1–6 (2013)

12. Mogelmose, A., Trivedi, M., Moeslund, T.: Vision-based traffic sign detection and analysis for intelligent driver assistance systems: perspectives and survey. IEEE Trans. Intell. Transp. Syst. **13**, 1484–1497 (2012)

13. Cirean, D., Meier, U., Masci, J., Schmidhuber, J.: Multi-column deep neural network for traffic sign classification. Neural Netw. **32**, 333–338 (2012). Selected Papers from IJCNN 2011

14. Sermanet, P., LeCun, Y.: Traffic sign recognition with multi-scale convolutional networks. In: 2011 International Joint Conference on Neural Networks (IJCNN), pp. 2809–2813 (2011)

15. Ferrari, V., Zisserman, A.: Learning visual attributes. In: Advances in Neural Information Processing Systems (2007)

16. Russakovsky, O., Fei-Fei, L.: Attribute learning in large-scale datasets. In: Kutulakos, K.N. (ed.) ECCV 2010 Workshops, Part I. LNCS, vol. 6553, pp. 1–14. Springer, Heidelberg (2012)

17. Cheng, K., Tan, X.: Sparse representations based attribute learning for flower classification. Neurocomputing **145**, 416–426 (2014)

18. Farhadi, A., Endres, I., Hoiem, D., Forsyth, D.: Describing objects by their attributes. In: 2009 IEEE Conference on Computer Vision and Pattern Recognition, CVPR 2009, pp. 1778–1785 (2009)

19. Rohrbach, M., Stark, M., Szarvas, G., Gurevych, I., Schiele, B.: What helps where – and why? Semantic relatedness for knowledge transfer. In: IEEE Conference on Computer Vision and Pattern Recognition (CVPR) (2010)

20. Lampert, C., Nickisch, H., Harmeling, S.: Learning to detect unseen object classes by between-class attribute transfer. In: 2009 IEEE Conference on Computer Vision and Pattern Recognition, CVPR 2009, pp. 951–958 (2009)

21. Lee, H., Battle, A., Raina, R., Ng, A.Y.: Efficient sparse coding algorithms. In: Schölkopf, B., Platt, J., Hoffman, T. (eds.) Advances in Neural Information Processing Systems 19, pp. 801–808. MIT Press, Cambridge (2007)

22. Lin, M., Chen, Q., Yan, S.: Network in network. CoRR abs/1312.4400 (2013)

23. Krizhevsky, A., Sutskever, I., Hinton, G.: Imagenet classification with deep convolutional neural networks. In: Advances in Neural Information Processing Systems, pp. 1097–1105 (2012)
24. Szegedy, C., Liu, W., Jia, Y., Sermanet, P., Reed, S., Anguelov, D., Erhan, D., Vanhoucke, V., Rabinovich, A.: Going deeper with convolutions. ArXiv e-prints (2014)
25. Simonyan, K., Zisserman, A.: Very deep convolutional networks for large-scale image recognition, pp. 1–13 (2015)

Novel Methods for Estimating Surface Normals from Affine Transformations

Daniel Barath[1,3], Jozsef Molnar[2], and Levente Hajder[1(✉)]

[1] Distributed Events Analysis Research Laboratory, MTA SZTAKI,
Kende utca 13-17, Budapest 1111, Hungary
{barath.daniel,hajder.levente}@sztaki.mta.hu
[2] Synthetic and Systems Biology Unit, Hungarian Academy of Sciences,
Szeged, Hungary
molnar.jozsef@brc.mta.hu
[3] Eötvös Loránd University, Budapest, Hungary
http://web.eee.sztaki.hu/

Abstract. The aim of this paper is to describe different estimation techniques in order to deal with point-wise surface normal estimation from calibrated stereo configuration. We show here that the knowledge of the affine transformation between two projections is sufficient for computing the normal vector unequivocally. The formula which describes the relationship among the cameras, normal vectors and affine transformations is general, since it works for every kind of cameras, not only for the pinhole one. However, as it is proved in this study, the normal estimation can optimally be solved for the perspective camera. Other non-optimal solutions are also proposed for the problem. The methods are tested both on synthesized data and real-world images. The source codes of the discussed algorithms are available on the web.

1 Introduction

Although computer vision has been an intensively researched area in computer science from many decades, several unsolved problems exist in the field. This paper proposes a novel optimal method for estimating the normal vector of a planar surface patch if the affine transformation of the patch between two calibrated (stereo) images is known.

The normal vector estimation problem itself can most accurately be solved by photometric stereo (PS) that was introduced many decades ago [15]. The main drawback of this method is that it can only be used in laboratories where light conditions are totally controlled. PS usually assumes that the object to be reconstructed is illuminated by directional light source(s) [15], but point-light sources can also be applied [5].

The image-based normal vector estimation is usually carried out by decomposing the homography of corresponding image patches between stereo setup [3, 13]. Unfortunately, these methods ambiguous as it was shown in several studies (e.g. in [12]). To the best of our knowledge, the problem of image-based normal

© Springer International Publishing Switzerland 2016
J. Braz et al. (Eds.): VISIGRAPP 2015, CCIS 598, pp. 316–337, 2016.
DOI: 10.1007/978-3-319-29971-6_17

vector estimation from affine transformation has not been solved yet. The first similar work was published in two papers by Habbecke and Kobbelt in [7,8]. They estimate the parameters of a flat patch based on photo-consistency. The plane is parameterized in 3D by the implicit parameters of a general plane. (These are three real values as the implicit parameters of a 3D plane are defined up to an arbitrary scale.) A very similar approach was proposed by Furukawa & Ponce [6] where the authors deal with the reconstruction of spatial patches. These patches are represented by both their locations and normals.

Our method only concentrates on the estimation of the spatial normal vector since the point of the plane can be estimated in 3D by triangulation if corresponding projections on two patches are known [9,10].

The closest work to our study is that of Megyesi et al. [17]. They compute a dense 3D reconstruction using normal vectors. The normal vectors themselves are calculated from the affine parameters between a rectified stereo image pair. For this reason, only two parameters of the affine transformation have to be estimated. The drawback of this work is that the rectification itself cannot be perfect due to noise and computational inaccuracy. Our method proposing here is more general as it works on arbitrary stereo image pairs. The only restriction is that the stereo images have to be taken by perspective cameras. (Or the non-perspective distortion of the images has to be undistorted.)

To the best of our knowledge, this is the first study that deals with surface normal computation from affine transformation using calibrated stereo images. *The main contribution of this paper is twofold: (i) We show here the general relationship among surface normal vector, affine transformation and camera parameters. The formulas proposing here is valid for every kind of cameras. (ii) Different surface normal estimators are proposed here including an optimal one that finds the optimal normal vector in the least squares sense if the affine parameters are contaminated with noise.*

The structure of the papers is as follows. The required geometric background in order to understand the proposed method is overviewed in Sect. 2. Then the novel methods are proposed in Sect. 3. The novel algorithms are tested both on synthesized data and real images as it is discussed in Sect. 4. Finally, the research is concluded in the final section. The appendix also contains very important details that are required to implement the algorithms proposed here. However, the Matlab/Octave implementations are also available at our webpage[1].

2   Geometric Background

Two projections of a 3D surface are given in stereo images. If the neighboring pixels are selected around image locations, these pixels form the so-called patches. The affine transformation between two corresponding patches are assumed to be known. The goal of this study is to show how the surface normal n can be estimated if the images are calibrated. The problem is visualized in Fig. 1.

---

[1] http://web.eee.sztaki.hu/home4/node/53.

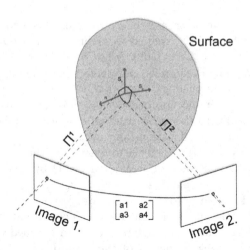

Fig. 1. 3D patch perspectively projected to stereo images.

The projected coordinates x and y is determined by the projective functions Π_x and Π_y as

$$x = \Pi_x(X, Y, Z), \quad y = \Pi_y(X, Y, Z),$$

where the surface point $[X, Y, Z]^T$ is written in parametric form

$$X = X(u, v), \quad Y = Y(u, v), \quad Z = Z(u, v).$$

As it is well-known in differential geometry [11], the tangent vectors of the plane are written by the partial derivatives of the spatial coordinates, while the surface normal is given by the cross product of the tangent vectors: $n = S_u \times S_v$ where

$$S_u = \left[\frac{\partial X(u,v)}{\partial u} \quad \frac{\partial Y(u,v)}{\partial u} \quad \frac{\partial Z(u,v)}{\partial u} \right],$$

$$S_v = \left[\frac{\partial X(u,v)}{\partial v} \quad \frac{\partial Y(u,v)}{\partial v} \quad \frac{\partial Z(u,v)}{\partial v} \right].$$

It is known that the 3D point $[X, Y, Z]^T$, and tangent vectors S_u and S_v determine the tangent plane. Locally, the surface can be approximated by its tangent plane. We assumed that we have images taken from the object. Now, a point of the surface close to the given 3D location $[X, Y, Z]^T$ is approximated by the first order Taylor-series:

$$\begin{bmatrix} x + \Delta x \\ y + \Delta y \end{bmatrix} \approx \begin{bmatrix} \Pi_x(X, Y, Z) \\ \Pi_y(X, Y, Z) \end{bmatrix} + \begin{bmatrix} \frac{\partial \Pi_x(X,Y,Z)}{\partial u} & \frac{\partial \Pi_x(X,Y,Z)}{\partial v} \\ \frac{\partial \Pi_y(X,Y,Z)}{\partial u} & \frac{\partial \Pi_y(X,Y,Z)}{\partial v} \end{bmatrix} \begin{bmatrix} \Delta u \\ \Delta v \end{bmatrix}.$$

Let us see that the partial derivatives of the projection functions give the affine transformation between 3D and 2D surface patches:

$$\begin{bmatrix} \Delta x \\ \Delta y \end{bmatrix} \approx A \begin{bmatrix} \Delta u \\ \Delta v \end{bmatrix},$$

where

$$A = \begin{bmatrix} \frac{\partial \Pi_x(X,Y,Z)}{\partial u} & \frac{\partial \Pi_x(X,Y,Z)}{\partial v} \\ \frac{\partial \Pi_y(X,Y,Z)}{\partial u} & \frac{\partial \Pi_y(X,Y,Z)}{\partial v} \end{bmatrix}.$$

The partial derivatives can be reformulated using the chain rule. For instance,

$$\frac{\partial \Pi_x(X,Y,Z)}{\partial u} = \frac{\partial \Pi_x(X,Y,Z)}{\partial X}\frac{X}{\partial u} + \frac{\partial \Pi_x(X,Y,Z)}{\partial Y}\frac{Y}{\partial u} + \frac{\partial \Pi_x(X,Y,Z)}{\partial Z}\frac{Z}{\partial u} = \nabla \Pi_x^T S_u,$$

where $\nabla \Pi_x$ is the gradient vector of the projection function w.r.t. the spatial coordinates X, Y, and Z of the surface patch. Similarly,

$$\frac{\partial \Pi_x}{\partial v} = \nabla \Pi_x^T S_v, \quad \frac{\partial \Pi_y}{\partial u} = \nabla \Pi_y^T S_u, \quad \frac{\partial \Pi_y}{\partial v} = \nabla \Pi_y^T S_v.$$

Therefore, the affine matrix can be written as

$$A = \begin{bmatrix} \nabla \Pi_x^T \\ \nabla \Pi_y^T \end{bmatrix} \begin{bmatrix} S_u & S_v \end{bmatrix}.$$

In stereo vision, two images are given. The affine transformation between the image patches is obtained by concatenating the inverse of affine transformation A_1 (between the patches of image #1 and the spatial patch), and the affine transformation A_2 (between 3D patch and that in image #2). Formally, it can be written as

$$\begin{bmatrix} \Delta x_2 & \Delta y_2 \end{bmatrix}^T = A_2 A_1^{-1} \begin{bmatrix} \Delta x_1 & \Delta y_1 \end{bmatrix}^T.$$

$A_2 A_1^{-1}$ is the affine transformation between the images. The inverse of the affine matrix A can be written as

$$A^{-1} = \frac{1}{det(A)} \begin{bmatrix} \Pi_x^T S_u & -\Pi_y^T S_u \\ -\Pi_x^T S_v & \Pi_y^T S_v \end{bmatrix},$$

where $det(A) = \Pi_x^T S_u \Pi_y^T S_v - \Pi_x^T S_v \Pi_y^T S_u$. If one makes elementary modification utilizing the fact that $S_v S_u^T - S_u S_v^T = [N]_\times$, then the affine transformation $A_2 A_1^{-1}$ can be written as

$$A_1^{-1} A_2 = \frac{1}{\Pi_x^{1^T} [N]_\times \Pi_y^1} \begin{bmatrix} \Pi_x^{2^T} [N]_\times \Pi_y^1 & \Pi_x^{1^T} [N]_\times \Pi_x^2 \\ \Pi_y^{2^T} [N]_\times \Pi_y^1 & \Pi_x^{1^T} [N]_\times \Pi_y^2 \end{bmatrix}.$$

Note that the scale of the normal is arbitrary since both the determinant and the matrix elements are multiplied with the scale of $[N]_\times$.

The expression $a^T [N]_\times b$ is also called the scalar triple product. Remark that $a^T [n]_\times b$ equals to $n^T (b \times a)$. Therefore, the final equation of the affine transformation is written as

$$\begin{bmatrix} a_1 & a_2 \\ a_3 & a_4 \end{bmatrix} = A_1^{-1} A_2 = \frac{1}{n^T w_5} \begin{bmatrix} n^T w_1 & n^T w_2 \\ n^T w_3 & n^T w_4 \end{bmatrix}, \tag{1}$$

where $w_1 = \nabla \Pi_y^1 \times \nabla \Pi_x^2$, $w_2 = \nabla \Pi_x^2 \times \nabla \Pi_x^1$, $w_3 = \nabla \Pi_y^1 \times \nabla \Pi_y^2$, $w_4 = \nabla \Pi_y^2 \times \nabla \Pi_x^1$, and $w_5 = \nabla \Pi_y^1 \times \nabla \Pi_x^1$. Equation 1 is a very important formula which shows the relations of the surface normal and the projection of the surface to the stereo image pair. *A very important advantage of this formula is that it is valid for every kind of camera since only the two projective equations must be known.* We show here that the above formula can be used to compute the surface normal if the perspective parameters are calibrated.

2.1 Pin-hole Camera Model

When the standard perspective camera model is used, the projection is written as

$$[x, y, 1]^T = \frac{1}{s} P_{persp}[X, Y, Z, 1]^T, \tag{2}$$

where $[x, y]$ are the projected coordinates in image space, s is the projective depth, P_{persp} is the so called projection matrix with size 3×4. Let us denote the rows of the projective matrix by p_1^T, p_2^T, and p_3^T. The projection formulas and their gradients can be written e.g. as

$$\frac{\partial x}{\partial X} = \frac{\partial \frac{p_1^T[X,Y,Z,1]^T}{p_3^T[X,Y,Z,1]^T}}{\partial X} = \frac{P_{11}s - P_{31}\left(p_1^T[X,Y,Z,1]^T\right)}{s^2} = \frac{1}{s}\left(P_{11} + xP_{31}\right).$$

where $s_i = p_3^T[X, Y, Z, 1]^T$ is the projective depth. Similarly,

$$\frac{\partial x}{\partial Y} = \frac{1}{s}\left(P_{12} + xP^{32}\right), \quad \frac{\partial x}{\partial Z} = \frac{1}{s}\left(P_{13} + xP_{33}\right),$$
$$\frac{\partial y}{\partial X} = \frac{1}{s}\left(P_{11} + yP^{31}\right), \quad \frac{\partial y}{\partial Y} = \frac{1}{s}\left(P_{12} + yP_{32}\right),$$
$$\frac{\partial y}{\partial Z} = \frac{1}{s}\left(P_{13} + yP_{33}\right).$$

Therefore, it can be written that

$$\nabla \Pi_x = \frac{1}{s}\begin{bmatrix} P_{11} + xP_{31} \\ P_{12} + xP_{32} \\ P_{13} + xP_{33} \end{bmatrix}, \quad \nabla \Pi_y = \frac{1}{s}\begin{bmatrix} P_{21} + yP_{31} \\ P_{22} + yP_{32} \\ P_{23} + yP_{33} \end{bmatrix},$$

where P_{ij} is the element in the ith row and j^{th} column in projection matrix P_{persp}. The projective depth is obtained as $s = p_3^T[X, Y, Z, 1]^T$. The affine transformation becomes

$$\begin{bmatrix} a_1 & a_2 \\ a_3 & a_4 \end{bmatrix} = \frac{1}{\alpha n^T w_5}\begin{bmatrix} n^T w_1 & n^T w_2 \\ n^T w_3 & n^T w_4 \end{bmatrix}, \tag{3}$$

where $\alpha = s^1/s^2$ is the ratio of the projective depths in the first and second images, and $w_1 = s^1 s^2 \left(\nabla \Pi_y^1 \times \nabla \Pi_x^2\right)$, $w_2 = s^1 s^2 \left(\nabla \Pi_x^2 \times \nabla \Pi_x^1\right)$, $w_3 = s^1 s^2 \left(\nabla \Pi_y^1 \times \nabla \Pi_y^2\right)$, $w_4 = s^1 s^2 \left(\nabla \Pi_y^2 \times \nabla \Pi_x^1\right)$, and $w_5 = s^2 s^2 \left(\nabla \Pi_x^1 \times \nabla \Pi_y^1\right)$.

A very important remark is that if the projective depth s^i is unknown, but the upper left 3×3 submatrices of the projection matrices P_1 and P_2 are known then the gradient vectors can be calculated up to an unknown scale. (This scale is the multiplicative inverse of the projective depth s^i.) Also note that the vectors $w_1 \ldots w_4$ are scaled by $s^1 s^2$ while w_5 by $s^2 s^2$.

Therefore, the normal vector is independent of the translation between the two cameras since the last columns of the projection matrices are the product of the camera intrinsic parameters and the translation. For this reason, the following two cases must be distinguished:

1. Both projection matrixes P^1 and P^2 are known. (In other words, the cameras are calibrated.)
2. Only the upper-left 3×3 submatrices of the projections are known. In this case, the projective depth of the points is notknown. However, the gradients can be computed up to a scale where this scale is the inverse of the projective depth.

Also remark that the normal vector cannot be estimated if $w_5 = 0$. This can only be true if $\nabla \Pi_x^1$ and $\nabla \Pi_y^1$ are parallel which is not a realistic case as it is only possible if the first and second rows of the 3×3 submatrix of projection matrix P_{persp} are parallel. If the camera calibration is valid it cannot be true. The problem itself is numerically stable if the angles between the vectors $w_1 \ldots w_4$ are relatively large. To our experiments, this is true for realistic reconstruction problems.

3   Normal Vector Estimation

This section shows different normal vector estimators. The first one is very fast and simple, later more sophisticated and accurate methods are introduced.

3.1   Fast Normal Estimation (FNE)

The base matrix equation (Eq. 3) consists of 4 elements. If two of those are selected and they are divided by each other, then an equation is obtained. If the same procedure is repeated for the rest of the matrix elements, then the second equation can be similarly yielded. For instance, two elements of the first and second rows give the equations

$$\frac{w_1^T n}{w_2^T n} = \frac{a_1}{a_2}, \quad \frac{w_3^T n}{w_4^T n} = \frac{a_3}{a_4}. \tag{4}$$

These equations can trivially be modified as

$$\left(a_2 w_1^T - a_1 w_2^T \right) n = 0, \tag{5}$$

$$\left(a_4 w_3^T - a_3 w_4^T \right) n = 0. \tag{6}$$

The normal vector n is perpendicular to both $a_2 w_1^T - a_1 w_2^T$ and $a_4 w_3^T - a_3 w_4^T$. Therefore, the normal can be obtained as the cross product of these vectors:

$$n = \left(a_2 w_1^T - a_1 w_2^T \right) \times \left(a_4 w_3^T - a_3 w_4^T \right). \tag{7}$$

Of course, the obtained vector n should be normalized, its length must be 1. A very nice property of this normal estimation is that it is independent of the scales appearing in vectors $w_1 \ldots w_4$. Therefore, the projective depths of the patch are not required to estimate the normal, because they influence only the length of n.

Remark, that the equation pairs (Eq. 4) can be selected in other two ways. In those cases, the normal vector is given by

$$n = \left(a_3 w_1^T - a_1 w_3^T\right) \times \left(a_4 w_2^T - a_2 w_4^T\right), \tag{8}$$

or

$$n = \left(a_4 w_1^T - a_1 w_4^T\right) \times \left(a_3 w_2^T - a_2 w_3^T\right). \tag{9}$$

The three variants of the fast estimator defined by Eqs. 7–9 can be unified into one homogeneous linear system of equations $Cn = 0$ since the cross products mean that the surface normal n has to be perpendicular to the vectors in the products Matrix C is defined as

$$C = \begin{bmatrix} a_2 w_1^T - a_1 w_2^T \\ a_4 w_3^T - a_3 w_4^T \\ a_3 w_1^T - a_1 w_3^T \\ a_4 w_2^T - a_2 w_4^T \\ a_4 w_1^T - a_1 w_4^T \\ a_3 w_2^T - a_2 w_3^T \end{bmatrix}. \tag{10}$$

The optimal solution of this problem is well-known: the optimal normal n is the eigenvector of matrix $C^T C$ corresponding to the smallest eigenvalue.

3.2 Optimal Normal Estimation with Known Projective Depth (OPT)

The aim of the optimal method is to minimize the error in the matrix base equation (Eq. 3). Formally, the estimation itself can be written as the minimization of Frobenius norm of Eq. 3 with respect to normal n. This is equivalent to

$$\arg\min_n \sum_{k=1}^{4} \left(\frac{n^T w_k}{n^T w_5} - a_k \right)^2. \tag{11}$$

It minimizes the normal vector in the least square sense assuming that the affine parameters are contaminated with noise. (This assumption is valid since the affine parameters are estimated as described later in Sect. 4.2 in short, and this estimation cannot be perfect since the images themselves contain noise.) The optimal solution is given in the Appendix with $\alpha = 1$.

3.3 Normal Estimation with Unknown Projective Depths Using Alternation (ALT)

If the projective depth is unknown then the base optimization equation (Eq. 11) cannot be applied since the parameter $\alpha = s_1/s_2$ is not known. The cost function defined in Eq. 11 has to be modified as

$$\arg\min_n \sum_{k=1}^{4} \left(\frac{n^T w_k}{\alpha n^T w_5} - a_k \right)^2 . \tag{12}$$

Unfortunately, this problem cannot be optimally solved to the best of our knowledge. We propose here an alternating-like approach which is overviewed in Algorithm 1. The alternation has two steps:

1. EstimateAlpha: The cost function (Eq. 12) is a linear one with respect to $1/\alpha$ since it can be written as $A\frac{1}{\alpha} = b$ where $A = \left[\frac{n^T w_1}{n^T w_5}, \ldots, \frac{n^T w_4}{n^T w_5} \right]^T$ and $b = [a_1, \ldots, a_4]^T$. The optimal solution of an overdetermined linear system can be solved optimally. In this case, that is obtained as

$$\frac{1}{\alpha} = \frac{n^T w_5}{\sum_j (n^T w_j)^2} \sum_j n^T w_j a_j. \tag{13}$$

2. EstimateNormal: The normal vector estimation is very similar to the optimal method described above, the only difference is that parameter α appears in the denominators. However, the method described in the Appendix can solve the subproblem optimally.

The alternation requires initial values for the parameters n and α to be optimized. We propose to use the linear methods described later in Sect. 3.4 in order to compute the initial values. The alternation converges to the closest (local) minimum since it optimizes a non-negative cost function and each step decreases (or does not increase) the cost. Unfortunately, we could not prove theoretically that the global optimum is reached in this way, however, to our practice, the method usually improves the initial normal n.

Algorithm 1. Alternation for Normal Estimation (ALT).

$n, \alpha \leftarrow$ Parameter Initialization by LNE-UPD
repeat
    $\alpha \leftarrow$ EstimateAlpha$(n, w_1, \ldots, w_5)$
    $n \leftarrow$ EstimateNormal$(\alpha, w_1, \ldots, w_5)$
until convergence

3.4 Linear Normal Estimation (LNE)

The base matrix equation (Eq. 3) is a nonlinear one. The elements can be linearized if they are multiplied with their common denominator $\alpha w_5^T n$. Then a cost function can be formed for the elements as

$$\arg\min_n \sum_{k=1}^{4} \left(n^T w_k - \alpha a_k n^T w_5\right)^2. \tag{14}$$

This is a usual trick, and the solution will not be optimal if this modification is carried out. However, the problem becomes linear, and it can be solved easily [2].

Linear Normal Estimation for Known Projective Depth (LNE-KPD). If the projective depth is known, then $\alpha = 1$ and the problem can be rewritten as an overdetermined homogeneous linear equation system $An = 0$ subject to $n^T n = 1$, where

$$A = \begin{bmatrix} w_1 - a_1 w_5 \\ w_2 - a_2 w_5 \\ w_3 - a_3 w_5 \\ w_4 - a_4 w_5 \end{bmatrix}. \tag{15}$$

The optimal solution of this system is the eigenvector of matrix $A^T A$ corresponding to the smallest eigenvalue [2].

Linear Normal Estimation for Unknown Projective Depth (LNE-UPD). If the projective depth is unknown, then the function to be optimized in Eq. 14 gives an overdetermined homogeneous linear system $Bb = 0$, similarly to the previous case (Sect. 3.4), but the matrix of coefficients B and the vector b differ as follows.

$$B^T = \begin{bmatrix} w_1^T, & -a_{11} \\ w_2^T, & -a_{12} \\ w_3^T, & -a_{21} \\ w_4^T, & -a_{22} \end{bmatrix}, \quad b = \begin{bmatrix} n \\ \alpha w_5^T n \end{bmatrix}.$$

The solution is given from the eigenvector of matrix $B^T B$ corresponding to the smallest eigenvalue [2]. If this vector is denoted by $\hat{b}$, then the estimation for the normal vector n is given by the first three coordinates of $\hat{b}$, but this vector should be normalized in order to fulfill the $n^T n = 1$ constraint. The parameter $\alpha = s_1/s_2$ can also be computed if n is known from the fourth coordinate of $\hat{b}$.

4 Experimental Results

The proposed normal vector estimators have been tested both on synthesized data and real world images.

4.1 Test on Synthesized Data

During synthesized tests, the main task was to generate different normal vectors with corresponding affine parameters. For this reason, a stereo image pair was first generated represented by their 3×4 projection matrices. Then a 3D object (sphere or cube) was generated. The sphere sampled by spherical coordinates. The normal vector of the locations on the sphere can easily be calculated as it is the direction pointing from the sphere center to the current surface points. The normal for the 3D cube was the perpendicular vector to the faces of the cube. The synthetic sphere and cube with ground truth normals are visualized in the images of Fig. 2. 72 and 150 different patches sampled for the synthetic sphere, and cube, respectively.

The affine parameters between the stereo images were calculated as follows: (i) The tangent plane of the sphere was determined first, (ii) then it was projected to the stereo images. (iii) The projections of the plane determine two homographies with respect to the 3D tangent plane. (iv) The homography between the two images were given by concatenating the two 3D→2D homographies. (v) The affine transformation is the first order approximation of the 2D→2D homography at the given locations.

The error values are defined as the average of the angular error between the estimated and ground truth normal vectors. We have tested all the methods described in this study. In the first test case, 72 patches of the sphere were generated, and the tests were repeated 50 times. Thus, the average error values come from $72 \cdot 50 = 3600$ run of the competitor methods. Two test cases were simulated: zero-mean Gaussian noise was added to the (i) affine parameters and (ii) to the elements of the projection matrices.

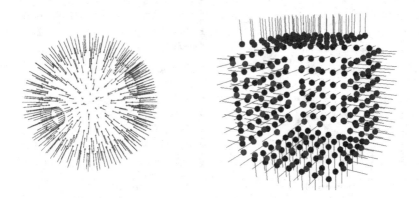

Fig. 2. Sphere and cube with normal vectors for synthesized test.

Then a synthetic cube was also generated as it is pictured in the right image of Fig. 2. We generated the points and normals of the cube for every side of the cube. Each side is uniformly sampled, 25 samples were generated per side.

Therefore, the synthetic cube dataset consisted of $6 \cdot 25 = 150$ feature points. Then the affine transformation calculated similarly to the spherical test sequence.

During the tests, we have compared the efficiency of the fast (FNE), linear (LIN-UPD and LIN-KPD), alternation (ALT), and optimal (OPT) normal estimators. The simplest (and more rapid) version of FNE is also compared, it is denoted by FNE-SIMP. The difference between FNE and FNE-SIMP is that the latter uses only the (minimally required) first two rows of matrix C defined in Eq. 10.

Remark that all the synthesized tests have been implemented in Octave[2].

Test with Contaminated Affine and Projective Parameters. In this test case, we compared the proposed methods except the linear estimators LIN-KPD and LIN-UPD. It is clear that the optimal estimator (OPT) outperforms the others as it is visualized in the plots of Fig. 3 since it optimally estimates the normal vector in the least square sense. It is also obvious that the fastest method FNE-SIMP is the less accurate one as the other three methods (OPT, ALT, and FNE) are significantly more sophisticated. However, it is interesting that although FNE and ALT differ, they yield approximately the same results. There are differences between their results, but they are very small.

In the first test case the affine parameters were contaminated by Gaussian noise. This is visualized in the left plots of Fig. 3. Hence, when noise were added to the elements of the projective matrices, the same conclusions can be drawn as the characteristics of the charts in the right plots of Fig. 3 are very similar.

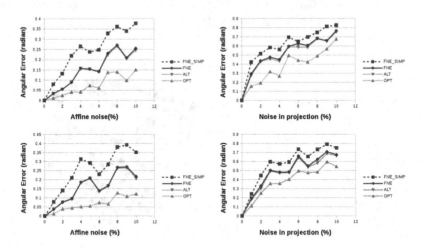

Fig. 3. Comparison of methods when affine (left) and projective (right) parameters and contaminated. Test results generated using synthesized cube (top) and sphere (bottom).

---

[2] www.octave.org.

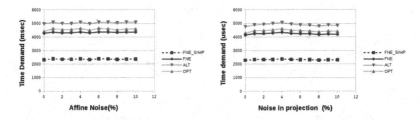

Fig. 4. Time demand of proposed methods when affine (left) and projective (right) parameters and contaminated. Test results generated using synthesized cube.

The time demand of the algorithms were also compared. Figure 4 shows how long the algorithms run. The fastest method is FNE-SIMP as it is expected. The application of this estimator is suggested to use in time critical applications. The slowest one is the alternating method (ALT) since it contains an iteration. It is also obvious that the time demand of the algorithms do not depend on the noise level. This statement is true both for contaminated affine and projective parameters.

We compared the linear methods to the corresponding non-linear ones. Namely, the linear method with unknown projective depth (LIN-UPD) algorithm is compared to the alternation and linear with known projective depth method (LIN-KPD) to the optimal one (Both charts are visualized in Fig. 5). The differences are significant only if the optimal (OPT) method is used instead of its linear version.

We also examined the expected values and the spreads of the five proposed methods. The expected value for the length of the difference between the ground truth and estimated vectors are close to zero as it is excepted. Therefore, the estimators are consistent. Their spreads are listed in Table 1. It shows that the optimal method has the lowest spread as it is expected, FNE is the highest one. It is interesting that the linear method with known projective depths gives significantly better result than the methods with unknown depths (LIN-UPD and ALT),

Table 1. Spread of error vector lengths.

FNE	LIN-UPD	ALT	LIN-KPD	OPT
0.55	0.449	0.433	0.352	**0.2919**

To conclude the synthetic tests, it can be declared that the optimal method is the best solution if the projective depth of the 3D point is known. If it is not, the alternation method serves the most efficient method, but its advantage over its linear version is very small. The alternation itself is an iterative algorithm, sometimes it can be very slow. Therefore, we propose the LIN-UPD method for

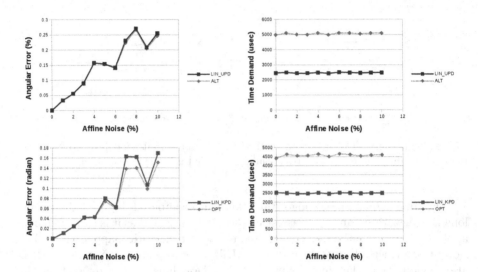

Fig. 5. Comparison of linear and corresponding nonlinear methods. Left: LIN-UPD vs. ALT. Right: LIN-KPD vs. OPT. Top: accuracy. Bottom: time demand.

time-critical application, ALT is the best selection for offline algorithms when the projective depths are unknown.

Normal Vector from Affine Parameters Versus Homography. The mainstream solution for computing the normal vector from two patches is to estimate the homography between the patches [13]. It has eight degrees of freedom, and it can be decomposed into the pose (3 DoF), the location (3 DoF), and the normal (2 DoF) of the plane.

We compare the accuracy of the homography-estimated normal vector with our optimal (OPT) estimator. The synthesized data is given by sampling the surface of a sphere similarly to the synthesized tests above. However, the homography and the affine transformation are both estimated from projected points: points are generated randomly on the 3D tangent plane (close to the location on sphere surface), and these points are projected to the image pair. Then noise is added to the projected coordinates in image space. The homography and the affine transformation are estimated using the corresponding points in image pairs. The estimation of the affine transformation is easier since it is trivial that affine parameter estimation is a linear problem. We estimate the homography via numerical optimization method, the initial parameters are computed by DLT (Direct Linear Transformation) algorithm [10]. Note that at least 4 points are required to estimate the homography, while 3 points are sufficient to compute the affine transformation.

The results of the comparison is visualized in Fig. 6. The transformations are estimated from the same point correspondences. The number of corresponding points are 4, 6, and 8 as it is seen in Fig. 6. The methods are denoted by 'HOM'

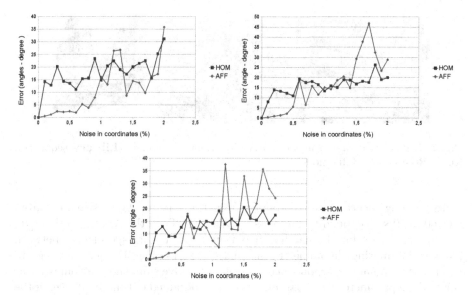

Fig. 6. Comparison of normal vector estimators. HOM: normal from homography decomposition AFF: normal from affine parameters by proposed OPT method. Top-left: transformation estimated from 4 points, Top-right: 6 points, Bottom: 8 points.

(normals from estimated homography) and 'AFF' (normals from affine transformation). It is interesting that AFF serves better results when the noise is low. It is true especially for the $P = 4$ case (left image in Fig. 6). It is because the homography is determined exactly by the given 4 points, while the affine transformation is overdetermined. When the number of points grows (center and right plots on Fig. 6), the normal from homography estimation becomes better than that from affine transformation since the projected coordinates are obtained via perspective projection, and homography represents the correct transformation between corresponding planes in image space.

4.2   Test on Real Image Pairs

Real Tests on Calibrated Images. The proposed optimal normal estimator has been also tested on real data. In order to use them, the projection matrices have to be known. We have downloaded building images and reconstruction data with camera parameters from the web page of the Visual Geometry Group at Oxford University[3] and Strecha Dense MVS database[4].

The Oxfordian data sets contain point correspondences, but we used ASIFT method [16] for this purpose instead of using the original data. The affine transformations for the pairs were computed as follows. (1) Two patches around the

---

[3] http://www.robots.ox.ac.uk/~vgg/data/.
[4] http://cvlabwww.epfl.ch/data/multiview/denseMVS.html.

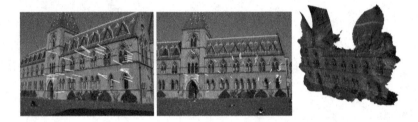

Fig. 7. Left, center: Stereo image pair with estimated normals (**Library** sequence). Right: Reconstructed 3D model.

corresponding locations were cropped from the images. Their size were from 60×60 to 80×80 depending on test sequences. (2) Then the ASIFT method [16] was applied again for the patch pair obtaining point correspondences between patches. Estimating the affine transformation is an affine 2D registration problem based on point correspondences. It is easy to solve since the problem is linear w.r.t. affine parameters, the parameters can be obtained optimally [2] even if the problem is overdetermined. Remark that the affine estimator should be robust since ASIFT can give false correspondences. We used a RANSAC [4]-like algorithm to discard the outliers. We were able to reconstruct the 3D surface of the estimated positions and corresponding normals using the Marching Cubes (APSS) filter of MeshLab[5]. It is visualized in the right plot of Fig. 7.

The proposed optimal method was carried out on two images of the **Fountain** sequence from the Strecha dataset. The affine transformation are computer by our unpublished affine matcher algorithm considering the calibration data of the sequence. The estimated normals are seen in Fig. 8, surface normals are visualized by white rods.

Fig. 8. Estimated normals on sequence **Fountain**.

---

[5] www.meshlab.org.

Fig. 9. Estimated normals on sequence **Bear**.

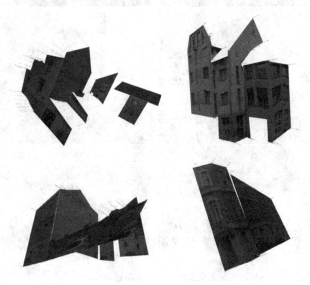

Fig. 10. Reconstructed 3D flat surfaces with estimated normals.

The proposed normal estimators are also tested on our own calibrated stereo image pair. The estimated surface normals of the plastic bear is pictured in Fig. 9.

Normal from Real Planar Surfaces. The proposed normal vector estimator (OPT) was also examined on images of buildings as it is pictured in Fig. 11. These objects mainly consist of planar walls and they can be matched by homography-based pairing methods especially when the images are rectified [14]. Thought the

Fig. 11. Estimated normals on planar surfaces.

homography itself can be decomposed [3] if the cameras are calibrated, and then the plane normal is obtained with the camera extrinsic parameters. However, the decomposition has ambiguity as it is discussed in [12] and two realistic normal vector can be achieved.

We reconstructed the plane normal via the affine transformation. The affine parameters can easily be calculated from homography as it is shown in the Appendix. The cameras were calibrated via point-based 3D reconstruction by bundle adjustment [1]. Then the normals were computed by the proposed optimal method. We tested the OPT algorithm on five different stereo pairs as it is visualized in Fig. 11. They were short baseline stereo images. The yielded normal vectors of the planes are drawn on the input images. The corresponding points on the wall surfaces are denoted by small dots, the normals are drawn both inside and outside the plane. The proposed method is robust enough, it computes very accurately the surface normals.

Figure 10 shows that the buildings can be reconstructed in 3D. The 3D coordinates of the plane corners were calculated by triangulation [9], the affine transformations came from the homography, and the plane normals estimated (visualized by small light-blue rods in the figure) using one of the proposed method.

It was found during the real tests that the proposed normal estimators can compute accurately the surface normals. This statement is especially true for the optimal (OPT) algorithm.

5   Conclusion and Future Work

In this paper, we discussed the geometric and algebraic issues of surface normal estimation of rigid objects that are taken by a standard perspective camera pair. Only the affine transformations between corresponding image patches are required to reconstruct the surface normal. We proposed several algorithms, one of those serves the optimal solutions in the least squares sense if the cameras are fully calibrated. It was shown here that the knowledge of the last row of perspective projection matrix is not necessary for the estimation. Moreover, we also proposed two linearized estimator that are significantly faster than their nonlinear version. The source codes of the proposed methods are available on the web[6]. In the testing section, the noise sensitivity of the algorithms were tested on synthesized data. We found the the optimal algorithm is suggested to use if the full perspective transformation is given, and the time demand is not crucial. The linearized versions can be used if fast algoritms are required. Finally, we demonstrated that the algorithms can cope with real image pairs for which the surface normals can efficiently be estimated.

In the future, the improvement of the affine transformation estimation is planned in order to get more accurate input for surface normal estimation. Novel reconstruction methods will also be developed using both point correspondences and affine transformations in order to obtain richer 3D model of real-world 3D

---

[6] http://web.eee.sztaki.hu/home4/node/53.

objects. If both surface normals and the corresponding 3D point cloud are given, more realistic 3D object reconstruction using continuous surface representation becomes possible.

A   Appendix

Algorithm for Optimal Normal Estimation. The task is to minimize the cost function defined in Eq. 12 with respect to normal vector n. The scale of the vector is arbitrary, only the direction of the normal is required. Such kind of problems are typically solved using Lagrange-multipliers, however, it cannot be applied here since the derivatives become very difficult. For this reason, we utilize another constraint for the normal: let the sum of the coordinates be 1. Thus, n is written as $n = [n_x, n_y, 1 - n_x - n_y]^T$. Equation 12 can be reformulated as,

$$\arg\min_m \sum_{k=1}^{4} \left(\frac{m^T q_k + w_{k,z}}{\alpha m^T q_5 + \alpha w_{5,z}} - a_k \right)^2,$$

where $m = [n_x, n_y]$, $q_i = [w_{i,x} - w_{i,z}, w_{i,y} - w_{i,z}]^T$. (Indices x, y, and z denote the first, second, and third coordinates of vectors, respectively.)

The minima/maxima can be obtained by taking the derivative with respect to vector m:

$$2 \sum_{k=1}^{4} \beta_k \gamma_k = 0,$$

where

$$\beta_k = \left(\frac{m^T q_k + w_{k,z}}{\alpha m^T q_5 + \alpha w_{5,z}} - a_k \right),$$

$$\gamma_k = \left(\alpha \frac{(m^T q_5 + w_{5,z}) q_k - (m^T q_k + w_{k,z}) q_5}{(\alpha m^T q_5 + \alpha w_{5,z})^2} \right).$$

After taking the lowest common multiple of the fractions, the left side should be equal to zero as $\sum_{k=1}^{4} \delta_k \kappa_k = 0$, where

$$\delta_k = \left(m^T q_k + w_{k,z} - a_k \alpha m^T q_5 - a_k \alpha w_{5,z} \right),$$

$$\kappa_k = \left((m^T q_5 + w_{5,z}) q_k - (m^T q_k + w_{k,z}) q_5 \right).$$

It can be simplified as $\sum_{k=1}^{4} e_k^1 e_k^2 = 0$, where

$$e_k^1 = \left(m^T (q_k - a_k \alpha q_5) + (w_{k,z} - a_k \alpha w_{5,z}) \right),$$

$$e_k^2 = \left((m^T q_5) q_k - (m^T q_k) q_5 + w_{5,z} q_k - w_{k,z} q_5 \right).$$

This is an equation with a 2D-vector:

$$\sum_{k=1}^{4} r \left(\begin{matrix} m^T (q_5 q_{k,x} - q_i q_{5,x}) + w_{5,z} q_{k,x} - w_{k,z} q_{5,x} \\ m^T (q_5 q_{k,y} - q_i q_{5,y}) + w_{5,z} q_{k,y} - w_{k,z} q_{5,y} \end{matrix} \right) = 0,$$

where $r = \left(m^T (q_k - a_k \alpha q_5) + (w_{k,z} - a_k \alpha w_{5,z}) \right)$.

By introducing the $m = [x, y]^T$ notation, the vector equation is modified as follows

$$\sum_{k=1}^{4} (\Omega_k x + \Psi_k y + \Gamma_k) \begin{pmatrix} \Omega_k^1 x + \Psi_k^1 y + \Gamma_k^1 \\ \Omega_k^2 x + \Psi_k^2 y + \Gamma_k^2 \end{pmatrix} = 0,$$

where

$$\Omega_k = q_{k,x} - \alpha q_{5,x} a_k, \qquad \Psi_k = q_{k,y} - \alpha q_{5,y} a_k,$$
$$\Gamma_k = w_{k,z} - a_k \alpha w_{5,z}, \qquad \Omega_k^1 = 0,$$
$$\Psi_k^1 = q_{5,y} q_{k,x} - q_{k,y} q_{5,x}, \quad \Gamma_k^1 = w_{5,z} q_{k,x} - w_{k,z} q_{5,x},$$
$$\Omega_k^2 = q_{5,x} q_{k,y} - q_{k,x} q_{5,y}, \qquad \Psi_k^2 = 0,$$
$$\Gamma_k^2 = w_{5,z} q_{i,y} - w_{i,z} q_{5,y}.$$

The rows of the vector equation give two special quadratic curves. They are written by their implicit equations as $\sum_{k=1}^{4} A_k^l x^2 + \sum_{k=1}^{4} B_k^l y^2 + \sum_{k=1}^{4} C_k^l xy + \sum_{k=1}^{4} D_k^l x + \sum_{k=1}^{4} E_k^l y + \sum_{k=1}^{4} F_k^l = 0$, where $A_k^l = \Omega_k \Omega_k^l$, $B_k^l = \Psi_k \Psi_k^l$, $C_k^l = \Omega_k^l \Psi_k + \Psi_k^l \Omega_k$, $D_k^l = \Omega_k^l \Gamma_k + \Gamma_k^l \Omega_k$, $E_k^l = \Psi_k^l \Gamma_k + \Gamma_k^l \Psi_k$ and $F_k^l = \Gamma_k \Gamma_k^l$, $l \in 1, 2$. They are special because $A_k^1 = 0$ and $B_k^2 = 0$.

The solution of the optimal method described in the study (within appendix) is given by the intersection of two quadratic equations.

$$B_1 y^2 + C_1 xy + D_1 x + E_1 y + F_1 = 0,$$
$$A_2 x^2 + C_2 xy + D_2 x + E_2 y + F_2 = 0.$$

Parameter y can be obtained from the latter equation as

$$y = -\frac{A_2 x^2 + D_2 x + F_2}{C_2 x + E_2}.$$

Substituting y into the first equation the following expression is obtained

$$B_1 \left(\frac{A_2 x^2 + D_2 x + F_2}{C_2 x + E_2} \right)^2 - C_1 x \frac{A_2 x^2 + D_2 x + F_2}{C_2 x + E_2} + D_1 x$$
$$- E_1 \frac{A_2 x^2 + D_2 x + F_2}{C_2 x + E_2} + F_1 = 0.$$

If both sides are multiplied with $(C_2 x + E_2)^2$, then the equation modifies as follows

$$B_1 (A_2 x^2 + D_2 x + F_2)^2 - C_1 x (A_2 x^2 + D_2 x + F_2)(C_2 x + E_2)$$
$$+ D_1 x (C_2 x + E_2)^2 - E_1 (A_2 x^2 + D_2 x + F_2)(C_2 x + E_2) + F_1 (C_2 x + E_2)^2 = 0$$

This is a fourth-order polynomial where the coefficients are as follows

$$x^4 : \qquad B_1 A_2^2 - C_1 A_2 C_2$$
$$x^3 : \quad 2B_1 A_2 D_2 - C_1 A_2 E_2 - C_1 D_2 C_2 + D_1 C_2^2 - E_1 A_2 C_2$$
$$x^2 : \quad B_1 D_2^2 + 2B_1 A_2 F_2 - C_1 D_2 E_2 - C_1 F_2 C_2 + 2D_1 C_2 E_2$$
$$\qquad \qquad - E_1 A_2 E_2 - E_1 D_2 C_2 + F_1 C_2^2$$
$$x^1 : \quad 2B_1 D_2 F_2 - C_1 F_2 E_2 + D_1 E_2^2 - E_1 D_2 E_2 - E_1 F_2 C_2$$
$$\qquad \qquad + 2F_1 C_2 E_2$$
$$x^0 : \qquad B_1 F_2^2 - E_1 F_2 E_2 + F_1 E_2^2$$

Remark that the equation $C_2x + E_2 = 0$ can also be considered. (In this case the first equation is independent from y.)

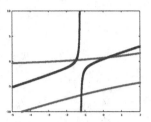

Fig. 12. Quadratic curves.

An example for two quadratic curves with three real intersections is visualized in Fig. 12. (The parameters of curves are $B1 = -1.9055$, $C1 = 2.2632$, $D1 = 2.8577$, $E1 = -9.4392$, $F1 = 7.7081$, and $A2 = -2.2632$, $C2 = 1.9055$, $D2 = -4.2074$, $E2 = 2.3903$, $F2 = -1.1190$).

Affine Parameters from Homography. The affine parameters can be obtained from the homography between the stereo image pairs. Let us assume that the homography H is given. Then the correspondence between the coordinates on the first (u and v) and second (u' and v') images is written as

$$u' = \frac{h_1^T [u, v, 1]^T}{h_3^T [u, v, 1]^T},$$

$$v' = \frac{h_2^T [u, v, 1]^T}{h_3^T [u, v, 1]^T},$$

where the 3×3 homography matrix H is written as

$$H = \begin{bmatrix} h_1^T \\ h_2^T \\ h_3^T \end{bmatrix} = \begin{bmatrix} h_{11} & h_{12} & h_{13} \\ h_{21} & h_{22} & h_{23} \\ h_{31} & h_{32} & h_{33} \end{bmatrix}.$$

The affine parameters come from the partial derivatives of the perspective plane-plane transformation. The top left element a_{11} of affine transformation matrix is as follows

$$a_{11} = \frac{\partial u'}{\partial u} = \frac{h_{11} h_3^T [u, v, 1]^T - h_{31} h_1^T [u, v, 1]^T}{\left(h_3^T [u, v, 1]^T \right)^2} = \frac{h_{11} - h_{31} u'}{s},$$

where $s = h_3^T[u, v, 1]^T$. The other components of affine matrix are obtained similarly

$$a_{12} = \frac{\partial u'}{\partial v} = \frac{h_{12} - h_{32}u'}{s},$$

$$a_{21} = \frac{\partial v'}{\partial u} = \frac{h_{21} - h_{31}v'}{s},$$

$$a_{22} = \frac{\partial v'}{\partial v} = \frac{h_{22} - h_{32}v'}{s}.$$

References

1. Triggs, B., McLauchlan, P.F., Hartley, R.I., Fitzgibbon, A.W.: Bundle adjustment – a modern synthesis. In: Triggs, B., Zisserman, A., Szeliski, R. (eds.) ICCV-WS 1999. LNCS, vol. 1883, pp. 298–372. Springer, Heidelberg (2000)
2. Björck, Å.: Numerical Methods for Least Squares Problems. SIAM, Philadelphia (1996)
3. Faugeras, O., Lustman, F.: Motion and structure from motion in a piecewise planar environment. Technical Report RR-0856, INRIA (1988)
4. Fischler, M., Bolles, R.: Random Sampling Consensus: a paradigm for model fitting with application to image analysis and automated cartography. Commun. Assoc. Comp. Mach. **24**, 358–367 (1981)
5. Fodor, B., Kazó, C., Zsolt, J., Hajder, L.: Normal map recovery using bundle adjustment. IET Comput. Vis. **8**, 66–75 (2014)
6. Furukawa, Y., Ponce, J.: Accurate, dense, and robust multi-view stereopsis. IEEE Trans. Pattern Anal. Mach. Intell. **32**(8), 1362–1376 (2010)
7. Habbecke, M., Kobbelt, L.: Iterative multi-view plane fitting. In: VMV06, pp. 73–80 (2006)
8. Habbecke, M., Kobbelt, L.: A surface-growing approach to multi-view stereo reconstruction. In: CVPR (2007)
9. Hartley, R.I., Sturm, P.: Triangulation. Comput. Vis. Image Underst. CVIU **68**(2), 146–157 (1997)
10. Hartley, R.I., Zisserman, A.: Multiple View Geometry in Computer Vision. Cambridge University Press, Cambridge (2003)
11. Kreyszig, E.: Differential Geometry. Dover Publications, New York (1991)
12. Liu, H.: Deeper Understanding on Solution Ambiguity in Estimating 3D Motion Parameters by Homography Decomposition and its Improvement. Ph.D. thesis, University of Fukui (2012)
13. Malis, E., Vargas, M.: Deeper understanding of the homography decomposition for vision-based control. Technical Report RR-6303, INRIA (2007)
14. Tanacs, A., Majdik, A., Molnar, J., Rai, A., Kato, Z.: Establishing correspondences between planar image patches. In: International Conference on Digital Image Computing: Techniques and Applications (DICTA) (2014)
15. Woodham, R.J.: Photometric stereo: A reflectance map technique for determining surface orientation from image intensity. In: Image Understanding Systems and Industrial Applications, Proceedings SPIE, vol. 155, pp. 136–143 (1978)
16. Yu, G., Morel, J.-M.: ASIFT: an algorithm for fully affine invariant comparison. Image Processing On Line, 2011 (2011)
17. Megyesi, Z., K'os, G., Chetverikov, D.: Dense 3d reconstruction from images by normal aided matching. Mach. Graph. Vis. **15**, 3–28 (2006)

Semi-automatic Hand Annotation
of Egocentric Recordings

Stijn De Beugher[1]($\boxtimes$), Geert Brône[2], and Toon Goedemé[1]

[1] EAVISE, ESAT, KU Leuven, Sint-Katelijne-Waver, Belgium
{stijn.debeugher,toon.goedeme}@kuleuven.be
[2] MIDI Research Group, KU Leuven, Leuven, Belgium
geert.brone@kuleuven.be

Abstract. We present a fast and accurate algorithm for the detection of human hands in real-life 2D image sequences. We focus on a specific application of hand detection, viz. the annotation of egocentric recordings. A well known type of egocentric camera is the mobile eye-tracker, which is often used in research on human-human interaction. Nowadays, this type of data is typically annotated manually for relevant features (e.g. visual fixations of gestures), which is a time-consuming and error-prone task. We present a semi-automatic approach for the detection of human hands in images. Such an approach reduces the amount of manual analysis drastically while guaranteeing high accuracy. In our algorithm we combine several well-known detection techniques together with an advanced elimination scheme to reduce false detections. We validate our approach using a challenging dataset containing over 4300 hand instances. This validation allows us to explore the capabilities and boundaries of our approach.

Keywords: Eye-tracking · Ego-centric · Annotation · Hand detection · Human-human interaction · (Semi-)automatic analysis

1 Introduction

Our motivation for developing a highly accurate hand detector comes from the wide applicability in a variety of disciplines including computer science, linguistics, sociology and psychology. Practical applications for such a technique include human-computer and human-robot interaction, gesture detection, automatic sign language translation, active gaming, etc. Detection of human hands in real-life images is an extremely challenging task due to their varying shape, orientation and position. Recently, several highly accurate hand detection algorithms were developed for 3D images [19]. Hand detection in 2D images, however, is far from a trivial task due the lack of depth context. Several attempts were made including skin-based detections [23], model-based detections [2,14,15] or pose estimation techniques [24]. Unfortunately when applied to real-life images, their performance drops significantly.

© Springer International Publishing Switzerland 2016
J. Braz et al. (Eds.): VISIGRAPP 2015, CCIS 598, pp. 338–355, 2016.
DOI: 10.1007/978-3-319-29971-6_18

On top of the challenging task we try to tackle, we aim to develop a generic method to achieve a high detection rate. It is well known that fully automatic approaches typically do not guarantee high accuracy in practical cases. However many applications could benefit from such a generic approach, e.g. the removal of privacy sensitive content such as faces in mobile mapping images, generation of ground-truth data, cartography by using object detection in aerial images, etc. To overcome this we expanded our framework with an intelligent mechanism which automatically demands for manual input when the confidence of a detection is below a threshold value. Using such an approach increases the detection rate significantly at the cost of a limited amount of manual interventions. For a certain target accuracy, our system computes the minimum amount of manual interactions.

In contrast to other techniques, we focus in this work on the detection of hands in video material. Using sequences of images gives us the opportunity to use the spatio-temporal relationship between consecutive frames to increase the detection rate. We use a 3-stage framework to generate the best possible result. First, we reduce the search space, using a human-torso detector. Second, we make a hypothesis using a sliding window approach of a hand model combined with a skin-based hand detection. Third, we use an advanced elimination approach to remove false detections in combination with a tracker resulting in reliable detections.

Fig. 1. Illustration of human-human interaction. Red dot is the position where the wearer of the mobile eye-tracker is looking at (Color figure online).

To validate our framework, we present a (semi-)automatic analysis of mobile eye-tracker data in the context of human-human interaction studies. The analysis of these data generally requires substantial manual annotation work [1,3,11,12]. The eye-tracking community would greatly benefit from the implementation of techniques that reduce the manual annotation load, like e.g. the detection of gesture strokes [11] and body language categorization [22]. The presented framework aims to contribute to these developments and proposes a technique to (semi-) automatically detect hands in video data recorded by a mobile eye-tracker. By mapping eye gaze data on interlocutors' body parts that are instrumental to face-to-face communication (like hands and faces), a first step in the analytical process is realized, as it allows for basic calculations of visual distribution. These data can then serve as the basis for further analytical work (e.g. the analysis of visual fixations on certain gesture types). An illustration of human-human interaction is given in Fig. 1, where an object is passed on. The red dot is the position where the giver is looking at, namely, the hand of the receiver, wearing an egocentric camera. An important research question that can be answered with this kind of experiments is if, when and how long persons look at their own hands when receiving a given object.

Next to a fast and accurate hand detection framework, an important contribution of this chapter is a generic (semi-)automatic detection approach. Furthermore, during our study, we noticed that it is hard to find fully annotated video material of human hands in real life recordings. Therefore we made our annotated dataset of eye-tracker recordings publicly available as described in [7]. The original dataset contained 1000 frames. In this paper we extended the original dataset with another challenging sequence of 1200 annotated frames. Thus the total dataset contains three sequences in which approximately 2200 frames were annotated[1].

This chapter is organised as follows: In Sect. 2, we discuss related work on hand detection. Section 3 clarifies our hand detection framework in detail. In Sect. 4 we discuss our novel (semi-)automatic approach in which a minimal manual intervention step enhances the detection rate. Finally in Sect. 5 we present the results on a pre-existing dataset and on our publicly available eye-tracker recordings that were performed to validate the approach.

2   Related Work

In recent years several attempts have been made to develop an accurate hand detector for 2D images, mostly by decreasing the complexity of the problem. Examples are the use of artificial markers e.g. coloured gloves [21] or using a static camera enabling the use of background segmentation [17]. In this chapter however, we focus on real-life applications where unmarked body parts need to be detected automatically, and therefore we only review the most popular methods that are applicable to natural settings.

A well known object detection technique is based on *Haar-like features* [20]. This technique combines a set of weak classifiers to build a final strong classifier and uses a sliding window approach to search for specific patterns in the image.

---

[1] http://www.eavise.be/insightout/Datasets/.

  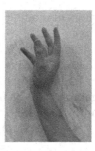

Fig. 2. Illustration of the hand models. The left image is the HOG representation of the hand model. The middle image illustrates the hand model, while the right image is an illustration of the context model (hand and its surrounding region including the background and wrist).

In [2] this technique is used as a basis for a hand detection algorithm, in combination with a skin detector to eliminate non-hand detections. Unfortunately the performance of this technique on unconstrained images is insufficient. Newer detectors outperform greatly Haar-based techniques.

A second approach is based on the *Deformable Parts Model* [10], which is an extension of *Histograms of Oriented Gradients* (HOG) [5]. This approach allows for the definition of a model of an object which is invariant to various postures or viewing angles. In [15] this technique is used to create two models of a hand, both with and without its surrounding region, e.g. the wrist (see Fig. 2). In addition, they use a skin detector based on the average skin colour of the face. This skin detector is used to improve the detection rate by searching for arms in the image. Finally a super-pixel based non-maxima suppression (NMS) is used in which overlapping bounding boxes are suppressed. A drawback of this method is the high computational cost: processing a frame of 360×640 pixels takes up to 4 min, of which the greater part is spent on superpixel calculations. Another hand detection approach was presented by [18]. In this chapter an invariant Hough Forest detector was used, resulting in a robust detection of the hand locations. Nevertheless, in our application the detection of the hand orientation is also of great importance on top of the location itself. Therefore we can not use such a basic approach.

In [9] the human pictorial structure is used. This approach searches for limbs in a human torso using the spatial relation between them. This method performs well on larger body parts (such as arms or heads), whereas smaller parts (e.g. a hand) are much more challenging. There are two major drawbacks of this method: (a) the requirement that all body parts are visible in the image and (b) they have a limited set of body poses that are detectable.

A pose estimation algorithm is proposed by [24]. This method is highly accurate since it has several parts for each limb and uses contextual co-occurrence relations between them. This method is designed for static images and its accuracy decreases drastically when motion blur is present, caused by moving body parts. The authors also admit their model has difficulties with some body poses (e.g. raised arms).

Based on a comparison of the previously described techniques, we opted for the work of [15] as a starting point for our algorithm. This approach achieves decent accuracy and its source code is publicly available so we can easily compare our method against it. In the next section we discuss the modifications we made in order to improve the detection results drastically, and how we extended to video.

3   Hand Detection Framework

An overview of our hand detection algorithm is given in Fig. 3. The general idea is that we first detect a human torso in the image, giving a robust reference for the detection of smaller body parts. Next we detect the face resulting in an indication of the hand sizes. After that, we detect hands using a model introduced by [15] in combination with a skin-based detection. Then we apply an advanced elimination scheme in order to remove false detections. Finally we use a Kalman filter to track left and right hand using the spatial relationship of consecutive frames.

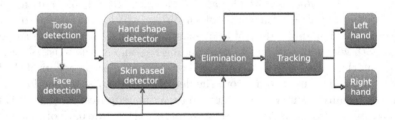

Fig. 3. Graphical representation of the proposed hand detection framework. The three stages: torso and face detection, hand detection and a combination of elimination and tracking.

3.1   Torso Detection

The first stage in our approach is the detection of a human torso, for which we use our own torso detector as we proposed in [6]. This torso detector is a part-based model [10], trained using only the upper 60 % of the labeled bounding boxes of human bodies of the standard PASCAL VOC dataset[2]. Using this model, rather than the more widely used full person detector, has the advantage that we can cope with images in which a person is not completely visible (from head to foot) such as, for example, in most of the images captured by a mobile eye-tracker (see Fig. 6) in a natural setting.

---

[2] The PASCAL Visual Object Classes Challenge 2009 (VOC2009) Dataset http://www.pascal-network.org/challenges/VOC/voc2009/workshop/index.html.

3.2 Face Detection

The next stage is a face detection step [20], which is used as a way to further improve the accuracy of the hand detections. In the work of [15], the face detection is only used for skin segmentation. If a face is detected, they apply a skin colour based proposal method to improve their detection results. In our approach on the other hand, we also make use of the proportions of the face by rejecting hand detections which have an abnormal size compared to the size of the face. This is based on the general rule that a human face has, about the same size as an outstretched human hand.

3.3 Hand Detection

When the torso and face location are known, we run our actual hand detection algorithm. Instead of searching for hands in the entire image, we define a search area by expanding the torso detection bounding box in both vertical and horizontal orientation. As mentioned before, we started from the work of [15]. This means we use the same part-based deformable model of a hand, as illustrated in the left part of Fig. 2. In their approach, an additional context model is used. However, the experiments we ran for this study showed that the addition of this model introduces a significant amount of false detections, so that we opted not to use it.

The hand model was developed to detect upstanding hands, but in real-life recordings any hand-orientation is possible. Therefore we rotate the enlarged region around the detected torso in steps of 10 degrees per rotation, as illustrated in Fig. 4, yielding an accurate detection of hands in any orientation. Using a larger step size decreases the computational cost, but also affects the accuracy of the detector as shown in Table 1. This table shows the performance of the hand model on a set of 100 annotated frames of 1280 × 720 pixels. To further decrease the computational cost related to this type of model evaluation, we used the acceleration approach of [8].

The hand model performs well as long as a hand is clearly visible in the image. However, when a hand is not visible or strongly deformed — for example due to motion blur caused by fast movements of the arms — these models show low detection rates. To overcome this problem, we developed an additional hand

Table 1. Accuracy of the hand model versus rotation angle of the images.

Step size	Precision	Recall	Time/frame
10 deg.	79,20 %	78,86 %	42 s
20 deg.	75,78 %	75,47 %	21 s
30 deg.	71,24 %	71,13 %	14 s
45 deg.	62,82 %	62,55 %	9,3 s
90 deg.	48,72 %	48,50 %	5 s

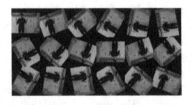

Fig. 4. Illustration of the rotation of our images in order to detect hands in any orientation. Left: step size is 10° per rotation. Right: step size is 20° per rotation.

detection technique as shown in Fig. 5. This technique segments the image in skin and no-skin based on three different colour spaces as introduced by [16]. In this work, skin colour is defined in both Red Green Blue (RGB), Hue Saturation Value (HSV) and Luma Chroma blue Chroma red (YCbCr) colour space resulting in a robust detection mechanism for skin, even under different lighting conditions. Using this approach is an improvement compared to the work of [15], because we no longer depend on the accuracy of the face detector for skin segmentation. We apply this segmentation to the stretched torso detection as shown in Fig. 5(b). Next, we skeletonize this result using a sequence of several erosion and dilation steps in order to get an accurate estimation of the skeleton, as illustrated in Fig. 5(c). In a following step, we apply the information obtained from the face detector. We use the correlation between the human body parts to classify the skeletonized image. If a skeletonized part has a length which is similar to the height of the face, we classify it as a hand (as illustrated by the top row in Fig. 5. Parts that are larger than a face are automatically treated as an arm (as illustrated by the bottom row in Fig. 5). For each part that is classified as an arm, we estimate a hand at both endpoints of the arm, as illustrated in Fig. 5(d). Estimated detections at the wrong endpoints are rejected using the elimination and tracking described in the next sections.

3.4   Elimination

After the above-mentioned steps, a large amount of hand detections is obtained, as seen in Fig. 6(a). The task of this elimination stage is to reject non-hand detections and to cluster overlapping detections. The output of this elimination operation is a reduced number of hand candidates as shown in Fig. 6(b). In our elimination process we apply the following steps:

– Remove hand detections which have an insufficient number of skin pixels, using the same skin detection algorithm as described in the previous step.
– Remove hand detections which have a divergent size with respect to the size of the face.

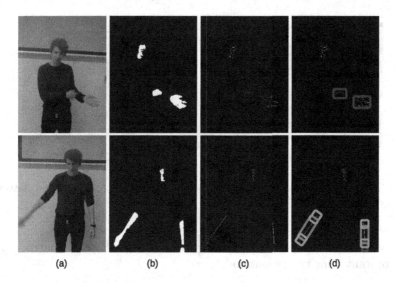

(a)   (b)   (c)   (d)

Fig. 5. From left to right: original image(a); binary image based on skin segmentation(b); skeletonization(c); arm and hand estimation(d). Purple boxes illustrate the hand classifications, blue boxes the arm detections and green boxes the estimated hands at the endpoints of the arm (Color figure online).

- Cluster overlapping detections based on their overlap and distance between their centers.
- Reduce the contribution of clusters that coincide with the face. We noticed that a face is often detected by the hand model. Only eliminating these detections is not a viable option since persons can hold their hands in front of the face. Therefore we reduce the score of those overlapping clusters by a predefined factor to minimize the impact.
- Remove hand detections which are too far from the predicted location by the Kalman trackers.

In the elimination step, we reduced the number of hand detections. Finally we classify the remaining detections in a left and right detection using the Kalman tracker information as explained in the next section.

3.5 Tracking

Our tracking stage is one of the most important contributions in order to improve the detection results. This is realized by steering the detections based on previous detections using a Kalman filter [13]. This mathematical filter is used to predict the position of the hands, which is needed when a detection is missing due to e.g. occlusions. A second advantage of using a Kalman filter is that the noise on the measured position of the detections is filtered out, resulting in more stable detections. For each torso detection we define two Kalman trackers: one for the left hand and one for the right hand in order to track each hand individually.

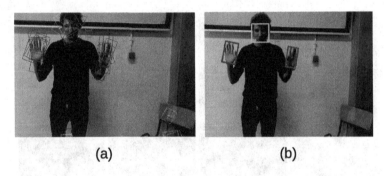

(a)   (b)

Fig. 6. Left: large amount of detections before elimination; Right: Final detections after elimination step.

We use a Kalman filter with the following state vector and update matrix, assuming a constant velocity motion model:

$$\mathbf{x} = \begin{bmatrix} x \\ y \\ v_x \\ v_y \end{bmatrix} \qquad A = \begin{bmatrix} 1\ 0\ 1\ 0 \\ 0\ 1\ 0\ 1 \\ 0\ 0\ 1\ 0 \\ 0\ 0\ 0\ 1 \end{bmatrix} \qquad (1)$$

where x and y are the position of the hand and v_x and v_y are the velocity of the hand. For each of the remaining clusters, as described in the previous section, we calculate the cost, based on the distance, to assign them to one of the Kalman trackers. By choosing the cluster with the lowest cost, we select the best candidate for each tracker.

To summarize this section we give an overview of our contributions as compared to the approach of [15]:

- Reduced computational footprint of our algorithm by avoiding both super-pixel calculation and the validation of the context model without loss in accuracy.
- Reduced search space by using a human-torso detector and only searching for hands in a region around the torso detection. This resulted in a reduced computational time and it reduced the number of false detections.
- Skin based detection is performed even when no face is detected, resulting in more detection candidates.
- Elimination of false-detections using the size of the face.
- Kalman tracker for both left and right hand that belongs to each torso detection.

4   Semi-automatic Analysis

As mentioned before, we aim to develop a framework that achieves a detection rate up to 100 %. Obviously it is unfeasible to develop an algorithm that achieves perfect accuracy on each dataset. Therefore we expanded our hand detection framework with a generic mechanism that allows for manual intervention resulting in a much higher accuracy. The key idea is that when the confidence drops under a specific (user-defined) threshold, our algorithm requests manual input. The user then has to manually annotate the missing detection. Relying only on the detection score results in a too large amount of manual interventions. To overcome this, we also take into account the distance between a detection and the predicted position (coming from the Kalman trackers). The formula of the confidence score is shown in Eq. 2:

$$M = \alpha log(D_{max} - D) + \beta S_i \qquad (2)$$

where:

$$D = \begin{cases} D_{max} - 1, & \text{if } d(C_i, C_{i-1}) \geq D_{max} \\ d(C_i, C_{i-1}), & \text{otherwise} \end{cases}$$

D_{max} stands for the maximum allowed distance between the current detection and a detection in the previous frame, C_i and C_{i-1} define respectively the center of the current and the previous detection. α and β are used to change the weight of the distance and detection score. In our experiments, we empirically determined the optimal value of those parameters: $\alpha = 0.5$ and $\beta = 1.0$.

The general concept of this approach is that a detection is likely to be valid if either the distance to the predicted location (based on previous detections) is low or if the detection score is high. If this value is below a user-defined threshold, manual input is requested. Thus by varying this threshold we can change the amount of manual interventions from zero (fully automatic detection) up to the number necessary to achieve full accuracy ((semi-)automatic detection). As illustrated in Fig. 7, the user is requested to manually annotate the missing detections when confidence score M is below a certain threshold. After this manual intervention the state vector of the corresponding kalman tracker is reset, thus resulting in a stable reference point for further detections.

5   Experimental Results

As mentioned in the introduction, we validate our hand detection framework using a data set of recordings. First we introduce our dataset, next we discuss the accuracy of our framework compared to other techniques.

5.1   Datasets

During our research we noticed that it is very hard to find video material containing hand annotations for each frame. In [15] a dataset of annotated movie frames is presented. Unfortunately, the available frames are not consecutive, which makes them unsuitable for our approach, designed for a sequence of frames. We also examined some video recordings from the MPI archive[3], but those were

---

[3] http://corpus1.mpi.nl.

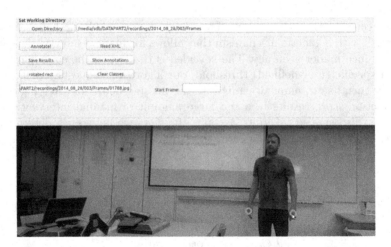

Fig. 7. Interface for manual intervention in which one can manually annotate the detection items.

annotated in terms of gestures (start and endpoint of the gesture) and contain no additional information of hand locations.

To overcome the lack of fully annotated video material, we set up a series of recordings. In each recording a mobile eye-tracker was used to capture the field of view of the test person. This eye-tracker records images at a resolution of 1280×720 pixels. In the first recording two persons stood face-to-face at a distance of 3 meters from each other. The person who wore the eye-tracker was told to look attentively at the interlocutor while this person made movements with his hands. The second recording was performed in a more natural setting. In this experiment, a PowerPoint presentation was given with the spectator wearing a mobile eye-tracker used as recording device. The third recording was conducted in the same setting as the second one, but another spectator was involved. This last sequence is a more challenging sequence since one of the hands are often occluded by furniture on the table. An illustration of these recordings is given in Fig. 8.

For each recording we manually annotated left and right hand in a sequence of consecutive frames. The annotations of the first two sequences consist of a bounding rectangle oriented with respect to the wrist. For practical reasons, the annotations of the last sequence consist of a single point for each hand. For the first sequence we chose a sequence in which a lot of arm- and hand movement is present. In total we have annotated a sequence of 403 consecutive frames. We chose a specific sequence of the second recording in which visual interaction between spectator and presenter was present, resulting in 491 consecutive frames. Finally, in the third sequence we annotated the first 1300 frames of the recording. This third sequence is an extremely difficult set for hand-tracking since the hands are often occluded by furniture. We specifically included this set since, due to its challenging nature, it fully exploits our algorithm and reveals it shortcomings.

Fig. 8. Illustration of all the datasets used in this chapter. First three images are frames from our own recorded sequences, last image is a frame from the 5-Signers dataset.

This results in a reference dataset of 4388 annotated hand instances, which can be used as reference dataset for benchmark tests. Since it is hard to find publicly available hand-annotated video material, we made our dataset publicly available[4] for other researchers. We plan to further expand this dataset in the future.

Next to our own datasets, we also found a small publicly available dataset, viz. the '5-signers' dataset [4]. This dataset contains time-series data of the hand positions collected from 5 signers during performance of sign language. Each of the signer sequences contains 39 frames resulting in 390 annotated hand-instances. An illustration of this dataset is given in the right image of Fig. 8.

5.2 Results

To validate our framework, we have performed a series of experiments. First we tested our hand detection algorithm without tracking of the hands nor manual intervention. We did this experiment on the first two sequences of our own dataset and the '5-signers' dataset. Examples of the detections on those datasets are shown in Figs. 9 and 10 respectively. The validation is done using the F-measure:

$$F = \frac{2TP}{2TP + FP + FN} \tag{3}$$

In each frame of our sequences one person and two hands are visible. Since our framework was designed to detect two hands for each torso instance, the number of false positives (FP) and false negatives (FN) are equal, hence the F-measure is reduced to the precision. A hand detection is considered valid if it is within half hand width from the ground-truth location of the hand. We compare the results to the performance of two state of the art techniques. The publicly available hand detection algorithm of [15] was used in which we use the two best detection scores as candidates for left and right hand. We also compare to the pose estimation proposal of [24] in which we classify the outermost bounding boxes of the arms as hands.

---

[4] http://www.eavise.be/insightout/Datasets/.

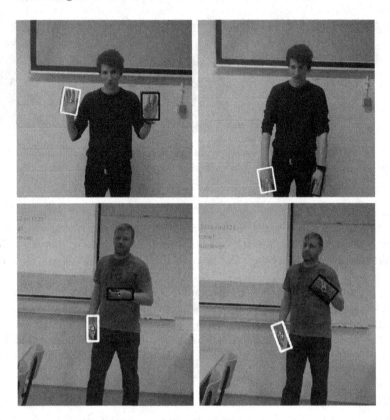

Fig. 9. Examples of hand detections on our own recorded sequences. Top row are images from our first sequence, bottom row are images from our second sequence.

Table 2. Accuracy of our hand detection algorithm compared to other techniques. Sequence 1 & 2 contains 1000 annotated hand-instances each, the '5-Signer' dataset contains 390 hand-instances.

	Mittal [15]	Yang [24]	Ours	Ours incl. tracking
Sequence 1	85 %	24.2 %	**83.4 %**	**88.2 %**
Sequence 2	48.9 %	46.5 %	**52.9 %**	**65.3 %**
5-Signers [4]	77.6 %	n.a.	**81.1 %**	**n.a.**

We compared our hand detection algorithm with tracking of the hands to the other techniques. Ours performs better than the other techniques in terms of accuracy. We outperfom the pose estimation technique, although a note on the bad performace of the approach of [24] should be made. The detection code we have used was developed to detect poses of persons from head to foot, whereas in the images of sequence 1 the legs of the person are not completely visible as shown in the first image of Fig. 8. The results of this comparison are shown in

Fig. 10. Examples of hand detections on the 5-Signer dataset.

Table 3. Execution times per frame averaged over all frames.

	Mittal [15]	Yang [24]	Ours
Avg time/frame	293.33 s	113 s	**36.67 s**

Table 2. We did those experiments on our own dataset, since we need sequences of frames. It is clear that the accuracy increases significantly when the tracking is applied, as shown in the right column of Table 2.

We also compared the execution speed of our algorithm, as shown in Table 3. It is clear that the execution time of our algorithm is drastically lower compared to the other techniques on the same hardware (Intel Xeon E5645). Our approach is much faster compared to the work of [15] since amongst others we no longer depend on the superpixel calculation. We also outperform the computational cost of [24] by a factor of 3.

Furthermore we present the extensive results of our (semi-)automatic approach on our own dataset as shown in Fig. 11. In this graph we plot the accuracy in function of the number of manual interventions expressed in a percent of the numbers of frames in the sequence. As mentioned before, by thresholding the result of Eq. 2, we can change the amount of necessary manual interventions. It is obvious that a higher amount of manual interventions results in a higher accuracy. We should also note the improvement in accuracy between no manual intervention and the lowest amount of manual interventions. For sequence 1, the accuracy increases from 90 % to 93 % at the cost of only 7 manual interventions, sequence 2 on the other hand has an accuracy improvement of 12 % at the cost of only 14 manual interventions. The result on sequence 3 indicates that a minimum amount of manual intervention is required in order to get a

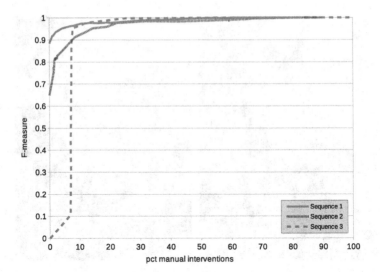

Fig. 11. Result of our (semi-)automatic approach in which accuracy is improved by manual interventions.

decent accuracy. This is caused by the complex setting of the recording. When we take a look at this sequence, we notice that the hands are often occluded as can be seen in Fig. 12. Those occlusions introduce wrong detections of the hand models. This is a perfect example in which we show the full potential of our semi-automatic approach. Without manual intervention, the accuracy on the third sequence is very bad, but when we introduce some manual intervention the accuracy increases significantly. The manual annotation of only 7.7 % of the frames results in an accuracy of 94.9 %. We observe that each manual

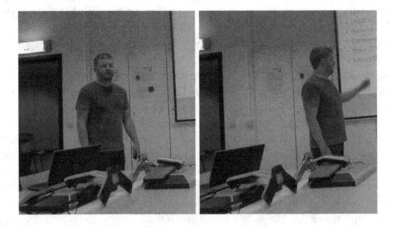

Fig. 12. Long standing occlusions in the third sequence resulting in a minimum amount of manual interventions in order to get a decent accuracy.

intervention restarts the tracker such that the hands in the following frames are again detected automatically.

6   Conclusion and Future Work

In order to provide an alternative for the manual annotation of human hands in videos, we present a semi-automatic approach. We build on the work of [15] and improved the accuracy and even decreased the computational cost. These improvements are realized by: (a) using a torso detector to reduce the search area (b) make use of a face detector whose detection is used to reject false hand detections (c) integration of an accurate skin-based hand and arm detection mechanism which is especially used in images where motion blur occurs (d) an advanced elimination scheme that is used to reject erroneous hand detections combined with a smart tracking mechanism for both left and right hands. We extended our approach with a generic mechanism that finds the optimal place to ask for manual intervention resulting in a much higher accuracy with minimal manual effort. This approach is based on a confidence score that is calculated using the detection score and the distance to the predicted detections. By thresholding this confidence score, we can change the amount of manual interventions. For validation we use three own recorded sequences, which we made publicly available, and one preexisting dataset. We report good accuracy and performance on several image sets as compared to state-of-the-art techniques. The third sequence is more complex than the others since hands are often occluded by furniture. When we apply our algorithm to that particular sequence, we show the full potential of our approach. The accuracy is very low without manual intervention, but when we apply a limited amount of manual interventions, the accuracy increases significantly.

Our future work concentrates on further reducing the computational cost of the hand detection algorithm. However we have realized a significant improvement in the processing time, our approach remains slow for practical use. We will investigate possibilities to reduce the computational cost of the model based detection. A first method is for example to apply a limited number of rotation angles around the angle of the previous detection. Next we will implement the possibility to indicate whether a hand is invisible (e.g. occluded by furniture). Such an indication could prevent that our system keeps searching for hands while no hands are visible. Furthermore we will work on the integration of the eye gaze data. Using such an approach enables the (semi-)automatic analysis of mobile eye-tracker data in terms of visual fixations on hands and reduce the manual workload related to this type of analysis.

Acknowledgements. This work is partially funded by KU Leuven via the projects Cametron and InSight Out. We also thank Raphael Den Dooven for his contributions.

References

1. Al Moubayed, S., Edlund, J., Gustafson, J.: Analysis of gaze and speech patterns in three-party quiz game interaction. In: Interspeech 2013 (2013)
2. Bo, N., Dailey, M.N., Uyyanonvara, B.: Robust hand tracking in low-resolution video sequences. In: Proceedings of the Third Conference on IASTED International Conference: Advances in Computer Science and Technology, Anaheim, CA, USA, pp. 228–233 (2007)
3. Brône, G., Oben, B.: Insight interaction. a multimodal and multifocal dialogue corpus. Lang. Resour. Eval. **49**(1), 195–214 (2014)
4. Buehler, P., Everingham, M., Huttenlocher, D., Zisserman, A.: Long term arm and hand tracking for continuous sign language tv broadcasts. In: Proceedings of the British Machine Vision Conference, pp. 110.1–110.10. BMVA Press (2008)
5. Dalal, N., Triggs, B.: Histograms of oriented gradients for human detection. In: CVPR, pp. 886–893 (2005)
6. De Beugher, S., Brône, G., Goedemé, T.: Automatic analysis of in-the-wild mobile eye-tracking experiments using object, face and person detection. In: Proceedings of the 9th International Joint Conference on Computer Vision, Imaging and Computer Graphics Theory and Applications (2014)
7. De Beugher, S., Brône, G., Goedemé, T.: A case study on real life mobile eye-tracker data. In: Proceedings of the 10th International Joint Conference on Computer Vision, Imaging and Computer Graphics Theory and Applications (2015)
8. Dubout, C., Fleuret, F.: Exact acceleration of linear object detectors. In: Fitzgibbon, A., Lazebnik, S., Perona, P., Sato, Y., Schmid, C. (eds.) ECCV 2012, Part III. LNCS, vol. 7574, pp. 301–311. Springer, Heidelberg (2012)
9. Eichner, M., Marin-Jimenez, M., Zisserman, A., Ferrari, V.: 2D articulated human pose estimation and retrieval in (almost) unconstrained still images. Int. J. Comput. Vis. **99**, 190–214 (2012)
10. Felzenszwalb, P.F., Girshick, R.B., McAllester, D., Ramanan, D.: Object detection with discriminatively trained part-based models. IEEE Trans. Pattern Anal. Mach. Intell. **32**(9), 1627–1645 (2010)
11. Gebre, B.G., Wittenburg, P., Lenkiewicz, P.: Towards automatic gesture stroke detection. In: The Eighth International Conference on Language Resources and Evaluation, pp. 231–235 (2012)
12. Jokinen, K.: Non-verbal signals for turn-taking and feedback. In: Proceedings of the Seventh International Conference on Language Resources and Evaluation (2010)
13. Kalman, R.: A new approach to linear filtering and prediction problems. Trans. ASME J. Basic Eng. **82**, 35–45 (1960)
14. Karlinsky, L., Dinerstein, M., Harari, D., Ullman, S.: The chains model for detecting parts by their context. In: 2010 IEEE Conference on Computer Vision and Pattern Recognition (CVPR), pp. 25–32 (2010)
15. Mittal, A., Zisserman, A., Torr, P.: Hand detection using multiple proposals. In: Proceedings of the British Machine Vision Conference, pp. 75.1–75.11. BMVA Press (2011)
16. Abdul Rahim, N.A., Wei, K.C., See, J.: RGB-H-CbCr skin colour model for human face detection. In: MMU International Symposium on Information and Communications Technologies (M2USIC), Petaling Jaya, Malaysia (2006)
17. Pfister, T., Charles, J., Everingham, M., Zisserman, A.: Automatic and efficient long term arm and hand tracking for continuous sign language TV broadcasts. In: British Machine Vision Conference (2012)

18. Spruyt, V., Ledda, A., Philips, W.: Real-time, long-term hand tracking with unsupervised initialization. In: Proceedings of the IEEE International Conference on Image Processing, pp. 3730–3734. IEEE (2013)
19. Van den Bergh, M., Van Gool, L.: Combining rgb and tof cameras for real-time 3d hand gesture interaction. In: Proceedings of the 2011 IEEE Workshop on Applications of Computer Vision (WACV), WACV 2011, pp. 66–72. IEEE Computer Society, Washington, DC (2011)
20. Viola, P., Jones, M.: Rapid object detection using a boosted cascade of simple features. In: IEEE Computer Society Conference on Computer Vision and Pattern Recognition, pp. 511–518 (2001)
21. Wang, R.Y., Popović, J.: Real-time hand-tracking with a color glove. In: ACM SIGGRAPH 2009 Papers, pp. 63:1–63:8 (2009)
22. Williams, G., Bregler, C., Hackney, P., Rosenthal, S., Mcdowall, I., Smolskiy, K.: Body signature recognition (2008)
23. Wu, Y., Liu, Q., Huang, T.S.: An adaptive self-organizing color segmentation algorithm with application to robust real-time human hand localization. In: Proceedings of Asian Conference on Computer Vision, pp. 1106–1111 (2000)
24. Yang, Y., Ramanan, D.: Articulated pose estimation with flexible mixtures-of-parts. In: 2011 IEEE Conference on Computer Vision and Pattern Recognition (CVPR), pp. 1385–1392. IEEE (2011)

Pedestrian Detection and Tracking in Challenging Surveillance Videos

Kristof Van Beeck[✉] and Toon Goedemé

EAVISE - Campus De Nayer, KU Leuven, De Nayerlaan 5,
2860 Sint-Katelijne-Waver, Belgium
{kristof.vanbeeck,toon.goedeme}@kuleuven.be

Abstract. In this chapter we propose a novel approach for real-time robust pedestrian tracking in surveillance images. Typical surveillance images are challenging to analyse since the overall image quality is low (e.g. low resolution and high compression). Furthermore often birds-eye viewpoint wide-angle lenses are used to achieve maximum coverage with a minimal amount of cameras. These specific viewpoints make it unfeasible to directly apply existing pedestrian detection techniques. Moreover, real-time processing speeds are required. To overcome these problems we introduce a pedestrian detection and tracking framework which exploits and integrates these scene constraints to achieve high accuracy results. We performed extensive experiments on publically available challenging real-life video sequences concerning both speed and accuracy. Our approach achieves excellent accuracy results while still meeting the stringent real-time demands needed for these surveillance applications, using only a single-core CPU implementation.

Keywords: Pedestrian detection · Tracking · Surveillance · Computer vision · Real-time

1 Introduction

Reliable pedestrian detection and tracking in surveillance images opens up a wide variety of applications (e.g. abnormal behaviour detection, path prediction, intruder detection, people safety on e.g. movable bridges and crowd counting). In recent years, tremendous advances concerning pedestrian detection were published. Current state-of-the-art detectors achieve excellent accuracy results on publicly available datasets (see Sect. 2). Unfortunately, directly applying these existing techniques on challenging surveillance images is not a trivial task. This is due to the inherent nature of these surveillance applications; often a large number of cameras are utilised since large areas need to be covered completely. Such scenarios impose severe constraints on the hardware: low-cost cameras are employed with wide-angle lenses, mounted high in a partly down-looking birds-eye view. Consequently image processing and analysis on these images is challenging.

Indeed, typical surveillance images are often captured at low-resolution and use high compression. Classic background subtraction based object detection

© Springer International Publishing Switzerland 2016
J. Braz et al. (Eds.): VISIGRAPP 2015, CCIS 598, pp. 356–373, 2016.
DOI: 10.1007/978-3-319-29971-6_19

Fig. 1. Example frame of one of the sequences of the CAVIAR dataset [8].

methods yield very noisy results at these high compression ratios. Moreover, these techniques do not differentiate between people and other objects. Due to their specific viewpoint (and wide-angle lens) standard pedestrian detectors - which are trained and evaluated on forward-looking images - are also unable to give accurate detection results on these images. Additionally, due to perspective effects some pedestrians to be detected appear very small in the image, which remains one of the most challenging tasks for current pedestrian detectors [15]. Furthermore, real-time processing speeds are required.

In this chapter we propose a flexible and fast pedestrian detection and tracking framework specifically addressing these challenging surveillance images. See Fig. 1 for a typical example frame of the publicly available surveillance dataset we used [8]. Our approach achieves excellent accuracy results at real-time processing speeds. We overcome the above mentioned challenges by the integration of three modalities: foreground segmentation approaches, the exploitation of scene constraints and an accurate pedestrian detector. This is done as follows. First, candidate regions in the image are generated. Using a calibrated scene distortion model, an early rejection of false patches is achieved. Next the candidate regions are warped to a standard viewing angle and used as input for a state-of-the-art pedestrian detector. As explained in Sect. 3 our approach allows for the use of a highly accurate pedestrian detector which would otherwise be too computationally intensive for real-time applications. Finally, the detections are employed in a *tracking-by-detection* approach to further increase the accuracy. Note that, using our approach the actual scene calibration is trivial and easily performed. We demonstrate the effectiveness of our approach on challenging surveillance video sequences, and present extensive accuracy and speed results. Our approach is generalisable to other object classes. The remainder of this chapter is structured as follows. In Sect. 2 we discuss related work on this topic, and distinguish our approach from existing work. Section 3 presents our framework in detail. Next we propose experimental results on challenging sequences in Sect. 4. Finally, in Sect. 5 we conclude our work and give final remarks on future work.

2   Related Work

Pedestrian detection and tracking in general is a very active research topic. Dalal and Triggs [10] initially proposed the use of *Histograms of Oriented Gradients* (HOG) for pedestrian detection. Their insights paved the way for numerous derived approaches; even today most state-of-the-art pedestrian detectors still rely on HOG features albeit in a more subtle manner (e.g. in combination with other features). A well-known example is the work of [17]. As opposed to the rigid model introduced by Dalal and Triggs they propose the inclusion of parts (representing e.g. the limbs or head of a pedestrian) to increase detection accuracy, coined the *Deformable Part Models* (DPM). In later work the authors tackled the inevitable increase in computational complexity by introducing a cascaded approach in which a fast rejection of negative detection windows is possible [16]. An extension was proposed using grammar models to cope with partial occlusion [19]. Girshick and Malik [21] published a new and fast training methodology for DPM models. As opposed to enriching the model with parts, [13] introduced the use of a rigid model with additional features, called *Channel Features* (ChnFtrs).

All previously mentioned approaches employ a sliding window paradigm: to cope with scale variations a scale-space feature pyramid is calculated and each layer is evaluated at each location. To speedup detection [12] proposed an approach which approximates features nearby to avoid a full feature pyramid calculation. Several other techniques have been proposed to speedup detection: using model scaling in stead of image scaling, GPU implementations [2] and search space minimisation techniques [1,9,27]. For several years, the DPM approaches remained among the top performing methods [14,15]. However, the need of parts for pedestrian detection remains unclear [4]. Indeed, recent work on optimised rigid models - e.g. *Roerei* [3] and *ACF* [11] - in fact outperform the DPM detectors.

In [18] the authors present the use of *convolutional neural networks* (R-CNN) for object detection, achieving unprecedented state-of-the-art accuracy results. This methodology existed for a long time, but its applicability to image classification tasks was highlighted by the work of [23]. Interestingly, their method steps away from the traditional sliding window approach, and utilises region proposals as input for the deep learning classifiers. Although currently not real-time, their framework is able to classify a large variety of classes simultaneously, making it ideal for large image database retrieval applications such as ImageNetimage [30]. Recently [20] presented a hybrid approach combining DPMs with CNNs, called *DeepPyramid DPM*.

Several pedestrian tracking algorithms exist. Due to recent advances in object detection techniques, *tracking-by-detection* has become increasingly popular. There, an object detector is combined with a reliable tracking algorithm (e.g. particle filtering); see for example [7]. Concerning existing work on pedestrian tracking in surveillance images many either operate on standard viewpoint and/or high-resolution images [6,31], or employ thermal cameras to facilitate segmentation to reduce the search area [24].

In previous work we presented a real-time pedestrian detection framework for similar viewpoint images which are captured with a blind-spot camera mounted on a real truck [32]. These images are - apart from the viewpoint - challenging since the camera is moving. However, in this work we can fully exploit and integrate foreground segmentation methods to increase both accuracy and speed. Furthermore, we work with images captured from genuine surveillance cameras. These images are of low-resolution, low-quality and, due to the use of wide-angle lenses show large amounts of distortion and contain non-trivial viewpoints.

Existing work on the same dataset either employs clustering algorithms with GPU optimisation [25], or focusses on motion analysis by matching trained silhouette models [28]. We differ significantly from these previous works: we developed an accurate tracking framework in which we can employ a highly accurate pedestrian detector on these challenging images, and thus perform much better than existing methods. We achieve real-time processing speeds on a single-core CPU implementation. Our approach easily lends itself for multi-threaded implementation if higher computational speeds are needed.

3   Algorithm Overview

Running standard pedestrian detectors such as the *Deformable Part Models* on surveillance images as shown in Fig. 1 is unfeasible. Current pedestrian detectors are only trained on upright pedestrians at a fixed height. Scale invariance is achieved using a scale-space pyramid. Thus in order to achieve decent detections on these surveillance images the detectors ought to run on multiple rotations and scales of the same surveillance image, using both dense rotation and scale steps. Evaluating the total 4D rotation-scale search space in real-time evidently is impossible. Nonetheless, the use of a pedestrian detector could significantly increase the accuracy, as opposed to standard techniques which only rely on e.g. background subtraction with blob analysis due to time constraints. Therefore, to overcome these challenges we propose the integration of a foreground segmentation approach with a scene model and a highly accurate pedestrian detector. Our approach allows for the detection of pedestrians in challenging viewpoints (e.g. rotated) under large lens distortion at low computational complexity, with very high accuracy. To retrieve the scene model, a simple one-time calibration procedure is performed, no explicit lens or camera calibration is needed. Our algorithm briefly works as follows. As seen in Fig. 1, pedestrians appear rotated and scaled based on their position in the image. We exploit this scene knowledge throughout our detection and tracking pipeline. For each input image, after a preliminary segmentation, we generate region proposals which potentially contain pedestrians. The scene model is used to reduce the number of region proposals. Next, based on the position in the image we warp each valid potential region to an upright and fixed-height image patch. These patches are given as input to a state-of-the-art pedestrian detector, which evaluates a pedestrian model on a single scale only. This is the key advantage of our work: since only one scale and position needs to be evaluated we can use a highly accurate pedestrian detector which would otherwise be too time-consuming. Furthermore this

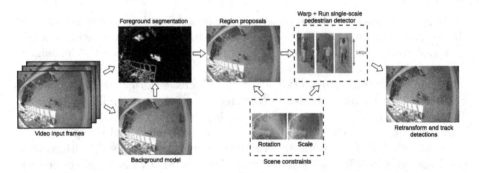

Fig. 2. Overview of our detection pipeline. After a first foreground segmentation step we extract region proposals which potentially contain pedestrians. Each region is warped to an upright fixed-height patch. Next, a highly accurate pedestrian detector is evaluated at a single scale. Finally, the detections are retransformed and tracked.

approach allows for the detection of extremely small pedestrians, if the detection model is powerful enough. The detections are retransformed to the original input image, and employed in a tracking-by-detection framework to associate pedestrian tracks and handle missing detections. Since each region can be evaluated independently, a fast multi-threaded implementation of this approach is trivial. Figure 2 shows an overview of our approach. In the next subsections we describe further details of each step in our pipeline, and motivate important design choices.

3.1 Foreground Segmentation

First we perform a foreground segmentation step to identify moving regions in the static camera images. Several segmentation approaches are applicable ranging from basic background subtraction methods to more advanced motion estimation methods. Since we employ scene constraints further on to reduce the number of region proposals, our approach allows for the use of a coarse segmentation. For this step we thus prefer low computational complexity over high accuracy, excluding time-consuming techniques (e.g. optical flow). Hence, we rely on background estimation techniques, which generate a statistical model of the scene. Several popular methods exist. Since a comprehensive comparison of these techniques is out of the scope of this work, we refer to [5,26] for a detailed overview. Concerning background subtraction, the main challenges in typical surveillance images arise from changing lightning conditions and camera shake. Based on these comparative works we opted for the method of [33], which employs *Gaussian Mixture Models* (GMM). These methods haven proven to cope well with (limited) background motion. Their proposed method is an extension of the original GMM where the number of Gaussian components per pixel is automatically selected. This effectively reduces memory requirements and increases the computation speed, making it ideal for this application. A qualitative segmentation output example is shown in the overview figure (Fig. 2).

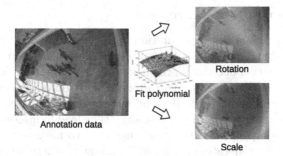

Fig. 3. A one-time calibration step is needed. The transformation parameters are extracted from the annotations.

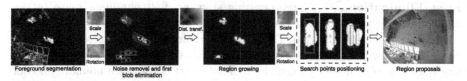

Fig. 4. Our region proposals pipeline. After foreground segmentation and noise removal a first blob elimination is performed. Next we perform region growing using a distance transform. Finally, we determine the optimal search points.

3.2 Modelling Scene Constraints

As previously mentioned, the pedestrians in the surveillance images appear rotated and scaled. Since the position of the surveillance camera is fixed with respect to the ground plane both parameters only depend on the position in the image. If we know the rotation and average pedestrian height for each pixel position $\mathbf{x} = [x, y]$ we can exploit this scene knowledge to achieve fast and accurate pedestrian detection, similar to [32]. During the generation of the region proposals this information can be used to reject regions which diverge too much from the expected region properties, thus limiting the search regions. For each valid proposed region, we use the transformation parameters to warp each patch to an upright, fixed-scale image patch, allowing the use of an accurate pedestrian detector whilst being real-time.

To retrieve these transformation parameters a one-time offline calibration needs to be performed (see Fig. 3). However, the scene calibration as proposed here is easy to perform and trivial. For this, we extracted the rotation and height of each annotated pedestrian from the dataset, giving the scale and rotation for that specific point. Next we interpolated the datapoints using a second order 2D polynomial function $f_i(\mathbf{x})$ for both parameters:

$$f_i(\mathbf{x}) = p_0 + p_1 x + p_2 y + p_3 x^2 + p_4 xy + p_5 y^2 \tag{1}$$

Both $f_{scale}(\mathbf{x})$ and $f_{rotation}(\mathbf{x})$ are used as *Lookup functions* (LUFs): at each position in the image they define the expected region properties and transformation parameters.

3.3   Generation of Region Proposals

In a next step we refine the segmentation and generate region proposals which need to be warped and evaluated using our single-scale pedestrian detector. Since we employ a pedestrian detector in the next stage to validate each region we are allowed to propose more regions than needed, i.e. regions without pedestrians. An accurate detector should indeed negatively classify such patches. However, it is important to early reject false patches, since they lead to useless computations and lower processing speeds. This stage thus tries to balance between optimal accuracy and speed, generating an optimal amount of search locations. Figure 4 gives an overview of our region proposal calculation. Let us now discuss each consecutive step in this pipeline.

First Elimination. As a preprocessing step, we first eliminate noise in the segmentation which remained after the background subtraction step (due to e.g. changing lightning conditions). This is simply done using *morphological opening*. Next, we perform a connected component analysis (using 8-connectivity), and test the local scene model for each blob. That is, we construct a bounding box of the expected scale and rotation around the centroid of each blob. We reject two types of regions: extremely small ones (25 pixels or less) due to the high SNR there (drawn in magenta in the second step of Fig. 4), and those that diverge from an area constraint (drawn in red). For this constraint, we require that the area of the connected component should be larger than a minimal percentage of the expected area (15 %). This step eliminates most invalid regions.

Region Growing. In the case of insufficient contrast, the foreground segmentation performs suboptimal (i.e. tends to split a valid pedestrian in multiple blobs, as seen for the largest pedestrian in Fig. 4). For each remaining valid region we therefore perform region growing based on the Euclidean distance transform, joining regions nearby. This has a second advantage: multiple pedestrians which are nearby are joined into a single detection region, even if one of them was removed after the first elimination. This is also illustrated in Fig. 4: after the first elimination only one of both small pedestrians is maintained. However, after region growing both are connected.

Defining Search Points. Finally, we define exact search locations where the pedestrian detector will be applied. This is done as follows. Each remaining region is again verified against the scene constraints since, due to the previous step, these regions could have grown significantly. This is the case when multiple (possibly previously invalid) regions are joined. Note that we do not reject regions at this stage. We locally evaluate each region and use the expected height and rotation to estimate the number of possible pedestrians. Based on the size of the region we first evaluate if multiple search points are necessary for this region. If so, we define a linear grid over the entire region of which the step size depends on the ratio of the expected and actual region parameters, and eliminate grid points which are located outside the segmented region.

The final region proposals are visualised as the green rectangles shown in the rightmost image in Fig. 4. As seen, our regions accurately predict possible pedestrians in the image. This is the power of this approach: by combining foreground segmentation and scene model constraints the search space for the computationally expensive pedestrian detector can be enormously restricted. Slight deviations from the exact pedestrian position are allowed since we employ a sliding-window approach in the final warped patch.

3.4 Warping Patches

Our scene model has another advantage: for each image location we know how a pedestrian is locally distorted. Each region proposal is warped to an upright pedestrian at a fixed-scale. Using this approach we are able to accurately detect even rotated and extremely small pedestrians, using a single-scale pedestrian detector only. The region proposals I are warped such that $I_{warp} = TI$ where transformation matrix T simply consists of a Euclidean transformation of which the parameters are extracted from the LUFs:

$$T = \begin{bmatrix} s\cos\theta & -s\sin\theta & t_x \\ s\sin\theta & s\cos\theta & t_y \\ 0 & 0 & 1 \end{bmatrix} \tag{2}$$

Note that the optimal scale to which the patches are warped highly depends on which pedestrian detector is used. This is dicussed in the next section, where we motivate the choice of pedestrian detector and determine the optimal scale.

3.5 Pedestrian Detector

The warped image patches are now classified by a pedestrian detector. In fact, the method described in the previous sections is generic and can be combined with each existing pedestrian detection algorithm. As discussed in Sect. 2, recent R-CNN based detection methods currently achieve top accuracy results concerning

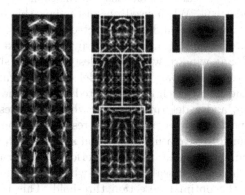

Fig. 5. Pedestrian model used in our implementation. (L) Root model. (M) Different parts. (R) Deformation costs.

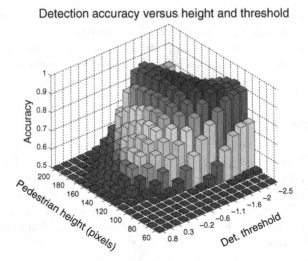

Fig. 6. The accuracy versus the pedestrian height and detection threshold for the single-scale cascaded DPM detector.

object detection in general. However, their performance is far from real-time, and they are more suited for multi-class large database retrieval tasks. Rigid pedestrian detectors (such as ChnFtrs) currently offer the best trade-off between speed and accuracy when a full-scale space pyramid needs to be constructed. However, since we need to evaluate a single scale only, no scale-space pyramid needs to be constructed. Therefore we are able to use an accurate pedestrian detector which would otherwise be too time-consuming, such as the *Deformable Part Models*. Moreover, since a rigid model does not allow for any deformation, using it in our single-scale approach is even unfeasible in a direct manner.

Since natural slight height variations exist between pedestrians (and due to small calibration errors), the detection accuracy significantly drops when using these models on a single-scale. Given this information, we opted to use the cascaded DPM model [16]. When used out-of-the-box this detector works as follows. First a scale-space pyramid is constructed in which for each layer HOG features are calculated resulting in a feature pyramid. Next, this pyramid is evaluated using a sliding window with the pedestrian model shown in Fig. 5.

A pedestrian is represented as a root model (left), several parts representing e.g. the limbs and head which are calculated at twice the resolution of the root model (middle), and a deformation cost which penalises large deviations from the expected part locations (right). The responses of both root filter and part filters are summed to give a final detection score. We altered this detector into a single-scale only implementation, and performed experiments to simultaneously determine the optimal scale factor to which the region proposals need to be warped, and the optimal detection threshold. This is done as follows. We extracted about 6000 annotated pedestrians from the CAVIAR dataset and warped them to different scales (heights). Combined with 6000 negative patches

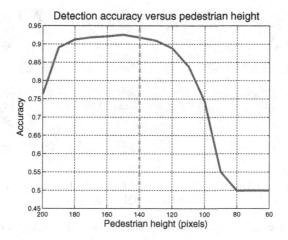

Fig. 7. The optimal threshold slice displaying the accuracy versus the pedestrian height.

we calculated the accuracy in function of the height and detection score threshold. The results are shown in Fig. 6.

As can be seen, at low resolutions the accuracy drops significantly, since only very limited spatial information is available. At high resolution similar behaviour is seen, since the pedestrians mismatch the detection model. Concerning the detection threshold, the detection accuracy is low at both high values (high *false negative rate*) and low values (high *false positive rate*). Figure 7 displays the optimal threshold slice extracted from Fig. 6. The accuracy is almost constant between 130–170 pixels. However, at larger pedestrian heights the detection time significantly increases. We therefore used 140 pixels as our optimal rescale height to which the region proposals will be warped such that a one-scale pedestrian model can be directly applied.

3.6 Tracking

The resulting detections are then retransformed to the input image coordinates. Next a *non-maxima suppression* step is performed, in which overlapping detections are filtered; only the highest scoring detection is kept. To link detections over multiple frames and to cope with occasional missing detections we integrate our approach in a *tracking-by-detection* framework. For this we employ the well-know Kalman filter [22]. For each new detection, a Kalman filter is initialised. We employ a constant velocity motion model. The state vector x_k thus consists of the centre of mass of each detection, the velocity and the scale: $x_k = \begin{bmatrix} x \; y \; v_x \; v_y \end{bmatrix}^T$. Our process matrix A thus equals:

$$A = \begin{bmatrix} 1 & 0 & 1 & 0 \\ 0 & 1 & 0 & 1 \\ 0 & 0 & 1 & 0 \\ 0 & 0 & 0 & 1 \end{bmatrix} \tag{3}$$

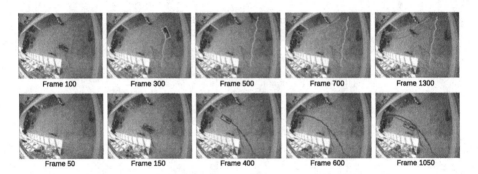

Fig. 8. Qualitative tracking example on two of the evaluation sequences (top and bottom row). See http://youtu.be/kWoKBPQoeQI for a video.

Table 1. Overview of the difference sequences of the CAVIAR dataset.

Scenario	# frames	Difficulty	Comments
Walk	3045	easy	Few people, low interaction
Browse	6654	medium	People browse at e.g. reception desk
Leave bags	5839	medium	Leaving objects behind
Rest	4220	medium	Resting on floor and in chairs
Fight	2492	difficult	People fighting. Difficult poses
Meet	4123	difficult	Group meetings, multiple occlusions

Using this motion model we predict the position of the pedestrians in the next frame. When a new frame is processed, we try to match each running tracker with a new detection as follows. We construct a circular region - based on the scale of that tracked detection - around the estimated new centroid. If the centroid of a new detection is found within that region, the detection is associated with this track, and the Kalman filter is updated. If multiple detections are found, we take the closest one based on the Euclidean distance. If no detection can be associated with a running track, we update the Kalman filter with its estimated position. If this occurs for multiple frames in a row, the track is discarded. For detections without an associated track, evidently a new Kalman filter is instantiated. Furthermore the exact size of the bounding boxes are averaged over multiple frames. See Fig. 8 for two qualitative tracking sequences of our proposed algorithm.

4   Experiments and Results

We performed extensive experiments concerning both speed and accuracy on the publicly available CAVIAR dataset [8]. This dataset was recorded at the entrance lobby of the INRIA labs with a wide-angle camera lens. The images are taken with a resolution of 384×288 at 25 frames per second, and are compressed using MPEG2. See Fig. 1 for an example frame. The dataset is divided into six

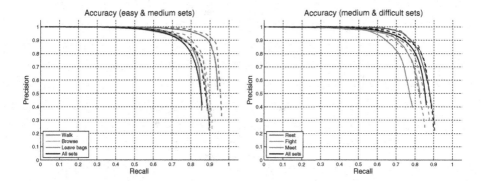

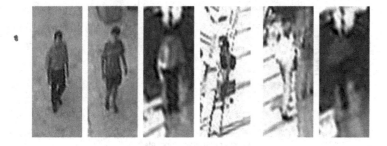

Fig. 9. The accuracy of our algorithm over the CAVIAR dataset. Solid lines indicate the results without tracking, dotted lines include tracking. The accuracy results for the six scenarios are divided over two graphs based on their difficulty for the sake of clarity. The black curve (*All sets*) indicates the average accuracy over all six scenarios.

Fig. 10. Example of warped annotations. Low-resolution and high-compression artifacts are noticeable.

different scenarios: *walk, browse, meet, leave bags, rest* and *fight*. Each scenario is again subdivided into multiple sequences, making a total of 28 sequences. We used all sets for testing. Note that some sequences contain pedestrians which are inherently undetectable with our proposed framework. For example, the *fight* sequences include scenarios with people in specific fighting poses, and the *rest* sequences contain scenarios where people fall on the floor or rest in e.g. chairs thus violating our scene constraints. Table 1 gives a textual overview of each scenario. For each scenario we give a *difficulty* measure, i.e. an indication of the complexity of the sequences of each scenario. Easy scenarios are composed of simple sequences with only few people and low interaction whereas difficult scenarios contain many occlusions and challenging poses. In total, our evaluation set consists of about 26400 frames, containing about 36200 annotations. Our algorithm is implemented in Matlab, with time-consuming parts (e.g. the detection and transformation) in C and OpenCV (using *mexopencv* as interface). Our test hardware consists of an Intel Xeon E5 CPU which runs at 3.1 GHz. All speed test are performed on a single CPU core. However, a multi-threaded CPU implementation to further increase the processing speed is trivial.

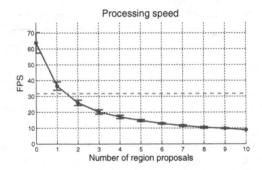

Fig. 11. The processing speed of our algorithm versus the number of region proposals (Color figure online).

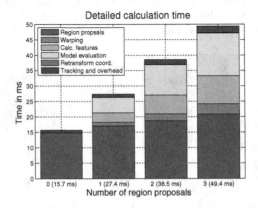

Fig. 12. An overview of the calculation time for each step in the algorithm versus the number of region proposals.

4.1 Accuracy

Figure 9 displays the accuracy results of our algorithm, using *precision-recall* curves. We give results for all scenarios mentioned above, ranging from easy (e.g. *walk* - limited number of persons) through difficult (e.g. *meet* - multiple persons with occlusions). For the sake of clarity we spread the accuracy results of the six sequences over two separate plots, based on their difficult. The left accuracy plot groups the easy and medium scenarios, the right plot gives the accuracy results for the more difficult scenarios. We exclude small pedestrians from the annotations (smaller than 20 pixels), and remove annotations in the top left corner of the image (on the balcony) and the bottom left corner of the image (people behind the covered reception desk). Furthermore we discard annotations close to the image border, since the pedestrians are not completely visible there (the annotation is strict and already starts when part of a pedestrian enters the frame). The solid lines in Fig. 9 indicate the accuracy without tracking, whereas the dotted lines show the accuracy with tracking. The black curves on

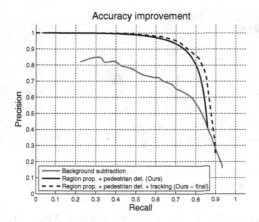

Fig. 13. The obtained accuracy improvement compared to a naive background subtraction approach.

both figures indicate the total average accuracy over the entire evaluation set (all six scenarios). To indicate the difficulty, in Fig. 10 we display some extracted annotations which are warped to an upright position.

As can be seen, these low-resolution output images contain severe compression artifacts. Even for humans they are sometimes difficult to recognize as a pedestrian. However, we achieve excellent accuracy results given these strict dataset annotations and challenging nature of these images. As observed, on some difficult scenarios (e.g. *Meet* and *Rest*) a lower accuracy is obtained.

This is mainly due to two reasons: these sets contain many long-term occlusions and poses which a standard pedestrian detector is unable to detect (e.g. sitting in a chair, lying on the floor). Since our tracker handles missing detections, the accuracy significantly improves.

4.2 Speed

The exact calculation time depends on the number of region proposals per image. Figure 11 therefore displays the speed of our algorithm (in frames per second), versus the number of region proposals.

Evidently, the processing speed decreases when multiple region proposals need to be evaluated. However, even at e.g. four region proposals we still achieve 17 fps. Over the entire evaluation set we achieve an average of 32 frames per second, indicated with the dotted red line. Note that all experimental results are performed on a single CPU core. In fact, each region proposal can be evaluated independently, thus allowing for an easy multi-threaded implementation.

Figure 12 visualises the individual calculation times for each important step in the entire algorithm pipeline, for a varying number of region proposals. As visualised, generation of the region proposals takes about 15–20 ms. The warping operation is very fast: on average 1 ms per region proposal is needed. Concerning the pedestrian evaluation step, the average feature calculation time per region

Fig. 14. Qualitative comparison between running a detector on all scales and rotations (left) versus the output of our algorithm (right).

is about 3 ms whereas the model evaluation takes 4 ms. The time needed to retransform the coordinates is negligible.

4.3   Comparative Evaluation

Figure 13 illustrates the accuracy improvement we achieved as compared to a basic background subtraction technique, i.e. interpreting the foreground blobs that are large enough as pedestrians. As seen, on these challenging images these naive methods yield poor results. The inclusion of our scene model and the application of a state-of-the-art pedestrian detector raises the accuracy enormously.

A quantitative comparison with other work using precision-recall curves on this dataset is difficult, since to the best of our knowledge no such accuracy results similar to our work exist. Existing work on these specific sequences of the CAVIAR dataset often focusses on activity recognition (e.g. *fight*) and anomaly detection. However, [29] present accuracy experiments using *tracking failure* measurements on 11 tracks of the CAVIAR dataset. For this, the authors consider a track lost if the tracking failed for 20 frames or more. In their work a multi-hypothesis tracking approach (particle filter) is used. They achieve a tracking failure percentage of 33.64 % with $N = 20$ particles and 16.82 % when $N = 50$. Using our approach we achieve a tracking failure of 9.1 % on the same sequences relying only on a single hypothesis tracker (Kalman filter). As a final qualitative analysis we compare our approach with a naive detection approach, that is running the standard deformable part model detector on all scales and all rotations.

For this, we need to upscale the image five times (the smallest pedestrian to be detected is only 25 pixels high, and the height of the detection model equals 120 pixels), and use a rotation step size of 10 degrees. Using this approach, the calculation time for a single frame increases to about 13 min. Figure 14 displays the detections found using this naive approach (left), and the output of our algorithm (right). As seen, the naive approach yields several false positives and fails to detect all pedestrians. Our algorithm achieves excellent accuracy results with minimal computational cost (89 ms for this frame).

5   Conclusions and Future Work

We presented a fast and accurate pedestrian detection and tracking framework targeting challenging surveillance videos. Our proposed algorithm integrates foreground segmentation methods with scene constraints to generate region proposals, which are then warped and evaluated by a single-scale pedestrian detector. Using this approach we can employ a highly accurate pedestrian detector for non-trivial camera-viewpoint images where existing pedestrian detectors fail, while still achieving real-time performance. We performed extensive evaluation experiments concerning both accuracy and speed on the publicly available CAVIAR dataset. This dataset consists of typical low-resolution high-compression surveillance images taken with a wide-angle lens from a challenging viewpoint. We show that our approach achieves both excellent accuracy and processing speeds using a single-core CPU implementation only. Furthermore, our proposed method easily lends itself for a multi-threaded implementation.

To improve the detection accuracy on very difficult scenarios (e.g. long-term occlusions, people in chairs or people lying on the floor) several further optimisations are possible. To cope with challenging poses an upperbody detector or an evaluation at multiple rotations could be employed. For this, the rotation should be included in the tracker. Additional features (e.g. color information) could be used to enable person reidentification. Furthermore, the scene calibration currently is based on annotation data. In the future we plan to investigate if an automated calibration method can be implemented (using e.g. an offline exhaustive search over all scales and rotations).

Acknowledgements. The authors would like to acknowledge that the dataset used here is from the EC Funded CAVIAR project/IST 2001 37540 [8].

References

1. Benenson, R., Mathias, M., Timofte, R., Van Gool, L.: Fast stixel computation for fast pedestrian detection. In: Fusiello, A., Murino, V., Cucchiara, R. (eds.) ECCV 2012 Ws/Demos, Part III. LNCS, vol. 7585, pp. 11–20. Springer, Heidelberg (2012)
2. Benenson, R., Mathias, M., Timofte, R., Van Gool, L.: Pedestrian detection at 100 frames per second. In: Proceedings of CVPR, pp. 2903–2910 (2012)
3. Benenson, R., Mathias, M., Tuytelaars, T., Van Gool, L.: Seeking the strongest rigid detector. In: Proceedings of CVPR, Portland, Oregon, pp. 3666–3673 (2013)
4. Benenson, R., Omran, M., Hosang, J., Schiele, B.: Ten years of pedestrian detection, what have we learned? In: Agapito, L., Bronstein, M.M., Rother, C. (eds.) ECCV 2014 Workshops. LNCS, vol. 8926, pp. 613–627. Springer, Heidelberg (2015)
5. Benezeth, Y., Jon, P.-M., Emile, B., Laurent, H., Rosenberger, C.: Review and evaluation of commonly-implemented background subtraction algorithms. In: 19th International Conference on Pattern Recognition, ICPR 2008, pp. 1–4. IEEE (2008)
6. Benfold, B., Reid, I.: Stable multi-target tracking in real-time surveillance video. In: CVPR, pp. 3457–3464 (2011)

7. Breitenstein, M.D., Reichlin, F., Leibe, B., Koller-Meier, E., Van Gool, L.: Online multiperson tracking-by-detection from a single, uncalibrated camera. IEEE PAMI **33**(9), 1820–1833 (2011)

8. CAVIAR project: Context aware vision using image-based active recognition. http://homepages.inf.ed.ac.uk/rbf/CAVIAR/

9. Cho, H., Rybski, P., Bar-Hillel, A., Zhang, W.: Real-time pedestrian detection with deformable part models. In: IEEE Intelligent Vehicles Symposium, pp. 1035–1042 (2012)

10. Dalal, N., Triggs, B.: Histograms of oriented gradients for human detection. In: Proceedings of CVPR, vol. 2, pp. 886–893 (2005)

11. Dollár, P., Appel, R., Belongie, S., Perona, P.: Fast feature pyramids for object detection (2014)

12. Dollár, P., Belongie, S., Perona, P.: The fastest pedestrian detector in the west. In: Proceedings of BMVC, pp. 68.1–68.11 (2010)

13. Dollár, P., Tu, Z., Perona, P., Belongie, S.: Integral channel features. In: Proceedings of BMVC, pp. 91.1–91.11 (2009)

14. Dollár, P., Wojek, C., Schiele, B., Perona, P.: A benchmark. In: Proceedings of CVPR, pp. 304–311 (2009)

15. Dollár, P., Wojek, C., Schiele, B., Perona, P.: Pedestrian detection: an evaluation of the state of the art. IEEE PAMI **34**, 743–761 (2012)

16. Felzenszwalb, P., Girschick, R., McAllester, D.: Cascade object detection with deformable part models. In: Proceedings of CVPR, pp. 2241–2248 (2010)

17. Felzenszwalb, P., McAllester, D., Ramanan, D.: A discriminatively trained, multi-scale, deformable part model. In: Proceedings of CVPR (2008)

18. Girshick, R., Donahue, J., Darrell, T., Malik, J.: Rich feature hierarchies for accurate object detection and semantic segmentation. In: Computer Vision and Pattern Recognition (2014)

19. Girshick, R., Felzenszwalb, P., McAllester, D.: Object detection with grammar models. In: Proceedings of NIPS, pp. 442–450 (2011)

20. Girshick, R.B., Iandola, F.N., Darrell, T., Malik, J.: Deformable part models are convolutional neural networks. CoRR, abs/1409.5403 (2014)

21. Girshick, R.B., Malik, J.: Training deformable part models with decorrelated features. In: IEEE International Conference on Computer Vision, ICCV 2013, Sydney, Australia, 1–8 December 2013

22. Kalman, R.: A new approach to linear filtering and prediction problems. Trans. ASME J. Basic Eng. **82**, 35–45 (1960)

23. Krizhevsky, A., Sutskever, I., Hinton, G.E.: Imagenet classification with deep convolutional neural networks. In: Advances in Neural Information Processing Systems, vol. 25, pp. 1097–1105. Curran Associates Inc. (2012)

24. Leykin, A., Hammoud, R.: Pedestrian tracking by fusion of thermal-visible surveillance videos. Mach. Vis. Appl. **21**(4), 587–595 (2010)

25. Orts-Escolano, S., Garcia-Rodriguez, J., Morell, V., Cazorla, M., Azorin, J., Garcia-Chamizo, J.M.: Parallel computational intelligence-based multi-camera surveillance system. J. Sens. Actuator Netw. **3**(2), 95–112 (2014)

26. Parks, D.H., Fels, S.S.: Evaluation of background subtraction algorithms with postprocessing. In: IEEE Fifth International Conference on Advanced Video and Signal Based Surveillance, AVSS 2008, pp. 192–199. IEEE (2008)

27. Pedersoli, M., Gonzalez, J., Hu, X., Roca, X.: Toward real-time pedestrian detection based on a deformable template model. IEEE Intell. Transp. Syst. **15**(1), 355–364 (2013)

28. Rogez, G., Orrite, C., Guerrero, J.J., Torr, P.H.S.: Exploiting projective geometry for view-invariant monocular human motion analysis in man-made environments. Comput. Vis. Image Underst. **120**, 126–140 (2014a)
29. Rogez, G., Rihan, J., Guerrero, J.J., Orrite, C.: Monocular 3D gait tracking in surveillance scenes. IEEE Trans. Syst. Man Cybern. Part B (Cybernetics) **44**(6), 894–909 (2014b)
30. Russakovsky, O., Deng, J., Su, H., Krause, J., Satheesh, S., Ma, S., Huang, Z., Karpathy, A., Khosla, A., Bernstein, M., Berg, A.C., Fei-Fei, L.: Imagenet large scale visual recognition challenge (2014)
31. Singh, V.K., Wu, B., Nevatia, R.: Pedestrian tracking by associating tracklets using detection residuals. In: IEEE Workshop on Motion and video Computing, WMVC 2008, pp. 1–8. IEEE (2008)
32. Van Beeck, K., Tuytelaars, T., Goedemé, T.: A warping window approach to real-time vision-based pedestrian detection in a truck's blind spot zone. In: Proceedings of ICINCO (2012)
33. Zivkovic, Z., van der Heijden, F.: Efficient adaptive density estimation per image pixel for the task of background subtraction. Pattern Recogn. Lett. **27**(7), 773–780 (2006)

Algorithmic Optimizations in the HMAX Model Targeted for Efficient Object Recognition

Ahmad W. Bitar, Mohamad M. Mansour, and Ali Chehab[✉]

Department of Electrical and Computer Engineering, American University of Beirut,
Beirut 1107 2020, Lebanon
{ab76,mmansour,chehab}@aub.edu.lb
http://www.aub.edu.lb

Abstract. In this paper, we propose various approximations aimed at increasing the accuracy of the S1, C1 and S2 layers of the original Gray HMAX model of the visual cortex. At layer S1, an image is convolved with 64 separable gabor filters in the spatial domain after removing some irrelevant information such as illumination and expression variations. At layer C1, some of the minimum scales values are exploited in addition to the maximum ones in order to increase the model's accuracy. By applying the embedding space in the additive domain, the advantage of some of the minimum scales values is taken by embedding them into their corresponding maximum ones based on a weight value between 0 and 1. At layer S2, we apply clustering, which is considered one the most interesting research areas in the field of data mining, in order to enhance the manner by which all the prototypes are selected during the feature learning stage. This is achieved by using the Partitioning Around Medoid (PAM) clustering algorithm. The impact of these approximations in terms of accuracy and computational complexity was evaluated on the Caltech101 dataset containing a total of 9,145 images split between 101 distinct object categories in addition to a background category, and compared with the baseline performance using support vector machine (SVM) and nearest neighbor (NN) classifiers. The results show that our model provides significant improvement in accuracy at the S1 layer by more than 10 % where the computational complexity is also reduced. The accuracy is slightly increased for both approximations at the C1 and S2 layers.

Keywords: HMAX · Support vector machine · Nearest neighbor · Caltech101

1 Introduction

The human visual system is quite powerful. It is perhaps not too surprising that the human brain has achieved, through millions of years of evolution, a remarkable ability to recognize and differentiate among very similar objects in a selective, robust and fast manner. Modern machines can perform many apparently

© Springer International Publishing Switzerland 2016
J. Braz et al. (Eds.): VISIGRAPP 2015, CCIS 598, pp. 374–395, 2016.
DOI: 10.1007/978-3-319-29971-6_20

complex tasks much faster, more efficiently and more precisely than humans. Some estimates indicate that the human visual system can discriminate at least tens of thousands of different object recognition. Therefore, it would be relatively easy to build a computer system that can be extremely selective by just memorizing all the pixels in several training images. Modern computers are able to translate the human ventral visual pathway (known as the "WHAT" stream) in order to achieve, in a similar manner to the human brain, an impressive trade-off between selectivity and invariance. Several scientists have attempted to model and mimic the human vision system [1].

The Hierarchical Model And X (HMAX) is an important model for object recognition in the visual cortex known for its high ability to achieve performance levels close to the human object recognition capability [2]. HMAX divides the human ventral stream into five layers: S1, C1, S2, C2 and View-TUned (VTU).

The first layer S1 of the HMAX model relies on the Gabor filter [3], which is a linear filter used for edge detection. It differs from other filters by its capability to highlight all the features that are oriented in the direction of the filtering. The features are therefore extracted from the images by tuning the gabor filter to several different scales and orientations using fine-to-coarse approach.

Several methods have been proposed in the literature in order to improve the efficiency of the original HMAX model. An extension of the original HMAX model has been proposed in [4], emphasizing the importance of shape selectivity in area V4. A simpler radial basis function (RBF) model for object recognition was proposed in [5] to maintain a good degree of translation and scale invariance. The proposed model was considered better than the original HMAX for translation and scale invariance by changing the point of attention and decreasing the amount of visual information to be processed. In [6], they developed a new set of receptive field shapes and parameters for cells in the S1 and C1 layers. The method serves to increase position invariance in contrast to scale invariance, which is decreased. In [7], they proposed a general framework for robust object recognition of complex visual scenes based on a quantitative theory of the ventral pathway of visual cortex. A number of improvements to the base model were proposed in [8] in order to increase the sparsity. The proposed model has shown a remarkable improvement on classification performance and the resulting model is found more economical in terms of computations. In [9], they proposed several approximations at the four HMAX layers (S1, C1, S2 and C2) in order to increase the efficiency of the model in terms of accuracy and computational complexity. A semi-supervised learning algorithm for visual object categorization was proposed in [10] by exploiting unlabelled data and employing a hybrid generative-discriminative learning scheme. The method achieved good performance in multi-class object discrimination tasks. In [11], they proposed a scheme based on a kernel function for discriminative classification. The method achieved improved accuracy and reduced computational complexity compared to the baseline model.

In this paper, the goal is to perform various optimizations at the S1, C1 and S2 layers of the original HMAX model. The results demonstrate that these optimizations increase the accuracy of the HMAX model as well as reduce its computational complexity at the S1 layer. The accuracy of the final model proves the advantage of exploiting only the important features for recognition and generating the prototypes in a more efficient way.

The remainder of this paper is organized as follows. In Sect. 2, the visual system of the human brain is briefly presented. In Sect. 3, a brief overview of the original HMAX model is explained. The proposed approximations at S1, C1 and S2 layers are presented in Sects. 4, 5 and 6, respectively. Experimental results are shown in Sect. 7. Finally, Sect. 8 gives concluding remarks and some directions for future work.

2   The Visual System of the Human Brain

The light enters our eye form the pupil to the retina through the Crystalline lens. The iris is considered the colored part of the eye and it controls the amount of light that enters our eye. The pupil is the central aperture of the iris and the retina sends images to the brain through the optic nerve [18,19].

The retina contains five types of neurons: Photoreceptors (95 % Rods and 5 % Cones), Horizontal neurons, Bipolar neurons, Amacrine neurons and Ganglion neurons. There is amazing collaboration among all 5 neuron types. Of the existing 120 millions of photoreceptors in each eye, 95 % are Rods. In fact, the Rods are located on the surface of the retina, only sensible to luminance, responsible for vision at low light and active in scotopic vision. The Cones are only sensible to chrominance, responsible for vision at normal light, active in photopic vision and located in the fovea which constitutes 1 % of the retina's surface. Interestingly, both Rods and Cones are active in mesopic vision which is considered as a combination between scotopic and photopic.

One of the most important problems in vision is that at low light, the pupil increases in size, the light reflects into the retina's surface where the Rods are present. When suddenly a high level of light comes to the eye and before the pupil decreases in size, it reflects directly into the Rods which are sensible to luminance not to chrominance.

The ganglion cells are the most important to study. The ganglion of type "P" are for Parvo (very small receptive field), the ganglion of type "M" are for Magno (large receptive field) and the ganglion of type "K" are for Konio or conio (also called non P-nonM). All ganglion cells' types have receptive field described as "Center ON-Peripheric OFF" or "Center OFF-Peripheric ON".

The optic nerve contains the ganglion fiber optic. Both left and right optic nerves of the left and right eyes, respectively, are crossed in a point called "Optic chiasm" which transmits the information received from the retina to the Lateral Geniculate Nucleus (LGN).

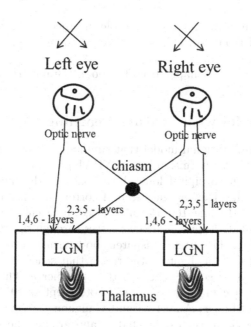

Fig. 1. Connections between the eye and LGN.

The LGNs are located in the thalamus of the brain and they are responsible for processing the information that arrives from the ganglion neurons. Each LGN is formed by six distinct layers numbered 1 to 6. The layers 1-2 contain neurons of type "M" and the layers 3-4-5-6 contain neurons of type "P". There is LGN in each hemisphere of the brain. The right LGN receives stimulus from the left visual field and vice versa as shown in Fig. 1. Layers 1, 4, 6 of both right and left LGNs receive axons from left part of the retina of each eye (nasal hemiretine) while the layers 2, 3, 5 of both right and left LGNs receive axons from right part of retina of each eye (temporal hemiretine).

The left and right primary visual cortex V1 receive information from the left and right LGN, respectively. The primary visual cortex is the part of the cerebral cortex responsible for processing visual information and it is located in the occipital lobe of the brain. The primary visual cortex V1 is also known as "Striate cortex" or "Broadman area 17 (BA17)". It is located in and around the calcarine fissure (or calcarine sulcus) in the occipital lobe of the brain. Importantly, it is divided into 6 distinct layers labeled 1 through 6. Layer 4, which receives the most visual input from the LGN is further divided into 4 layers: 4A, 4B, 4Cα (receives most Magnocellular inputs from the LGN) and 4Cβ (receives most Parvocellular inputs from the LGN). The V1 of each hemisphere transmits information to two primary pathways: Dorsal Stream and Ventral Stream. The object recognition in cortex is thought to be mediated by the ventral visual pathway running from visual cortex V1, over extrastriate visual areas V2 and V4 to Inferotemporal cortex IT. Based on physiological experiments in monkeys,

IT has been postulated to play a central role in object recognition. IT in turn is a major source of input to PFC, "the center of cognitive control" involved in linking perception to memory.

For further details, see Chap. 7 from the book "Brains: How They Seem To Work" [20].

3   HMAX Model with Feature Learning

HMAX [12] is a computational model that summarizes the organization of the first few stages of object recognition in the WHAT pathway of the visual cortex, which is located in the occipital lobe at the back of the human brain. It is considered a primordial part of the cerebral cortex responsible for processing visual information in the first 100–150 ms. Indeed, light enters our eye from the central aperture, called "Pupil", and then passes through the "Crystalline lens" which is considered the biconvex transparent body situated behind the iris into the eye and aiming to focus light on the retina that sends images to a specific part of the brain (visual cortex) through the optic nerve. The retina contains five different types of connected neurons: Photoreceptors (95 % rods and 5 % cones), Horizontal, Bipolar, Amacrine and Ganglion through which the light leaves the eye. The visual cortex, located in and around the calcarine sulcus, refers to the striate cortex V1, anatomically equivalent to Brodmann area 17 ($BA17$), connected to several extrastriate visual cortical areas (V2, V4, V5, etc.), anatomically equivalent to Brodmann area 18 and Brodmann area 19. The right and left V1 receive information from the right and left Lateral Geniculate Nucleus (LGN), respectively. The LGNs are located in the thalamus of the brain and they receive information directly from the ganglion cells of the retina via the optic nerve and optic chiasm.

3.1   Computational Complexity

The operations of the five layers of the HMAX model are briefly summarized.

S1 Layer: All the responses of the S1 units are summarized here by simply performing 2-D convolution between 64 Gabor filters (16 scales in steps of two pixels and 4 orientations) shown in Fig. 2 and the input images in the spatial domain.

Firstly, each Gabor filter of a specific scale and orientation can be initialized as:

$$G(x,y) = \exp^{-\left(\frac{u^2+\gamma^2 v^2}{2\sigma^2}\right)} \times \cos\left(\frac{2\pi}{\lambda}u\right), \tag{1}$$

where:

$$u = x\cos\theta + y\sin\theta,$$
$$v = -x\sin\theta + y\cos\theta,$$
$$\gamma = 0.0036 \times \rho^2 + 0.35 \times \rho + 0.18,$$
$$\lambda = \frac{\gamma}{0.8}.$$

The parameter γ is the aspect ratio at a particular scale, θ is the orientation $\in$ $[0°, 45°, 90°, 135°]$, σ represents the effective width ($=0.3$ in our case), λ is the wavelength at a particular scale, and ρ represents the scale.

Secondly, all the S1 image responses are computed by applying a two dimensional convolution between the initialized Gabor filters and the input images in the spatial domain. The S1 image responses are so-called: the Gabor features.

In fact, all the filters are arranged in 8 bands. There are two filter scales with four orientations at each band.

The S1 layer has a computational complexity of $O(N^2 M^2)$ where $M \times M$ is the size of the filter and $N \times N$ is the size of the image.

C1 Layer: The C1 units are considered to have larger receptive field sizes and a certain degree of position and scale invariance. For each band, each C1 unit response (image response) is computed by taking the maximum pooling between the gabor features of the two scales at the same orientation. The main role of the maximum pooling function is to subsample the number of the S1 image responses and increase tolerence to stimulus translation and scaling. Then, the pooling over local neighborhood using a grid of size $n \times n$ is performed. From band 1 to 8, the value of n starts from 8 to 22 in steps of two pixels, respectively. Furthermore, a subsampling operation can also be performed by overlapping between the receptive fields of the C1 units by a certain amount Δ_s ($= 4_{\text{band1}}, 5_{\text{band2}}, \cdots, 11_{\text{band8}}$), given by the value of the parameter C1Overlap. The value C1Overlap $= 2$ is mostly used, meaning that half the S1 units feeding into a C1 unit were also used as input for the adjacent C1 unit in each direction. Higher values of C1Overlap indicate a greater degree of overlap. This layer has a computational complexity of $O(N^2 M)$.

S2 Layer: The original version of HMAX was the *standard model* in which the connectivity from C1 to S2 was considered *hard-coded* to generate several combinations of C1 inputs. The model was not able to capture discriminating features to distinguish facial images from natural images. To improve that, an extended version was proposed [1], and is called *HMAX with feature learning*. In this model, each S2 unit acts as a Radial Basis Function (RBF) unit, which serves to compute a function of the distance between the input and each of the stored prototypes learned during the feature learning stage. That is, for an image patch X from the previous C1 layer at a particular scale, the S2 response (image response) is given by:

$$S2_{\text{out}} = \exp^{\left(-\beta \|X - P_i\|^2\right)}, \tag{2}$$

where β represents the sharpness of the tuning, P_i is the ith prototype and $\| \cdot \|$ represents the Euclidean distance. This layer has a computational complexity of $O\left(PN^2 M^2\right)$, where P is the number of prototypes.

C2 Layer: It is considered the layer at which the final invariance stage is provided by taking the maximum response of the corresponding S2 units over all

Fig. 2. 64 Gabor filters (16 scales in steps of two pixels $[7 \times 7$ to $37 \times 37] \times 4$ orientations $[0°, 45°, 90°, 135°]$).

scales and orientations. The C2 units provide input to the VTUs. This layer has a computational complexity of $O(N^2MP)$.

VTU Layer: At runtime, each image in the database is propagated through the four layers described above. The C1 and C2 features are extracted and further passed to a simple linear classifier. Typically, support vector machine (SVM) and nearest neighbor (NN) classifiers are employed.

The learning stage: The learning process aims to randomly select P prototypes used for the S2 units. They are selected from a random image at the C1 layer by extracting a patch of size 4×4, 8×8, 12×12, or 16×16 at random scale and position (Bands 1 to 8). For an 8×8 patch size for example, it contains $8 \times 8 \times 8 = 512$ C1 unit values instead of 64. This is expected since for each position, there are units representing each of the four orientations $[0°, 45°, 90°, 135°]$.

4   S1 Layer Approximations

At the S1 layer, several approximations are investigated in order to increase the efficiency of the original HMAX model in terms of accuracy and computational complexity. Each approximation has been evaluated independently using SVM and NN classifiers.

4.1   Combined Image-Based HMAX Using 2-D Gabor Filters

In this approximation, all unimportant information such as illumination and expression variations are eliminated from the image and hence its salient features become richer [13]. To achieve this, four main steps are applied to the original image A of size $h \times a$:

Step 1 – *Adaptive Histogram Equalization:* In order to handle the large intensity values to some extent, adaptive histogram equalization is applied to the original image A:

$$\text{Adapted\_Image} = \text{AdaptHistEq}(A) \tag{3}$$

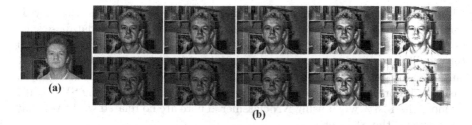

Fig. 3. (a) The original image and (b) Combined images using $\alpha = 0.25,\ 0.5,\ 0.75,\ 1$ and 1.25, respectively. c is equal to 0.25 and 0.75 on the top and bottom, respectively.

Step 2 – *SVD Decomposition:* Singular value decomposition (*SVD*) is applied to the image after equalization. The concept behind *SVD* is to break down the image into the product of three different martices as:

$$SVD(\text{Adapted\_Image}) = \mathbf{L} \times \mathbf{D} \times \mathbf{R^T} \tag{4}$$

where $\mathbf{L}$ is the orthogonal matrix of size $h \times h$, $\mathbf{R^T}$ is the transpose of an orthogonal matrix $\mathbf{R}$ of size $a \times a$ and $\mathbf{D}$ is the diagonal matrix of size $h \times a$. This decomposition helps the computations to be more immune to numerical errors, as well as to expose the substructure of the original image more clearly and orders their elements from most amount of variation to the least.

Step 3 – *Reconstruction Image:* According to the values of $\mathbf{L}$, $\mathbf{D}$ and $\mathbf{R}$, the reconstructed image is computed as follows:

$$\text{Reconstructed\_Image} = \mathbf{L} * \mathbf{D}^\alpha * \mathbf{R^T}, \tag{5}$$

where α is a magnification factor that varies between 1 and 2. The idea to have the value of α vary between one and two in order to magnify the singular values of $\mathbf{D}$ is to make them invariant to illumination changes. When α equals to 1, the reconstructed image is equivalent to the equalized image. When α is chosen between $]1\ 2]$, then the singular values greater than unity will be magnified. Thus, the combination between the reconstructed image and the equalized image will be a fruitful step to making the model more robust against illumination and expression variations.

Interestingly, when the singular values are scaled in the exponent, a nonlinearity is introduced. Therefore for a specific database (Caltech101 for example), scaling down the magnification factor α may be helpful.

Step 4 – *Combined Image:* The combined image is produced by simply combining the reconstructed image and the equalized image as shown in Fig. 3, using a combination parameter c which varies between 0 and 1.

$$\text{I}_{\text{Comb}} = \frac{\text{Adapted\_Image} + (c * \text{Reconstructed\_Image})}{1 + c} \tag{6}$$

By applying this approximation, the computations in this layer become faster as shown in Fig. 6 since only the significant information are used for recognition. In addition, the approximation can significantly improve the model's accuracy. It can be explained by the fact that when the model uses a challenge database such as Calech101 or Caltech256 in which there are a lot of unimportant information such as illumination and expression variations, it will be interesting to exploit only the most important features in the images in order to make the recognition easier and more robust where the accuracy is increased by 10 % using SVM while by more than 13 % when using NN classifier. There are no related works yet that approximate the S1 layer.

4.2 Combined Image-Based HMAX Using Separable Gabor Filters

In this approximation, all the combined images of the previous approximation are convolved with 64 Gabor filters in a separable manner $(G(x, y) = f(x)g(y))$, instead of just performing the 2-D convolution. In this case, the Gabor features are computed using two 1-D convolutions corresponding to convolution by $f(x)$ in the x-direction and $g(y)$ in the y-direction. Based on the definition of separable 2-D filters, the Gabor filters are parallel to the image axes $(\theta = k\pi/2)$. In order to be applied to an image along diagonal directions, they have been extended to further work with $\theta = k\pi/4$. The main issue of these techniques is that they will not work with any other desired direction. To handle this problem, Eq. (1) can be rewritten using the isotropic version $(\gamma = 1, \text{ circular})$ in the complex domain [9]. In this case, $u^2 + v^2 = (x\cos\theta + y\sin\theta)^2 + (-x\sin\theta + y\cos\theta)^2 = x^2 + y^2$.

$$G(x, y) = e^{-\frac{x^2+y^2}{2\sigma^2}} \times \cos\left(\frac{2\pi}{\lambda}(x\cos(\theta)+y\sin(\theta))\right)$$
$$= \text{Re}(f(x)g(y))$$

where

$$f(x) = e^{-\frac{x^2}{2\sigma^2}} \times e^{ix\cos(\theta)},$$
$$g(y) = e^{-\frac{y^2}{2\sigma^2}} \times e^{iy\sin(\theta)}.$$

Finally, the convolution using this approximation can therefore be expressed as:

$$\text{I}_{\text{Comb}} * \text{G}(x, y) = \text{I}_{\text{Comb}}(x, y) * f(x) * g(y) \tag{7}$$

By exploiting the separability of Gabor filters and convolving them with the original image, the computational complexity is reduced from $O(N^2M^2)$ to $O(tN^2M)$ where $t = 8$ due to complex valued arithmetic. But since in this approximation, the separable Gabor filters are convolved with the combined image I_{Comb}, the complexity is being more reduced since only the significant information are used for recognition. The accuracy is not increased by more than 10.5 % for SVM (between 10.4 % and 10.5 %) while is increased by more than 14 % for the NN classifier.

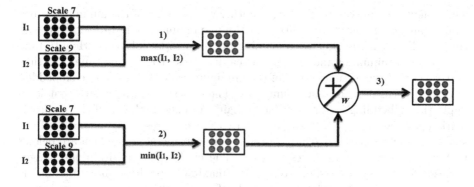

Fig. 4. Scheme example of the C1 approximation.

4.3 Combined Image-Based HMAX Using Haar Wavelet Transform

The foundation of the discrete wavelet transform (DWT) goes back to 1976 when Crochiere et al. for the first time introduced sub-band coding [15]. In 1983, Burt defined a technique very similar to sub-band coding and named it pyramidal coding which is also known as multi resolution analysis [16]. Later in 1989, Vetterli and Le Gall made some improvements to the sub-band coding scheme and removed the existing redundancy in the pyramidal coding scheme [17]. DWT definition is based on sub-band coding and multi-resolution analysis.

In this approach, we add one more step to the Subsect. 4.1 in order to have a total of 5 steps. Hence, to handle efficiently the condition variations, the wavelet transform DWT of LEVEL1 decomposition can be used to segment the image into four sub-bands: Low frequency component (LL), and High frequency components (LH, HL and HH). Thus, to help the recognition process to fully focus on important features, the LL sub-band has been considered ineffective with illumination changes and expression variations.

4.4 Baseline Model Using Haar Wavelet Transform

In this approach, we only added the Multi-resolution approach step to the baseline model.

5 C1 Layer Approximations

Concerning the C1 layer, a pooling between the S1 responses over scales within each band is performed by simply taking the maximum response between them. By testing what can be the result of the minimum pooling that has not been exploited at this layer, it was noticed that all the minimum scales values are very close to their corresponding maximum ones. Some of them equal, otherwise the most of minimum scales values are not smaller more than 6 or 7 %. As such, it will be important to further consider some of the minimum scales values when taking the maximum pooling. In other words, some of the minimum scales values can be exploited in addition to the maximum ones in order to increase the model's accuracy. But the remaining question to be solved is "How to take advantage of minimum and maximum scales values at the same time". So that under a specific conditions, some of the minimum scales values can be embedded into their corresponding maximum ones. The easiest way to achieve that is to apply the embedding in the additive domain. A general scheme of this approximation is shown in Fig. 4. In this figure, two S1 image responses I_1 and I_2 of the same orientation at the first band (band1) are considered and which are belong to the filter scale 7 and 9, respectively. The circles shown within the images correspond to their pixels. In step 1, the maximum pooling (max function) is performed between I_1 and I_2. The pixels of the resulting image correspond to the maximum scales values (shown with blue circles). In step 2, the minimum pooling (min function) is performed between I_1 and I_2. The pixels of the resulting image correspond to the minimum scales values (shown with violet circles) that are then embedded into their corresponding maximum ones in the additive domain under specific conditions as shown in step 3. In other words, each minimum scale value is added into the maximum one that has the same (x, y) coordinates. w is the weight of the embedding.

Embedding in the Additive Domain: This kind of embedding is very straightforward to implement since the minimum scales values (after applying the minimum pooling over scales within each band) can be directly embedded into their corresponding maximum values by simply using the addition operator.

Generally, the embedding process at a particular pixel coordinate (x, y) in the additive domain can be expressed as:

$$I_{\text{Embed}}(x, y) = \max_{\text{scale}}(x, y) + w * \min_{\text{scale}}(x, y), \tag{8}$$

where $I_{\text{Embed}}(x, y)$ represents the final result after the embedding process, $\max_{\text{scale}}$ is the maximum scale value, $\min_{\text{scale}}$ is the minimum scale value, and $w \in [0, 1]$ represents the weight of the embedding.

Two different conditions are considered to embed the minimum scales values into their corresponding maximum ones:

Condition 1: At each band, after computing the maximum pooling over scales of the same orientation, the minimum pooling is also performed and then all the minimum scales values are embedded into their corresponding maximum ones. In this case, w is set to 1.

Condition 2: Each minimum scale value is embedded if and only if its corresponding maximum value belongs to the interval $[0\,\%\ 5\,\%[$. The values within the interval specifies how much a maximum scale value is greater than its corresponding minimum one. In fact, the interval $[0\,\%\ 5\,\%[$ is divided into two groups: $[0\,\%\ 2\,\%[$ and $[2\,\%\ 5\,\%[$, and two distinct sub-conditions are thus considered:

- *Sub-condition 1:* The embedding is performed by setting w to 1 for $[0\,\%\ 2\,\%[$ and 0.5 for $[2\,\%\ 5\,\%[$.
- *Sub-condition 2:* The embedding is performed by setting w to 0.5 for $[0\,\%\ 2\,\%[$ and 0.1 for $[2\,\%\ 5\,\%[$.

The accuracy is not increased by more than 1 % in all conditions when SVM is used, while the opposite for NN classifier. However, the computational complexity at this layer is slightly increased due to the embedding process.

6   S2 Layer Approximations

At the S2 layer, the focus is to enhance the manner by which all the prototypes are selected during the feature learning stage. In the original model, P prototypes are randomly selected from the training images at the C1 layer. If more than P prototypes are used, the model's accuracy will increase at the expense of additional computational complexity. That is why our motivation is to learn the same number of prototypes P but in an efficient way in order to decrease the model's false classification rate while keeping the same computational complexity.

In order to achieve this, clustering is exploited, which is considered one of the most important research areas in the field of data mining. It aims to divide the data into groups, (clusters) in such a way that data of the same group are similar and those in other groups are dissimilar. Clustering is considered useful to obtain interesting patterns and structures. That is why, one of the existing clustering algorithms, more specifically the Partitioning Around Medoid (PAM) clustering algorithm [14] has been exploited in this approximation to generate the prototypes.

Furthermore, one of the important issues to consider, is the redundancy of some prototypes especially those selected from the homogeneous areas of the image (prototypes' pixels are being equal to zero). That is why, our contribution also aims to generate a non-redundant P prototypes and force the model not to generate any unimportant prototype. Accordingly, each of the selected prototypes will be important and aims to increase the model's accuracy.

PAM is characterized by its robustness to the presence of noise and outliers. Its complexity is defined by $O(i(b-q)^2)$ where i is the number of iterations, q is the number of clusters, and b represents the total number of objects in the data set.

To generate 2000 prototypes in a more efficient way and use them in our model instead of the traditional ones, the PAM algorithm is performed and it consists of 6 different steps:

Step 1 –5 medoids of 4 × 4 pixels at four orientations of each training category (total of 30 images) from the total 102 categories are randomly initialized.

Step 2 –For each category, the Frobenius distance between each of the C1 response of each image with all the selected medoids is then computed in order to associate each data image to the closest medoid.

Step 3 –For a random cluster, a non-medoid image patch is randomly selected in order to be swaped with the original medoid of the cluser in which the non-medoid is selected.

Step 4 –steps 2 and 3 are repeated until the total cost of swapping becomes greater than zero. The total cost of swapping can be defined as follows:

$$\text{Cost}_{\text{swapping}} = \text{Current Total Cost} - \text{Past Total Cost}$$

Step 5 –All the previous steps are also performed for all the other remaining three sizes of the medoids (8×8, 12×12 and 16×16) in order to have a total of 20 medoids in each category.

Step 6 –Finally, a total of 2040 medoids are being selected to be used as prototypes. 10 prototypes are dropped from each size in order to end up with only 2000 prototypes.

This algorithm is complex since there are six steps to perform in order to generate the prototypes. But in fact, the run of the HMAX model relies on two parts. The first part is responsible to generate and reserve all the necessary prototypes by only running the first two layers S1 and C1. The second part consists of running the whole model and use the prototypes that have been generated and reserved for the S2 layer. Interestingly, the complexity of the model depends only on the second part, which means that the large complexity of our algorithm does not affect the computational complexity of the model, more precisely, of the S2 layer. That is why, the computational complexity at the S2 layer of our model remains $O\left(PN^2M^2\right)$, where P is the number of prototypes. By applying this approximation, the accuracy of the model increases by 0.68% approximately using the SVM classifier.

7   Experimental Results

The proposed optimizations at the S1, C1 and S2 layers were implemented using MATLAB in order to evaluate their accuracy and computational complexity using experimental simulations. The S1, C1 and S2 approximations were evaluated using the Caltech101 database, which contains a total of 9,145 images split between 101 distinct object categories in addition to a background category. All the results of our approximations were the average of 3 independent runs. For each run, the following steps were performed:

1. A set of 30 images are randomly chosen from each category for training, while all the remaining images are used for testing. All the images are normalized to 140 pixels in height and the width is rescaled accordingly so that the image aspect ratio is preserved.

Table 1. Simulation results for SVM and NN on face category.

Positive training	SVM	NN
50	92.325 %	52.903 %
100	95.678 %	79.578 %
150	96.656 %	88.240 %
200	97.018 %	90.658 %
250	97.075 %	92.181 %
300	97.165 %	92.634 %
350	96.019 %	91.480 %
400	94.558 %	87.649 %

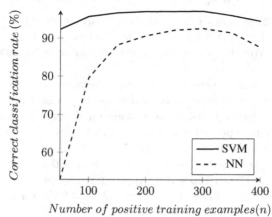

Fig. 5. SVM and NN accuracies on face category.

2. C1 sub-sampling ranges do not overlap in scales.
3. The prototypes are learned at random scales and positions. They are extracted from all the eight bands.
4. C2 vectors are built using the training set.
5. Training applied using both SVM and Nearest-Neighbor classifiers.
6. C2 vectors for the test set are built, and then the test images are classified.

7.1   Performance of SVM and NN

The performance of both SVM and NN are performed on the face category extracted from the caltech101 database and which contains 435 face images.

The results show that the accuracy decreases when the number of training becomes greater than 300. This is expected because the data becomes unbalanced.

The images were rescaled to 160×160 pixels, the C1 sub-sampling ranges overlap in scale (C1Overlap $= 2$) and the prototypes are chosen only from Bands

1 and 2. The classifiers were trained with $n = 50, 100, 150, 200, 250, 300, 350$ and 400 positive examples and 50 negative examples from the background class, while they are tested with all the remaining positive examples and 50 examples from the negative set as shown in Table 1 and Fig. 5. 1000 prototypes (250 patches) $\times$ (4 sizes) are used in the S2 layer.

7.2   Evaluations at the S1 Layer - Part 1

At this layer, the computational complexity and correct classification rates (accuracies) for each of the proposed approximations (**Approx**) are compared to the baseline model.

- **Approx1:** Combined Image-based HMAX using 2-D Gabor filters.
- **Approx2:** Combined Image-based HMAX using separable Gabor filters.

Interestingly, to avoid any confusion, the performances of the two approaches "Combined Image Based HMAX using Haar Wavelet Transform" and "Baseline model using Haar Wavelet Transform" are tested in an independent subsection (see Subsect. 7.3).

In order to compute the speed of the approximations at this layer, the total time complexity of the S1 layer is measured on a specific face image from the face category. All the evaluations were done on a core i7 2.4 GHZ machine. The simulations were repeated five times. Figure 6 illustrates an average of the results. It shows that both Approx1 and Approx2 are faster than the baseline (blue curve) for all the tested image sizes. It has been noticed that for an image of size between 100×100 and 160×160, Approx1 is always faster than Approx2. For example, Approx1 is faster than Approx2 by 3.23 % for an image of size 100×100.

For other image sizes greater than or equal to 160×160, Approx2 always shows lower timing than Approx1. For example, for an image of size 160×160, Approx1 is faster than the baseline by 2.95 % while by 3.42 % for Approx2.

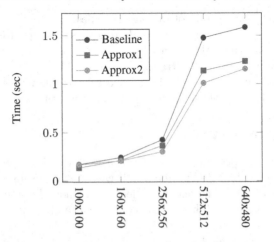

Fig. 6. Timing comparison (in sec).

Table 2. Classification accuracies of Approx1 approximation.

Approx1	Classifier	$c = 0.25$	$c = 0.75$
$\alpha = 0.25$	SVM	34.36 %	33.59 %
	NN	20.28 %	20.28
$\alpha = 0.5$	SVM	45.20 %	45.16 %
	NN	31.23 %	31.23
$\alpha = 0.75$	**SVM**	**49.02 %**	48.27 %
	NN	**35.01 %**	34.45 %
$\alpha = 1$	SVM	47.74 %	47.74 %
	NN	31.36 %	31.36 %
$\alpha = 1.25$	SVM	39.35 %	40.74
	NN	23.99 %	24.08 %

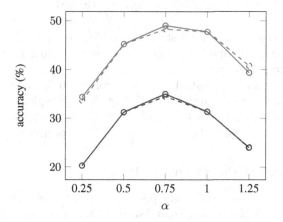

Fig. 7. Approx1 accuracies under different values of α and c (Color figure online).

For an image of size 256×256, Approx1 is faster than the baseline by 6.29 % while by 12.56 % for Approx2.

In order to assess the correct classification rates, both SVM and NN classifiers were used. The average accuracies of Approx1 under different values of α and c are shown in Table 2. From all the following experiments, 2000 prototypes (500 patches) $\times$ (4 sizes) are used and all the images were rescaled to 140 in height. Recall that C1 sub-sampling ranges do not overlap in scales and the prototypes are extracted from all the eight bands. The performance of the original model reaches 39 % and 21.2 % when using 30 training examples per class averaged over 3 repetitions under SVM and NN, respectively. Table 2 proves our significant contribution at the S1 layer especially for $\alpha = 0.75$ and $c = 0.25$ where the accuracy is increased by 10.02 % and 13.811 % using SVM and NN, respectively.

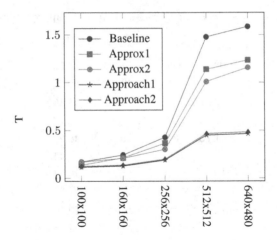

Fig. 8. Timing comparison (in sec).

Figure 7 illustrates the results shown in Table 2. It shows 4 different curves. The red and blue solid curves represent the accuracy values for $c = 0.25$ under SVM and NN, respectively.

While the red and blue dashed curves are for $c = 0.75$ under SVM and NN, respectively. Finally, the separability of Gabor filters is exploited and applied to the combined image with $\alpha = 0.75$ and $c = 0.25$. Approx2 shows an accuracy equal to 49.471 % and 35.372 % for SVM and NN, respectively.

7.3   Evaluations at the S1 Layer - Part 2

In this subsection, we aim to test the performance of the two approaches "Combined Image-Based HMAX using Haar Wavelet Transform" and "Baseline Model using Haar Wavelet Transform" in terms of speed and correct classification rates. To facilitate the notations, we name the first approach by "Approach1" while the second by "Approach2".

- **Approach1:** Combined Image-Based HMAX using Haar Wavelet Transform.
- **Approach2:** Baseline Model using Haar Wavelet Transform.

In order to compute the speed of the two approaches, we perform exactly the same computations as we did in Fig. 6. In other words, we just added the two approaches to Fig. 6 without any modification to the inputs of the model in order to get Fig. 8 that contains a total of 5 curves.

The simulations were also repeated five times and Fig. 8 illustrates the average of the results. It shows that both Approach1 and Approach2 are significantly faster than the baseline, Approx1 and Approx2 for all the tested images. From Fig. 8, we also notice that Approach1 is always approximately faster than Approach2. For more details about the numerical results of Fig. 8, refer to Table 3.

By normalizing the images to 140 pixels in height where the width is rescaled accordingly so that the image aspect ratio is preserved, some images become too

Table 3. Timing comparison (in sec) for all the proposed approximations at the S1 layer versus the baseline model.

Size	Baseline	Approx1	Approx2	Approach1	Approach2
100 × 100	0.170	0.136	0.168	0.115	0.124
160 × 160	0.242	0.213	0.208	0.128	0.135
256 × 256	0.427	0.364	0.301	0.189	0.198
512 × 512	1.476	1.138	1.009	0.450	0.465
640 × 480	1.584	1.235	1.156	0.465	0.483

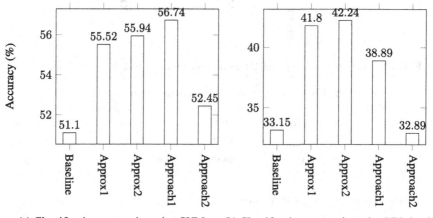

(a) Classification accuracies using SVM classifier.

(b) Classification accuracies using NN classifier.

Fig. 9. Classification accuracies.

small after applying Level1 Haar wavelet decomposition. Hence, choosing the prototypes from all bands becomes impossible. That is why, our contribution is to rescale only in this part all the images to 160 × 160. As in Subsect. 7.2, the correct classification rates are measured in the same way. The performance of the original model reaches 51.1 % and 33.1 % when using 30 training examples per classaveraged over 3 repetitions under SVM and NN, respectively. Figure 9 (a) shows that by using the SVM classifier, Approach1 is the best and that Approach2 is only better than the baseline by 1.35 %. Approach2 is worse than (Approx1, Approx2 and Approach1) by (3.07 %, 3.49 % and 4.29 %), respectively.

Fig. 9(b) shows that by using the NN classifier instead of SVM, Approx2 becomes the best and Approach2 the worst. Approx2 reaches 42.24 % while Approach1 and Approach2 reach 38.89 % and 32.89 %, respectively.

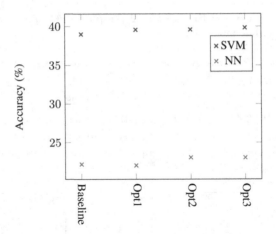

Fig. 10. Average accuracies of C1 approximations (Color figure online).

Table 4. Accuracy of S2 approximation.

Approximation	SVM
Baseline	39 %
PAM	39.68 % (+0.68 %)

7.4 Evaluations at the C1 Layer

Figure 10 shows the average accuracies of the SVM (blue x points) and NN (red x points) on several C1 optimization options (**Opt**). The accuracy of the model is increased a little bit when three cases of the additive method are applied. For example, using SVM, the accuracy is increased by 0.577 %, 0.607 %, 0.843 % on Opt1, Opt2 and Opt3, respectively. On the other hand, the increase is 0.846 %, 1.88 %, 1.85 % using NN.

– **Opt1:** Embedding all pixels ($\alpha = 1$).
– **Opt2:** $[0\,\%, 2\,\%[$, ($\alpha = 0.5$); $[2\,\%, 5\,\%[$, ($\alpha = 0.1$)
– **Opt3:** $[0\,\%, 2\,\%[$, ($\alpha = 1$); $[2\,\%, 5\,\%[$, ($\alpha = 0.5$)

7.5 Evaluations at the S2 Layer

Table 4 shows the average accuracy of the SVM classifier based on the S2 approximation.

This approximation has a big advantage on the model since the selected prototypes are non-redundant and generated in more intelligent way. Therefore, each prototype serves to slightly increase the accuracy. The accuracy of the model is increased approximately by 0.68 %.

7.6   Combined Classification Accuracies

Figure 11 shows the average accuracies of SVM on the combination of the approximations "Approx2" + "Opt3" + "PAM". Our model shows an accuracy equal to 51 % when using only 2000 prototypes while it shows 53.8 % when using higher number of protoypes (4080) as used in [1, 8, 10, 11].

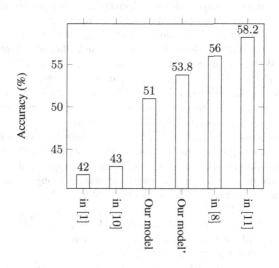

Fig. 11. The accuracies of the final models.

8   Discussion and Future Work

In this work, the complexity of all the five different layers of the original model of object recognition in the visual cortex, HMAX, is presented. Different approximations were added to the first three layers S1, C1 and S2.

The results have shown that removing all unimportant information such as illumination, expression variations and occlusions, to be a fruitful approach to improving performance. The idea behind separability of Gabor filters has been also exploited in order to be applied on the combined images generated after keeping only the important features for recognition. The change of the main concept at the C1 layer is further applied by exploiting the advantage of some of the minimum scales values and using them to be embedded into the extracted maximum scales values. The accuracy was slightly increased when the embedding process has been applied using the additive method. Our model serves also to always use an intelligent version of selected prototypes at the S2 layer in order to remove all the possibilities of having an unimportant prototype aiming to decrease the model's accuracy.

As for future enhancements, a natural extension would be to adapt our work to the HMAX model in color mode. In addition, several new approximations will be applied and tested on more challenging databases.

References

1. Serre, T., Wolf, L., Poggio, T.: Object recognition with features inspired by visual cortex. In: IEEE Conference on Computer Vision and Pattern Recognition (CVPR 2005), pp. 994–1000 (2005b)
2. Serre, T., Kouh, M., Cadieu, C., Knoblich, U., Kreiman, G., Poggio, T.: A theory of object recognition: computations and circuits in the feedforward path of the ventral stream in primate visual cortex. CBCL Paper #259/AI Memo #2005-036, Massachusetts Institute of Technology, Cambridge, MA (2005a)
3. Amayeh, G., Tavakkoli, A., Bebis, G.: Accurate and efficient computation of gabor features in real-time applications. In: Bebis, G., et al. (eds.) ISVC 2009, Part I. LNCS, vol. 5875, pp. 243–252. Springer, Heidelberg (2009)
4. Cadieu, C., Kouh, M., Riesenhuber, M., Poggio, T.: Shape representation in v4: Investigating position-specific tuning for boundary conformation with the standard model of object recognition. J. Vis. 5(8), 671 (2005)
5. Bermudez-Contreras, E., Buxton, H., Spier, E.: Attention can improve a simple model for object recognition. Image Vis. Comput. 26, 776–787 (2008)
6. Serre, T., Riesenhuber, M.: Realistic modeling of simple and complex cell tuning in the hmax model, and implications for invariant object recognition in cortex. Massachusetts Institute of Technology, Cambridge, MA. CBCL, Paper 239/AI Memo 2004-017 (2004)
7. Serre, T., Wolf, L., Bileschi, S., Riesenhuber, M., Poggio, T.: Robust object recognition with cortexlike mechanisms. In: IEEE Conference on Pattern Analysis and Machine Intelligence, vol. 29, pp. 411–426 (2007b)
8. Mutch, J., Lowe, D.G.: Multiclass object recognition with sparse, localized features. In: IEEE Conference on Computer Vision and Pattern Recognition (CVPR), vol. 1, pp. 11–18 (2006)
9. Chikkerur, S., Poggio, T.: Approximations in the hmax model. MIT-CSAIL-TR-2011-021, CBCL-298, p. 12 (2011)
10. Holub, A., Welling, M.: Exploiting unlabelled data for hybrid object classification. In: Advances in Neural Information Processing Systems (NIPS 2005) Workshop in Inter-Class Transfer (2005)
11. Grauman, K., Darrell, T.: The pyramid match kernel: Discriminative classification with sets of image features. In: Proceedings of the IEEE International Conference on Computer Vision (ICCV), vol. 2, pp. 1458–1465 (2005)
12. Serre, T., Kreiman, G., Kouh, M., Cadieu, C., Knoblich, U., Poggio, T.: A quantitative theory of immediate visual recognition. Prog. Brain Res. Comput. Neurosci. Theor. Insights Brain Funct. 165, 33–56 (2007a)
13. Sharif, M., Anis, S., Raza, M., Mohsin, S.: Enhanced SVD based face recognition. J. Appl. Comput. Sci. Math. 12, 49 (2012)
14. Kumar, P., Wasan, S.K.: Comparative study of k-means, pam and rough k-means algorithms using cancer datasets. In: Proceedings of CSIT: 2009 International Symposium on Computing, Communication, and Control (ISCCC) Singapore, 2011, pp. 136–140 (2011)
15. Crochiere, R., Webber, S., Flanagan, J.: Digital coding of speech in sub-bands. In: IEEE International Conference on Acoustics, Speech, and Signal Processing, pp. 233–236 (1976)
16. Burt, P., Adelson, E.: The Laplacian pyramid as a compact image code. IEEE Trans. Commun. 31(4), 532–540 (1983)

17. Vetterli, M., Le Gall, D.: Perfect reconstruction FIR filter banks: Some properties and factorizations. IEEE Trans. Acoust. Speech Sig. Process. **37**(7), 1057–1071 (1989)
18. Hubel, D.H., Freeman, W.H.: The Human Eye: Structure and Function. Sinauer Associates, Sunderland (1999)
19. Oyster, C.W.: Eye, Brain and Vision. vol. 12(1), pp. 40–41 (1989)
20. Purves, D.: Brains: How They Seem To Work. FT Press, Upper Saddle River (2010)

Estimating Visual Motion Using an Event-Based Artificial Retina

Luma Issa Abdul-Kreem[1,2]($\boxtimes$) and Heiko Neumann[1]

[1] Institute for Neural Information Processing, Ulm University, Ulm, Germany
{luma.issa,heiko.neumann}@uni-ulm.de
http://www.uni-ulm.de/in/neuroinformatik.html
[2] Control and Systems Engineering Department, University of Technology, Baghdad, Iraq

Abstract. Biologically inspired computational models of visual processing often utilize conventional frame-based cameras for data acquisition. Instead, the Dynamic Vision Sensor (DVS) emulates the main processing sequence of the mammalian retina and generates spike-trains to encode temporal changes in the luminance distribution of a visual scene. Based on such sparse input representation we propose neural mechanisms for initial motion estimation and integration functionally related to the dorsal stream in the visual cortical hierarchy. We adapt the spatio-temporal filtering scheme as originally suggested by Adelson and Bergen to make it consistent with the input representation generated by the DVS. In order to regulate the overall activation of single neurons against a pool of neighboring cells, we incorporate a competitive stage that operates upon the spatial as well as the feature domain. The impact of such normalization stage is evaluated using information theoretic measures. Results of optical flow estimation were analyzed using synthetic ground truth data.

Keywords: Event-based vision · Optic flow · Neuromorphic sensor · Neural model · Motion integration

1 Introduction

A frame-based imager transmits moving scenes into a series of consecutive frames. These frames are constructed at a fixed time rate, which generates an enormous amount of redundant information. In contrast, a Dynamic Vision Sensor (DVS) (see [1,2]) reduces this redundancy using a new technology inspired by visual systems. The functionality of this sensor is similar to the biological retina, where a stream of spike events are generated as a polarity format ON (+1) or OFF (−1) if a positive or negative luminance change is detected. Luminances that do not change over time, on the other hand, do not produce any output, and as a consequence, any such redundant information sampled by frame-based cameras is reduced.

A DVS has high temporal resolution, where the events are generated asynchronously and sent out almost instantaneously on the address bus. Thus, subtle

© Springer International Publishing Switzerland 2016
J. Braz et al. (Eds.): VISIGRAPP 2015, CCIS 598, pp. 396–415, 2016.
DOI: 10.1007/978-3-319-29971-6_21

and fast motions can be detected. In addition, a DVS has low latency and a large dynamic range due to the pixels locally responding to relative changes in intensity. A DVS's ability to produce an event at 1 μs time precision and a latency of 15 μs with bright illumination were illustrated in [3].

The new sensor technology has led to several recent applications in many fields to exploit the advantages of DVSs compared with traditional frame-based imagers, where several application-oriented studies have capitalized on those features. Such works include [1,4], where Litzenberger and co-authors introduced an algorithm that used the silicon retina imager to estimate vehicle speed based on the slope of the events cloud. Delbruck and co-authors presented a hybrid neuromorphic procedural system for object tracking via an event-driven cluster tracker algorithm. The authors showed how a moving ball can be detected, tracked and successfully blocked by a robot goalie despite a low contrast object and complex background. The event-cluster algorithm was introduced by [5,6], where a first study considered a real world application, namely vehicle tracking for traffic monitoring in real time, and a second study addressed microrobotics tracking.

In our previous work reported in [7], we have proposed a bio-inspired model using the energy model of [8]. In this paper, we extend the previous work by introducing a new set of temporal filters which are compatible with the vision sensor functionality and accurately transcribed the biphasic temporal filters that were suggested by Adelson-Bergen [8]. To achieve balanced activities of individual cells against the neighborhood activities, a normalization process is carried out. Here, we investigate the impact of such normalization process to enhance the initial responses of filtering stage. We tested our model using different kinds of stimuli that were moved via translatory and rotatory motions. The results highlight an accurate flow estimation compared with synthetic ground truth. In order to show the robustness of our model, we examined the model by probing it with synthetically generated ground-truth stimuli and realistic complex motions, e.g. biological motions and a bouncing ball, with satisfactory results. The following sections details our methodology and results.

2   Previous Models

Motion estimation is an advanced topic in automated visual processing and has been investigated widely using conventional cameras (see, e.g., [9–12]). Few studies have been published using the new vision sensor technology of an address-event silicon retina. Benosman and co-authors [13] implemented the energy minimization method introduced in [14] to calculate motion flow using an event-based retina. Since the vision sensor generates a stream of events (ON or OFF) and does not provide gray levels, the authors suggested using pixel activities by integrating events within a short temporal window. Gradients were estimated by comparing active pixels over one temporal window to calculate the spatial gradient, and two temporal windows to calculate the temporal gradient. A least squares error minimization technique was used to calculate the local optic flow based on such pixel neighborhoods. Benosman and co-authors showed beneficial results, however, their method to approximate local gradients of the luminance function from event-sequences has its limitations and in some cases leads to inconclusive results (see [15]).

Recently [16] presented an algorithm for motion estimation where the authors utilized spatiotemporal filters of the type suggested by findings of [17] to estimate a local motion flow calculated for each event occurring in the scene. The spatiotemporal filters were implemented over a spatial buffer of (11×11) which stores the timestamp of the events. This method is characterized as a neuroscience approach and showed adequate results. In [18], the authors systematically investigated the implications of event based sensing in the visual flow estimation. They discussed different principal approaches for motion estimation and showed that gradient-based methods for local motion suffer from the sparse encoding in address-event representations (AER). While approaches exploiting the local plane-like structure of the event cloud are shown to be well suited.

The motion estimation using address event representation thus requires further investigation and development. Our model differs from these methods in the initial process of motion estimation. We propose neural mechanisms for motion estimation which are inspired by the dorsal stream of the visual system and are consistent with the vision sensor functionality. Here, we adapted the spatiotemporal filtering scheme as originally suggested by Adelson and Bergen [8] to be consistent with the functionality of AER principles. In addition, we incorporate normalization responses in the spatial domain as well as in feature domain to regulate the overall activity of single neuron against the neighboring activations.

3   Methodology

3.1   Initial Input Representation from ON/OFF Events

High temporal resolution, low latency and large dynamic range visual sensing are key features of the address-event-representation (AER) principle, where each pixel of the vision sensor responds independently and almost instantaneously translates local changes in the intensity (log I) into events (ON or OFF), see Fig. 1. This principle is used in our study to profit from the advantage of the event-based technology instead of using standard frame-based camera technology. Since a single event in spatiotemporal domain yields an ambiguity for motion estimation using spatiotemporal filtering, pixel activities, ON (+1) and OFF (−1), are accumulated during a temporal window. The accumulated ON/OFF events are denoted by

$$e(\mathbf{p}, t) = e^{on}(\mathbf{p}, t) + e^{off}(\mathbf{p}, t), \tag{1}$$

where $e^{on}(\mathbf{p}, t)$ and $e^{off}(\mathbf{p}, t)$ are ON and OFF events, respectively, which occurred at position $\mathbf{p} = (x, y)$ and time t.

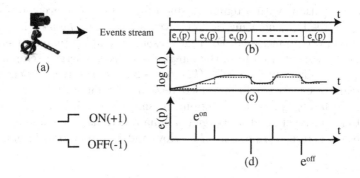

Fig. 1. (a) DVS sensor. (b) Event stream which is represented as a sequence of events e at a position **p** and time t. (c) Local changes in intensity (log I) excite (ON or OFF) events. (d) e^{on} and e^{off} identify the event activity (+1) ON and (−1) OFF, respectively.

3.2   Detection of Motion Energy from Event Input

Motion estimation using spatiotemporal filters emulates motion detection processing of the primary visual cortex [8], where the space-time filters are 3D and can here be decomposed into separable products of two 2D spatial and two 1D temporal kernels. The two spatial filters consist of different phases (even and odd) while the temporal filters consist of two different temporal integration windows (fast and slow). The spatial receptive fields (RFs) of odd and even filters can be implemented using Gabor functions, which provide a close description of the receptive fields in primary visual cortex area (V1) [19]. We thus used these functions to build even and odd spatial filters as in Eqs. (2) and (3), respectively namely

$$F_{even}(\mathbf{p}, \theta_k, f_s) = \frac{1}{2\pi\sigma_s^2} \cdot exp(-\frac{\breve{x}^2 + \breve{y}^2}{2\sigma_s^2}) \cdot cos(2\pi f_s\varsigma \mathbf{p}), \qquad (2)$$

$$F_{odd}(\mathbf{p}, \theta_k, f_s) = \frac{1}{2\pi\sigma_s^2} \cdot exp(-\frac{\breve{x}^2 + \breve{y}^2}{2\sigma_s^2}) \cdot sin(2\pi f_s\varsigma \mathbf{p}), \qquad (3)$$

where $\begin{pmatrix} \breve{x} \\ \breve{y} \end{pmatrix} = \begin{pmatrix} cos\theta_k & -sin\theta_k \\ sin\theta_k & cos\theta_k \end{pmatrix} \cdot \begin{pmatrix} x \\ y \end{pmatrix}$, θ_k is the spatial filter orientation with N different orientations where $k = \{1, 2, 3...N\}$, σ_s is the standard deviation of the spatial filters and f_s represents the spatial frequency tuning, $\varsigma = \begin{pmatrix} cos\theta_k \\ -sin\theta_k \end{pmatrix}$, the index **p** represents the spatial location $\mathbf{p} = (x, y)$.

In the model of [8] the authors suggested to utilize temporal gamma functions of different duration in order to accomplish temporal smoothing and differentiation, leading to a temporally biphasic response shape. In order to transcribe this functionality to the AER output of the sensor, we make use of the following approximation: The biphasic Adelson-Bergen temporal filters can be decomposed into a convolution of numerical difference kernel (to approximate a first-order derivative operation) with a temporal smoothing filter. The event-based sensor already operates by generating discrete events based on *changes*, i.e. temporal derivatives, in the input signal. For that reason, we employ temporal smoothing filters which are calculated from the integral of Adelson-Bergen temporal filters ($f(t) = \int (kt)^2 \cdot exp(-kt^2) \cdot [1/n! - (kt)^2/(n+2)!]$) and convolve them with the input stream of events to obtain scaled versions of temporally smoothed derivatives of the input luminance function. To simplify the integral operation, we suggest to reconstruct Adelson-Bergen gamma functions by combining two temporally offset Gaussian functions. The slow and fast temporal filters are thus given as

$$g^{slow,fast}(t) = [\Lambda_1(t) - \Lambda_2(t)]/c, \tag{4}$$

$$\Lambda_{1,2}(t) = 1/(\sigma_{1,2}\sqrt{2\pi}) \cdot exp(-(t - \mu_{1,2})^2/(2\sigma_{1,2}^2)) \tag{5}$$

$\Lambda_{1,2}$ represent Gaussian functions that are parametrized by the standard deviations ($\sigma_{1,2}$) and mean values ($\mu_{1,2}$), the symbol c denotes a scalar factor of Gaussian combination to closely resemble the shape of the Adelson-Bergen temporal filters. The results are scaled versions of temporally smoothed derivatives of the input luminance function as suggested by [8]. We chose $\sigma_1 = 1$, $\mu_1 = 2.5$, $\sigma_2 = 2$, $\mu_2 = 7$, $c = 2.6$ to generate the fast temporal filter and $\sigma_1 = 1.3$, $\mu_1 = 4$, $\sigma_2 = 2.3$, $\mu_2 = 9.2$, $c = 3.1$ to generate the slow temporal filter, see Fig. 2. Subsequently, the smoothing temporal filters are calculated by

$$T^{slow,fast}(t) = \frac{1}{\int_0^\infty (H_1 - H_2)dt} \cdot (H_1 - H_2)(t), \tag{6}$$

$$H_1 = \left[\frac{1}{2}(1 + erf\left(\frac{(t - \mu_1)}{\sigma_1\sqrt{2}}\right)\right]/c, \tag{7}$$

$$H_2 = \left[\frac{1}{2}(1 + erf\left(\frac{(t - \mu_2)}{\sigma_2\sqrt{2}}\right)\right]/c. \tag{8}$$

The slow and fast temporal filters ($T^{slow,fast}(t)$) are scaled to 1 to prevent any biases in calculating responses.

Figure 3 (a) and (b) show the spatial and temporal filters, respectively. The spatiotemporal separable responses are calculated according to the scheme proposed in [8]. The products of two spatial and two temporal filters responses are shown in the first row of Fig. 3 (c). These responses are combined in a linear fashion to get the oriented selectivity as shown in the second row of Fig. 3 (c). The oriented linear combinations are denoted by

$$F_a^{v1}(\mathbf{p}, \theta_k, f_s, t) = F_{even}(\mathbf{p}, \theta_k, f_s) \cdot T_{slow}(t) + F_{odd}(\mathbf{p}, \theta_k, f_s) \cdot T_{fast}(t), \tag{9}$$

$$F_b^{v1}(\mathbf{p}, \theta_k, f_s, t) = F_{even}(\mathbf{p}, \theta_k, f_s) \cdot T_{fast}(t) - F_{odd}(\mathbf{p}, \theta_k, f_s) \cdot T_{slow}(t). \tag{10}$$

The spatiotemporal response for a stream of events input e can be achieved through nonlinear combinations of contrast invariant responses:

$$r_\theta^{v1} = [F_a^{v1} * e]^2 + [F_b^{v1} * e]^2, \tag{11}$$

where $*$ indicates convolution, θ indicates motion directions (left vs. right relative to the orientation axis). The local spatial coordinate and feature selectivities are omitted for better readability.

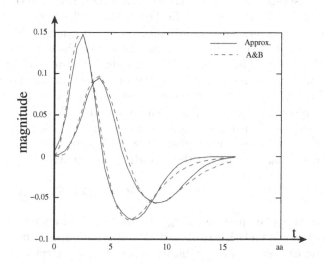

Fig. 2. Fast and slow temporal filters. Dashed lines show Adelson-Bergen (A&B) filters and solid lines show the approximated shapes (Approx.) derived by two shifted Gaussian envelopes as filters.

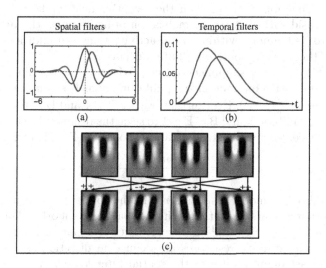

Fig. 3. Spatiotemporal filter construction. (a) Spatial filters. (b) Temporal filters. (c) The first row represents the products of two spatial and two temporal filters. The second row represents the sum and difference of the spatio-temporal product filters.

3.3 Response Normalization

The activity in neurons show significant nonlinearities depending on spatio-temporal activity distribution in the space-feature domain surrounding a target cell [20]. Such response nonlinearities have been demonstrated in the LGN, early visual cortex (area V1), and beyond. In theoretical studies (e.g. [21]) it has been proposed that such compression of stimulus responses can be achieved through the normalization of the target cell response defined by the weighted integration of activities in a neighborhood defined over the space-feature domain of a target cell. In other words, the normalization operation utilizes contextual information from a local neighborhood that is defined in space as well as feature domains relevant for the current computation. Such a normalization can be generated at the neuronal level by divisive, or shunting, inhibition (see [22–24]).

Given the activity of a neuron (defined by its membrane potential) the rate of change can be characterized by the following rate equation [25]

$$\tau \frac{dv(t)}{dt} = -A \cdot v(t) + (B - C \cdot v(t)) \cdot net_{ex} - (D + E \cdot v(t)) \cdot net_{in}, \quad (12)$$

given A representing the passive leakage, B and D are parameters to denote the saturation potentials (relative to C and E, respectively), and net_{ex} and net_{in} denote generic excitatory and inhibitory inputs to the target cell and impose the conductances of the membrane potential. In order to achieve balanced cell activations against the pool of neighboring cells, a normalization is calculated, following [21,26]. We employ a spatial weighting function Λ_σ^{pool} which realizes a distance-dependent weighting characteristics, e.g., Gaussian. The size of this neighborhood function is larger than the receptive field, or kernel, size of the cells under consideration. After normalization of activations, the responses are guaranteed to be bounded within a local activity range. In addition, a spectral whitening of the local response distribution occurs [27].

We realized a slightly simplified version of the scheme described in [21] and solve the normalization interaction at equilibrium, namely evaluating the steady-state response for $\frac{dv(t)}{dt} = 0$. Ever further, we set $C = 1$ and $D = 0$ in Eq. (12) to define shunting inhibition, and $B = E = 1$ to scale the response levels accordingly. As a consequence, we get the steady state response for Eq. (12) which reads

$$v_\infty = \frac{net_{ex}}{A + net_{ex} + net_{in}}. \quad (13)$$

We normalize the model responses r^{v1} in the spatial domain using an integration field that weights the activities in the spatial neighborhood of the target. We propose a spatial weight fall-off in accordance to a Gaussian weighting function. The motion selective responses are defined in direction space relative to the local contrast orientation θ of the spatial filter kernels used. We take the direction feature space into account as well by calculating the average activity over all directions. In all, we can denote the overall pool activation by

$$r^{pool}(\mathbf{p}) = \frac{1}{2N} \sum_\theta \{r_\theta^{v1} * \Lambda_\sigma\}_\mathbf{p}, \quad (14)$$

with θ denoting motion direction, '$*$' denotes the (spatial) convolution operator, $\mathbf{p} = (x, y)$ represents the spatial position of the cell, N is the number of contrast filter orientations and Λ is the weighting function of the spatial pooling operation. The latter is parametrized by the parameter σ to control the width of the spatial extent. The resulting normalized response is finally calculated by

$$r_\theta^{v1nor}(\mathbf{p}) = \frac{r_\theta^{v1}(\mathbf{p})}{A + r_\theta^{v1}(\mathbf{p}) + r^{pool}(\mathbf{p})}, \tag{15}$$

A denotes the passive decay of the dynamic mechanism and prevents from zero division in the steady-state.

4 Experimental Setup and Results

4.1 Ground Truth Data

To evaluate our method, a set of different stimuli with translatory and rotational motions were recorded using the DVS128 sensor. The rotational and translatory motions were generated using linear and rotational actuators, in which the linear actuator's speed is 20 cm/sec and the rotational actuator's speed is 5.23 rad/sec. The DVS sensor was mounted on a tripod and placed 23 cm away from the stimulus.

Table 1. Parameters used in our model.

Definition	variable	value
Spatial filter frequency	f_s	0.25
Motion directions	θ	0°,45°,90°,135°
Standard deviation of spatial filters	σ_s	2 pixel
Number of motion directions	N	4
Standard deviation of Gaussian function normalization	σ	15 pixel
leakage activities	A	0.01

The model parameters used for the illustrated results are shown in Table 1. The estimated results of the optic flow were based on the summed responses of r which generates a confidence for the motion direction $(u_e(\mathbf{p})\, v_e(\mathbf{p}))^T = \sum_\theta r(\mathbf{p}, \theta, f_s, t) \cdot (\cos\theta \quad - \sin\theta)^T$. Figures 4 and 5 present the translatory and rotational motion results respectively, where the stimulus image, ON/OFF events and ground truth are presented in the first row. The second row shows the estimated flow and the direction selectivity which is depicted in polar plot. In order to measure the accuracy of our approach, we calculated the angular error $\Phi(\mathbf{p}) = cos^{-1}(\mathbf{V}_e(\mathbf{p}) \cdot \mathbf{V}_g(\mathbf{p}))/(|\mathbf{V}_e(\mathbf{p})||\mathbf{V}_g(\mathbf{p})|)$, where $\mathbf{V}_e(\mathbf{p})^T = (u_e(\mathbf{p}), v_e(\mathbf{p}))$ and $\mathbf{V}_g(\mathbf{p})^T = (u_g(\mathbf{p}), v_g(\mathbf{p}))$ represent the estimated and ground truth flow vectors at position $\mathbf{p} = (x, y)$, respectively. The error values in the range of $[0°, 180°]$ are depicted as a histogram which is shown in the third column.

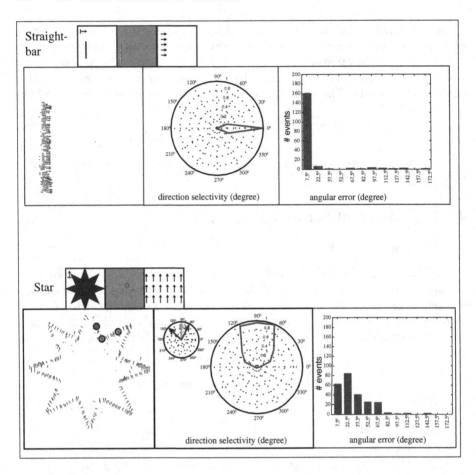

Fig. 4. Processing results of translatory motion stimuli, straight-bar and star. The first row of each stimulus contains the input image, the ON/OFF accumulated events (3 *msec.*) and a sketch of the ground truth optical flow field. The second row shows the estimated motion, the polar plot of the direction selectivity and the histogram of the angular error between the estimated motion and their respective ground truth. In the star stimulus, the smaller polar plot represents the direction selectivity at corner and bar contours while the larger polar plot represents the direction selectivity of the whole stimulus. The abscissa of the histogram represents the binning in the range of the angular errors Φ that are combined into single bins $[\theta-7.5°, \theta+7.5°)$. The ordinate represents the number of pixels.

In case of *translatory* motion, we used two stimuli: a straight bar and a star in which different directions were selected to move these stimuli. The straight bar stimulus was moved to the left while the star was moved vertically up the field of view. For the straight bar, the polar plot shows that the spatiotemporal filters are selective to the right direction (0°). According to the histogram, the angular error between the estimated flow and the ground truth flow reveals that

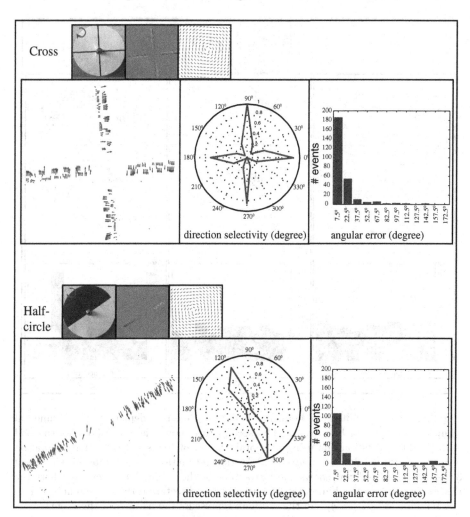

Fig. 5. Processing results of rotational motion stimuli, cross and half-circle. The first row of each stimulus contains the input image, the ON/OFF accumulated events (3 *msec.*) and the ground truth optical flow field. The second row represents the estimated motion, the polar plot of the direction selectivity and the histogram of the angular error between the estimated motion and their respective ground truth. The abscissa of the histogram represents the binning in the range of the angular errors Φ that are combined into single bins $[\theta - 7.5°, \theta + 7.5°)$. The ordinate represents the number of pixels.

motions were estimated with correct directions in which the error values are accumulated within a small range of $[0°, 15°)$. In the star stimulus, different slanted bars are connected to form the star shape. Since each bar generates motion components that suffer locally from the aperture problem the initial estimation and subsequent integration impose a challenge. The smaller polar plot

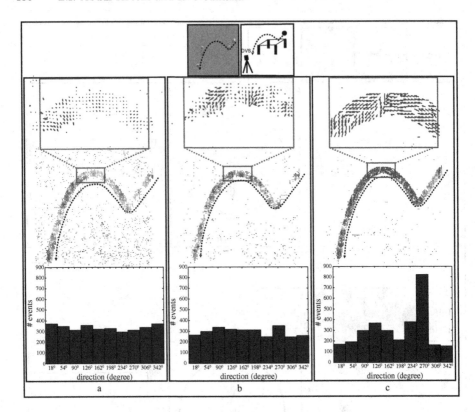

Fig. 6. Estimated motion of bouncing ball: The first row depicts the experimental setup and samples of ball events. The second row shows the optic flow for the ball path. (a) Motion estimation using temporal window of $20\,msec$. (b) Motion estimation using temporal window of $10\,msec$. (c) Motion estimation using temporal window of $3\,msec$. The second row shows the histograms of the estimated directions where the abscissa represents the binning in the rang of the direction θ that are combined into one bin $[\theta - 18°, \theta + 18°)$, and the ordinate represents the number of the events.

shows the direction selectivity at a particular position (corner and bar contours) while the larger polar plot shows the whole direction selectivity. Although an ambiguous motion was estimated along bar contours, real motion was estimated at the corner regions (aperture problem, see Sect. 5). The aperture problem can be resolved via the feedback of larger integration receptive field MT cells (for more details see [28]). The angular error histogram showed some spurious flow in 45° due to the small slanted lines. Moreover, a smaller spurious flow occurred $\Phi > 60°$ due to the low resolution of DVS (128×128) which gives rise to spatial aliasing.

In the case of *rotational* motion, again two stimuli were used: cross and half-circle. These stimuli were rotated clockwise. In the cross stimulus, the motion estimation showed a flow pattern on the stimulus blades in which motion was estimated with preferred direction of (0°, 90°, 180°, 270°). In the half-circle stimulus, the results showed a flow motion over the stimulus diagonal along the elongated contrast edge contour. Here, the direction selectivity of the spatiotemporal filters is tuned to 290° and 110°. In both stimuli, the results revealed appropriate flow estimation compared with the ground truth of the clockwise rotation in which most of the angular error was sandwiched between 15° and 30°.

4.2   Complex Realistic Movements

To demonstrate the usefulness of our model under realistic acquisition conditions, we extended our evaluation to include bouncing ball and articulated biological motions, Figs. 6 and 7. In the case of the bouncing ball, different projected velocities occur since the ball moves from a distant position towards the camera. This leads to the additional challenge for our model to estimate the motion when different velocities occur. We tested the influence of the temporal window size of events integration on the motion estimation. We used three different window sizes, 20 $msec$, 10 $msec$ and 3 $msec$, in which the first accumulated temporal period is equivalent to the typical sampling rate of a conventional frame-based imager.

Figure 6 shows that the motion estimation can be improved with decreasing the temporal window size. This referred to the higher sampling rate interval can capture small number of events and instantaneously transcribe their motion in contrast to the larger interval window that integrates more events over time space which leads to lose the intermediate motion details. In general, the result of smaller number of events acquisition, Fig. 6 (c), shows a proper estimation of flow direction comparing with other sampling cases.

Our model was tested using articulated movements in which real body motions are represented. Figure 7 shows two actions (jumping-jack and two hands waving) of an actor, where different movements and speeds were generated from body and limbs motions. The estimated motions for the two actions have been determined using a sampling rate of 3 $msec$ in which flow motions were obtained.

Fig. 7. Estimated flow of articulated motion of an actor. (a) Jumping-jack. (b) Two hands waving. The first column of each action represents original image input and the integrated ON/OFF events. The second column of each action represent the estimated motions.

5   Aperture Problem

Neurons in the primary visual cortical area V1 that are selective to spatio-temporal stimulus features have small RFs, or filter sizes. Consequently, they can only detect local motion components that occur within their RFs. That means that along elongated contrasts only ambiguous motion information can be detected locally. It is the normal flow component that can be measured along the local contrast gradient of the luminance function (aperture problem). In our test scenarios this has been investigated with input shown in Fig. 4. The aperture problem can be resolved either by utilizing local feature responses at corners, line ends, or junctions that belong to a single surface to be tracked. Another approach is to integrate several normal flow estimates at distant locations.

The integration strategy might be either based on vector integration (VA) [29] or on the intersection-of-constraints (IOC) mechanism, as suggested by [30]. The latter approach can be demonstrated to calculate the exact movement of, e.g., two overlapping gratings which move translatory behind a circular aperture in distinct directions (plaid), see Fig. 8. If the two gratings have same contrast and spatial frequency the plaid appears as a single pattern that moves in the direction of the intersecting normal flow constraint lines defined by the component gratings. This direction correspond to the feature motions generated by the grating intersections.

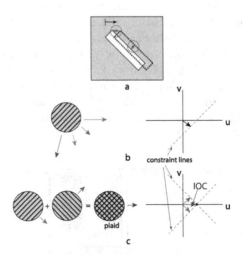

Fig. 8. Aperture problem and the velocity-space diagram of the IOC. (a) Elongated bar moves to the right direction, the circles represent the detector cells at the corner and center of the bar. (b) Single grating and a large family of motions identified by one constraint line (the dash line). (c) Plaid composes by two gratings and the IOC prediction for the motion of the plaid.

The IOC method could be used here as well by utilizing a voting scheme that is initialized by the normal flow components derived from the spatio-temporal filter responses as described above (in Sect. 3.2). Since the spatio-temporal weights of the filters already take into account the uncertainty of the detection and estimation process the IOC approach could be formulated within the Bayesian framework [31]. To implement this mechanism in our model, we could combine the local estimated motions from spatio-temporal filters and the likelihoods for the corresponding constraint lines. The IOC solution would then be the maximum likelihood response of the multiplied constraint component likelihoods. The integration of normal flow motions in the IOC is valid under the assumption that the contributions from component flows are generated by translatory motions. For rotational flows of an extended object, such as the ones shown in Fig. 5, the IOC (as well as the VA) does not yield the correct integrated motion estimation (compare [32]). The high temporal of input events delivered by the DVS sensor leads to motion components that can be considered as to mainly represent motion components tangential to a rotational sweep. However, since those local measures are noisy and need to be integrated over a temporal window, the rotational components become more prominent and gradually deteriorate the IOC solution.

In order to account for integrating local motion responses of unknown components and compositions, we further pursue a biologically inspired motion integration which is motivated by our own previous work reported in [26,28]. In this framework we utilize model mechanisms of cortical area MT that integrate initial V1 cell responses. The RF of cells in MT are larger in their size by up to an order of magnitude. In other words, such cells operate at a much larger spatial context to properly integrate localized responses, similar to the VA method. As a consequence, localized feature responses at line ends or corners lead to stronger responses in the integration process. In our model, this has been accomplished by weighting the responses of area V1 (Sect. 3.2) with larger RFs via area MT (V1:MT 1:3). This integration process can be denoted by

$$r_{\theta,\mathbf{p}}^{MT} = r_{\theta,\mathbf{p}}^{v1nor} * \Lambda_{\sigma_{MT}}, \tag{16}$$

where '$*$' defines the (spatial) convolution operator, θ denoting motion direction, $\mathbf{p} = (x,y)$ represents the spatial position of the cell. Λ is Gaussian weighting function of the spatial pooling operation. The latter is parametrized by the parameter σ_{MT} to denote the width of the spatial extent. Figure 9 shows how the aperture problem of star stimulus in Fig. 4 (star) is reduced after integrating the localize estimation of the spatiotemporal filters. Here, the direction selectivity in area MT, Fig. 9, is more tuned toward the correct direction ($90°$) comparing with Fig. 4 (star). According to the histograms, the error values of Fig. 9 are accumulated at range of $[0°, 15°)$ which refers that motion was estimated more accurately than in Fig. 4 (star).

In order to investigate the impact of the normalization process (Sect. 3.3) on motion estimation and aperture problem, a slanted bar stimulus ($45°$) was used as shown in Fig. 10. Many studies (see e.g., [33,34]) showed that firing strength of some visual cortex cells is decreased with increasing the length of the bar which is centered on its RF. This property is called end-stopping. In [35], the effect of the

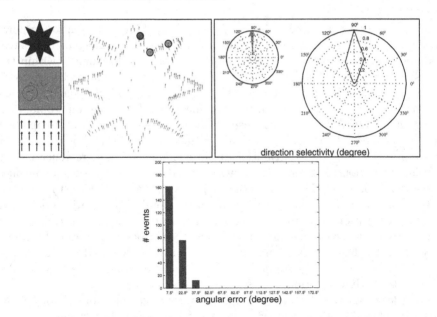

Fig. 9. Motion integration of star stimulus. The localize responses of the spatiotemporal filters are integrated with larger RFs (V1:MT 1:3). The size of the MT cell is depicted over the accumulated events. The direction selectivity is depicted via polar plots in which the larger polar plot shows the direction selectivity of the whole responses while the smaller polar plot shows the direction selectivity at corner and bar contours. The histogram demonstrates the angular error between the estimated motion and the ground truth of upward motion direction. The abscissa of the histogram represents the binning in the range of the angular errors Φ which are combined into one bar $[\theta - 7.5°, \theta + 7.5°)$, and the ordinate represents the number of events.

end-stopping property was emulated using an interaction between a center unit and six surround units and immediately after the filtering with the separated spatial filters only. In our model, we simulate the property of the end-stopping by a uniform surround integration that operates upon the weighted activities in the spatial and feature domain (see Sect. 3.3). To reduce the responses along the bar contour compared with the responses at the endpoints, we set the parameters A and E of Eq. 12 to 1 and 10, respectively. In Fig. 10, the bar was moved in a direction that differs from its normal flow as highlighted in the top-left of the input stimulus. Here, the responses of the filtering stage (Sect. 3.2) before and after the normalization process are shown in second row while the direction selectivity is shown in the third row. The results reveal that the activities at the endpoints of the bar are higher than the activities along the bar contour. As a consequences, normalization process enhances the response selectivity in which the direction selectivity is more tuned toward correct direction (180°).

We used multi-information (MI) [36] to evaluate the influence of the normalization stage to reduce the dependency of the motion selective cells. The MI function can be defined as the Kullback-Leibler divergence (see [18,37,38]) between the joint distribution and the product of its marginals

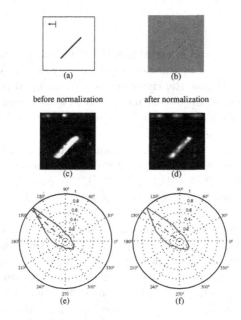

before normalization   after normalization

Fig. 10. Effect of normalization process on the motion estimation for slanted bar. (a) The input image. (b) The integrated ON/OFF events. (c) The responses of the filtering stage before normalization process. (d) The responses of the filtering stage after normalization process. (e) Polar plot of the direction selectivity of the filtering stage before normalization process. (f) Polar plot of the direction selectivity of the filtering stage after normalization process. The dash line denotes the median of the direction selectivity.

$$MI(I) = D_{KL}\left(p(I)\|\prod_k p(I_k)\right) = \sum_{x_1 \in I_1}\sum_{x_2 \in I_2}\cdots\sum_{x_d \in I_n} p(x_1, x_2, ..., x_n)$$
$$\cdot log\frac{p(x_1, x_2, ..., x_n)}{p(x_1)p(x_2)...p(x_n)}, \quad (17)$$

where $p(x_1, x_2, ..., x_n)$ represents the joint distribution, $p(x_k)$, $k = 1..n$ denotes a marginal distribution, with $n = 8$ defining the movement directions. The results show that $MI(I)$ is reduced from $MI(I) = 0.0509$ before normalization to $MI(I_{nor}) = 0.0112$ after the normalization stage. This indicates that the normalization process tends to reduce the dependency of the representation of the movement estimates, although it does not completely decorrelate the motion estimates (in the latter case the mutual information should reach $MI(I_{nor}) = 0$). Further investigations are necessary to indicate whether the reduction in $MI(I_{nor})$ as observed above is significant. Since we employ binary variables for the instances of the random variables, the effects of threshold definition and the number of discrete direction estimates need to be evaluated.

6   Discussion

In this paper, we introduced a neural model for motion estimation using neuro-morphic vision sensors. The neural model processing was inspired by the low-level filtering at the initial stage of the visual system. We adopted the spatio-temporal filtering model suggested in [8] and integrated new temporal filters to fit with AER principles. In addition, a normalization mechanism over the space-feature domain have been incorporated.

Many works have addressed motion estimation using the frame-based imager, which can be characterized as computer vision approaches, [9,10,14] and bio-inspired related models [8,39–41]. Recently, [13,16] carried out motion estima-tion using retina sensors in which the first article adopted a computer vision approach, while the second considered a bio-inspired model. Our approach con-tributed to bio-inspired motion estimation using DVS sensors by suggesting tem-poral filters consistent with polarity responses of the retina sensors. According to [8], the temporal filters in bio-inspired models defined as a smoothing functions with biphasic shape responses, in which temporal gamma functions of different duration were used to achieved temporal smoothing and differentiation. These functions can be approximately decomposed into a convolution of a numerical difference kernel with a temporal smoothing filters. Since already the represen-tation AER uses the first order temporal derivative, where the discrete events generated based on the *changes* in the input. Thus, we suggest to employ tem-poral smoothing filters and convolve them with the input stream of events to obtain scaled versions of temporally smoothed derivatives of the input luminance function.

In order to reduce the aperture problem, the local motion estimation of area V1 is integrated in model area MT. The RFs of cells in area MT are larger in their size by up to an order of magnitude. In other words, such cells operate at a much larger spatial context to properly integrate localized responses, similar to the VA method. As a consequence, localized feature responses at line ends or corners lead to stronger responses in the integration process.

Our model was tested using different kinds of stimuli. In many cases, the model shows accurate results for translatory motion estimation compared with synthetic ground truth. The error value increased in rotational motion cases due to the limited number of estimated directions compared with the ground truth. This drawback could be overcome by increasing the estimated directions of our model. The spatial low resolution of the DVS sensor has unfavorable impact on several locations of the star image case due to the spatial aliasing problem which leads to spurious estimations.

The size of the temporal sampling interval can affect the motion estima-tion results in which smaller temporal window size gives better estimation. This smaller window can capture more motion details since it accumulates the occurred events immediately and transcribes their motions instantaneously. As a consequence, the subtle information can be maintained.

Balancing the activations of the individual cells is achieved by the normal-ization process. This process operates in the spatial and directional domain.

Consequently, the overall cells activities are adjusted in a local region. The impact of such normalization stage is evaluated using information theoretic measures. The results show that the normalization process can enhance the responses of the endpoints of the bar that contains unambiguous motion. As a consequence, the direction selectivity of the filtering stage is more tuned toward the real motion direction.

Our model can be extended by considering feedback processing sweep from higher area (MT) to lower area (V1). The feedback projection modulates the initial responses of the lower areas and then uncertain flow estimation can be enhanced and improved.

Acknowledgements. LIAK. has been supported by grants from the Ministry of Higher Education and Scientific Research (MoHESR) Iraq and from the German Academic Exchange Service (DAAD). HN. acknowledges support from DFG in the Collaborative Research Center SFB/TR (A companion technology for cognitive technical systems). The authors would like to thank M. Schels for his help in recording biological motion.

References

1. Litzenberger, M., Belbachir, A.N., Donath, N., Gritsch, G., Garn, H., Kohn, B., Posch, C., Schraml, S.: Estimation of vehicle speed based on asynchronous data from a silicon retina optical sensor. In: IEEE Intelligent Transportation Systems Conference Toronto, Canada, pp. 17–20 (2006)
2. Liu, S., Delbruck, T.: Neuromorphic sensory systems. Neurobiology **20**, 288–295 (2010)
3. Lichtsteiner, P., Posch, C., Delbruck, T.: A 128 × 128 120 db 15 μs latency asynchronous temporal contrast vision sensor. IEEE J. Solid-State Circuits **43**, 566–576 (2008)
4. Delbruck, T., Lichtsteiner, P.: Fast sensory motor control based on event-based hybrid neuromorphic-procedural system. In: IEEE International Symposiom on Circuit and System, pp. 845–848 (2007)
5. Litzenberger, M., Posch, C., Bauer, D., Belbachir, A.N., Schon, P., Kohn, B., Garn, H.: Embedded vision system for real-time object tracking using an asynchronous transient vision sensor. In: 12th - Signal Processing Education Workshop. IEEE DSPW, pp. 173–178 (2006)
6. Ni, Z., Pacoret, C., Benosman, R., Ieng, S., Regnier, S.: Asynchronous event-based high speed vision for microparticle tracking. J. Microsc. **43**, 1365–2818 (2011)
7. Abdul-Kreem, L.I., Neumann, H.: Bio-inspired model for motion estimation using address event representation. In: 10th International Conference on Computer Vision Theory and Application, VISIGRAPP, Berlin, Germany, 11–14 March 2015
8. Adelson, E., Bergen, J.: Spatiotemporal energy models for the perception of motion. J. Opt. Soc. Am. **2**, 90–105 (1985)
9. Brox, T., Bruhn, A., Papenberg, N., Weickert, J.: High accuracy optical flow estimation based on a theory for warping. In: Pajdla, T., Matas, J.G. (eds.) ECCV 2004. LNCS, vol. 3024, pp. 25–36. Springer, Heidelberg (2004)
10. Drulea, M., Nedevschi, S.: Motion estimation using the correlation transform. IEEE Trans. Image Process. **22**, 1057–7149 (2013)

11. Fleet, D., Jepson, A.: Computation of component image velocity from local phase information. Int. J. Comput. Vis. **5**, 77–104 (1990)
12. Horn, B., Schunck, B.: Determining optical flow. Artif. Intell. **17**, 185–203 (1981)
13. Benosman, R., Leng, S., Clercq, C., Bartolozzi, C., Srinivasan, M.: Asynchronous framless event-based opticlal flow. Neural Netw. **27**, 32–37 (2012)
14. Lucas, B.D., Kanade, T.: An iterative image registration technique with and application to stereo vision. In: Proceedings of Imaging Understanding Workshop, pp. 121–130 (1981)
15. Tschechne, S., Brosch, T., Sailer, R., Egloffstein, N., Abdul-Kreem, L.I., Neumann, H.: On event-based motion detection and integration. In: Proceedings of 8th International Conference on Bio-inspired Information and Communication Technologies, BICT, December 1–3, Boston, MA, USA. ACM digital library (2014)
16. Tschechne, S., Sailer, R., Neumann, H.: Bio-inspired optic flow from event-based neuromorphic sensor input. In: El Gayar, N., Schwenker, F., Suen, C. (eds.) ANNPR 2014. LNCS, vol. 8774, pp. 171–182. Springer, Heidelberg (2014)
17. De Valois, R., Cottarisb, N.P., Mahonb, L.E., Elfara, S.D., Wilsona, J.A.: Spatial and temporal receptive fields of geniculate and cortical cells and directional selectivity. Vis. Res. **40**, 3685–3702 (2000)
18. Brosch, T., Tschechne, S., Neumann, H.: On event-based optical flow detection. Front. Neurosci. **9**, Article No. 137, 1–15 (2015)
19. Ringach, D.L.: Spatial structure and symmetry of simple-cell receptive fields in macaque primary visual cortex. Neurophysiology **88**, 455–463 (2002)
20. Carandini, M., Heeger, D.J.: Normalization as a canonical neural computation. Nat. Rev. Neurosci. **13**, 51–62 (2012)
21. Brosch, T., Neumann, H.: Computing with a canonical neural circuits model with pool normalization and modulating feedback. Neural Comput. **26**, 2735–2789 (2014)
22. Blomfield, S.: Arithmetical operations performed by nerve cells. Brain Res. **69**, 115–124 (1974)
23. Dayan, P., Abbot, L.F.: Theoretical Neuroscience. MIT Press, Cambridge (2001)
24. Silver, R.A.: Neuronal arithmetic. Nat. Rev. Neurosci. **11**, 474–489 (2010)
25. Grossberg, S.: Nonlinear neural networks: principles, mechanisms, and architectures. Neural Netw. **1**, 17–61 (1988)
26. Bouecke, J., Tlapale, E., Kornprobst, P., Neumann, H.: Neural mechanisms of motion detection, integration, and segregation: from biology to artificial image processing systems. EURASIP J. Adv. Signal Process. **2011**, Article ID 781561, 22 (2010). doi:10.1155/2011/781561
27. Lyu, S., Simoncelli, E.P.: Nonlinear extraction of independent components of natural images using radial gaussianization. Neural Comput. **21**, 1485–1519 (2009)
28. Bayerl, P., Neumann, H.: Disambiguating visual motion through contextual feedback modulation. Neural Comput. **16**, 2041–2066 (2004)
29. Yo, C., Wilson, H.: Perceived direction of moving two-dimensional patterns depends on duration, contrast and eccentricity. Vis. Res. **32**, 135–147 (1992)
30. Adelson, E., Movshon, J.: Phenomenal coherence of moving visual pattern. Nature **300**, 523–525 (1982)
31. Simoncelli, E.: Bayesian multiscale differential optical flow. In: Handbook of Computer Vision and Applications, Chap. 14. Academic Press (1999)
32. Caplovitz, G., Hsieh, P., Tse, P.: Mechanisms underlying the perceived angular velocity of a rigidly rotating object. Vis. Res. **46**, 2877–2893 (2006)
33. Hubel, D.H., Wiesel, T.N.: Receptive fields and functional architecture in two nonstriate visual areas (18 and 19) of the cat. J. Neurophysiol. **28**, 229–289 (1965)

34. Pack, C.C., Livingstone, M.S., Duffy, K.R., Born, R.T.: End-stopping and the aperture problem: two-dimensional motion signals in macaque v1. Neuron **39**, 671–680 (2003)
35. Tsui, J.M.G., Hunter, N., Born, R.T., Pack, C.C.: The role of v1 surround suppression in mt motion integration. J. Neurophysiol. **24**, 3123–3138 (2010)
36. Studený, M., Vejnarová, J.: The multiinformation function as a tool for measuring stochastic dependence. In: Jordan, M.I. (ed.) Learning in Graphical Models. NATO ASI Series, vol. 89, pp. 261–297. Springer, Heidelberg (1998). (Kluwer Academic Publishers)
37. Cover, T.M., Thomas, J.A.: Elements of Information Theory, 2nd edn. Wiley, Hoboken (2006)
38. Lyu, S., Simoncelli, E.P.: Nonlinear extraction of independent components of natural images using radial gaussianization. Neural Comput. **21**, 1485–1519 (2009)
39. Strout, J.J., Pantle, A., Mills, S.L.: An energy model of interframe interval effects in single-step apparent motion. Vis. Res. **34**, 3223–3240 (1994)
40. Emerson, R.C., Bergen, J.R., Adelson, E.H.: Directionally selective complex cells and the computation of motion energy in cat visual cortex. Vis. Res. **32**, 203–218 (1992)
41. Challinor, K.L., Mather, G.: A motion-energy model predicts the direction discrimination and mae duration of two-stroke apparent motion at high and low retinal illuminance. Visi. Res. **50**, 1109–1116 (2010)

CURFIL: A GPU Library for Image Labeling with Random Forests

Hannes Schulz[✉], Benedikt Waldvogel, Rasha Sheikh, and Sven Behnke

Autonomous Intelligent Systems, Computer Science Institute VI, University of Bonn,
Friedrich-Ebert-Allee 144, 53113 Bonn, Germany
schulz@ais.uni-bonn.de, mail@bwaldvogel.de, rasha@uni-bonn.de,
behnke@cs.uni-bonn.de

Abstract. Random forests are popular classifiers for computer vision tasks such as image labeling or object detection. Learning random forests on large datasets, however, is computationally demanding. Slow learning impedes model selection and scientific research on image features. We present an open-source implementation that significantly accelerates both random forest learning and prediction for image labeling of RGB-D and RGB images on GPU when compared to an optimized multi-core CPU implementation. We further use the fast training to conduct hyper-parameter searches, which significantly improves on earlier results on the NYU depth v2 dataset. Our flexible implementation allows to experiment with novel features, such as height above ground, which further increases classification accuracy. CURFIL prediction runs in real time at VGA resolution on a mobile GPU and has been used as data term in multiple applications.

Keywords: Random forest · Computer vision · Image labeling · GPU · CUDA

1 Introduction

Random forests are ensemble classifiers that are popular in the computer vision community. Random decision trees are used when the hypothesis space at every node is huge, so that only a random subset can be explored during learning. This restriction is countered by constructing an ensemble of independently learned trees—the random forest.

Variants of random forests were used in computer vision to improve e.g. object detection or image segmentation. One of the most prominent examples is the work of Shotton et al. (2011), who use random forests in Microsoft's Kinect system for the estimation of human pose from single depth images. Here, we are interested in the more general task of image labeling, i.e. determining a label for every pixel in an RGB or RGB-D image (Fig. 1).

The real-time applications such as the ones presented by Lepetit et al. (2005) and Shotton et al. (2011) require fast prediction in few milliseconds per image.

© Springer International Publishing Switzerland 2016
J. Braz et al. (Eds.): VISIGRAPP 2015, CCIS 598, pp. 416–432, 2016.
DOI: 10.1007/978-3-319-29971-6_22

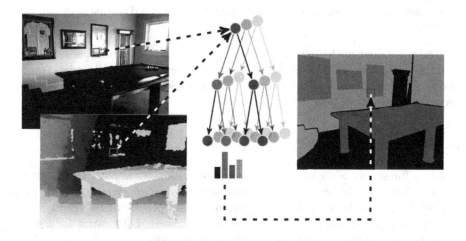

Fig. 1. Overview of image labeling with random forests: Every pixel (RGB and depth) is classified independently based on its context by the trees of a random forest. The leaf distributions of the trees determine the predicted label (Color figure Online).

This is possible with parallel architectures such as GPUs, since every pixel can be processed independently. Random forest training for image labeling, however, is not as regular—it is a time consuming process. To evaluate a randomly generated feature candidate in a single node of a single tree, a potentially large number of images must be accessed. With increasing depth, the number of pixels in an image arriving in the current node can be very small. It is therefore essential for the practitioner to optimize memory efficiency in various regimes, or to resort to large clusters for the computation. Furthermore, changing the visual features and other hyper-parameters requires a re-training of the random forest, which is costly and impedes efficient scientific research.

This work describes the architecture of our open-source GPU implementation of random forests for image labeling (CURFIL). CURFIL provides optimized CPU and GPU implementations for the training and prediction of random forests. Our library trains random forests up to 26 times faster on GPU than our optimized multi-core CPU implementation. Prediction is possible in real-time speed on a single mobile GPU.

In short, our contributions are as follows:

1. we describe how to efficiently implement random forests for image labeling on GPU,
2. we describe a method which allows to train on horizontally flipped images at significantly reduced cost,
3. we show that our GPU implementation is up to 26 times faster for training (up to 48 times for prediction) than an optimized multi-core CPU implementation,
4. we show that simply by the now feasible optimization of hyper-parameters, we can improve performance in two image labeling tasks, and

5. we make our documented, unit-tested, and MIT-licensed source code publicly available[1].

The remainder of this paper is organized as follows. After discussing related work, we introduce random forests and our node tests in Sects. 3 and 4, respectively. We describe our optimizations in Sect. 5. Section 6 analyzes speed and accuracy attained with our implementation.

2   Related Work

Random forests were popularized in computer vision by Lepetit et al. (2005). Their task was to classify patches at pre-selected keypoint locations, not—as in this work—all pixels in an image. Random forests proved to be very efficient predictors, while training efficiency was not discussed. Later work focused on improving the technique and applying it to novel tasks.

Lepetit and Fua (2006) use random forests to classify keypoints for object detection and pose estimation. They evaluate various node tests and show that while training is increasingly costly, prediction can be very fast.

The first GPU implementation for our task was presented by Sharp (2008), who implements random forest training and prediction for Microsoft's Kinect system that achieves a prediction speed-up of 100 and training speed-up factor of eight on a GPU, compared to a CPU. This implementation is not publicly available and uses Direct3D which is only supported on the Microsoft Windows platform.

An important real-world application of image labeling with random forests is presented by Shotton et al. (2011). Human pose estimation is formulated as a problem of determining pixel labels corresponding to body parts. The authors use a distributed CPU implementation to reduce the training time, which is nevertheless one day for training three trees from one million synthetic images on a 1,000 CPU core cluster. Their implementation is also not publicly available.

Several fast implementations for general-purpose random forests are available, notably in the *scikit-learn* machine learning library (Pedregosa et al., 2011) for CPU and *CudaTree* (Liao et al., 2013) for GPU. General random forests cannot make use of texture caches optimized for images though, i.e., they treat all samples separately. GPU implementations of general-purpose random forests also exist, but due to the irregular access patterns when compared to image labeling problems, their solutions were found to be inferior to CPU (Slat and Lapajne, 2010) or focused on prediction (Van Essen et al., 2012).

The prediction speed and accuracy of random forests facilitates applications interfacing computer vision with robotics, such as semantic prediction in combination with self localization and mapping (Stückler et al., 2012) or 6D pose estimation (Rodrigues et al., 2012) for bin picking.

CURFIL was successfully used by Stückler et al. (2013) to predict and accumulate semantic classes of indoor sequences in real-time, and by Müller and Behnke (2014) to significantly improve image labeling accuracy on a benchmark dataset.

---

[1] https://github.com/deeplearningais/curfil/.

Since this library was developed, convolutional neural networks (CNN) have been shown to outperform random forests in terms of accuracy. Similar to random forests, CNN can also profit from height above ground and depth normalization (Schulz et al., 2015), but even more from transfer learning (e.g. Eigen and Fergus, 2014; Gupta et al., 2014) using the ImageNet dataset. The purpose of CURFIL, however, is to provide a library for fast training and prediction. This allows us to conduct hyper-parameter searches and to employ the trained model on mobile GPUs in realtime.

3   Random Forests

Random forests—also known as random decision trees or random decision forests—were independently introduced by Ho (1995) and Amit and Geman (1997). Breiman (2001) coined the term "random forest". Random decision forests are ensemble classifiers that consist of multiple decision trees—simple, commonly used models in data mining and machine learning. A decision tree consists of a hierarchy of questions that are used to map a multi-dimensional input value to an output which can be either a real value (regression) or a class label (classification). Our implementation focuses on classification but can be extended to support regression.

To classify input x, we traverse each of the K decision trees T_k of the random forest $\mathcal{F}$, starting at the root node. Each inner node defines a test with a binary outcome (i.e. true or false). We traverse to the left child if the test is positive and continue with the right child otherwise. Classification is finished when a leaf node $l_k(x)$ is reached, where either a single class label or a distribution $p(c \mid l_k(x))$ over class labels $c \in \mathcal{C}$ is stored.

The K decision trees in a random forest are trained independently. The class distributions for the input x are collected from all leaves reached in the decision trees and combined to generate a single classification. Various combination functions are possible. We implement majority voting and the average of all probability distributions as defined by

$$p(c \mid \mathcal{F}, x) = \frac{1}{K} \sum_{k=1}^{K} p(c \mid l_k(x)). \tag{1}$$

A key difference between a decision tree and a random decision tree is the training phase. The idea of random forests is to train multiple trees on different random subsets of the dataset and random subsets of features. In contrast to normal decision trees, random decision trees are not pruned after training, as they are less likely to overfit (Breiman, 2001). Breiman's random forests use CART as tree growing algorithm and are restricted to binary trees for simplicity. The best split criterion in a decision node is selected according to a score function measuring the separation of training examples. CURFIL supports information gain and normalized information gain (Wehenkel and Pavella, 1991) as score functions.

A special case of random forests are random ferns, which use the same feature in all nodes of a hierarchy level. While our library also supports ferns, we do not discuss them further in this paper, as they are neither faster to train nor did they produce superior results.

4   Visual Features for Node Tests

Our selection of features was inspired by Lepetit et al. (2005)—the method for visual object detection proposed by Viola and Jones (2001). We implement two types of RGB-D image features as introduced by Stückler et al. (2012). They resemble the features of Sharp (2008); Shotton et al. (2011)—but use depth-normalization and region averages instead of single pixel values. Shotton et al. (2011) avoid the use of region averages to keep computational complexity low. For RGB-only datasets, we employ the same features but assume constant depth. The features are visualized in Fig. 2.

For a given query pixel $\mathbf{q}$, the image feature f_θ is calculated as the difference of the average value of the image channel ϕ_i in two rectangular regions R_1, R_2 in the neighborhood of $\mathbf{q}$. Size w_i, h_i and 2D offset $\mathbf{o}_i$ of the regions are normalized by the depth $d(\mathbf{q})$:

$$f_\theta(\mathbf{q}) := \frac{1}{|R_1(\mathbf{q})|} \sum_{\mathbf{p} \in R_1} \phi_1(\mathbf{p}) - \frac{1}{|R_2(\mathbf{q})|} \sum_{\mathbf{p} \in R_2} \phi_2(\mathbf{p})$$

$$R_i(\mathbf{q}) := \left(\mathbf{q} + \frac{\mathbf{o}_i}{d(\mathbf{q})}, \frac{w_i}{d(\mathbf{q})}, \frac{h_i}{d(\mathbf{q})} \right). \tag{2}$$

Fig. 2. Sample visual feature at three different query pixels. Feature response is calculated from difference of average values in two offset regions. Relative offset locations $\mathbf{o}_i$ and region extents w_i, h_i are normalized with the depth $d(\mathbf{q})$ at the query pixel $\mathbf{q}$.

CURFIL optionally fills in missing depth measurements. We use integral images to efficiently compute region sums. The large space of eleven feature parameters—region sizes, offsets, channels, and thresholds—requires to calculate feature responses on-the-fly since pre-computing all possible values in advance is not feasible.

5   CURFIL Software Package

CURFIL's speed is the result of careful optimization of GPU memory throughput. This is a non-linear process to find fast combinations of memory layouts, algorithms and exploitable hardware capabilities. In the following, we describe the most relevant aspects of our implementation.

User API. The CURFIL software package includes command line tools as well as a library for random forest training and prediction. Inputs consist of images for RGB, depth, and label information. Outputs are forests in JSON format for training and label-images for prediction. Datasets with varying aspect ratios are supported.

Our source code is organized such that it is easy to improve and change the existing visual feature implementation. It is developed in a test-driven process. Unit tests cover major parts of our implementation.

CPU Implementation. Our CPU implementation is based on a refactored, parallelized and heavily optimized version of the *Tuwo Computer Vision Library*[2] by

---

Algorithm 1. Training of random decision tree

Require: $\mathcal{D}$ training instances
Require: F number of feature candidates to generate
Require: P number of feature parameters
Require: T number of thresholds to generate
Require: stopping criterion (e.g. maximal depth)
  1: $D \leftarrow$ randomly sampled subset of $\mathcal{D}$ $(D \subset \mathcal{D})$
  2: $N_{\text{root}} \leftarrow$ create root node
  3: $C \leftarrow \{(N_{\text{root}}, D)\}$ ▷ initialize candidate nodes
  4: **while** $C \neq \emptyset$ **do**
  5:     $C' \leftarrow \emptyset$ ▷ initialize new set of candidate nodes
  6:     **for all** $(N, D) \in C$ **do**
  7:         $(D_{\text{left}}, D_{\text{right}}) \leftarrow$ EVALBESTSPLIT(D)
  8:         **if** $\neg$STOP(N, D_{left}) **then**
  9:             $N_{\text{left}} \leftarrow$ create left child for node N
10:             $C' \leftarrow C' \cup \{(N_{\text{left}}, D_{\text{left}})\}$
11:         **if** $\neg$STOP(N, D_{right}) **then**
12:             $N_{\text{right}} \leftarrow$ create right child for node N
13:             $C' \leftarrow C' \cup \{(N_{\text{right}}, D_{\text{right}})\}$
14:     $C \leftarrow C'$ ▷ continue with new set of nodes

---

[2] http://www.nowozin.net/sebastian/tuwo/.

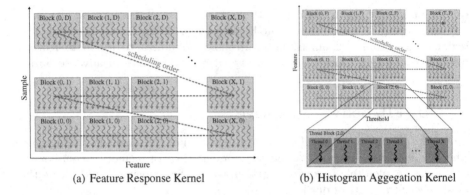

(a) Feature Response Kernel   (b) Histogram Aggegation Kernel

Fig. 3. (a) Two-dimensional grid layout of the feature response kernel for D samples and F features. Each block contains n threads. The number of blocks in a row, X, depends on the number of features. $X = \lceil F/n \rceil$. Feature responses for a given sample are calculated by the threads in one block row. The arrow (red dashes) indicates the scheduling order of blocks. (b) Thread block layout of the histogram aggregation kernel for F features and T thresholds. One thread block per feature and per threshold. X threads in block aggregate histogram counters for D samples in parallel. Every thread iterates over at most $\lceil D/x \rceil$ samples (Color figure online).

Nowozin. Our optimizations make better use of CPU cache by looping over feature candidates and thresholds in the innermost loop, and by sorting the dataset according to image ID before learning. Since feature candidate evaluations do not depend on each other, we can parallelize over the training set and make use of all CPU cores even when training only a single tree.

GPU Implementation. Evaluation of the optimized random forest training on CPU (Algorithm 1) shows that the vast majority of time is spent in the evaluation of the best split feature. This is to our benefit when accelerating random forest training on GPU. We restrict the GPU implementation efforts to the relatively short feature evaluation algorithm (Algorithm 2) as a drop-in replacement and leave the rest of the CPU computation unchanged. We use the CPU implementation as a reference for the GPU and ensure that results are the same in both implementations.

Split evaluation can be divided into the following four phases that are executed in sequential order:

1. random feature and threshold candidate generation,
2. feature response calculation,
3. histogram aggregation for all features and threshold candidates, and
4. impurity score (information gain) calculation.

Each phase depends on results of the previous phase. As a consequence, we cannot execute two or more phases in parallel. The CPU can prepare data for the launch of the next phase, though, while the GPU is busy executing the current phase.

Algorithm 2. CPU-optimized feature evaluation

Require: D samples
Require: $\mathbf{F} \in \mathbb{R}^{F \times P}$ random feature candidates
Require: $\mathbf{T} \in \mathbb{R}^{F \times T}$ random threshold candidates
 1: initialize histograms for every feature/threshold
 2: **for all** $d \in D$ **do**
 3:     **for all** $f \in 1 \ldots F$ **do**
 4:         calculate feature response
 5:         **for all** $\theta \in \mathbf{T}_f$ **do**
 6:             update according histogram
 7: calculate impurity scores for all histograms
 8: **return** histogram with best score

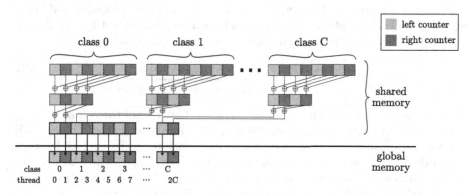

Fig. 4. Reduction of histogram counters. Every thread sums to a dedicated left and right counter (indicated by different colors) for each class (first row). Counters are reduced in a subsequent phase. The last reduction step stores counters in shared memory, such that no bank conflicts occur when copying to global memory (Color figure online).

5.1   GPU Kernels

Random Feature and Threshold Candidate Generation. A significant amount of training time is used for generating random feature candidates. The total time for feature generation increases per tree level since the number of nodes increases as trees are grown.

The first step in the feature candidate generation is to randomly select feature parameter values. These are stored in a $F \times 11$ matrix for F feature candidates and eleven feature parameters of Eq. (2). The second step is the selection of one or more thresholds for every feature candidate. Random threshold candidates can either be obtained by randomly sampling from a distribution or by sampling feature responses of training instances. We implement the latter approach, which allows for greater flexibility if features or image channels are changed. For every feature candidate generation, one thread on the GPU is used and all T thresholds for a given feature are sampled by the same thread.

In addition to sorting samples according to the image they belong to, feature candidates are sorted by the feature type, channels used, and region offsets. Sorting reduces branch divergence and improves spatial locality, thereby increasing the cache hit rate.

Feature Response Calculation. The GPU implementation uses a similar optimization technique to the one used on the CPU, where loops in the feature generation step are rearranged in order to improve caching.

We used one thread to calculate the feature response for a given feature and a given training sample. Figure 3(a) shows the thread block layout for the feature response calculation. A row of blocks calculates all feature responses for a given sample. A column of blocks calculates the feature responses for a given feature over all samples. The dotted red arrow indicates the order of thread block scheduling. The execution order of thread blocks is determined by calculating the Block ID *bid*. In the two-dimensional case, it is defined as

$$bid = \text{blockIdx}.x + \underbrace{\text{gridDim}.x}_{\text{blocks in row}} \cdot \underbrace{\text{blockIdx}.y}_{\text{sample ID}}. \tag{3}$$

The number of features can exceed the maximum number of threads in a block; therefore, the feature response calculation is split into several thread blocks. We use the x coordinate in the grid for the feature block to ensure that all features are evaluated before the GPU continues with the next sample. The y coordinate in the grid assigns training samples to thread blocks. Threads reconstruct their feature ID f using block size, thread and block ID by calculating

$$f = \text{threadIdx}.x + \underbrace{\text{blockDim}.x}_{\text{threads in block row}} \cdot \underbrace{\text{blockIdx}.x}_{\text{block index in grid row}}. \tag{4}$$

After sample data and feature parameters are loaded, the kernel calculates a single feature response for a depth or color feature by querying four pixels in an integral image and carrying out simple arithmetic operations to calculate the two regions sums and their difference.

Histogram Aggregation. Feature responses are aggregated into class histograms. Counters for histograms are maintained in a four-dimensional matrix of size $F \times T \times C \times 2$ for F features, T thresholds, C classes, and the two left and right children of a split.

To compute histograms, the iteration over features and thresholds is implemented as thread blocks in a two-dimensional grid on GPU; one thread block per feature and threshold. This is depicted in Fig. 3(b). Each thread block slices samples into partitions such that all threads in the block can aggregate histogram counters in parallel.

Histogram counters for one feature and threshold are kept in the shared memory, and every thread gets a distinct region in the memory. For X threads and C classes, $2XC$ counters are allocated. An additional reduction phase is then required to reduce the counters to a final sum matrix of size $C \times 2$ for every feature and threshold.

Figure 4 shows histogram aggregation and sum reduction. Every thread increments a dedicated counter for each class in the first phase. In the next phase, we iterate over all C classes and reduce the counters of every thread in $O(\log X)$ steps, where X is the number of threads in a block. In a single step, every thread calculates the sum of two counters. The loop over all classes can be executed in parallel by $2C$ threads that copy the left and right counters of C classes.

The binary reduction of counters (Fig. 4) has a constant runtime overhead per class. The reduction of counters for classes without samples can be skipped, as all counters are zero in this case.

Impurity Score Calculation. Computing impurity scores from the four-dimensional counter matrix is the last of the four training phases that are executed on GPU.

In the score kernel computation, 128 threads per block are used. A single thread computes the score for a different pair of features and thresholds. It loads $2C$ counters from the four-dimensional counter matrix in global memory, calculates the impurity score and writes back the resulting score to global memory.

The calculated scores are stored in a $T \times F$ matrix for T thresholds and F features. The matrix is then finally transferred from device to host memory space.

Undefined Values. Image borders and missing depth values (e.g. due to material properties or camera disparity) are represented as NaN, which automatically propagates and causes comparisons to produce *false*. This is advantageous, since no further checks are required and the random forest automatically learns to deal with missing values.

5.2   Global Memory Limitations

Slicing of Samples. Training arbitrarily large datasets with many samples can exceed the storage capacity of global memory. The feature response matrix of size $D \times F$ scales linearly in the number of samples D and the number of feature candidates F. We cannot keep the entire matrix in global memory if D or F is too large. For example, training a dataset with 500 images, 2000 samples per image, 2000 feature candidates and double precision feature responses (64 bit) would require $500 \cdot 2000 \cdot 2000 \cdot 64\,\text{bit} \approx 15\,\text{GB}$ of global memory for the feature response matrix in the root node split evaluation.

To overcome this limitation, we split samples into partitions, sequentially compute feature responses, and aggregate histograms for every partition. The maximum possible partition size depends on the available global memory of the GPU.

Image Cache. Given a large dataset, we might not be able to keep all images in the GPU global memory. We implement an image cache with a last recently used (LRU) strategy that keeps a fixed number of images in memory. Slicing samples ensures that a partition does not require more images than can be fit into the cache.

Memory Pooling. To avoid frequent memory allocations, we reuse memory that is already allocated but no longer in use. Due to the structure of random decision trees, evaluation of the root node split criterion is guaranteed to require the largest amount of memory, since child nodes always contain less or equal samples than the root node. Therefore, all data structures have at most the size of the structures used for calculating the root node split. With this knowledge, we are able to train a tree with no memory reallocation.

5.3   Extensions

Hyper-Parameter Optimization. Cross-validating all the hyper-parameters is a requirement for model comparison, and random forests have quite a few hyper-parameters, such as stopping criteria for splitting, number of features and thresholds generated, and the feature distribution parameters.

To facilitate model comparison, CURFIL includes support for cross-validation and a client for an informed search of the best parameter setting using Hyperopt (Bergstra et al., 2011). This allows to leverage the improved training speed to run many experiments serially and in parallel.

Image Flipping. To avoid overfitting, the dataset can be augmented using transformations of the training dataset. One possibility is to add horizontally flipped images, since most tasks are invariant to this transformation. CURFIL supports training horizontally flipped images with reduced overhead.

Table 1. Comparison of random forest *training* time on a quadcore CPU and two non-mobile GPUs. Random forest parameters were chosen for best accuracy.

	NYU		MSRC	
Device	time [min]	speed-up [×]	time [min]	speed-up [×]
i7–4770K	369	1.0	93.2	1.0
Tesla K20c	55	6.7	5.1	18.4
GTX Titan	24	15.4	3.4	25.9

Table 2. Random forest *prediction* time in milliseconds, on RGB-D images at original resolution, comparing speed on a recent quadcore CPU and various GPUs. Random forest parameters are chosen for best accuracy.

	NYU		MSRC-21	
Device	time [min]	speed-up [×]	time [min]	speed-up [×]
i7-440K	477	1	409	1
GTX 675M	28	17	37	11
Tesla K20c	14	34	10	41
GTX Titan	12	39	9	48

Instead of augmenting the dataset with flipped images and doubling the number of pixels used for training, we horizontally flip each of the two rectangular regions used as features for a sampled pixel. This is equivalent to computing the feature response of the same feature for the same pixel on an actual flipped image. Histogram counters are then incremented following the binary test of both feature responses. The implicit assumption here is that the samples generated through flipping are independent.

The paired sample is propagated down a tree until the outcome of a node binary test is different for the two feature responses, indicating that a sample and its flipped counterpart should split into different directions. A copy of the sample is then created and added to the samples list of the other node child.

This technique reduces training time since choosing independent samples from actually flipped images requires loading more images in memory during the best split evaluation step. Since our performance is largely bounded by memory throughput, dependent sampling allows for higher throughput at no cost in accuracy.

6   Experimental Results

We evaluate our library on two common image labeling tasks, the NYU Depth v2 dataset and the MSRC-21 dataset. We focus on the processing speed, but also discuss the prediction accuracies attained. Note that the speed between datasets is not comparable, since dataset sizes differ and the forest parameters were chosen separately for best accuracy.

6.1   Datasets

The NYU Depth v2 dataset by Silberman et al. (2012) contains 1,449 densely labeled pairs of aligned RGB-D images from 464 indoor scenes. We focus on the

Table 3. Segmentation accuracies on NYU Depth v2 dataset of our random forest compared to state-of-the-art methods trained only on this dataset. We used the same forest as in the training/prediction time comparisons of Tables 1 and 2.

	Accuracy [%]	
Method	Pixel	Class
Silberman et al. (2012)	59.6	58.6
Couprie et al. (2013)	63.5	64.5
Our random forest[a]	68.1	65.1
Our random forest[a] (with height, cf. Sect. 6.5)	69.6	66.5
Stückler et al. (2013)[b]	70.6	66.8
Hermans et al. (2014)	68.1	69.0
Müller and Behnke et al. (2014)[b]	72.3	71.9

[a]see main text for hyperparameters used.
[b]based on our random forest prediction (without height).

semantic classes *ground, furniture, structure*, and *props* defined by Silberman et al. (2012).

To evaluate our performance without depth, we use the MSRC-21 dataset[3]. Here, we follow the literature in treating rarely occuring classes *horse* and *mountain* as *void* and train/predict the remaining 21 classes on the standard split of 335 training and 256 test images.

6.2   Training and Prediction Time

Tables 1 and 2 show random forest training and prediction times, respectively, on an Intel Core i7-4770K (3.9 GHz) quadcore CPU and various NVidia GPUs. Note that the CPU version is using all cores.

For the RGB-D dataset, training speed is improved from 369 min to 24 min, which amounts to a speed-up factor of 15. Dense prediction improves by factor of 39 from 477 ms to 12 ms.

Training on the RGB dataset is finished after 3.4 min on a GTX Titan, which is 26 times faster than CPU (93 min). For prediction, we achieve a speed-up of 48 on the same device (9 ms vs. 409 ms).

Prediction is fast enough to run in real time even on a mobile GPU (GTX 675M, on a laptop computer fitted with a quadcore i7-3610QM CPU), with 28 ms (RGB-D) and 37 ms (RGB).

Fig. 5. Image labeling examples on NYU Depth v2 dataset. Left to right: RGB image, depth visualization, ground truth, random forest segmentation.

---

[3] http://jamie.shotton.org/work/data.html.

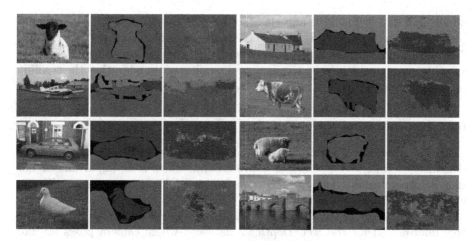

Fig. 6. Image labeling examples on the MSRC-21 dataset. In groups of three: input image, ground truth, random forest segmentation. Last row shows typical failure cases.

6.3 Classification Accuracy

Our implementation is fast enough to train hundreds of random decision trees per day on a single GPU. This fast training enabled us to conduct an extensive parameter search with five-fold cross-validation to optimize segmentation accuracy of a random forest trained on the NYU Depth v2 dataset (Silberman et al., 2012). Table 3 shows that we outperform other state-of-the art methods simply by using a random forest with optimized parameters. The resulting model and the fast CURFIL prediction were used in two publications which improved the results further by 3D accumulation of predictions in real time (Stückler et al., 2013) and superpixel CRFs (Müller and Behnke, 2014). This shows that efficient hyper-parameter search is crucial for model selection. Example segmentations are displayed in Figs. 5 and 6.

Methods on the established RGB-only MSRC-21 benchmark are so advanced that their accuracy cannot simply be improved by a random forest with better hyperparameters. Our pixel and class accuracies for MSRC-21 are 59.2 % and 47.0 %, respectively. This is still higher than other published work using random forests as the baseline method, such as 49.7 % and 34.5 % by Shotton et al. (2008). However, as Shotton et al. (2008) and the above works show, random forest predictions are fast and constitute a good initialization for other methods such as conditional random fields.

6.4 Image Mirroring Training Speed and Accuracy

Finally, we trained the MSRC-21 dataset by augmenting the dataset with horizontally flipped images using the naïve approach and our proposed method. The naïve approach doubles both the total number of samples and the number of images, which quadruples the training time to 14.4 min. Accuracy increases to

60.6 % and 48.6 % for pixel and class accuracy, respectively. With paired samples (introduced in Sect. 5.3), we reduce the runtime by a factor of two (to now 7.48 min) at no cost in accuracy (60.9 % and 49.0 %). The remaining difference in speed is mainly explained by the increased number of samples, thus the training on flipped images has very little overhead.

6.5 Incorporating Novel Features

With few changes in code, CURFIL allows to incorporate novel features. To demonstrate this, we chose height above ground, which is an important cue for indoor scene classification, and has been used in multiple other studies (Gupta et al., 2014; Müller and Behnke, 2014; Schulz et al., 2015). On a robot with known camera pose, height above ground can be inferred directly. To generate this information for the NYU Depth v2 dataset—where camera poses are not available—we proceed as suggested by Müller and Behnke (2014). We extract normals in the depth images, find ten clusters in normal space with k-means and determine the cluster that is most vertical. We then project all points to this normal and subtract the height of the lowest point.

We add the height image as an additional depth channel. Instead of computing region differences as in Eq. (2), we determine the average height above ground in R_1, such that

$$f_{\text{height},\theta}(\mathbf{q}) := \frac{1}{|R_1(\mathbf{q})|} \sum_{\mathbf{p}\in R_1} \phi_{\text{height}}(\mathbf{p}). \tag{5}$$

Using the same hyperparameters as without height, the classification accuracy improves significantly by 1.5 and 1.3 percentage points for class and pixel accuracy, respectively. Analysis of the learned forest shows that overall, height above ground is used in roughly 12 % of the split nodes, followed by depth differences (38 %) and color (50 %). These numbers reflect the statistics of the feature proposal distribution.

6.6 Random Forest Parameters

The hyper-parameter configurations for which we report our timing and accuracy results were found with global parameter search and cross-validation on the training set. The cross-validation outcome varies between datasets.

For the NYU Depth v2 dataset, we used three trees with 4537 samples / image, 5729 feature candidates / node, 20 threshold candidates, a box radius of 111 px, a region size of 3, tree depth 18 levels, and minimum samples in leaf nodes 204.

For MSRC-21 we found 10 trees, 4527 samples / image, 500 feature candidates / node, 20 threshold candidates, a box radius of 95 px, a region size of 12, tree depth 25 levels, and minimum samples in leaf nodes 38 to yield best results.

7 Conclusion

We provide an accelerated random forest implementation for image labeling research and applications. Our implementation achieves dense pixel-wise classification of VGA images in real-time on a GPU. Training is accelerated on GPU by a factor of up to 26 compared to an optimized CPU version. The experimental results show that our fast implementation enables effective parameter searches that find solutions which outperform state-of-the art methods. CURFIL prepares the ground for scientific progress with random forests, e.g. through research on improved visual features.

References

Amit, Y., Geman, D.: Shape quantization and recognition with randomized trees. Neural Comput. **9**(7), 1545–1588 (1997)

Bergstra, J., Bardenet, R., Bengio, Y., Kégl, B., et al.: Algorithms for hyper-parameter optimization. In: Neural Information Processing Systems (NIPS) (2011)

Breiman, L.: Random forests. Mach. Learn. **45**(1), 5–32 (2001)

Couprie, C., Farabet, C., Najman, L., LeCun, Y.: Indoor semantic segmentation using depth information. The Computing Resource Repository (CoRR) abs/1301.3572 (2013)

Eigen, D., Fergus, R.: Predicting depth, surface normals and semantic labels with a common multi-scale convolutional architecture. arXiv preprint arxiv:1411.4734 (2014)

Gupta, S., Girshick, R., Arbeláez, P., Malik, J.: Learning rich features from RGB-D images for object detection and segmentation. In: Fleet, D., Pajdla, T., Schiele, B., Tuytelaars, T. (eds.) ECCV 2014, Part VII. LNCS, vol. 8695, pp. 345–360. Springer, Heidelberg (2014)

Hermans, A., Floros, G., Leibe, B.: Dense 3D semantic mapping of indoor scenes from RGB-D images. In: International Conference on Robotics and Automation (ICRA). IEEE, Hong Kong, May 2014

Ho, T.: Random decision forests. In: International Conference on Document Analysis and Recognition (ICDAR), vol. 1, pp. 278–282. IEEE (1995)

Lepetit, V., Fua, P.: Keypoint recognition using randomized trees. IEEE Trans. Pattern Anal. Mach. Intell. (PAMI) **28**(9), 1465–1479 (2006)

Lepetit, V., Lagger, P., Fua, P.: Randomized trees for real-time keypoint recognition. In: Conference on Computer Vision and Pattern Recognition (CVPR), vol. 2, pp. 775–781 (2005)

Liao, Y., Rubinsteyn, A., Power, R., Li, J.: Learning random forests on the gpu. In: NIPS Workshop on Big Learning: Advances in Algorithms and Data Management (2013)

Müller, A.C., Behnke, S.: Learning depth-sensitive conditional random fields for semantic segmentation of rgb-d images. In: International Conference on Robotics and Automation (ICRA). IEEE, Hong Kong, May 2014

Pedregosa, F., Varoquaux, G., Gramfort, A., Michel, V., Thirion, B., Grisel, O., Blondel, M., Prettenhofer, P., Weiss, R., Dubourg, V., Vanderplas, J., Passos, A., Cournapeau, D., Brucher, M., Perrot, M., Duchesnay, E.: Scikit-learn: machine learning in python. J. Mach. Learn. Res. **12**, 2825–2830 (2011)

Rodrigues, J., Kim, J., Furukawa, M., Xavier, J., Aguiar, P., Kanade, T.: 6D pose estimation of textureless shiny objects using random ferns for bin-picking. In: International Conference on Intelligent Robots and Systems (IROS), pp. 3334–3341. IEEE (2012)

Schulz, H., Höft, N., Behnke, S.: Depth and height aware semantic RGB-D perception with convolutional neural networks. In: European Symposium on Artificial Neural Networks (2015)

Sharp, T.: Implementing decision trees and forests on a GPU. In: Forsyth, D., Torr, P., Zisserman, A. (eds.) ECCV 2008, Part IV. LNCS, vol. 5305, pp. 595–608. Springer, Heidelberg (2008)

Shotton, J., Fitzgibbon, A., Cook, M., Sharp, T., Finocchio, M., Moore, R., Kipman, A., Blake, A.: Real-time human pose recognition in parts from single depth images. In: Conference on Computer Vision and Pattern Recognition (CVPR), pp. 1297–1304 (2011)

Shotton, J., Johnson, M., Cipolla, R.: Semantic texton forests for image categorization and segmentation. In: Conference on Computer Vision and Pattern Recognition (CVPR) (2008)

Silberman, N., Hoiem, D., Kohli, P., Fergus, R.: Indoor segmentation and support inference from RGBD images. In: Fitzgibbon, A., Lazebnik, S., Perona, P., Sato, Y., Schmid, C. (eds.) ECCV 2012, Part V. LNCS, vol. 7576, pp. 746–760. Springer, Heidelberg (2012)

Slat, D., Lapajne, M.: Random Forests for CUDA GPUs. Ph.D. thesis, Blekinge Institute of Technology (2010)

Stückler, J., Birešev, N., Behnke, S.: Semantic mapping using object-class segmentation of RGB-D images. In: International Conference on Intelligent Robots and Systems (IROS), pp. 3005–3010. IEEE (2012)

Stückler, J., Waldvogel, B., Schulz, H., Behnke, S.: Dense real-time mapping of object-class semantics from RGB-D video. J. Real-Time Image Process. 10, 599–609 (2013)

Van Essen, B., Macaraeg, C., Gokhale, M., Prenger, R.: Accelerating a random forest classifier: Multi-core, GP-GPU, or FPGA? In: International Symposium on Field-Programmable Custom Computing Machines (FCCM). IEEE (2012)

Viola, P., Jones, M.: Rapid object detection using a boosted cascade of simple features. In: Conference on Computer Vision and Pattern Recognition (CVPR) (2001)

Wehenkel, L., Pavella, M.: Decision trees and transient stability of electric power systems. Automatica 27(1), 115–134 (1991)

C-EFIC: Color and Edge Based Foreground Background Segmentation with Interior Classification

Gianni Allebosch$^{(\boxtimes)}$, David Van Hamme, Francis Deboeverie,
Peter Veelaert, and Wilfried Philips

Department of Telecommunications and Information Processing,
Image Processing and Interpretation, Ghent University - iMinds, Gent, Belgium
gianni.allebosch@telin.ugent.be

Abstract. The detection of foreground regions in video streams is an essential part of many computer vision algorithms. Considerable contributions were made to this field over the past years. However, varying illumination circumstances and changing camera viewpoints provide major challenges for all available algorithms. In this paper, a robust foreground background segmentation algorithm is proposed. Both Local Ternary Pattern based edge descriptors and RGB color information are used to classify individual pixels. Furthermore, camera viewpoints are detected and compensated for. We will show that this algorithm is able to handle challenging conditions and achieves state-of-the-art results on the comprehensive ChangeDetection.NET 2014 dataset.

Keywords: Foreground background segmentation · Moving edges · Illumination invariance · Camera motion compensation

1 Introduction

The extraction of foreground regions is often the first step in video analysis. For many applications, it is vital that only interesting parts are analysed further, while the other (background) regions are ignored. Notable examples include the tracking of moving objects for traffic or surveillance applications, or fine grained motion analysis in sports or medical rehabilitation. Most often, as in the examples described above, the foreground regions are related to moving objects.

Foreground background segmentation algorithms first create a model of the background appearance. Then, every new input frame is compared with this model. Regions in the image that differ significantly from the background model, are considered to be foreground. Individual algorithms distinguish themselves in the way they represent the background appearance, and how the background model is maintained.

Many foreground background segmentation algorithms can be considered as variations on the classical Gaussian Mixture Model (GMM) [1]. It is assumed that the frequency of a certain appearance level (intensity, RGB . . .) occurring

© Springer International Publishing Switzerland 2016
J. Braz et al. (Eds.): VISIGRAPP 2015, CCIS 598, pp. 433–454, 2016.
DOI: 10.1007/978-3-319-29971-6_23

at a certain pixel can be modelled statistically as a mixture of Gaussian distributions. Simple, static backgrounds can be modelled locally by a single narrow Gaussian, while more complex (dynamic) backgrounds will result in multiple and/or wider Gaussians. The more recent ViBe algorithm [2] and its successor ViBe+ [3] are built on similar principles as the GMM, but store the distributions as a collection of samples rather than by the model parameters. The authors of [4] further elaborated on these methods, by adding a framework for automatic local parameter tuning, which alters the detection and updating mechanisms locally. The algorithm models the dynamic changes for each pixel and raises both the detection threshold and the learning rate when necessary. This avoids an important source of false positives, without sacrificing high detection rates in static regions.

Pixel appearances can change drastically under changing lighting conditions [5]. The previously described methods still handle light changes poorly. Some methods adapt the background updating mechanism at runtime if a large illumination change is detected [6]. Illumination invariant regional descriptors can also provide a robust solution. In the method of Heikkilä and Pietikäinen [7], Local Binary Patterns (LBPs) are used to construct histograms of regional pixel variations. A more recent and high performing method, coined SuBSENSE [8] further extends this by using the more advanced Local Binary Similarity Patterns [9], additional color information, and a framework for automatic local parameter tuning, based on [4].

Alternatively, in the method of [10], strong edges are used as features in both a short and long term background model. Assuming a static camera viewpoint, the location of these edges remains static under changing lighting conditions. This makes edges based foreground masks more reliable in these situations. However, since moving objects are generally contiguous, additional contour filling strategies are required. Foreground edges are prone to gaps, which classical boundary filling techniques are unable to cope with. So, in previous work [11], we developed an edge based segmentation framework, using Local Ternary Patterns and successive robust interior classification.

Dynamic camera viewpoints pose another issue in foreground background segmentation. Most foreground background segmentation models strictly build a local model of the scene, which is no longer relevant once the viewpoint changes. Methods that do not make a static camera assumption, prove to be more robust in such situations. Sajid and Cheung [12] combine multiple local and global change statistics. Each mechanism results in a separate foreground mask. The final foreground mask is reached through majority voting between individual masks. The global statistics are less prone to small viewpoint changes, which generally results in more reliable foreground detection. However, once the camera viewpoint strays to far from the original one, the observed scene is often totally different and the global statistics also become irrelevant. For this reason, in [13] we developed a detection mechanism based on optical flow that is able to handle these more challenging situations.

Our previous work already showed state-of-the-art performance in these difficult conditions, but some issues still remained. If there are no lighting or camera

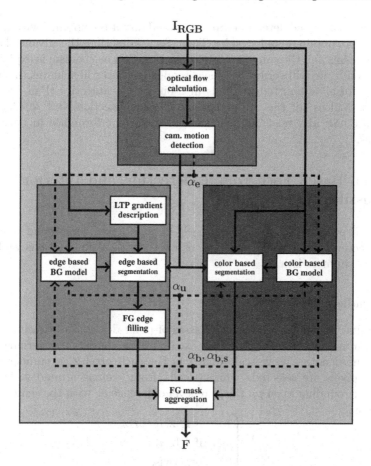

Fig. 1. General overview of the components of our method. Solid lines represent images being fed forward, dashed lines fed back or fed forward parameters. Note that our algorithm consists of three large parts: camera motion compensation (red), an edge based detection framework (green) and a color based one (blue) (Color figure online).

viewpoint changes, less advanced strategies could often yield better results. Most notable, the lack of color information in our previous frameworks limited the performance in some (easier) categories.

In this paper, we describe an extension to these previously proposed foreground background segmentation algorithms, coined C-EFIC (Color and Edge based Foreground background segmentation with Interior Classification). Since the benefits of these methods were shown to be so significant, we extend them with additional RGB color information to mitigate remaining limitations. We will show that this is possible without sacrificing robustness in difficult illumination conditions and for changing camera viewpoints. An overview of the interactions between the principal parts of this method is shown in Fig. 1.

In Sect. 2, the edge based foreground detection mechanisms and successive interior filling mechanism are explained in depth. Afterwards, in Sect. 3, we explain the color based detection framework and how it is combined with the edge based mask to obtain reliable results. Then, Sect. 4 describes the advanced updating mechanisms and the camera motion compensation mechanism is explained in Sect. 5. Finally, we will verify that the extended method still achieves the highest F-measure of all submitted methods in 2 out of 11 categories (Pan-Tilt-Zoom and Night Videos) of the comprehensive ChangeDetection.NET 2014 dataset [14], whilst now also reaching near state-of-the-art performance in most other categories.

2   Edge Based Foreground Detection and Interior Classification

In [11], we presented a robust approach to edge based foreground detection. In this section, we give an overview of its theoretical foundations and most important characteristics.

2.1   Gradient Description

Local Ternary Patterns (LTPs) are classes of point descriptors which represent the local intensity or color variations. Each pattern consists of a ternary string [15]. Let us denote a gray scale image I. For each pixel $\mathbf{z}$, a number of surrounding pixels are selected, symmetrically from a circle around $\mathbf{z}$. For each selected surrounding pixel p_i, the gray level difference d_s with the center pixel $\mathbf{z}$ is coded as

$$d_s[\mathbf{z}, i] = \begin{cases} 1 & \text{if } I_s[\mathbf{z}, i] - I[\mathbf{z}] > T \\ -1 & \text{if } I_s[\mathbf{z}, i] - I[\mathbf{z}] < -T \\ 0 & \text{otherwise} \end{cases} \tag{1}$$

with T a fixed, typically low threshold (e.g. 3 with a total intensity range of 255) and $I_s[\mathbf{z}, i]$ the gray scale value of p_i. $d_s[\mathbf{z}, i]$ represents whether the selected surrounding pixel p_i has a significantly different intensity value from $\mathbf{z}$. When

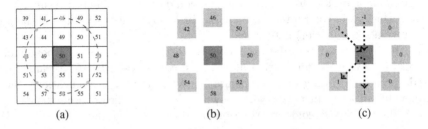

(a)   (b)   (c)

Fig. 2. An example of a Local Ternary Pattern with 8 neighborhood points: (a) Intensity levels in a certain image region, with circular neighborhood points denoted. (b) Interpolated neighborhood intensity levels. (c) LTP representation ($T = 3$), with our binary gradients representation superimposed.

(a)

(b)

Fig. 3. (a) Input image. (b) Gradient descriptors, where the colors represent the direction and more saturated means more likely. Note that all clear edges in the image are represented by similarly strong gradient confidences (Color figure online).

$d_s[\mathbf{z}, i] = 0$, the thresholded difference is considered to be below noise level. An example is given in Fig. 2.

It can be observed that every $d_s[\mathbf{z}, i]$ can also be considered as binary evidence for a gradient $\mathbf{g_b}[\mathbf{z}, i]$, either with unit or zero length, and with a direction either from $\mathbf{z}$ towards p_i, or from p_i towards $\mathbf{z}$ (Fig. 2c). Formally,

$$\mathbf{g_b}[\mathbf{z}, i] = d_s[\mathbf{z}, i]\mathbf{e_i} \ , \tag{2}$$

where $\mathbf{e_i}$ is the unit vector pointing from $\mathbf{z}$ towards p_i.

If there is an edge located at the center pixel $\mathbf{z}$, we expect the individual binary gradients in the LTP descriptor to roughly point in the same direction in that neighborhood. By summing over all $\mathbf{g_b}$s in the pattern, the gradient evidence can be combined:

$$\mathbf{G}[\mathbf{z}] = \sum_{i=0}^{N-1} \mathbf{g_b}[\mathbf{z}, i] \ . \tag{3}$$

This vector's direction can be regarded as the most likely gradient direction. Its length is a measure for the confidence of the calculated gradient direction, rather than its strength. A large vector means the binary evidence is supported by that of the other surrounding pixels. Note that $\mathbf{G}[\mathbf{z}]$ should thus be considered a gradient descriptor as opposed to the original LTP texture descriptor [15].

Since edges in most real video images locally tend to continue in the same direction (perpendicular to the gradient), we also apply an adaptive smoothing strategy. The gradient vectors are filtered along both directions with rotated, anisotropic 2D Gaussian kernels. The Gaussians have significantly different standard deviations along both dimensions, so smoothing is mainly performed along the chosen orientation. The orientation of the kernels is adjusted locally, depending on the direction of $\mathbf{G}[\mathbf{z}]$.

Our strategy of thresholding the individual intensity differences first and then combining the evidence over a larger area, is what makes our descriptor robust in different illumination conditions. In our approach, all significant intensity differences in the descriptor are treated as being equally important. What matters is that the evidence is supported in the surrounding region. This stabilizes the gradient confidence levels and also diminishes the possibility that isolated noisy pixels sharply disturb the descriptor. A result of the gradient calculation and successive smoothing step of our algorithm is shown in Fig. 3.

2.2   Foreground Edges

In this section, we discuss the construction of the temporal LTP gradient model. Our model is based on the model used in the LBP based foreground detection framework developed by Heikkilä and Pietikäinen [7]. In [11], we replaced the fixed learning rate with an adaptive one, in order to handle more challenging situations properly. This mechanism will be explained in Sect. 4.

At each pixel $\mathbf{z}$, a fixed number M of the gradient vectors described in Sect. 2.1 are stored. This results in a temporal gradient descriptor model denoted $\mathbf{G}_j$, with $1 \leq j \leq M$. At every input frame (time t), the new gradient vector $\mathbf{G}[\mathbf{z}, t]$ is compared to each vector in the temporal model at that location and the vector $\mathbf{G}_k[\mathbf{z}, t]$ with the lowest L_2-distance to the new vector is updated as follows:

$$\mathbf{G}_k[\mathbf{z}, t] = \alpha[\mathbf{z}, t]\mathbf{G}[\mathbf{z}, t] + (1 - \alpha[\mathbf{z}, t])\mathbf{G}_k[\mathbf{z}, t - 1] \qquad (4)$$

with $\alpha[\mathbf{z}, t]$ the learning rate, which satisfies $0 \leq \alpha[\mathbf{z}, t] \leq \frac{1}{M}$ and will be discussed in depth in Sect. 4.

Furthermore, for each vector $\mathbf{G}_j[\mathbf{z}, t]$, a weight factor $W_j[\mathbf{z}, t]$ is kept, which is a measure for the likelihood that the vector represents the background. W_j is updated as follows:

$$W_j[\mathbf{z}, t] = \begin{cases} \alpha[\mathbf{z}, t] + (1 - \alpha[\mathbf{z}, t])W_j[\mathbf{z}, t - 1] & \text{if } j = k \\ (1 - \alpha[\mathbf{z}, t])W_j[\mathbf{z}, t - 1] & \text{otherwise} \end{cases} . \qquad (5)$$

Thus, $W_k[\mathbf{z}, t]$ is boosted, while the other weights are decreased. To find the background, the temporal vectors are sorted according to their weights. The most weighted vector always represents background. Consecutive vectors are added to the background vector set, until their cumulative weight exceeds a chosen threshold T_w. So, the background set $\mathbf{B}_{LTP}[\mathbf{z}, t]$ is always a subset of $\{\mathbf{G}_1, \mathbf{G}_2 \ldots \mathbf{G}_M\}$ at $\mathbf{z}$ and t. The background in static regions will likely be represented by a single vector, while the set will contain multiple vectors in more dynamic environments. If the lowest L_2-distance between the input vector and the background set exceeds a threshold $T_{e,LTP}$, this pixel is considered to be foreground.

The remaining vectors can be considered 'pre-background'. They are not used in the detection step, but they can become part of the background set later on, if their weights become higher. If the lowest squared difference with all vectors in the temporal model exceeds $T_{e,LTP}$, the temporal vector with the lowest weight is replaced by the current vector $\mathbf{G}[\mathbf{z}, t]$ and its weight is reset to $\alpha[\mathbf{z}, t]$. When $\alpha[\mathbf{z}, t]$ is high, new vectors can be quickly added to the background set in that way.

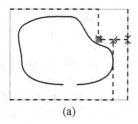 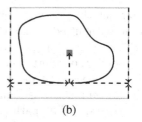

(a)   (b)

Fig. 4. Examples of the path distance calculation from all 4 image corners to a chosen pixel. Black solid lines represent strong edges, dotted lines the individual paths. In (a), the path from 1 image corner is larger than the Manhattan distance. In (b), the paths from 2 image corners are longer.

To conclude this section, we observe that the foreground gradients often occupy more than 1 pixel. This is a direct consequence of the nature of the Local Ternary Patterns, i.e. the local features aggregate information from a larger region around the center pixel, and also non-minimal gradient levels might be detected as foreground. So, we incorporated the thinning algorithm described in [16] to raise the detection accuracy. The main benefit of this algorithm is that it avoids the creation of gaps. Furthermore, to further eliminate the possibility of (small) gaps occurring in the contours, morphological closing is executed with a 3 by 3 circular kernel.

2.3   Shortest Path Based Interior Filling

In the previous sections, we described our strategy to create a foreground image, which only contains edges. Here, we describe how interior points can be added to the silhouette.

To determine the entire foreground object, we must fill its contour, consisting of foreground edge points. Topologically, a point can be classified as interior if it is completely enclosed by contour points. If this would be the case for all foreground objects, classical contour filling strategies (e.g. floodfill) would suffice. However, if the foreground object passes in front of objects with similarly oriented edges, gaps can still occur in the foreground edges.

In [11], we proposed a filling algorithm which for a single object produces a silhouette $S(P)$ that satisfies

$$C(P) \subseteq S(P) \subseteq H_o(P) \subseteq H_c(P) , \qquad (6)$$

where the set $C(P)$ is obtained by filling the closed contours in the pixel set P and leaving the other edges intact. $H_o(P)$ and $H_c(P)$ are the orthogonal hull [17] and convex hull of P respectively.

This mechanism is based on the following idea: *The probability of a point being interior, is proportional to how difficult it is to reach that point from the outside, where the strong gradient points act as obstacles.* If an object's foreground edges consist of a closed contour, these edges block every possible path from the outside to one of the interior points. If there is a single gap in the contour, it would be possible to reach the interior points, but a path to it would likely be longer than in a scenario without edge pixels. An example is given in Fig. 4.

Now consider the foreground gradient image to be an originally fully 4-connected input graph. Vertices coinciding with foreground pixels and their links are removed. The shortest paths from the four image corners c_i ($1 \leq i \leq 4$) to each of the image pixels z is found by using Breadth First Search (BFS) in the graph. The corresponding path length is denoted $D_P[z, i]$. We define the excess distance D_E as the difference between the actual path length and the minimal length. In particular, we have

$$D_E[z, i] = D_P[z, i] - D_M[z, i] , \tag{7}$$

where $D_M[z, i]$ is the Manhattan distance or L_1-distance from c_i to z, i.e. the path length ignoring foreground edge pixels. If z cannot be reached, $D_E[z, i]$ is set to ∞.

If the foreground object is orthogonally convex, it is possible to reach any point on the outside of the contour from at least 3 image corners with minimal distance if there is a free path along the image borders. However, this cannot be guaranteed for the distance from the 4th corner. So, we exclude the corner c_l with the largest excess distance from further analysis in each pixel (Fig. 4) such that Eq. 6 holds. The total excess distance $D_{E,tot}$ for a pixel z is defined as:

$$D_{E,tot}[z] = \sum_{i \neq l} D_E[z, i] . \tag{8}$$

The interior pixels are now classified by imposing a threshold $T_d \geq 0$. The resulting binary, LTP edge based foreground image F_{LTP} is constructed as follows:

$$F_{LTP}[z] = \begin{cases} 1 & \text{if } D_{E,tot}[z] > T_d \\ 0 & \text{otherwise} \end{cases} . \tag{9}$$

If there is one moving object, and the filled silhouette is constructed as described above, Eq. 6 is always satisfied. Higher thresholds remove more pixels from the silhouette, but never more than contour filling and never less than the orthogonal hull. If there are multiple objects moving in the scene, Eq. 6 is satisfied for the individual objects as long as their rectangular bounding boxes do not overlap.

Hence, our filling algorithm may fill in too many pixels when there are many moving objects close to each other, or when the objects are not convex. One way to avoid this problem is to classify the foreground edges into different object classes, and to correct the silhouettes accordingly. Since this classification requires reasoning at a much higher level, we will not solve the classification problem, but instead combine filling with color based segmentation. This will be discussed in the next section.

3   Combined Foreground Detection

In the previous sections, we described a framework for edge based foreground detection. However, as already noted, even though the resulting foreground

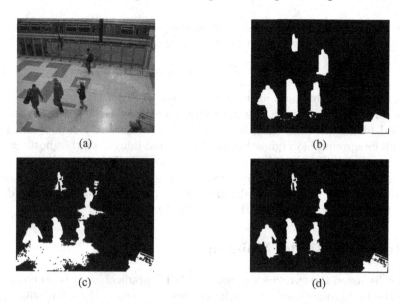

Fig. 5. Creation of the aggregated foreground mask (a) Input. (b) Edge based mask. (c) Color based mask. (d) Bitwise AND of b and c. Note that the edge based mask suffers from unnecessarily filled concavities, while the color bask mask has more issues with shadows.

silhouettes satisfy Eq. 6, their shapes can still differ from the objects themselves. Appearing concavities (e.g. pedestrians, space between the arms and torso) can result in a large excess distance $D_{E,tot}$ (see Eq. 8). Intensity or color based algorithms work in a totally different way, so they are less prone to these effects. However, as discussed before, these features are highly unstable when the illumination changes. In order to generate the best possible silhouettes, our algorithm combines both color features and the LTP features described above.

The R,G and B values of the input frames are fed into a similar temporal model as described in Sect. 2.2. Instead of 2-dimensional gradient descriptors, this model holds 3-dimensional color vectors. The same mechanisms as described in Sect. 2.2 are used. Only the detection threshold, which is dynamic and denoted $T_{e,RGB}$, differs:

$$T_{e,RGB}[\mathbf{z}, t] = k_0 + k_1 U_L[\mathbf{z}, t] \qquad (10)$$

where k_0 and k_1 are constants and $U_L[\mathbf{z}, t]$ is the local unreliability rate, which will be defined in Sect. 4. For now, we note that the local unreliability rate is higher when the temporal model is unreliable, e.g. in dynamic regions. Since dynamic regions are more likely to produce false positive foreground pixels, raising the detection threshold helps avoid this phenomenon. The resulting RGB color based foreground mask is denoted F_{RGB}.

The filled edged based mask and the color based mask will generally produce distinct errors, due to concavities and illumination changes respectively. Concavities do not influence the color based mask, while changing illumination and weak shadows leave the edge based mask untouched. So, a logical AND operation between the two masks will sharply reduce both unwanted effects. Thus, the resulting foreground mask F is determined as

$$F[\mathbf{z}] = F_{LTP}[\mathbf{z}] \& F_{RGB}[\mathbf{z}]. \tag{11}$$

An example of the mask aggregation is show in Fig. 5.

To conclude this section, we note that objects which are static when the temporal model is being built, but later move to another location, can leave behind a foreground blob (ghost) in the foreground image. So, the ghost removal methodology described in [18] is added the framework. This algorithm compares the location of the contours of foreground objects with the location of edges in the input image. If the Chamfer distance between them exceeds a threshold, the object is entirely removed from the foreground mask.

4   Temporal Model Maintenance

Having discussed the two kinds of features (LTP gradients and RGB colors) on which the background modelling is based, we now turn to the adaptivity of the model.

In the simplest scenarios, the background is entirely visible in the first frame, and remains unchanged throughout the rest of the sequence. However, in many realistic scenario's, this assumption is invalid. So, the temporal model needs to be maintained and updated over time. In C-EFIC, the maintenance mechanism utilizes three kinds of learning: fixed, exponentially decreasing and related to the background unreliability.

4.1   Fixed Learning Rate

In background regions, a constant amount of learning is necessary throughout the sequence. This allows the temporal models to adapt to small gradual changes. So, the learning rate is set $\geq \alpha_c$, a very small constant. To avoid foreground objects getting learned into the background, α_c is not used in foreground regions.

4.2   Exponential Learning Rate

The learning rate must be higher in the beginning of the sequence. This ensures that foreground objects present in the first few frames get replaced quickly by background vectors in the model once they move. We impose that part of the learning rate α_e decreases exponentially over time:

$$\alpha_e[t] = \frac{1}{M} e^{-\tau t}, \tag{12}$$

where t is the number of frames since the background model was initialized and τ is a user settable parameter, which determines the rate at which the function decreases. The multiplication by $\frac{1}{M}$ avoids that the weights of the current model drop too drastically as soon as all temporal vectors have been initialized.

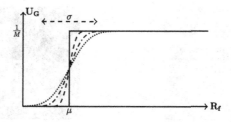

Fig. 6. U_G as a function of the foreground (gradient) pixel percentage R_f for different values of σ. The solid line represents the case where $\sigma = 0$, and behaves like a regular (hard) threshold or step function.

4.3   Unreliability Learning Rate

If there are large or quickly occurring changes in the image, not due to foreground objects, the temporal model becomes unreliable. We denote a background unreliability rate α_u, which is increased in two different scenario's:

1. The model should be corrected globally (e.g. a small camera displacement).
2. The model should be corrected locally (e.g. the wind deforming a tree).

The first scenario with global unreliability can be identified fairly easily. If there is a large disagreement between the current frame and the model (i.e. there are too many foreground pixels), the camera position has likely shifted or a significant change has occurred in the background. The model should adapt to these changes quickly. For this purpose, we impose a 'global unreliability' term U_G, that is sigmoid-shaped (Fig. 6):

$$U_G[t] = \frac{0.5}{M} \left(1 + \text{erf} \left(\frac{R_f[t] - \mu}{\sigma} \right) \right) \tag{13}$$

$$\text{with erf}(x) = \frac{2}{\sqrt{\pi}} \int_0^x e^{-y^2} dy . \tag{14}$$

erf is the error function, $R_f[t]$ is the percentage of foreground pixels in frame t and σ and μ are user settable parameters. It behaves like a soft threshold at μ. σ determines how steep the function is. If $\sigma = 0$, U_G behaves like a sharp threshold, like in the method of Wang and Suter [19]. The minimum of $U_G[t]$ is 0 and the maximum is again $\frac{1}{M}$. As will be shown in the results section, this method works very well when the camera viewpoint constantly shifts around the initial position (camera jitter). Unlike in [11,13], we only use the global unreliability rate when camera jitter is detected. This avoids large foreground objects to needlessly cause a large learning rate.

When the camera keeps moving away from that point (e.g. panning), the connection to the original model is quickly lost, and quick updating only has limited potential. In Sect. 5, a method to distinguish between panning and jitter and to compensate for panning motions is presented.

(a)

(b)

Fig. 7. (a) Input frame with dynamic background below a busy street [14]. (b) Visualization of U_G for the entire image, lighter means more unreliable. Note that the region corresponding to the water is correctly classified as more unreliable background, while the busy street above has a much smaller unreliability term.

Local unreliability is less trivial to identify. Simply looking at the amount of foreground pixels is not recommended, since dynamic background can be easily confused with regions that simply contain a lot of foreground (e.g. a busy street). However, experiments show that dynamic background regions will often produce many isolated foreground pixels (Fig. 7). In our algorithm, this phenomenon is exploited by building a local isolated pixel rate R_i over time:

$$R_i[\mathbf{z}, t] = \alpha_i[t]R_{i,curr}[\mathbf{z}, t] + (1 - \alpha_i[t])R_i[\mathbf{z}, t-1] , \tag{15}$$

$$\text{where } \alpha_i[t] = C_i + (1 - C_i)M\alpha_e[t] \tag{16}$$

with C_i a small constant. $R_{i,curr}[\mathbf{z}, t] = 1$ if $\mathbf{z}$ is the only foreground gradient pixel in his 8-connected neighborhood, otherwise $R_{i,curr}[\mathbf{z}, t] = 0$. Note that α_i is higher in the beginning of the sequence, since the accumulated evidence of dynamic background gets more reliable over time. The 'local unreliability' U_L at $\mathbf{z}$ in frame t is now

$$U_L[\mathbf{z}, t] = R_i[\mathbf{z}, t] * K_G , \tag{17}$$

where K_G is an isotropic Gaussian kernel, used to combine evidence of unreliability in the neighborhood of $\mathbf{z}$.

The background unreliability rate $\alpha_u[\mathbf{z}, t]$ is now defined as the maximum of the global and local unreliability:

$$\alpha_u[\mathbf{z}, t] = \max[U_G[t], U_L[\mathbf{z}, t]] . \tag{18}$$

The total learning rate $\alpha[\mathbf{z}, t]$, taking into account all of the considerations described above, is

$$\alpha[\mathbf{z}, t] = \begin{cases} \min\left[\frac{1}{M}, \alpha_c + \alpha_e[t] + \alpha_u[\mathbf{z}, t]\right] & \text{if } F[\mathbf{z}] = 0 \\ \min\left[\frac{1}{M}, \alpha_e[t] + \alpha_u[\mathbf{z}, t]\right] & \text{otherwise.} \end{cases} \tag{19}$$

5   Camera Motion Compensation

So far, we have addressed many issues in foreground background segmentation, all related to the scene. However, as soon as the camera viewpoint changes, the appearance of all pixels might change, even though the background itself does not. In this section we describe how camera motion can be detected and compensated for, based on optical flow estimation through the framework originally presented in [13]. We distinguish two scenarios: one where the camera is undergoing a fairly constant motion away from the original position (e.g. panning) and one where the camera keeps moving around the same position (jitter). A notable change w.r.t. the latter category was made to the previous algorithm.

5.1 Camera Motion Detection

Optical flow is an image feature which essentially represents the motion of individual pixels between subsequent frames. By assuming a constant brightness, optical flow calculation boils down to the estimation of the displacement $(\Delta x, \Delta y)$ of a pixel at $\mathbf{z} = (x, y)$ [20]. This vector can only be obtained by introducing additional constraints, which is where optical flow methods described in literature differ from one another [21].

In most video sequences, the amount of pixels that represent dynamically moving objects is relatively small compared to the static ones. However, if the camera viewpoint changes, most pixels in the image will also appear to move. So, camera motion can be detected by the occurrence of significant optical flow vectors in the majority of the image.

As in [13], (dense) optical flow vectors are calculated by using the efficient algorithm described in [22]. Let $\mathbf{V}[\mathbf{z}]$ be the flow vector image at pixel $\mathbf{z}$. Now, we define the optical flow mask F_f as follows:

$$F_f[\mathbf{z}] = \begin{cases} 1 & \text{if } ||\mathbf{V}[\mathbf{z}]|| > T_f \\ 0 & \text{otherwise ,} \end{cases} \tag{20}$$

where T_f is a typically low threshold (e.g. 1 pixel). So, F_f represents a significance classification of all flow vectors. If the ratio of significant flow vectors is larger than a second threshold T_n, camera motion is detected.

5.2 Distinction Between Panning/Tilting and Jitter

In order to compensate for camera motion, the effect of this motion should be mitigated at every pixel location. To compare a new image with a background model, the image should first be transformed such that coinciding pixels also represent the same objects in the model. Since the distances to the objects in the scene are not known a priori in most applications, the effects of potential perspective changes on the image formation are difficult to model. Luckily, when a scene's relief is small, relative to the average distance from the objects to the camera, the weak-perspective image formation model can be used to describe the image formation [23]. As proven by [24], arbitrary projection transformation matrices can be written in the form of an affine matrix, assuming a weak-perspective image formation model.

The affine transformation between two images can be estimated by the Pyramidal implementation of the Lucas Kanade Feature Tracker [25]. The algorithm first detects interest points and then calculates sparse optical flow vectors to detect their individual motion. In the second phase of the algorithm, the robust affine transformation matrix M_{tf} is selected through a RANSAC framework. From this matrix, the expected flow $\mathbf{V_e}$ can now be directly determined. For a certain pixel $\mathbf{z} = (x, y, 1)^T$ in homogeneous coordinates:

$$\mathbf{V_e}[\mathbf{z}] = M_{tf}\mathbf{z} \ . \tag{21}$$

If the center pixel is also the origin of the camera's coordinate system, the expected flow at the center pixel $\mathbf{V_e}[\mathbf{0}]$ can be represented by the third column of M_{tf}, also known as a translation vector $\mathbf{t_r} = (t_{r,x}, t_{r,y})$. If the camera viewpoint changes between successive frames, $\mathbf{t_r}$ will be a nonzero vector whose orientation (arctan $\frac{t_{r,y}}{t_{r,x}}$) represents the direction of the camera shift. If this direction is more or less constant for a longer period, the camera viewpoint obviously moves away from its original position, and a panning (or tilting) camera motion can be detected. Conversely, if the direction of $\mathbf{t_r}$ changes a lot, it is more likely that the camera is jittery, but not necessarily moving away from the original position.

In the proposed method, the distinction between panning and jittery camera's is derived from the reasoning above. Let us define two accumulators: acc_p for panning and acc_j for jitter, both initialized to 0 in the beginning of the sequence. Every time a camera viewpoint change is detected, the current direction of $\mathbf{t_r}$ is compared to the previous one. If the angles differ by more than 90, acc_j is incremented. Otherwise, acc_p is incremented. The camera motion compensation is then executed with regard to the highest corresponding accumulator value.

5.3 Jitter Compensation

If the camera viewpoint is shifting around the same position, it is likely that it moves in a small region around it. Many of these viewpoints will be reached multiple times over a relatively short time span. This means that a (multimodal) background model can be learned from past observations.

In [13], we proposed a compensation framework based on affine transformation of the input image with regard to the temporal modal. When a correct image transformation is found, this method proved to work very well. However, experiments showed that such obtained transformation matrices are sometimes erroneous, most notably caused by motion blur in the input frames. A faulty image transformation results in comparison with the wrong background pixels. To further increase the robustness of our method, we deviate from that approach.

Early in the sequence, the background model will not yet be fully adapted, and will produce many false positive foreground pixels. However, as described in Sect. 2.2, the global unreliability rate is designed to learn more rapidly as the percentage of foreground pixels rises. So, if the camera reaches the same position later again, the model will already be (partially) adjusted, so the false detection rate will be much lower. Detecting jitter first and then selecting the global unreliability rate later on has proven very beneficial (Sect. 6).

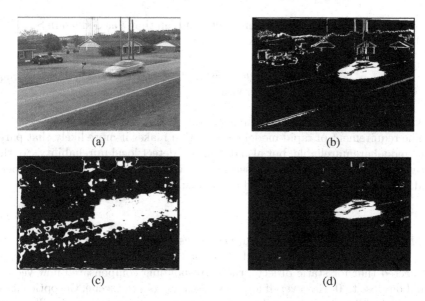

Fig. 8. Creation of the foreground RGB image in a sequence with panning camera. (a) Input image, (b) Short-term foreground mask, (c) compensated flow mask, (d) bitwise AND of short-term RGB foreground mask and compensated flow mask. Similar results are obtained for the foreground LTP image.

5.4   Panning Compensation

If a (slow) panning or tilting camera change is detected, the original background image is no longer usable if the new camera viewpoint has deviated too far from the original one. The benefits of the global unreliability rate are limited here. By the time the camera returns to its original position, the background model will already have been replaced by other, more recent observations.

However, if the camera motion is relatively slow, such that the spatial relation between successive frames can be established, it becomes possible to build a (unimodal) short term LTP gradient and a RGB color background model, denoted $\mathbf{B}_{LTP,s}$ and $\mathbf{B}_{RGB,s}$ respectively. A new frame is compared to this model after using the affine transformation as discussed in the previous sections. Here, $\mathbf{B}_{LTP,s}$ and $\mathbf{B}_{RGB,s}$ are also transformed after every frame as long as the panning motion continues. They are updated by using a large fixed learning rate $\alpha_{c,s}$ in background regions. Regions where the background model is not yet initialized, are copied directly from the LTP gradient and RGB color input images.

Comparing the LTP features determined from the input with $\mathbf{B}_{LTP,s}$ results in the short term foreground mask $F_{LTP,s}$:

$$F_{LTP,s}[\mathbf{z},t] = \begin{cases} 1 & \text{if } ||\mathbf{G}[\mathbf{z},t] - \mathbf{B}_{LTP,s}[\mathbf{z},t]|| > T_{LTP,s} \\ 0 & \text{otherwise} , \end{cases} \tag{22}$$

where $\mathbf{G}[\mathbf{z}]$ is the input LTP gradient vector at $\mathbf{z}$ an time t as defined in Sect. 2.1. Similarly, for the color based mask:

$$F_{RGB,s}[\mathbf{z}, t] = \begin{cases} 1 & \text{if } ||\mathbf{I}_{RGB}[\mathbf{z}, t] - \mathbf{B}_{RGB,s}[\mathbf{z}, t]|| > T_{RGB,s} \\ 0 & \text{otherwise}, \end{cases} \quad (23)$$

where I_{RGB} is the color input image.

The requirement of rapid model construction makes it more likely that parts of the model are unreliable, but also difficult to detect local unreliability. So, the flow vector image $\mathbf{V}$ is used as a secondary decision mechanism. The compensated flow image $\mathbf{V_c}$ is now defined as follows:

$$\mathbf{V_c} = \mathbf{V} - \mathbf{V_e} . \quad (24)$$

Thus, the original flow vectors are compensated with regard to the expected affine image transformation, resulting from the camera viewpoint change. It is expected that for static objects, the corresponding compensated flow vectors should be close to $\mathbf{0}$. However, if an object is moving in the scene, the optical flow vectors will differ from the globally calculated transformation and will locally coincide with nonzero compensated flow vectors. Thus, utilizing a final flow compensation threshold $T_{f,c}$, the compensated foreground flow mask $F_{f,c}$ is now defined as

$$F_{f,c}[\mathbf{z}, t] = \begin{cases} 1 & \text{if } ||\mathbf{V_c}[\mathbf{z}, t]|| > T_{f,c} \\ 0 & \text{otherwise} . \end{cases} \quad (25)$$

The resulting foreground RGB and LTP masks now consist of the pixelwise bitwise AND operation between $F_{f,c}$ and the respective short term foreground masks. Figure 8 shows a visual example of foreground detection in a sequence with a panning camera. Note that the contour filling step described in Sect. 2.3 is still executed after the foreground edge detection step. Finally, once the panning motion has stopped, the exponential learning rate (Sect. 4) is reset, such that the temporal model is quickly rebuilt at the new position.

6   Results

We compared the proposed algorithm to other the state-of-the-art, through the comprehensive ChangeDetection.NET 2014 dataset [14]. This dataset contains a total of 53 videos, spread across 11 categories. For all videos, we submitted our binary masks to the website, which automatically calculates a total of 7 different performance measures for each video and category, as well as the overall performance. Note that the same parameter set has to be used for all videos in the dataset, such that optimizing for one particular category is discouraged.

As a preprocessing step, all input images were smoothed with a 3 by 3 Gaussian kernel. Post processing was done with a 5 by 5 median filter on the binary foreground images. The parameters were tuned manually, and their values can be found in Table 1. The proposed algorithm was developed using C++

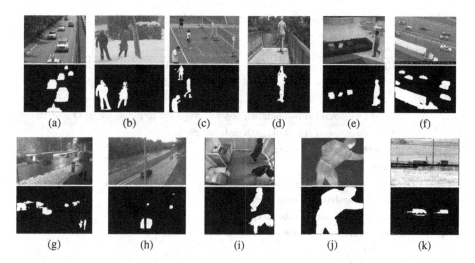

Fig. 9. Performance of our proposed method on a few hand picked frames from the ChangeDetection.NET 2014 dataset [14]. (a) Baseline, (b) Bad Weather, (c) Camera Jitter, (d) Dynamic background, (e) Intermittent Object Motion, (f) Low Framerate, (g) Night Videos, (h) Pan-Tilt-Zoom, (i) Shadows, (j) Thermal, (k) Turbulence.

programming code and runs at $16\,fps$ for 320 by 240 pixel videos on a desktop with an Intel® Xeon® E5 Quad Core processor.

Visible results for all categories are shown in Fig. 9. The calculated measures are given in Table 2. Since the video sequences differ strongly between the categories (e.g. many or few foreground objects, overall degree of difficulty ...), these measures show considerable variance. The F-measures for all categories and overall are compared with those of other methods in Table 3. As explained in [8], the F-measure can be regarded as the most unbiased performance measure, so this will be the focus of our analysis.

Including our own previous work as a competitor [13], the newly proposed method significantly achieves the highest F-measure of all methods in 2 of the categories significantly: Night and Pan-Tilt-Zoom. The night video category holds the most difficult illumination conditions, with generally low lighting, but changing appearances caused by traffic lights, street illumination and head and tail lights of cars. Given all robustness measures described above, it is not surprising that the method performs so well.

The top performance in the Pan-Tilt-Zoom category can be attributed to the robust camera motion compensation framework. It should be stated however that the proposed method does not detect the camera zooming in one of the four videos, and thus has limited performance in this sequence. Since the method uses affine transformations to compensate for camera viewpoint changes, this framework could also be used if zooming could be detected. In the Camera Jitter sequences, the method is close to the best performing methods, thanks to the global unreliability rate, which is now only triggered when necessary.

Table 1. Fixed parameters used in the experiments.

Parameter	Description	Value
T	LTP threshold	3 (intensity range: $[0, 255]$)
T_w	Sum of temporal model weights threshold	85 (% of total weight)
$T_{e,LTP}$	LTP foreground threshold	25 (% of max. vector length)
k_0	RGB foreground threshold, 0th order coeff	4 (% of range for R, G and B)
k_1	RGB foreground threshold, 1th order coeff	4
M	# vectors in the temporal models	5 (per pixel)
α_c	Fixed learning rate	0.0002
τ	Exponential learning rate	0.04
μ	Global unreliability rate threshold	0.50
σ	Global unreliability rate steepness	0.10
C_i	Local unreliability rate constant factor	0.05
T_d	Filling threshold	3
T_f	Optical flow threshold	1
T_n	Flow mask threshold	70 (% of # pixels)
$T_{f,c}$	Compensated flow threshold	2
$T_{e,LTP,s}$	Short term LTP foreground threshold	40 (% of max. vector length)
$T_{e,RGB,s}$	Short term RGB foreground threshold	8 (% of range for R, G and B)
$\alpha_{c,s}$	Short term fixed learning rate	0.1

Table 2. Results on the ChangeDetection.NET 2014 dataset using all 7 measures [14].

	Recall	TNR	FPR	FNR	PBC	F-Measure	Precision
Bad Weather (BW)	0.7352	0.9977	0.0023	0.2648	0.6600	0.7867	0.8719
Low Framerate (LF)	0.8077	0.9976	0.0024	0.1923	0.5532	0.6806	0.7135
Night Videos (NV)	0.7223	0.9866	0.0134	0.2777	2.5899	0.6677	0.6636
Pan-Tilt-Zoom (PTZ)	0.8686	0.8947	0.1053	0.1314	10.597	0.6207	0.6144
Turbulence (TB)	0.6494	0.9990	0.0010	0.3506	0.2542	0.6275	0.7047
Baseline (BL)	0.9455	0.9970	0.0030	0.0545	0.5201	0.9309	0.9170
Dynamic background (DB)	0.6556	0.9952	0.0048	0.3444	1.0825	0.5627	0.5674
Camera Jitter (CJ)	0.8458	0.9890	0.0110	0.1542	1.6653	0.8248	0.8157
Intermittent Object Motion (IOM)	0.8107	0.9172	0.0828	0.1893	8.4615	0.6229	0.5823
Shadow (SH)	0.9191	0.9920	0.0080	0.0809	1.1933	0.8778	0.8453
Thermal (TH)	0.8131	0.9943	0.0057	0.1869	1.3706	0.8349	0.8690
Overall (Over.)	0.7976	0.9782	0.0218	0.2024	2.6316	0.7307	0.7543

Furthermore, the method also shows near state-of-the-art performance for Baseline, Low Framerate and Shadow and sequences. The Baseline sequences can be considered to be the easiest ones, with near constant illumination and empty frames in the beginning of the sequences. The Shadow sequences are comparable to the Baseline ones, but the moving objects cause more pronounced shadows. Our method is able to deal with weak shadows very well, since they do not cause edges in the LTP image. Stronger shadows obviously pose a more difficult

Table 3. Comparison of the F-Scores of all methods applied to the ChangeDetection.NET 2014 database per category and overall. The highest score is denoted in bold faced numbers. References to the other methods can be found on the ChangeDetection.NET website [14].

Method	BW	LF	NV	PTZ	TB	BL	DB	CJ	IOM	SH	TH	Over
C-EFIC	78.67	68.06	**66.77**	**62.07**	62.75	93.09	56.27	82.48	62.29	87.78	83.49	73.07
EFIC [13]	77.86	66.32	65.48	58.42	67.13	91.72	57.79	71.25	57.53	82.02	**83.88**	70.88
IUTIS-3	80.32	**73.27**	49.48	39.21	78.57	**95.46**	**89.60**	81.39	71.36	85.85	82.10	**75.43**
SuBSENSE	86.19	64.45	55.99	34.76	77.92	95.03	81.77	81.52	65.69	89.86	81.71	74.08
MBS	79.80	63.50	51.58	55.20	58.58	92.87	79.15	**83.67**	75.68	79.68	81.94	72.88
FTSG	82.28	62.59	51.30	32.41	71.27	93.30	87.92	75.13	**78.91**	88.32	77.68	72.83
SaliencySubsense	85.93	65.15	53.48	33.99	75.12	94.83	81.57	80.71	60.12	**89.94**	68.57	71.76
MBS V0	77.30	62.79	51.58	51.18	56.98	92.87	79.04	**83.67**	70.92	77.84	81.15	71.39
Superpixel Strengthen BG Subtraction	85.80	69.10	53.84	33.37	71.39	94.10	83.91	80.04	54.00	89.58	69.06	71.29
M4CD Version 1.0	78.67	61.19	49.77	23.48	**80.26**	92.04	68.11	80.51	63.93	89.13	71.61	69.16
CwisarDH	68.37	64.06	37.35	32.18	72.27	91.45	82.74	78.86	57.53	85.81	78.66	68.12
Spectral-360	75.69	64.37	48.32	36.53	54.29	93.30	77.66	71.42	56.09	85.19	77.64	67.32
Bin Wang Apr 2014	76.73	46.89	38.02	13.48	75.45	88.13	84.36	71.07	72.11	81.28	75.97	65.77
AAPSA	77.42	49.42	41.61	33.02	46.43	91.83	67.06	72.07	50.98	79.53	70.30	61.79
IUTIS-2	74.01	60.34	51.54	21.98	71.45	79.13	57.41	71.65	48.36	86.21	53.06	60.26
SC_ SOBS	66.20	54.63	45.03	4.09	48.80	93.33	66.86	70.51	59.18	77.86	69.23	59.61
KNN	75.87	54.91	42.00	21.26	51.98	84.11	68.65	68.94	50.26	74.68	60.46	59.37
SOBS_ CF	63.70	51.48	44.82	21.26	47.02	92.99	65.19	71.50	58.10	77.21	71.40	58.83
CP3-online	74.85	47.42	39.19	26.60	37.43	88.56	61.11	52.07	61.77	70.37	79.17	58.05
IUTIS-1	67.05	56.94	47.70	4.53	58.29	92.98	41.89	59.97	50.73	84.94	71.74	57.89
RMoG	68.26	53.12	42.65	24.70	45.78	78.48	73.52	70.10	54.31	72.12	47.88	57.35
GMM - Stauffer and Grimson	73.80	53.73	40.97	15.22	46.63	82.45	63.30	59.69	52.07	73.70	66.21	57.07
KDE - ElGammal	75.71	54.78	43.65	3.65	44.78	90.92	59.61	57.20	40.88	76.60	74.23	56.88
GraphCutDiff	**87.87**	51.27	46.88	37.23	51.43	71.47	53.91	54.89	40.19	72.28	57.86	56.84
GMM - Zivkovic	74.06	50.65	39.60	10.46	41.69	83.82	63.28	56.70	53.25	73.22	65.48	55.66
Euclidean dist.	67.01	50.15	38.59	3.95	41.35	87.20	50.81	48.74	48.92	67.86	63.13	51.61
Multiscale Spatio-Temporal BG	63.71	33.65	41.64	3.64	52.91	84.50	59.53	50.73	44.97	79.18	51.03	51.41
Mahalanobis dist	22.12	7.97	13.74	3.74	33.59	46.42	17.98	33.58	22.90	33.53	13.83	22.67

problem, which requires more advanced modelling of either the structure of the scene or the colormetric nature of shadows in general. The good performance in the Low Framerate category can be attributed to the intelligent updating mechanism, which is not fooled by the rapidly changing foreground masks.

The proposed method comes a close second to only our previous work in the Thermal category, mostly since the edge based approach is also able to handle thermal reflections well. Since the videos in this category all consist of a single channel, the additional color mask does not benefit the detection. The F-measure is slightly less, since the same parameters have to be used everywhere, which are now optimized for the majority of the (mostly color) videos in the dataset. The same reason for slight performance loss can be found in the Turbulence category.

A limitation of the edge based method shows up in the Dynamic Background category. Here, the background set vectors are spread across the entire feature space, which results in many false negatives in two sequences with boats on rivers. An adjusted mask aggregation framework in dynamic regions could improve these results.

Overall, our method ranks near the top of the classification, with the third highest F-Measure of all methods. There is also an improvement of more than 2 % with regard to our previous work. This gain can be attributed to the additional color mask and the improved handling of camera jitter. Also note that the top ranking method (IUTIS-3 [26]) actually combines the top 3 performing methods at its time of submission (SuBSENSE [8], FTSG [27] and CwisarDH [28]). So, by utilizing a similar framework which incorporates our method, even better results may be achieved.

7   Conclusion

In this paper, we presented a combined color and edge based foreground background estimation algorithm. Gradient orientations and confidence measures are calculated by using Local Ternary Patterns. These gradients serve as input in a background modelling framework, which can dynamically adjust the learning rate. Then, interior points are added to the edge based foreground mask, based on a Breadth First Search strategy. The combined foreground mask consists of the bitwise AND operation between the edge based image and a color based foreground image.

Furthermore, possible camera motion is detected and a distinction is made between jitter and panning situations. Jitter is handled by the global unreliability rate, while panning motions are compensated through an optical flow based framework.

We have shown that our method performs especially well in the presence of difficult lighting conditions, e.g. at night, and for panning camera's compared to state-of-the-art methods. Furthermore, the method also provides near state-of-the- art performance for Baseline and Low Framerate sequences, Thermal videos, Shadows and Camera Jitter, which results in an overall third highest F-measure to date on the challenging ChangeDetection.NET 2014 dataset.

Acknowledgements. We would like to thank the creators of ChangeDetection.NET and all those responsible for providing the means to evaluate our foreground background estimation algorithm on this very comprehensive dataset.

References

1. Stauffer, C., Grimson, W.E.L.: Adaptive background mixture models for real-time tracking. In: CVPR, pp. 2246–2252 (1999)
2. Barnich, O., Droogenbroeck, M.V.: ViBe: a universal background subtraction algorithm for video sequences. IEEE Trans. Image Process. **20**, 1709–1724 (2011)

3. Droogenbroeck, M.V., Paquot, O.: Background subtraction: experiments and improvements for ViBe. In: CVPR Workshops, pp. 32–37. IEEE (2012)
4. Hofmann, M., Tiefenbacher, P., Rigoll, G.: Background segmentation with feedback: the pixel-based adaptive segmenter. In: 2012 IEEE Computer Society Conference on Computer Vision and Pattern Recognition Workshops (CVPRW), pp. 38–43 (2012)
5. Cristani, M., Farenzena, M., Bloisi, D., Murino, V.: Background subtraction for automated multisensor surveillance: a comprehensive review. EURASIP J. Adv. Sig. Process. **2010**, 43:1–43:24 (2010)
6. Porikli, F., Tuzel, O.: Human body tracking by adaptive background models and mean-shift analysis. In: IEEE International Workshop on Performance Evaluation of Tracking and Surveillance (2003)
7. Heikkila, M., Pietikainen, M.: A texture-based method for modeling the background and detecting moving objects. IEEE Trans. Pattern Anal. Mach. Intell. **28**, 657–662 (2006)
8. St-Charles, P.L., Bilodeau, G.A., Bergevin, R.: Flexible background subtraction with self-balanced local sensitivity. In: IEEE Conference on Computer Vision and Pattern Recognition (CVPR) Workshops (2014)
9. Bilodeau, G.A., Jon, J.P., Saunier, N.: Change detection in feature space using local binary similarity patterns. In: CRV, pp. 106–112. IEEE (2013)
10. Gruenwedel, S., Van Hese, P., Philips, W.: An edge-based approach for robust foreground detection. In: Blanc-Talon, J., Kleihorst, R., Philips, W., Popescu, D., Scheunders, P. (eds.) ACIVS 2011. LNCS, vol. 6915, pp. 554–565. Springer, Heidelberg (2011)
11. Allebosch, G., Van Hamme, D., Deboeverie, F., Veelaert, P., Philips, W.: Edge based foreground background estimation with interior/exterior classification. In: Proceedings of the 10th International Conference on Computer Vision Theory and Applications, vol. 3, pp. 369–375. SCITEPRESS (2015)
12. Sajid, H., Cheung, S.C.S.: Background subtraction for static and moving camera. In: IEEE International Conference on Image Processing (2015)
13. Allebosch, G., Deboeverie, F., Veelaert, P., Philips, W.: EFIC: edge based foreground background segmentation and interior classification for dynamic camera viewpoints. In: Battiato, S., et al. (eds.) ACIVS 2015. LNCS, vol. 9386, pp. 130–141. Springer, Heidelberg (2015). doi:10.1007/978-3-319-25903-1_12
14. Wang, Y., Jon, P.M., Porikli, F., Konrad, J., Benezeth, Y., Ishwar, P.: Cdnet 2014: an expanded change detection benchmark dataset. In: IEEE Conference on Computer Vision and Pattern Recognition (CVPR) Workshops (2014)
15. Tan, X., Triggs, B.: Enhanced local texture feature sets for face recognition under difficult lighting conditions. IEEE Trans. Image Process. **19**, 1635–1650 (2010)
16. Zhang, T.Y., Suen, C.Y.: A fast parallel algorithm for thinning digital patterns. Commun. ACM **27**, 236–239 (1984)
17. Ottmann, T., Soisalon-Soininen, E., Wood, D.: On the definition and computation of rectilinear convex hulls. Inf. Sci. **33**, 157–171 (1984)
18. Evangelio, R., Sikora, T.: Complementary background models for the detection of static and moving objects in crowded environments. In: 2011 8th IEEE International Conference on Advanced Video and Signal-Based Surveillance (AVSS), pp. 71–76 (2011)
19. Wang, H., Suter, D.: A re-evaluation of mixture of Gaussian background modeling [video signal processing applications]. In: 2005 Proceedings of the IEEE International Conference on Acoustics, Speech, and Signal Processing (ICASSP 2005), vol. 2, pp. ii/1017–ii/1020 (2005)

20. Fleet, D.J., Weiss, Y.: Optical flow estimation. In: Paragios, N., Chen, Y., Faugeras, O. (eds.) Handbook of Mathematical Models in Computer Vision, pp. 237–257. Springer US, London (2006)
21. Fortun, D., Bouthemy, P., Kervrann, C.: Optical flow modeling and computation: a survey. Comput. Vis. Image Underst. **134**, 1–21 (2015). Image Understanding for Real-world Distributed Video Networks
22. Farnebäck, G.: Two-frame motion estimation based on polynomial expansion. In: Bigun, J., Gustavsson, T. (eds.) SCIA 2003. LNCS, vol. 2749, pp. 363–370. Springer, Heidelberg (2003)
23. Forsyth, D.A., Ponce, J.: Geometric camera models. In: Computer Vision: A Modern Approach, 2nd edn, pp. 33–61. Pearson. International edn. (2012)
24. Faugeras, O.D., Luong, Q.T., Papadopoulo, T.: The Geometry of Multiple Images - The Laws that Govern the Formation of Multiple Images of a Scene and Some of Their Applications. MIT Press, Cambridge (2001)
25. Bouguet, J.Y.: Pyramidal implementation of the Lucas Kanade feature tracker description of the algorithm. Technical report, Intel Corporation Microprocessor Research Labs (2000)
26. Bianco, S., Ciocca, G., Schettini, R.: How far can you get by combining change detection algorithms? IEEE Transactions on Image Processing (2015, Submitted)
27. Wang, R., Bunyak, F., Seetharaman, G., Palaniappan, K.: Static and moving object detection using flux tensor with split Gaussian models. In: IEEE Conference on Computer Vision and Pattern Recognition Workshops (CVPR) (2014)
28. De Gregorio, M., Giordano, M.: Change detection with weightless neural networks. In: 2014 IEEE Conference on Computer Vision and Pattern Recognition Workshops (CVPRW), pp. 409–413 (2014)

Choosing the Best Embedded Processing Platform for On-Board UAV Image Processing

Dries Hulens[1]([✉]), Jon Verbeke[2], and Toon Goedemé[1]

[1] EAVISE, KU Leuven, Sint-Katelijne-Waver, Leuven, Belgium
{dries.hulens,jon.verbeke,toon.goedeme}@kuleuven.be
[2] Department of Engineering, KU Leuven, Leuven, Belgium

Abstract. Nowadays, complex image processing algorithms are a necessity to make UAVs more autonomous. Currently, the processing of images of the on-board camera is often performed on a ground station, thus severely limiting the operating range. On-board processing has numerous advantages, however determining a good trade-off between speed, power consumption and weight of a specific hardware platform for on-board processing is hard. Many hardware platforms exist, and finding the most suited one for a specific vision algorithm is difficult. We present a framework that automatically determines the most-suited hardware platform given an arbitrary complex vision algorithm. Our framework estimates the speed, power consumption and flight time of this algorithm for multiple hardware platforms on a specific UAV. We demonstrate this methodology on two real-life cases and give an overview of the present top performing CPU-based platforms for on-board UAV image processing.

Keywords: UAV · Vision · On-board · Real-time · Speed estimation · Power estimation · Flight time estimation

1 Introduction

Nowadays UAVs (Unmanned Aerial Vehicles) are used in a variety of tasks such as surveillance, inspection, land surveying,... They are mostly manually controlled remotely or follow a predefined flight path, while collecting interesting images of the environment. These images are often analyzed offline since the processing power of these UAVs is limited. Otherwise a wireless link is provided to do the processing of the images on a ground station giving the instructions to the UAV. To be more autonomous and operate more robustly, UAVs should be equipped with processing power so that images can be processed on-board. This will ensure that UAVs can analyze and react in real-time on the images and that they can fly much further since a wireless link is not necessary. Recent advances concerning embedded platforms show an ongoing increase in processing power at reasonable power consumption and weight. Currently, it even becomes possible to employ these complex hardware platforms under UAVs. However, since various parameters need to be taken into account, finding an optimal hardware platform

© Springer International Publishing Switzerland 2016
J. Braz et al. (Eds.): VISIGRAPP 2015, CCIS 598, pp. 455–472, 2016.
DOI: 10.1007/978-3-319-29971-6_24

for a specific algorithm is not trivial. Example applications that need on-board complex image processing are e.g. visual SLAM for 3D sense and avoid, the detection and tracking of people for surveillance purposes, navigating through the corridor between trees in an orchard for counting fruit, the automation of a film crew by UAVs, a vision-based navigation system to automatically clean solar panels,... Determining the optimal trade-off between the processing capabilities and the physical constraints is a daunting task because of their variety. Therefore, in this paper we answer the question: *Which hardware platform is best suited to perform a particular image processing task on a UAV?* A hardware platform can be a simple embedded processor (e.g. a Raspberry PI) or even a small computer like a laptop, depending on the processing power that is needed. Using these under a UAV impose severe constraints on the hardware platforms: they should be lightweight, small and have adequate processing power at low power consumption to maintain long flight times. To determine the effective processing speed of a particular algorithm on a specific hardware platform, one should implement the algorithm on each specific platform. Acquiring a large variety of test platforms to determine the most suitable one evidently is not time nor cost efficient. Therefore, in this paper we present a framework that, given a specific algorithm, estimates the processing speed, power consumption and flight time on a large set of hardware platforms, without the need to acquire any of them. For this we rely on two benchmark algorithms. This paper provides data for a number of hardware platforms only restricted in the fact that they are CPU-based. However since our framework is generic, new platforms can easily be added to the framework. An overview of the platforms that we have included can be found in Table 1. The framework will be evaluated on two real cases. In the first case we track a person with a UAV using a face detection algorithm [2]. For this, we search for a hardware platform that can run the face detector at *4 fps* while minimizing the power consumption (e.g. maximum flight time). In our second case the UAV should visually navigate through a fruit orchard corridor, running a vantage point detection algorithm [1] on-board at *10 fps* (Fig. 1).

Fig. 1. Parrot AR Drone (left) and XBird 250 (right) carrying an Odroid hardware platform for real-time image processing.

The main contributions of this paper are:

- State-of-the-art overview of the current best CPU-based processing platforms for complex image processing on-board a UAV.
- Present experimental results of benchmark computer vision experiments on each of these state-of-the-art platforms.
- We propose a generic model to estimate the processing speed, power consumption and UAV flight time of any given image processing algorithm on a variety of hardware platforms.
- Validation of the proposed generic model on two real cases (people detection/tracking and vision-based navigation).

This paper is structured as follows: in the next section we give an overview of the related work on this topic. In Sect. 3 we briefly discuss the hardware platforms that we used in the framework. In Sect. 4 we present our framework and in Sect. 5 we verify our framework with some experiments and show our results.

2   Related Work

Currently, UAVs are often used to capture images of the environment which are then processed afterwards e.g. surveying [12]. For this the UAVs are controlled manually or by means of GPS. However, our main focus is on autonomously flying UAVs. To enable this, UAVs mainly rely on vision algorithms. Therefore, algorithms like path planning and obstacle avoidance (e.g. object detection) are used to steer the UAV to a certain position [7,13,14]. Due to their computational complexity, on-board UAV processing is often practically unfeasible. Therefore, in these approaches, a ground station (with desktop computer) is used to process the images and steer the UAV. However this severely limits their operating range. In cases where on-board processing currently is employed, only light-weight algorithms are used. For example [10] use sky segmentation (color segmentation), running on a Pentium III processor, to detect and avoid objects in the sky. [8] use a marker detection system to follow a predefined path. [17] use line detection, running on a Cortex-A9, for the inspection of pole-like structures. [9] track an IR-LED-pattern mounted on a moving platform, using a ATmega 644P controller and [15] filters laser scanner data on an Atom-based processing platform to estimate crop height. However, our real-life test case algorithms are much more complex. To implement more complex algorithms on a UAV often FPGAs or ASICs are used since they offer an optimal trade-off between weight, power consumption and processing power. [11] designed an FPGA based path planning algorithm, and [6] evaluate other hardware like ASICs as on-board vision processing platform. However, translating e.g. OpenCV code (C, C++ or python) to hardware (using e.g. VHDL) is a tedious and time consuming task. [16] use a high-end processing platform for on-board path planning and obstacle avoidance. This is possible since, in their case, power consumption or weight is less relevant because they use an octacopter with a large carrying capacity. Currently, work exists which achieves real-time performance of complex vision algorithms on UAV mounted embedded platforms [18–20]. However, their algorithms are specifically adapted or designed to perform real-time performance on a targeted hardware platform. We aim to develop a framework that performs the opposite operation; i.e. given a specific algorithm we determine the most suited hardware platform. To resolve all problems mentioned above, in this paper we present a framework that automatically determines the most suitable hardware platform given a user's computer vision algorithm from state-of-the-art, affordable (from $30 to $800), embedded platforms. Our framework enables the use of complex computer vision algorithms which run in real-time on-board of the UAV, directly programmed in OpenCV.

3   State-of-the-Art Image Processing Platforms

Nowadays, a number of CPU-based processing platforms are available which are lightweight and powerful and therefore suited for the task at hand. An overview is given in Table 1. We will describe them briefly, in order of ascending processing power (and thus increasing weight). A well-known lightweight processing platform is the Raspberry PI. The PI is a small, low-cost 1 GHz ARM11 based hardware platform developed for educational purposes. The main advantage of this small platform is that it runs a linux-based distribution, which allows the compilation and usage of well-known vision libraries e.g. OpenCV. Of course, the processing speed is limited, but simple vision algorithms, like e.g. face detection based on skin color segmentation, run at real-time performance. The PI is equipped with a Broadcom GPU which recently became open-source. A more powerful alternative for the PI is the family of Odroid platforms. One of those platforms is the U3 that is even smaller than the PI and has an ARM based 1.7 GHz Quad-Core Samsung processor that is also used in smartphones. Speed tests on the U3 indicated that this platform is *20 times* faster than the Raspberry PI. The XU3 is another Odroid platform which has a Samsung Exynos5422 Cortex-A15 2.0 GHz quad core and a Cortex-A7 quad core processor making him two times faster as the U3. The XU3 has a fan to cool the processor where the U3 is passively cooled. Both the U3 and XU3 are equipped with an eMMC slot which is a much faster alternative for the SD card. Another novel and promising platform is the Jetson TK1 Development Kit with an on-board NVIDIA GPU and a quad-core ARM15 CPU, making the platform especially useful for GPU based vision algorithms. In this paper we only perform experiments on the CPU but in future work the GPU will also be evaluated. The Jetson has several IO ports making it easy to communicate with sensors or inertial measurement units (IMUs), it even has a sata connection for a hard-drive. The CPU speed is comparable with the U3, but when GPU vision algorithms are used this platforms really shines. A more powerful family of hardware platforms are the Mini-ITX platforms. Mini-ITX platforms all have the same dimensions ($17 \times 17cm$) but can be equipped with different processors and IO. They are basically small computers with the same IO as a normal desktop computer. The mini-ITX platforms can be classified into two categories: the Atom platforms that can be compared

Table 1. Overview hardware platforms that we have tested for our framework.

Name	Processor	Memory	Weight (gram)	Power (Watt)	Volume (cm^3)
Desktop	Intel I7-3770	20 GB	740	107	4500
Raspberry PI	ARM1176JZF-S	512 MB	69	3,6	95
Odroid U3	Samsung Exynos	2 GB	52	6,7	79
Odroid XU3	Samsung Exynos	2 GB	70	11	131
Jetson	Cortex A15	2 GB	185	12,5	573
mini-ITX atom	Intel Atom D2500	8 GB	427	23,5	1270
mini-ITX I7	Intel I7-4770S	16 GB	684	68	1815
Nuc	Intel I5-4250U	8 GB	550	20,1	661
Brix	Intel I7-4500	8 GB	172	26	261

with netbooks and the I7-3000 platforms that can be compared with desktops. The Atom Mini-ITX platform has a 1.86 GHz Fanless Dual Core processor like in many netbooks computers. Its speed is comparable with the U3 and therefore less interesting due to its larger size, power consumption and weight. Unlike the previous, the Intel i7-3770 platform has a quad core processor and is much faster. This platform is one of the fastest platforms we have tested in this paper. It is five times faster than the XU3 and even faster than our reference system that we used (normal desktop computer). Together with a power supply that can be connected to a LiPo battery and a SSD hard drive, this platform can handle complex image processing algorithms on-board a UAV. The disadvantage of this platform is its power consumption and weight. The next family of platforms are the Brix and Nuc barebone mini-computers. These computers are designed to be mounted on the back of a monitor and have a size of 11 × 11cm. These platforms consume less power than the Mini-ITX I7 platform but are twice as slow, which is still very fast for such a small computer. The Brix has an Intel I7-4500 quad-core processor and is comparable in speed with the Nuc that has an Intel I5-4250U processor. When stripping down the casing of these two platforms, the Brix only weighs 172 g (SSD included) compared to the Nuc that still weigh 550 g, giving the Brix the most interesting specs to mount on a UAV for complex image processing algorithms. Section 5.1 gives an overview of the tests we have performed on these platforms.

4   Approach

The goal of our framework is to find the best hardware platform to run a user's new vision algorithm on a UAV. The main criterion we try to optimize is the amount the processing platform reduces the UAV's flight time. Indeed, both because of the hardware platform's own weight and of its electrical power consumption it drains the battery during flight. The best platform is found when

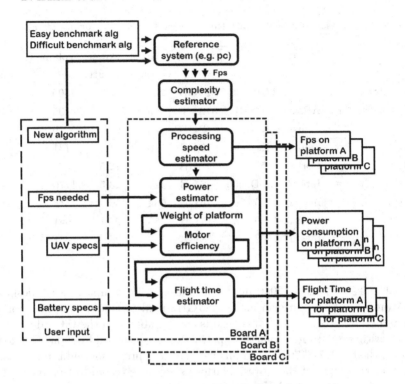

Fig. 2. Overview of our framework.

a vision algorithm can run on it at a certain *required speed (in fps frames per second)*, while it consumes as little as possible and the weight of the platform is as low as possible in order to extend flight time. The *required speed* can be much lower than the maximum speed that the algorithm can run on a certain platform, e.g. a face detector that runs at *100 fps* but only *20 fps* is required for a certain application. The power consumption reduces dramatically when reducing the frame rate of the algorithm on the same platform.

We propose a generic calculation model that estimates the flight time reduction for an arbitrary vision algorithm on a specific embedded processing platform. As seen in Fig. 2 this model consists of six blocks. In the first block the user's new algorithm and two benchmark algorithms are executed on a reference system (e.g. the user's desktop computer) and their frame rate is given to the next block where the relative complexity of the new algorithm is estimated. With this, for each hardware platform, its speed is estimated in the next block. Then the power consumption of every platform, while running the new algorithm at a certain required speed, is estimated. In the next block the power consumption of the UAV carrying each hardware platform is calculated. Finally, in the last block the flight-time of the UAV, carrying each hardware platform running the new algorithm at a certain speed, is estimated. In the next subsections these blocks are discussed in detail.

4.1   Complexity and Processing Speed Estimator

To estimate the speed of a new algorithm on every hardware platform we first estimate the complexity of this algorithm. For the sake of simplicity, we assume a linear relation between the processing speed and the complexity of the algorithm. We will validate this linearity assumption in Sect. 5. The speed of the algorithm ($f_{alg} = \frac{1}{T_{alg}}$) on the reference system, e.g. the user's desktop PC, is used as measurement for the complexity (C_{alg}). We empirically measure the relative complexity of the new algorithm with respect to two reference (benchmark) algorithms. The first benchmark algorithm is an easy algorithm that we let correspond with 0 % complexity (C_1). For this algorithm we chose the OpenCV implementation of a 3×3 median filter on a color image of 640×480 pixels. The second algorithm is a more difficult algorithm that corresponds to a complexity of 100 % (C_2), where OpenCV's HOG person detector is applied to an image of 640×426 pixels. Our *Complexity estimator* uses the execution time of these two benchmark algorithms (T_1 and T_2) and the user's new algorithm (T_{alg}) running on the reference system to calculate the complexity of the new algorithm (see Fig. 3). The complexity is then calculated as:

$$C_{alg} = \frac{T_{alg} - T_1}{T_2 - T_1} C_2 + C_1 \tag{1}$$

We assume a linear relation between the computational complexity and the speed of these vision algorithms because they all do mainly the same operations, like applying filters on an image and extracting features. Vision algorithms are always data intensive but most of the time not computationally intensive. Note that code optimizations for specific architectures evidently affect the results. Details like memory usage are not taken into account in this simple model, because the memory on the advanced hardware platforms is easy expandable. Moreover, in our model we only assume CPU-based processing platforms, no other architectures such as GPU or FPGA for which a code translation step would be necessary. In Sect. 5 the validity of this linear relation is verified.

Now that the complexity of the new algorithm (C_{alg}) is known, the speed of the algorithm can be estimated on every platform by following Fig. 3 in the other direction, as demonstrated in Fig. 4 for two fictitious platforms. The simple and difficult algorithm is run on every platform what results in a T_1 and T_2 for each platform. Because C_{alg} is known from the previous step, T_{alg} can now be calculated for each platform:

$$T_{alg} = \frac{C_{alg} - C_1}{C_2}(T_2 - T_1) + T_1 \tag{2}$$

At this point the speed ($f_{alg} = \frac{1}{T_{alg}}$) of a new algorithm can be estimated for each hardware platform, hence in the next step we can estimate the power consumption of the new algorithm on each platform.

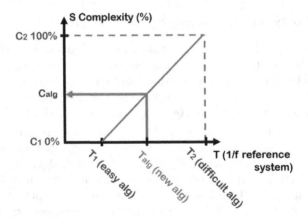

Fig. 3. Linear complexity model. Complexity C_{alg} (red) is estimated with T_1, T_2 and T_{alg} as input (green) (Color figure online).

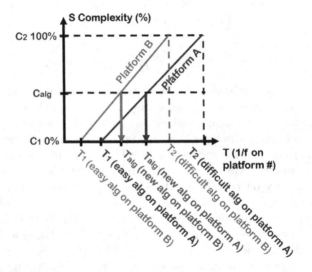

Fig. 4. Calculating T_{alg} (red) for each processing platform (blue and orange) with known C_{alg}, $T1$ and $T2$ (Color figure online).

4.2 Power Estimator

In UAV applications flight time is of utmost importance. Therefore our framework estimates the power consumption of each hardware platform running the new algorithm at the required speed. We performed experiments to determine the relation between processing speed and power consumption, indicating that a linear model is again a good approximation (see Sect. 5). When the maximum speed of the algorithm is not required, the power consumption can be lower than when the algorithm is running at full speed. By taking the *required fps* as an input of the *Power Estimation Block* we can estimate the power consumption more precisely for each platform.

To calculate the power consumption P_{alg} of a certain algorithm, the power consumption of each platform is measured when in idle state P_{idle} (doing nothing) and when running all cores at full speed P_{max} (algorithm running at full speed). Together with the required speed (in frames per second) f_{req} and the maximum speed of the algorithm f_{max} the power consumption of the platform can be linearly interpolated as follows:

$$P_{alg} = \frac{P_{max} - P_{idle}}{f_{max}} f_{req} + P_{idle} \tag{3}$$

In this step we also have to eliminate hardware platforms which do not reach the *required fps* (when $\frac{1}{T_{alg}} < f_{req}$). At this point the power consumption of every remaining platform, running the user's new algorithm at a certain speed, is known. In the next step the power consumption of the UAV itself, carrying the platform as payload, is calculated.

4.3 Motor Efficiency

In [4] a model has been developed that enables the user to estimate the power consumption of a multicopter at hover. The performance estimates are based on momentum disk theory and blade element theory of helicopters combined with empirically determined correction factors for multicopters [3]. The model requires the user to input several parameters such as weight, number of propellers n_{props} and propeller radius R. The model uses some empirical parameters such as the Figure of Merit FM (basically the propeller efficiency), the motor efficiency η_{motor} (including the electronic speed controller efficiency) and an installed-to-hover power ratio $\frac{P_{installed}}{P_{hover}}$ of 2 (based on industry standards). The empirical parameters were determined with actual tests on several motors and propellers which are middle grade RC components. The user can (slightly) change these as their multicopter might have higher or lower grade components. We will use this model to estimate the power consumption of the UAV carrying the hardware platform. During hover and slow forward flight it can be assumed that thrust T_{hov}(approximately) equals the total weight force W_{tot} in Newton ($W_{tot} = m_{tot}g = (m_{UAV} + m_{platform})g$) and the hover power per propeller can be calculated through the disk loading DL, induced velocity v_i and air density ρ:

$$DL = \frac{W_{tot}}{\pi R^2 n_{props}} \tag{4}$$

$$P_{hov_{theo}} = T_{hov} v_{i_{hov}} = W_{tot} v_{i_{hov}} = W_{tot} \sqrt{\frac{DL}{2\rho}} \tag{5}$$

$$P_{hov_{real}} = \frac{P_{hov_{theo}}}{FM \eta_{motor}} \tag{6}$$

Calculating the power consumption of the multicopter based on hover conditions is a rather safe method as during slow forward flight the required power actually decreases by 10 % and most multicopter operations take place in this regime [5].

Together with the hardware power consumption P_{alg}, the total electrical power consumption P_{tot} can be calculated as:

$$P_{tot} = \frac{W_{tot}\sqrt{\frac{DL}{2\rho}}}{FM\eta_{motor}} + P_{alg} \qquad (7)$$

At this stage the total power consumption of the UAV, carrying the hardware platform that is running a certain algorithm, is known. In the next subsection the flight time is estimated.

4.4   Flight Time Estimator

The flight time for every platform can be estimated since the power consumption of every platform running an algorithm at a certain speed together with the power consumption of the UAV itself carrying each of the platforms is known now. These two values together with the capacity of the batteries are the inputs of this block. Nowadays most UAVs are using lithium polymer batteries because of their good capacity vs weight ratio. Nevertheless the capacity mentioned on the batteries applies only as long as the remaining battery voltage is above a certain value. Therefore most of the time 75 % of the battery's capacity is taken as a more fair value to calculate the flight time. Flight time is subsequently calculated as follows:

$$T_{flight}(h) = \frac{0.75V_{bat}C_{bat}}{P_{tot}} \qquad (8)$$

where C_{bat} is the capacity mentioned on the battery in Ah, V_{bat} is the voltage of the battery and P_{tot} is the total power consumption of the UAV at hover (Eq. 7).

At this point the main question *"Which hardware platform is best suited to perform a particular image processing task on a UAV?"* can be answered, which we will demonstrate in the next section for our two example algorithms.

5   Experiments and Results

We performed extensive experiments to validate our framework using a wide variety of platforms and multiple algorithms. In the first subsection we performed multiple speed tests of two algorithms to compare the different hardware platforms. In the next subsection we proof that the assumption of a linear complexity and power model holds. Finally we present validation experiments on two computer vision-based real-life cases: face detection and tracking on a UAV for people following and visual navigation to find the corridor in an orchard for fruit counting/inspection.

5.1 Speed Tests of Two Algorithms on Each Hardware Platform

In our first test the processing speed of the OpenCV implementation of a HOG person detector and a Viola and Jones face detector is measured on all platforms. Thereby speed can be compared for every hardware platform. The result can be seen in Fig. 5. In Fig. 6 we display the ratio of the measured speed of these two algorithms and the power consumption of every platform while running the two algorithms. Figure 7 displays the ratio of the speed and the volume of the hardware platforms and in Fig. 8 the ratio of the processing speed and the weight of the platforms is shown.

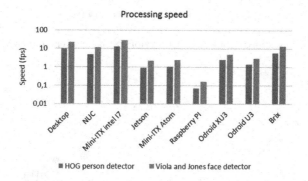

Fig. 5. Speed (logarithmic) of HOG person detector (blue) and Viola and Jones Face detector (orange) for every platform (Color figure online).

As seen in Figs. 5, 6, 7, and 8, the Mini ITX Intel I7 platform is one of the fastest but also very heavy. The Jetson and Atom platforms score below average compared to the other platforms because the Jetson is a processing platform designed for GPU implementations and the Atom is already an older generation of CPUs. The Nuc and Brix have a similar speed and power consumption, but the Brix is much lighter and smaller. The two Odroid platforms are similar in power consumption, volume and weight but the XU3 is twice as fast as the U3 platform. Overall, the Brix scores best when all test are taken into account.

5.2 Validation of Models

In Sect. 4.1 we assumed a linear relation between the complexity of a vision algorithm and the execution speed (the higher the execution time of the algorithm the more complex it is). The linearity is validated by estimating the speed of our two real-case-algorithms, on a desktop computer, for every platform and comparing it with the real speed of these algorithms on every platform. In Fig. 10 the percentage deviation between *estimated fps* and *measured fps* is given for the two algorithms. As seen, the error is not greater than 10 % which is indicating that the assumption of a linear model for the estimation of the complexity can be taken as valid.

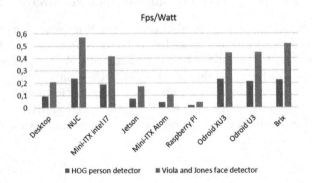

Fig. 6. Processing speed / power consumption ratio for every hardware platform.

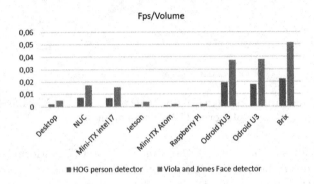

Fig. 7. Processing speed / volume ratio for every hardware platform.

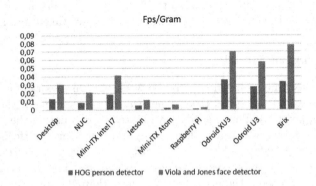

Fig. 8. Processing speed / gram ratio for every hardware platform.

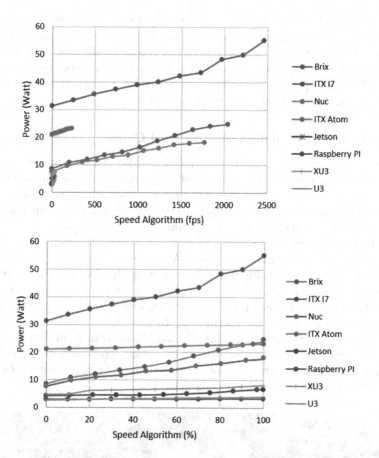

Fig. 9. Power consumption of each platform measured while increasing the speed (top: in fps, bottom: in %) of the easy (Median filter) algorithm.

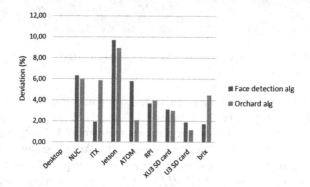

Fig. 10. Deviation between *estimated fps* and *measured fps* of our two real-case-algorithms.

As mentioned in Sect. 4.2 there is also a linear relation between the power consumption and the processing speed of an algorithm running on a hardware platform. To verify this statement the power consumption of each hardware platform is measured while incrementally increasing the processing speed. As seen in Fig. 9, the power consumption increases indeed practically linear with the processing speed for each processing platform.

5.3 Framework Validation on Two Real Cases

For two application cases, we demonstrated the use of the proposed model to find out which hardware platform is best suited for on-board computer vision processing. For both cases a Parrot AR Drone and a XBird 250 is used. On the Parrot the forward looking camera is used to capture images for our algorithms and the XBIRD is equipped with a webcam to capture images.

In the first real case, the UAV should follow a single person. The detection of the person is done by using the OpenCV implementation of Viola and Jones face detector [2] as seen in Fig. 12. This algorithm should run at least at *4 fps*. In the second case the UAV should navigate through a fruit orchard as seen in Fig. 13. Therefore an orchard-path-detection algorithm is used to find the middle and the vanishing point of the corridor [1]. In this algorithm, filters are applied on the image for preprocessing, followed by a Hough transform to find straight lines (the corridor) and a Kalman filter to predict and track the middle and vanishing point of the corridor. This algorithm should run at least at *10 fps* to fly smoothly through the orchard (Fig. 11).

Fig. 11. Left: AR Drone with forward looking camera. Right: XBird with webcam attached on the frame.

We ran both algorithms on a normal desktop computer to know their speed with which their complexity is estimated (Table 2). When their complexity is known their speed on every hardware platform is estimated (Eq. 2), together with their power consumption (Eq. 3) on every platform. At this stage some hardware platforms are discarded because they do not reach the required speed. Thereafter, the total power consumption of the UAV (AR Drone) carrying every hardware platform, running the algorithm, is calculated (Eq. 7). Finally, flight time is estimated with Eq. 8. Results can be seen in Tables 3 and 4. Table 4 indicates that the power consumption of the algorithm can't be ignored when using small UAVs.

Fig. 12. The detection of the persons head is used to track the person with a UAV.

Fig. 13. The first three pictures are the preprocessing steps, in the third picture lines are detected with Hough Transform and displayed on the fourth picture. The intersections of the red lines are the vanishing point (pink) of the corridor. The intersection of the green cross is where the vanishing point should be to steer the drone through the orchard (Color figure online).

Secondly, we verified the estimated flight time by attaching the proposed hardware platform on both the AR Drone and XBird while running the specific algorithm. Flight time is measured while hovering, as seen in Table 5 the deviation between estimated and measured data is very small (less than 7 %) indicating that our framework indeed finds the best hardware platform for a specific vision algorithm and estimates the speed and flight time very precisely. Note that, when the UAV runs the orchard or face algorithm the flight time reduces with 30.21 % and 39.58 % as compared to the flight time without payload.

Table 2. Algorithm complexity estimation results.

Algorithm	Speed (fps) Desktop	Complexity (%) Desktop
Benchmark 1 (median)	2040	0
Benchmark 2 (HOG)	9,91	100
Orchard	388	2,08
Face	22,44	43,9

Table 3. Results of our framework for the face detection and orchard algorithm. Platforms in red are eliminated because they do not reach the required speed. The platform in green is the best platform to run this algorithm on, on the specific UAV (AR Drone).

Algorithm	Face			Orchard		
Platform	Est. speed (fps)	Est. power consump. (Watt)	Est. flight time (min)	Est. speed (fps)	Est. power consump. (Watt)	Est. flight time (min)
Desktop	22,44			388		
Nuc	10,75	11,48	4,6	199	8,04	4,8
ITX i7	28,88	34,6	3,3	483,9	31,8	3,4
Brix	13,21	13,44	8,3	243,14	9,17	8,9
ITX atom	2,34	24,76		40,28	21,55	5
Jetson	1,98	9,25		16,52	5,71	9,3
RPI	0,16	10,39		1,86	4,61	
XU3	5,06	7,39	11,6	19,5	6,44	11,9
U3	2,95	4,5		14,8	3,55	13,4

Table 4. Power consumption of each part of the system.

Power consumption	Face (Watt)	(%)	Orchard (Watt)	(%)
Algorithm	7,39	17,2	3,55	9,6
Board weight	7	16,3	5	13,5
UAV weight	26	60,6	26	70
IMU	2,55	5,9	2,55	6,9

Table 5. Estimated and measured data.

Alg.	Est. speed (fps)	Measured speed (fps)	Estimated flight time AR Drone (min)	Measured flight time AR Drone (min)	Estimated flight time XBird (min)	Measured flight time XBird (min)
Face	5,06	4,9	11,6	12,4	7,4	7,12
Orchard	14,8	14,97	13,4	12,7	8,1	7,54

6 Conclusion and Future Work

We developed a framework that finds the best hardware platform for a specific vision processing algorithm that should run at a certain speed on-board a UAV. Furthermore the speed of the algorithm running on each platform is estimated. Thanks to this framework researchers can find a suitable hardware platform without buying them all to test their algorithm on. A second novelty of our framework is that flight time can be estimated for the user's UAV, carrying the proposed platform. We validated the framework with success on two real test cases allowing us to find a suitable hardware platform for our application and to estimate the flight time with our AR Drone and XBird carrying this platform.

Also, we made this model available via an online front end that other researchers can use to find the best platform for their algorithm and even add their own hardware to the framework and expand the database of hardware platforms (www.eavise.be/VirtualCameraman.html). In the future we will keep adding new state-of-the-art platforms and extend the framework with GPU platforms.

Acknowledgements. This work is funded by KU Leuven via the CAMETRON project.

References

1. Dries, H., Maarten, V.: UAV autonoom laten vliegen in een boomgaard. College University Lessius, Department of Industrial Engineering (2012)
2. Paul, V., Michael, J.: Rapid object detection using a boosted cascade of simple features. Comput. Vis. Pattern Recogn. 1, 511–518 (2001)
3. Prouty, R.W.: Helicopter Performance, Stability, and Control. Krieger, Malabar (1995)
4. Verbeke, J., Hulens, D., Ramon, H., Goedemé, T., De Schutter, J.: The design and construction of a high endurance hexacopter suited for narrow corridor (2014)
5. Theys, B., Dimitriadis, G., Andrianne, T., Hendrick, P., De Schutter, J.: Wind tunnel testing of a VTOL MAV propeller in tilted operating mode. In: ICUAS (2014)

6. Shoaib, E., McDonald-Maier, K.D.: On-board vision processing for small UAVs: time to rethink strategy. In: NASA/ESA Conference on Adaptive Hardware and Systems (2009)
7. Taro, S., Yoshiharu, A., Takumi, H.: Development of a SIFT based monocular EKF-SLAM algorithm for a small unmanned aerial vehicle. In: SICE Annual Conference (SICE) (2011)
8. Lorenz, M., Petri, T., Friedrich, F., Marc, P.: Pixhawk: a system for autonomous flight using onboard computer vision. In: IEEE International Conference on Robotics and Automation (ICRA) (2011)
9. Wenzel, K.E., Masselli, A., Zell, A.: Automatic take off, tracking and landing of a miniature UAV on a moving carrier vehicle. Journal of intelligent & robotic systems **61**, 221–238 (2011)
10. McGee, T.G., Raja, S., Karl, H.: Obstacle detection for small autonomous aircraft using sky segmentation. In: Proceedings of the 2005 IEEE International Conference on Robotics and Automation, ICRA 2005 (2005)
11. Kok, J., Gonzalez, L.F., Kelson, N.: FPGA implementation of an evolutionary algorithm for autonomous unmanned aerial vehicle on-board path planning. In: IEEE Transactions on Evolutionary Computation (2013)
12. Siebert, S., Teizer, J.: Mobile 3D mapping for surveying earthwork projects using an Unmanned Aerial Vehicle (UAV) system. Autom. Constr. **41**, 1–14 (2014)
13. Allen, F., Jesse, F., Edward, V., Lee Gregory, S.: UAV obstacle avoidance using image processing techniques. In: IEEE International Conference on Technologies for Practical Robot Applications (TePRA) (2012)
14. Yucong, L., Srikanth, S.: Path planning using 3D dubins curve for unmanned aerial vehicles. In: International Conference on Unmanned Aircraft Systems (ICUAS) (2014)
15. David, A., Sebastian, E., Aaron, L., Carrick, D.: On crop height estimation with UAVs. In: IEEE/RSJ International Conference on Intelligent Robots and Systems (IROS 2014) (2014)
16. Matthias, N., Sven, B.: Hierarchical planning with 3D local multiresolution obstacle avoidance for micro aerial vehicles. In: Proceedings of the Joint International Symposium on Robotics (ISR) and the German Conference on Robotics (ROBOTIK) (2014)
17. Inkyu, S., Stefan, H., Peter, C.: Inspection of pole-like structures using a vision-controlled VTOL UAV and shared autonomy. In: IEEE/RSJ International Conference on Intelligent Robots and Systems (IROS 2014) (2014)
18. Shaojie, S., Yash, M., Nathan, M., Vijay, K.: Vision-based state estimation and trajectory control towards high-speed flight with a quadrotor. Robotics: Science and Systems (2013)
19. Christophe, D.W., Sjoerd, T., Remes Bart, D.W., de Croon Guido, C.H.E.: Autonomous flight of a 20-gram flapping wing MAV with a 4-gram onboard stereo vision system. In: IEEE International Conference on Robotics and Automation (ICRA) (2014)
20. Christian, F., Matia, P., Daviden, S.: SVO: fast semi-direct monocular visual odometry. In: Proceedings of the IEEE International Conference on Robotics and Automation (2014)

Author Index

Printed in the United States
By Bookmasters